Texts and Monographs in Physics

Series Editors: R. Balian   W. Beiglböck   H. Grosse   E. H. Lieb
N. Reshetikhin   H. Spohn   W. Thirring

T0237925

Springer

Berlin
Heidelberg
New York
Barcelona
Hong Kong
London
Milan
Paris
Singapore
Tokyo

Texts and Monographs in Physics

Series Editors: R. Balian   W. Beiglböck   H. Grosse   E. H. Lieb
N. Reshetikhin   H. Spohn   W. Thirring

Malte Henkel

Conformal Invariance and Critical Phenomena

With 65 Figures and 63 Tables

 Springer

Professor Dr. Malte Henkel

Laboratoire de Physique des Matériaux CNRS-UMR 7556
Université Henri Poincaré Nancy I
B. P. 239
F-54506 Vandœuvre lès Nancy, France

ISSN 0172-5998
ISBN 978-3-642-08466-9

Library of Congress Cataloging-in-Publication Data.
Henkel, M. (Malte), 1960–, Conformal invariance and critical phenomena / Malte Henkel. p. cm. –
(Texts and monographs in physics, ISSN 0172-5998) Includes bibliographical references and index.
1. Conformal invariants. 2. Critical phenomena (Physics)
I. Title. II. Series. QC174.52.C66H46 1999 530.14'3–dc21 99-11044 CIP

© Springer-Verlag Berlin Heidelberg 2010
Printed in Germany

Dem Gedenken an meinen Vater, Dr. Rudolf Henkel
und für meine Mutter, Ilse Henkel

Preface

Critical phenomena arise in a wide variety of physical systems. Classical examples are the liquid–vapour critical point or the paramagnetic–ferromagnetic transition. Further examples include multicomponent fluids and alloys, superfluids, superconductors, polymers and fully developed turbulence and may even extend to the quark–gluon plasma and the early universe as a whole. Early theoretical investigators tried to reduce the problem to a very small number of degrees of freedom, such as the van der Waals equation and mean field approximations, culminating in Landau's general theory of critical phenomena. Nowadays, it is understood that the common ground for all these phenomena lies in the presence of strong fluctuations of infinitely many coupled variables. This was made explicit first through the exact solution of the two-dimensional Ising model by Onsager. Systematic subsequent developments have been leading to the scaling theories of critical phenomena and the renormalization group which allow a precise description of the close neighborhood of the critical point, often in good agreement with experiments. In contrast to the general understanding a century ago, the presence of fluctuations on all length scales at a critical point is emphasized today. This can be briefly summarized by saying that at a critical point a system is *scale invariant*.

In addition, *conformal invariance* permits also a non-uniform, local rescaling, provided only that angles remain unchanged. This has been familiar for almost a century; the invariance of Maxwell's equations in the vacuum under both scale and conformal transformations is the classic example. In general, conformal invariance holds for any massless relativistic quantum field theory with a local energy-momentum tensor. In consequence, any statistical system described in the continuum limit by such a quantum field theory should be conformally invariant, where the requirement of masslessness translates into the condition of criticality.

For *two-dimensional systems*, conformal invariance has considerably extended our knowledge of the nature of a critical point. In more than two dimensions, the consequences of conformal invariance for critical phenomena, first discussed by Polyakov in 1970, are rather weak. This is completely different in two dimensions, however, as pointed out in the seminal work by Belavin, Polyakov and A. Zamolodchikov in 1984. Scale invariance alone is capable of casting systems into universality classes dependent only on a few

selected properties such as the global symmetry, the dimension of the space
and the number of components of the order parameter. Two-dimensional
conformal invariance yields, in addition, a classification of the critical point
partition functions and thereby furnishes exact values of the critical exponents. Furthermore, the critical multipoint correlation functions of the local
variables of the system can be determined exactly.

This book provides an introduction to the basic techniques and results
of conformal invariance applied to the study of critical phenomena. At first
sight, the field-theoretical techniques required, to understand a central notion
such as the *central charge*, for example, might appear to be quite formidable
and more adapted to the realm of a string theorist than a condensed-matter
physicist. I believe that this point of view is at least overly simplistic and
that a lot of very physical problems in condensed-matter physics could be
brought to a better understanding with the help of conformal invariance
methods. Conformal invariance techniques are now developed to a point that
they should be useful for applications on a wide and varied front.

This book goes back to joint lectures with Philippe Christe on conformal
invariance and their applications to critical phenomena, held at the University of Fribourg in 1991/92. Already six years ago, we believed that the
basic techniques of conformal invariance could be presented in a form accessible to the condensed-matter physicist. The text grown from these joint
lectures was published in the Springer Lectures Notes in Physics, New Series
m: monographs, under the title *Introduction to Conformal Invariance and Its
Applications to Critical Phenomena*, which appeared in 1993. Regrettably, in
summer 1992 Philippe had reoriented his career, away from fundamental research, for a job in industry, bringing this fruitful collaboration to an end.
The present book takes up the same topic, and tries to take more recent
developments into account. Since Philippe is no longer active in research, we
thought it better that I should perform the task of rewriting alone, and he
kindly let me use the entire material of our common Lecture Notes volume
as I felt best.

Lecturing on the subject subsequently in Porto (1994) and Nancy (1997),
I could try out improvements in the presentation of several points. When
Springer kindly proposed to republish the text, I decided to rewrite the book
considerably. A new book, rather than a slightly updated second edition,
appears justified, since many new developments (including new experimental advances) have occurred since the Lecture Notes volume was written.
Although I used much of the material as presented in the Lecture Notes
volume, much new material has been added.

The text is written for a reader with some experience in scaling and the
phenomenology of the renormalization group without necessarily being an
expert in quantum field theory, but who has a specific physical problem at
hand and is curious about what new insights the application of conformal
invariance might provide to the understanding of his problem. The constant
emphasis in this book on lattice models and explicit numerical calculations

should be helpful here. I have somewhat downplayed the continuum field theory and also the more formal aspects of conformal invariance (in any case, these are superbly presented in several existing reviews and books). I have not tried to approach anything even nearly reaching encyclopedic completeness; rather, the material selected is what I believe will be encountered first when trying to understand the basics of theory. Besides that, the set of topics retained is certainly subjective. Regrettably, many of the more advanced applications of conformal invariance had to be skipped completely. I mention here as keywords string theory (which stands at the origin of conformal invariance), field theories on a fluctuating metric and quantum gravity and the related matrix models, the deep connections with integrable systems (only an outline of A. Zamolodchikov's theory of relevant perturbations of conformally invariant models and the application to the 2D Ising model in magnetic field will be presented here), the mathematical theory of knots and links, extensions of the conformal algebra like superconformal algebras, W-algebras, Temperley–Lieb algebras and generalizations thereof (at least some of the simplest results are quoted), the Peierls instability, weakly disordered systems or the Kondo and quantum Hall effects. Discussion of these fascinating topics must be left to more specialized articles.

On the other hand, percolation and polymers are among the conformal invariance applications which are now included. Many new tables and figures with explicit model results were added, which confront the predictions of conformal field theory with precise simulational data. The chapter on numerical methods has been supplemented by sample Fortran codes and an introduction to the new density matrix renormalization group technique. In the last few years important advances in experimental techniques, in particular of molecular beam epitaxy, have made two-dimensional systems widely available for careful precision studies. I could not resist the temptation to give at least an overview of the applications of these to two-dimensional critical phenomena. Furthermore, the same technological advance has made it possible for the first time to check finite-size scaling theory experimentally. Exercises were added for most of the chapters. In the main, these are simply 'Fingerübungen' to encourage the reader to play with the material further, but some of those for the later chapters may become quite formidable. Even if I normally did not formulate exercises of the type 'prove equation so-and-so', it is understood that the first exercise for the reader is to fill in any gaps in the derivations in the text; it was of course impossible to spell out all the calculational details, but I think sufficiently many signposts are provided.

The scheme indicated below should give a rough idea of the relationship between the different chapters and is meant to help the reader find a fast path towards the applications without necessarily having to digest all of the preceeding chapters. Chapters 1–3 contain basic material, parts of which may already be familiar to some readers, which serves as the foundation for the remainder of this book. Chapters 4–7 discuss quantum field theory techniques and the Virasoro algebra on a more abstract level. In particular, the calculation of critical four-point correlators is discussed in detail. Chapters 8–10 describe lattice Hamiltonians and their numerical treatment, which at a first reading may only require some of the notions presented in Chap. 4. Of course, the Ising model will serve as the master example where the conformal techniques can be tried out at length. The material presented in Chaps. 4–6,10 is complementary and the reader is encouraged to go back and forth between these parts. The later chapters (11–16) are written in more of a review style and I have tried to provide the flavour of a variety of advanced problems, some of which are still open, where conformal invariance methods have been useful. Chapters 11 and 12 discuss further applications at criticality. The vicinity of the critical point is studied in Chaps. 13 and 14. Surface criticality in various forms is studied in Chap. 15. Chapter 16 reviews the present state of the art for extensions of conformal invariance suitable for a discussion of dynamical scaling. For a reader in a hurry, one possibility would be to go through Chaps. 2 and 3, briefly review the representation theory results of Chap. 4 and the Hamiltonian limit in Chap. 8 before looking in detail at the Ising model applications in Chap. 10 (especially Sec. 10.6) and then going on to more advanced topics.

I admit that I feel overwhelmed by the exploding literature in the field. In the bibliography I have compiled those works I happened to come across before and during writing the text, but it is clear that the selection of references is highly incomplete and does to a large part reflect my own focus and/or my lack of knowledge of some directions in the field. I sincerely apologize to any authors who might feel that I did not give proper credit to their contribution to the field. To some extent, this lapse might be made up by the bibliographies compiled in reviews/books cited in the general references.

Quite a few people might believe at this point that critical phenomena, let alone conformal invariance, were a purely academic, or even theoretical, game. Not so. Material properties close to criticality may dramatically change in response to only slight variations in external parameters. In supercritical fluids, for example, the unusual physical properties near criticality may be exploited in new, environmentally benign, kinds of chemical and materials processing, as reviewed in [E26]. The theoretical setting for this (in 2D) is discussed in Chaps. 13,14.

Finally, I have the pleasure to express (albeit in this imperfect way) my gratitude to friends and colleagues who in the past few years provided support or advice, shared with me their insight, or simply participated in the thrill of pursuing a problem in collaboration. In one way or other, they have influenced my views on physics and the life around it. I have greatly appreciated this, whether or not this becomes explicit in this book. To F.C. Alcaraz, S. Andrieu, P. Bauer, B. Berche, J.L. Cardy, M. Cieplak, R. Dickman, M. Droz, L. Frachebourg, A.B. Harris, W. Hofstetter, A. Honecker, D. Karevski, E.S. Lage, M.C. Marques, J.F.F. Mendes, N. Mousseau, E. Orlandini, M. Piecuch, R. Peschanski, I. Peschel, J. Richert, V. Rittenberg, J. Santos, G.M. Schütz, F. Seno, L. Turban and J.M. Yeomans: Herzlichen Dank, merci beaucoup! Of course, without P. Christe the precursor volume would have never been written and I remain greatly indebted to him. C. Chatelain and H. Rieger kindly provided Figs. 9.3 and 15.6. Je remercie en particulier L. Turban pour son aide constante et ses multiples remarques. Dem Springer-Verlag danke ich für die erfreuliche Zusammenarbeit.

Nancy, January 1999   Malte Henkel

The following sketch is intended to give an idea on the relationships and dependencies of the various chapters.

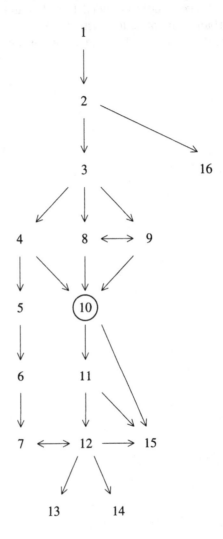

Contents

1. **Critical Phenomena: a Reminder** 1
   1.1 Phase Diagrams and Critical Exponents 1
   1.2 Scale Invariance and Scaling Relations 6
   1.3 Some Simple Spin Systems 12
       1.3.1 Ising Model.. 13
       1.3.2 Tricritical Ising Model 14
       1.3.3 q-States Potts Model 16
       1.3.4 Vector Potts Model 19
       1.3.5 XY Model 21
       1.3.6 Yang–Lee Edge Singularity 23
       1.3.7 Percolation 24
       1.3.8 Linear Polymers................................. 26
       1.3.9 Restricted Solid-On-Solid Models 27
   1.4 Some Experimental Examples 30
   1.5 Correspondence Between Statistical Systems and Field Theory 37
   1.6 Correspondence of Physical Quantities 39
       1.6.1 Free Energy Density 40
       1.6.2 Correlation Functions 40
       1.6.3 Correlation Lengths 41

2. **Conformal Invariance** 43
   2.1 From Scale Invariance to Conformal Invariance 43
   2.2 Conformal Transformations in d Dimensions 44
   2.3 Conformal Transformations in Two Dimensions 46
   2.4 Conformal Invariance in Two Dimensions 49
   2.5 Correlation Functions of Quasi-primary Operators 51
   2.6 The Energy–Momentum Tensor 53

3. **Finite-Size Scaling** 63
   3.1 Statistical Systems in Finite Geometries 63
   3.2 Finite-Size Scaling Hypothesis 64
   3.3 Universality 68
   3.4 Phenomenological Renormalization 72
   3.5 Consequences of Conformal Invariance 74

3.6   Comparison with Experiments 78

4. **Representation Theory of the Virasoro Algebra** 83
    4.1   Verma Module 84
    4.2   Hilbert Space Structure 88
    4.3   Null Vectors 90
    4.4   Kac Formula and Unitarity 92
    4.5   Minimal Characters 97

5. **Correlators, Null Vectors and Operator Algebra** 101
    5.1   Null Vectors and Correlation Functions 101
    5.2   Operator Algebra and Associativity 104
    5.3   Analyticity and the Monodromy Problem 110
    5.4   Riemann's Method 112

6. **Ising Model Correlators** 117
    6.1   Spin-Density Four-Point Function 117
    6.2   Energy-Density Four-Point Function 121
    6.3   Mixed Four-Point Functions 123
    6.4   Semi-Local Four-Point Functions 124

7. **Coulomb Gas Realization** 127
    7.1   The Free Bosonic Scalar Field 127
    7.2   Screened Coulomb Gas 132
    7.3   Minimal Correlation Functions 134
    7.4   Minimal Algebras and OPE Coefficients 137

8. **The Hamiltonian Limit and Universality** 141
    8.1   Hamiltonian Limit in the Ising Model 141
    8.2   Hubbard–Stratonovich Transformation 144
    8.3   Hamiltonian Limit of the Scalar ϕ^4 Theory 146
    8.4   Hamiltonian Spectrum and Conformal Invariance 148
    8.5   Temperley–Lieb Algebra 150
    8.6   Laudau–Ginzburg Classification 154

9. **Numerical Techniques** 157
    9.1   Simple Properties of Quantum Hamiltonians 157
    9.2   Some Further Physical Quantities
           and their Critical Exponents 160
    9.3   Translation Invariance 162
    9.4   Diagonalization 163
    9.5   Extrapolation 170
           9.5.1   VBS Algorithm 174
           9.5.2   BST Algorithm 174
    9.6   The DMRG Algorithm 177

10. Conformal Invariance in the Ising Quantum Chain 183
   10.1 Exact Diagonalization 183
      10.1.1 General Remarks 183
      10.1.2 Jordan–Wigner Transformation 184
      10.1.3 Diagonalization of a Quadratic Form 185
      10.1.4 Eigenvalue Spectrum and Normalization 187
   10.2 Character Functions 189
   10.3 Finite-Size Scaling Analysis 191
      10.3.1 Ground State Energy 191
      10.3.2 Operator Content 194
      10.3.3 Finite-Size Corrections 197
      10.3.4 Finite-Size Scaling Functions 197
   10.4 The Spin-1 Quantum Chain 198
   10.5 The Virasoro Generators 201
   10.6 Recapitulation 203

11. Modular Invariance 205
   11.1 The Modular Group 205
   11.2 Implementation for Minimal Models 206
   11.3 Modular Invariance at $c = 1$ 211
      11.3.1 Circle or Coulomb Models 212
      11.3.2 Orbifold Models 213
   11.4 Lattice Realizations 216

12. Further Developments and Applications 219
   12.1 Three-States Potts Model 219
   12.2 Tricritical Ising Model 221
      12.2.1 Operator Content 221
      12.2.2 Supersymmetry and Superconformal Invariance 224
   12.3 Yang–Lee Edge Singularity 227
   12.4 Ashkin–Teller Model 230
      12.4.1 Relation with the XXZ Quantum Chain 231
      12.4.2 Global Symmetry and Boundary Conditions 231
      12.4.3 Phase Diagram 233
      12.4.4 Operator Content on the $c = 1$ Line 234
   12.5 XY Model 236
   12.6 XXZ Quantum Chain 238
   12.7 Ising Correlation Functions on Cylinders 242
   12.8 Alternative Realizations of the Conformal Algebra 242
      12.8.1 Logarithmic Conformal Theories 243
      12.8.2 Lattice Two-Point Functions 244
   12.9 Percolation 245
   12.10 Polymers 247
      12.10.1 Linear Polymers 247
      12.10.2 Lattice Animals 251

12.11 A Sketch of Conformal Turbulence 254
12.12 Some Remarks on 3D Systems 258

13. Conformal Perturbation Theory 261
13.1 Correlation Functions in the Strip Geometry 261
13.2 General Remarks on Corrections to the Critical Behaviour .. 263
13.3 Finite-Size Corrections................................. 265
      13.3.1 Tower of the Identity 266
      13.3.2 Application to the Ising Model 267
      13.3.3 Application to the Three-States Potts Model 268
      13.3.4 Checking the Operator Content
             from Finite-Size Corrections 270
13.4 Finite-Size Scaling Functions 270
      13.4.1 Ising Model: Thermal Perturbation 271
      13.4.2 Ising Model: Magnetic Perturbation 273
13.5 Truncation Method 275

14. The Vicinity of the Critical Point 279
14.1 The c-Theorem 280
      14.1.1 Application to Polymers.......................... 284
14.2 Conserved Currents Close to Criticality 285
14.3 Exact S-Matrix Approach.............................. 289
14.4 Phenomenological Consequences 298
      14.4.1 Integrable Perturbations 298
      14.4.2 Universal Critical Amplitude Ratios 304
      14.4.3 Chiral Potts Model 306
      14.4.4 Oriented Interacting Polymers 307
      14.4.5 Non-integrable Perturbations 311
14.5 Asymptotic Finite-Size Scaling Functions 316

15. Surface Critical Phenomena 321
15.1 Systems with a Boundary 321
15.2 Conformal Invariance Close to a Free Surface 326
15.3 Finite-Size Scaling with Free Boundary Conditions 330
15.4 Surface Operator Content 332
      15.4.1 Ising Model..................................... 332
      15.4.2 Three-States Potts Model 337
      15.4.3 Temperley–Lieb Algebra
             and Relation with the XXZ Chain 338
      15.4.4 Tricritical Ising Model 340
      15.4.5 Yang–Lee Edge Singularity 340
      15.4.6 Ashkin–Teller Model............................. 340
      15.4.7 XXZ Quantum Chain 342
      15.4.8 Percolation 343
      15.4.9 Polymers 346

15.5 Profiles 346
15.6 Defect Lines 350
       15.6.1 Aperiodically Modulated Systems 362
       15.6.2 Persistent Currents in Small Rings 363

16. Strongly Anisotropic Scaling 369
16.1 Dynamical Scaling 369
16.2 Schrödinger Invariance 372
16.3 Towards Local Scale Invariance for General θ 377
16.4 Some Remarks on Reaction–Diffusion Processes 383

Anhang/Annexe 385

List of Tables 388

List of Figures 390

References 391

Index 411

1. Critical Phenomena: a Reminder

We begin by recalling some basic facts about critical phenomena. This is not intended to provide a complete and exhaustive introduction to the field. We rather put emphasis on those standard concepts which will form the basis for the extension to conformal invariance in later chapters. Besides this, we establish our notation.

After a brief discussion of the phenomenology of second-order phase transitions, the conventional bulk critical exponents are defined. Scale invariance, as implemented via the renormalization group, will be used to derive the well-known scaling relations between the critical exponents. Some of the most conventional model systems used in the study of critical phenomena will be introduced. We also quote some experimental results for the critical exponents of two-dimensional systems. Finally, we recall the analogy between statistical mechanics and quantum theory.

The material of this chapter is fairly standard and further information can be found in the general overviews quoted. For the novice, a good, non-technical starting point is provided in [G1, G2, G3].

1.1 Phase Diagrams and Critical Exponents

The classical (and, historically, first) example of a spin system with a continuous phase transition is the celebrated **Ising model**[1] [360]. It is defined in terms of spin variables $\sigma_i = \pm 1$ attached to the sites i of some lattice and the Hamiltonian is

$$\mathcal{H} = -J \sum_{\langle i,i' \rangle} \sigma_i \sigma_{i'} - B \sum_i \sigma_i, \qquad (1.1)$$

where B is the external magnetic field. In this simplest form of the model, interactions occur only between pairs of nearest-neighbour sites and for our purposes, it is almost always easier to consider ferromagnetic interactions with an exchange coupling $J > 0$, rather than antiferromagnetic long-range order. While the model is initially defined on a lattice with a finite number

---

[1] Since the model was proposed by Lenz in 1920, it is sometimes referred to as Lenz–Ising model. Ising (1925) solved the model only in 1D, where no phase transitions occur.

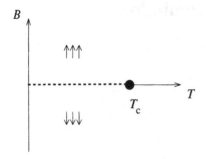

Fig. 1.1. Phase diagram of an Ising ferromagnet. Along the dotted line $B = 0$, there is a first-order transition between the two indicated ordered states. These two become indistinguishable at the second-order transition $T = T_c$, marked by the black dot

of sites $\mathcal{N}$, phase transitions can strictly occur only in the thermodynamic limit $\mathcal{N} \to \infty$. One is interested in the phase diagram as a function of both the temperature T and the magnetic field B. The generic Ising model phase diagram, displayed in Fig. 1.1, has a very simple structure. At low temperature, there are, for $B \neq 0$, two ordered ferromagnetic phases, where all the spins σ_i tend to have the same value, resulting in a macroscopically observable total magnetization M. There is a first-order phase transition along the line $B = 0$, where M changes its sign. At high temperatures, there is a single disordered phase where M vanishes for vanishing magnetic field B. Finally, the two ordered phases become physically indistinguishable precisely *at* the **critical point** located at $T = T_c, B = 0$. In this book, we shall be interested in the behaviour of spin systems like the Ising model close to these second-order phase transitions.

Consider further macroscopic thermodynamic quantities like the specific heat C or the magnetic susceptibility χ, as well as the magnetization M. In general, all these observables are *analytic* functions of both T and B, but that is no longer true when the line of first-order transitions $B = 0, T \leq T_c$ is crossed. We illustrate this in Fig. 1.2, where a sketch of the magnetization M as a function of B is given for several values of the temperature. At low temperatures, where macroscopically large domains of ordered spins are stable, M has a jump discontinuity across the line of first-order transitions at $B = 0$. In the high temperature phase, there is no macroscopic long-range order and M is continuous across the $B = 0$ line. However, even for $T > T_c$, there occur still large clusters of spins of the same orientation. The size of these clusters defines the **correlation length** ξ. If the distance r between two spins is small compared to ξ, we say that the two spins are correlated, since there is a large probability that they have the same value. On the other hand, if $r \gg \xi$, they are not and in this case one expects that the correlation probability should decrease roughly exponentially with r, for r sufficiently large. The mean cluster size of correlated spins grows with decreasing T (for $B = 0$ fixed) leading to an increase of ξ as T approaches the critical point T_c,[2] until, precisely at the critical point $T = T_c$, $B = 0$, the correlation

---

[2] For a nice experimental demonstration of this, see [E89].

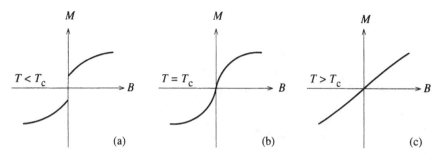

Fig. 1.2. Magnetization M of a spin system as a function of B, for (**a**) $T < T_c$, (**b**) $T = T_c$ and (**c**) $T > T_c$

length diverges. By definition, a second-order **critical point** is characterized by a diverging correlation length.[3] That does not mean, however, that at a critical point, all spins belong to a single, ordered, cluster; rather, the correlation length sets the maximal scale across which long-range correlations are dynamically created. Indeed, close to a critical point, fluctuations occur not only on length scales of the order of ξ, but on all length scales $\rho \leq \xi$ (we neglect here that the continuum description implicitly used is certainly inadequate on length scales of order $\rho \sim \mathcal{O}(a)$ of the lattice constant a). At the critical point, where $\xi \to \infty$, fluctuations on *all* length scales become important. Graphically, the presence of fluctuations on *all* length scales is usually visualized through Kadanoff's **droplet picture** [379]. We sketch this in Fig. 1.3 for the Ising model, which shows a nested hierarchy of correlated droplets with spin "up" inside larger correlated droplets with spin "down" and so forth.[4] Of course, the fractal nature of the clusters at criticality can only be schematically represented. As a consequence, microscopic details such as the precise lattice structure or precise form of the interactions, are no longer discernible and become close to criticality completely dominated by the intrinsic, dynamic fluctuations. Therefore the behaviour of the observables close to the critical point, which typically can be formulated in terms of simple algebraic relationships and described by critical exponents (which we shall define below), is independent[5] of them. In that sense, the microscopic details become unimportant for the critical behaviour.

The Ising model (1.1) may be conveniently defined on a hypercubic lattice and we had only taken nearest-neighbour interactions into account. However,

---

[3] On the other hand, ξ remains *finite* at a first-order transition, which is thus not critical.

[4] While many textbooks on statistical mechanics now contain reproductions of Monte Carlo simulations illustrating the droplet picture, it has recently also become possible to demonstrate this experimentally. See [E89] for an example occurring in the disordering of the Si(113)-(3 × 1) reconstruction.

[5] Experimentally, this is demonstrated very convincingly for binary liquids and the liquid–gas transition for a large number of substances in [E20].

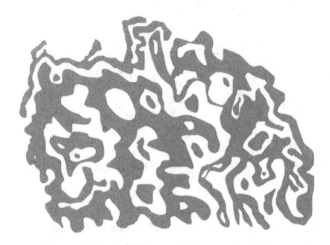

Fig. 1.3. Droplet picture for the Ising model. The different shading corresponds to the two possible states. The correlation length ξ is of the order of the size of the largest droplets

if we had also included next-nearest-neighbour interactions and taken all couplings to be ferromagnetic, the critical exponents would be independent of the ratio of the two couplings. The same values of the critical exponents are found when the model is defined on a hexagonal lattice, say, instead of a hypercubic lattice. The independence of quantities like critical exponents from such microscopic details is referred to as **universality**. Phenomenologically, the critical behaviour of systems near to their critical points can be cast into **universality classes** characterized by the values of the critical exponents. Put otherwise, the droplet picture means that at a critical point, a snapshot of the fluctuating spins taken at the length scale ρ would on average be independent of the value of ρ. That is the physical basis of **scale invariance** of a many-body system at criticality. Scale invariance is required to be able to describe the behaviour of a complicated system in terms of a few simple numbers, e.g. the critical exponents. One would of course like to be able to present a systematic classification of the possible universality classes. While that has in general not yet been achieved, for two-dimensional systems, **conformal invariance** will lead to a (partial) classification of the possible universality classes. How this comes about will be a major topic in this book.

For an analysis of the problem, the basic variables (in the context of the Ising model example (1.1), but also for more general systems) are the **reduced temperature** t and the **reduced magnetic field** h, defined as

$$t := \frac{T - T_c}{T_c} \qquad \text{and} \qquad h := \frac{B}{k_B T_c}, \tag{1.2}$$

which describe the deviation of the system from the critical point. We shall use units such that the Boltzmann constant $k_B = 1$ throughout.

The relationship of the thermodynamic observables to a statistical mechanics description is standard and we recall it here for the Ising model example. The basic quantity is, in the canonical ensemble, the **partition function**

$$\mathcal{Z} = \sum_{\{\sigma\}} \exp\left(-\frac{1}{T}\mathcal{H}\right),$$ (1.3)

where the sum is over all spin configurations. The free energy is

$$F = -T \ln \mathcal{Z},$$ (1.4)

from which the thermodynamics is recovered as usual, namely

$$C = -\frac{T}{\mathcal{N}}\frac{\partial^2 F}{\partial T^2} \ , \quad M = -\frac{1}{\mathcal{N}}\frac{\partial F}{\partial B} \ , \quad \chi = \frac{\partial M}{\partial B},$$ (1.5)

where C is the specific heat, M is the magnetization per site and χ is the susceptibility per site (where $\mathcal{N}$ is the number of sites). Averages $\langle X \rangle$ for any observable are calculated as, where $X[\{\sigma\}]$ is the value of X for the spin configuration $\{\sigma\}$

$$\langle X \rangle = \frac{1}{\mathcal{Z}} \sum_{\{\sigma\}} \left(X[\{\sigma\}] e^{-\mathcal{H}/T} \right).$$ (1.6)

In order to describe the correlation between clusters of spins, we shall need correlation functions like the **two-point correlation function** $G(i, j)$ of the spins σ_i and σ_j at sites i and j. It is defined as

$$G(r) = G(i, j) = \langle \sigma_i \sigma_j \rangle - \langle \sigma_i \rangle \langle \sigma_j \rangle,$$ (1.7)

where $r = |\boldsymbol{i} - \boldsymbol{j}|$ is the distance between the sites i and j. Also, thermodynamic observables can be rewritten in terms of averages. As an example, consider the magnetization per site, M, rewritten in the form

$$M = \langle \sigma_i \rangle = \frac{T}{\mathcal{N}\mathcal{Z}}\frac{\partial \mathcal{Z}}{\partial B} = \frac{T}{T_c}\frac{1}{\mathcal{N}\mathcal{Z}}\frac{\partial \mathcal{Z}}{\partial h}.$$ (1.8)

Returning to the two-point function $G(\boldsymbol{r})$, it is known that away from criticality and at large (compared to the microscopic lattice spacing a) distances $r/a \gg 1$, $G(r)$ decays exponentially:

$$G(r) \approx r^{-\tau} e^{-r/\xi} \ , \qquad T \neq T_c, \ B = 0,$$ (1.9)

where $\xi = \xi(T, B)$ is the **correlation length**. The value of the exponent τ depends on whether T is larger or smaller than T_c. In the two-dimensional Ising model one has, for example, $\tau = 1/2$ for $T > T_c$ and $\tau = 2$ for $T < T_c$ [G22, p. 306].

For the *definition of the critical exponents*, we write $X(t) \sim |t|^{x_\pm}$ if

$$x_\pm := \lim_{t \to \pm 0} \frac{\ln X(t)}{\ln |t|}$$ (1.10)

exists. This defines the critical exponents $x_\pm$ associated to the quantity $X(t)$. In most cases, the exponent obtained is the same for $t < 0$ and $t > 0$, that is $x_+ = x_-$. We shall henceforth assume this to be the case.

The conventional **critical exponents** α, β, γ, and δ are now defined by

$$
\begin{aligned}
C &\sim |t|^{-\alpha} \quad ; \quad h = 0 \\
M &\sim (-t)^\beta \quad ; \quad t < 0,\, h = 0 \\
\chi &\sim |t|^{-\gamma} \quad ; \quad h = 0 \\
M &\sim h^{1/\delta} \quad ; \quad t = 0.
\end{aligned}
\tag{1.11}
$$

However, the associated *scaling amplitudes* (not detailed in equation (1.11)) may depend on the sign of t. The available information on these is excellently reviewed in [521]. We shall briefly discuss a few conformal invariance predictions for them in Chap. 14, but otherwise, scaling amplitudes will not be considered in this book. We shall also need the correlation length critical exponent ν defined by

$$
\xi \sim |t|^{-\nu} \quad ; \quad h = 0
\tag{1.12}
$$

and, finally, the large-distance behaviour of $G(r)$ at criticality defines[6] the critical exponent η

$$
G(r) \sim r^{-d+2-\eta} \quad ; \quad t = 0,\, h = 0,
\tag{1.13}
$$

which is the last of the conventional bulk critical exponents we define.[7]

Critical exponents describe phenomenologically the behaviour of many-body systems close to criticality. It will be one of our main tasks to develop techniques for their explicit computation. Two-dimensional critical phenomena are peculiar in the sense that the conformal invariance of correlation functions at criticality allows one to find the exponents, and much more, *exactly* (see Table 1.1 below for numerical values of the exponents).

1.2 Scale Invariance and Scaling Relations

Having defined the conventional bulk critical exponents, we now show that scale invariance implies certain relations between them.

A popular way of deriving these is through renormalization group arguments. In view of the later extension of scale invariance to conformal invariance, we prefer here to follow a different route. Rather than relying on

---

[6] Sometimes, it is attempted to combine (1.9,1.13) into a single equation of the form $G(r) \sim r^{-d+2-\eta} \exp(-r/\xi)$, supposedly valid for all temperatures T close to T_c and for $B = 0$. However, the example of the 2D Ising model mentioned above [G22, p. 306] shows that such attempts have in general no meaning. This example shows as well that the Ornstein-Zernicke (mean field) form $G(r) \sim r^{-(d-1)/2} \exp(-r/\xi)$ cannot be generally valid in the immediate neighbourhood of the critical point either.

[7] η may be negative. For example, in the 3D cubic model, $\eta = -0.39 \pm 0.22$ [41].

heuristic arguments to derive scaling, we take here the point of view that a model at a critical point can be *axiomatically* characterized through the **covariance** of the entire set of all n-point correlation functions of all observables (below, we shall refer to these as *scaling operators*) under scale transformations. This idea will be extended to different forms of conformal invariance in later chapters. However, at the end of this section we briefly recall the renormalization group derivation of scaling.

To be specific, we shall appeal to the Ising model example, but the following discussion is completely general. The simplest examples of scaling operators are the order parameter density $\sigma(r)$ and the energy density $\varepsilon(r)$. We are interested in their two-point functions

$$G_\sigma(r_1 - r_2) = \langle \sigma(r_1)\sigma(r_2) \rangle - \langle \sigma(r_1) \rangle \langle \sigma(r_2) \rangle,$$
$$G_\varepsilon(r_1 - r_2) = \langle \varepsilon(r_1)\varepsilon(r_2) \rangle - \langle \varepsilon(r_1) \rangle \langle \varepsilon(r_2) \rangle \tag{1.14}$$

and we have already taken advantage of translation invariance, which normally holds for distances $a \ll |r| \ll \xi$, where a is a microscopic length and ξ the correlation length. In terms of the reduced variables t and h defined above, our starting point are the relations

$$G_\sigma(r; t, h) = b^{-2x_\sigma} G_\sigma(r/b; tb^{y_t}, hb^{y_h}), \tag{1.15}$$
$$G_\varepsilon(r; t, h) = b^{-2x_\varepsilon} G_\varepsilon(r/b; tb^{y_t}, hb^{y_h}), \tag{1.16}$$

where x_σ and x_ε are called the **scaling dimensions** of the **scaling operators** σ and ε.[8] Throughout this book, h and t will be called the **scaling fields** canonically conjugate to σ and ε. The numbers y_t and y_h will be referred to as **renormalization group eigenvalues** and b is the dilatation factor. Equations (1.15,1.16) express at the critical point $t = h = 0$ the covariance of G_σ and G_ε under scale transformations.

These correlators are related to the susceptibility and specific heat, through the fluctuation-dissipation theorems

$$\chi = \frac{1}{T} \sum_r G_\sigma(r) \simeq \frac{1}{T} \int d^d r \, G_\sigma(r)$$
$$C = \frac{1}{T^2} \sum_r G_\varepsilon(r) \simeq \frac{1}{T^2} \int d^d r \, G_\varepsilon(r). \tag{1.17}$$

Integrating (1.15), we get for the scaling of the susceptibility per site

$$\chi(t, h) = b^{d-2x_\sigma} \chi(tb^{y_t}, hb^{y_h}). \tag{1.18}$$

Recall that $\chi = -\partial^2 f/\partial h^2$. Therefore, integrating twice with respect to h, we arrive at the scaling of the **free energy density** $f = F/\mathcal{N}$

---

[8] The reason why the classical observables σ and ε are referred to as scaling operators will become clear later, when these will be related to operators in a quantum field theory.

$$f(t, h) = b^{d-2x_\sigma - 2y_h} f(tb^{y_t}, hb^{y_h}). \tag{1.19}$$

We remark that the scaling form just obtained for f merely describes the scaling of the *singular* part f^{sin} of the free energy. In general, the free energy density $f = f^{\text{reg}} + f^{\text{sin}}$ does contain a regular and a singular part. The regular part is analytic at a second-order critical point. It merely acts as a background term and is of little interest for us. The singular part contains the terms which produce the singular behaviour of the specific heat close to criticality. Unless stated explicitly otherwise, we shall always restrict attention to the singular part f^{sin} and shall normally simply write f for short. In the same way as for χ, we get for the scaling of the specific heat

$$C(t, h) = b^{d-2x_\varepsilon} C(tb^{y_t}, hb^{y_h}) \tag{1.20}$$

and consequently, since $C = -\partial^2 f/\partial t^2$, up to factors which are constant and non-zero close to criticality, we have

$$f(t, h) = b^{d-2x_\varepsilon - 2y_t} f(tb^{y_t}, hb^{y_h}). \tag{1.21}$$

Comparing the two forms for the scaling of f, we see that $x_\varepsilon + y_t = x_\sigma + y_h$. In fact, a similar relationship in the scaling of the free energy could be derived for any pair of scaling operators coupled to their canonically conjugate scaling fields. Since the above constant is independent of the scaling operators, it should have simple value. A plausible choice,[9] in view of the scaling of the singular free energy density $f = F/\mathcal{N}$ is

$$x_\varepsilon + y_t = x_\sigma + y_h = d \tag{1.22}$$

and we finally have for the scaling of the (singular) free energy density

$$f(t, h) = b^{-d} f(tb^{y_t}, hb^{y_h}). \tag{1.23}$$

The factor b^{-d} can be understood in a simple way from the fact that the number of sites $\mathcal{N}$ which appears in the definition of f is reduced by a factor b^{-d} if all distances are rescaled by a factor b. Below, we shall recover (1.22) from a renormalization group argument.

One may repeat this dilatation n times to arrive at

$$f(t, h) = b^{-dn} f(b^{ny_t} t, b^{ny_h} h). \tag{1.24}$$

From this, an important classification of the scaling fields through the values of their *renormalization group eigenvalues* becomes apparent. To be generic, we call the scaling field u and the associated renormalization group eigenvalue y. Now consider the argument $\mathcal{U} = b^{ny} u$ of f, where criticality corresponds to

---

[9] Recall that perturbing the system away from criticality amounts to add terms $\sum_i (t\varepsilon_i + h\sigma_i)$ to the Hamiltonian $\mathcal{H}$. Since x is the scaling dimension of a scaling operator and y the one of its conjugate scaling field, (1.22) follows from dimensional counting.

$u = \mathcal{U} = 0$. If $y > 0$, repeated application of a dilatation will drive the system away from the critical point, for any arbitrarily small initial perturbation $u \neq 0$ around the fixed point $u = 0$. Such a scaling field is called **relevant**. On the other hand, if $y < 0$, repeated dilatations will drive the system towards the critical value $\mathcal{U} = 0$. Such a scaling field is called **irrelevant**. Finally, if $y = 0$, the scaling field u is called **marginal**.[10] In general, both t and h are relevant scaling fields. The notions of relevance, irrelevance and marginality are also carried over to the conjugate scaling operators.

In principle, there are infinitely many irrelevant scaling operators in a given model. Close to criticality, their effect can be made arbitrarily small by considering the system on sufficiently large scales. It follows that the long-distance behaviour of the system is completely determined from the relevant scaling operators and it will be only those which will enter into scaling forms for quantities like the free energy. That is the reason why we left out the irrelevant scaling fields already in (1.15,1.16).

Formally, the scaling form (1.23) of the (singular) free energy density is a homogeneity relation. The dilatation factor b can be eliminated by fixing $b^{y_t} t = K$ and we obtain

$$f^{\text{sin}}(t, h) = \left|\frac{t}{t_0}\right|^{d/y_t} W_\pm \left(\left(\frac{h}{h_0}\right)\left(\frac{t}{t_0}\right)^{-y_h/y_t}\right), \tag{1.25}$$

where $W_\pm$ are **scaling functions** and the index $\pm$ refers to $t > 0$ or $t < 0$, respectively. The scaling functions $W_\pm$ are **universal**, that is, independent of the irrelevant scaling operators and scaling fields. Irrelevant scaling fields only determine the non-universal constants t_0 and h_0.

We now discuss the relation of the scaling of the free energy with the critical exponents defined in the previous section. Recall the definition of the exponent α from the zero-field specific heat

$$C = -\partial^2 f/\partial t^2 \sim |t|^{d/y_t - 2} \; ; \; h = 0 \tag{1.26}$$

leads to $\alpha = 2 - d/y_t$. The exponents β, γ and δ are found analogously from $M = -\partial f/\partial h$ and $\chi = -\partial^2 f/\partial h^2$, with the result

---

[10] In this discussion, we have taken (1.15,1.16) literally and simply used the reduced variables t and h to represent the scaling fields. That is only correct in the immediate vicinity of the critical point. More precisely, the scaling fields conjugate to the scaling operators σ, ε are functions u_t, u_h which transform under dilatations as $u'_t = b^{y_t} u_t(t, h)$, $u'_h = b^{y_h} u_h(t, h)$ and only for t, h sufficiently small, one actually has $u_t \sim t, u_h \sim h$. These distinctions become important when carrying out a real-space renormalization group procedure explicitly. Then the above relations between u'_t and u_t can be calculated explicitly as a result from a renormalization group transformation.

Also, calling the y's 'renormalization group eigenvalues' amounts to a certain abuse of language, because from the preceeding remark is it clear that $\lambda_t = b^{y_t}$, and not y_t, is an eigenvalue under a renormalization group transformation at a fixed point. But we shall follow the common (bad) habit in the sequel.

$$\alpha = 2 - d/y_t$$
$$\beta = (d - y_h)/y_t$$
$$\gamma = (2y_h - d)/y_t \qquad (1.27)$$
$$\delta = y_h/(d - y_h).$$

From these, y_t and y_h can be eliminated and we obtain the first two of the celebrated **scaling relations**

$$\boxed{\begin{aligned} \alpha + 2\beta + \gamma &= 2 \\ \alpha + \beta(1 + \delta) &= 2 \end{aligned}} \qquad (1.28)$$

We reconsider the order parameter correlation function scaling (1.15). For simplicity, let $h = 0$. Then, eliminating the dilatation factor as before by fixing $b^{y_t} t = K$, we have

$$G_\sigma(r; t, 0) = t^{2x_\sigma/y_t} G_\sigma \left(r/(K/t)^{1/y_t}; K, 0 \right) \qquad (1.29)$$

and it follows that the r-dependence of $G_\sigma(r)$ is always expressed in terms of the variable r/ξ. Close to, but not precisely at, criticality, $G_\sigma(r)$ decays exponentially with $|r|$ and we can identify ξ with the correlation length, which consequently must show the scaling

$$\xi \sim t^{-1/y_t} \ ; \ h = 0, \qquad (1.30)$$

which gives $\nu = 1/y_t$. Considering the limit $t \to 0$, a finite limit for G_σ is only possible if $G_\sigma(r; 0, 0) \sim r^{-2x_\sigma}$, or $\eta = 2x_\sigma + 2 - d$. From the scaling (1.18) of the susceptibility, we have $\gamma = (d - 2x_\sigma)/y_t$. Thus

$$\boxed{\gamma = \nu(2 - \eta)} \ . \qquad (1.31)$$

The last of the scaling relations involves **hyperscaling**. From the scaling behaviour of the singular free energy density (1.25), we read off

$$\boxed{\alpha = 2 - d\nu} \ . \qquad (1.32)$$

This is only valid provided that the scaling functions $W_\pm$ remain finite at the critical point. One can show that no singularities occur in the scaling functions below the **upper critical dimension** d^*.[11] For $d \geq d^*$, on the other hand, singularities of the free energy scaling function occur generically and are due to so-called **dangerous irrelevant** scaling fields (Fisher [231]). While dangerous irrelevant variables are absent for $d < d^*$, the breaking of hyperscaling is still possible if the scaling functions vanish precisely at the critical point.

We remark that the following exponent *in*equalities can be derived rigorously and depend only on the convexity of the free energy for their derivation, see e.g. [G5]

---

[11] For dimensions $d > d^*$ the critical exponents coincide with the results of a mean field calculation. For the Ising model with short-range interactions, $d^* = 4$.

$$\alpha + 2\beta + \gamma \geq 2 \ , \quad 2 - \alpha \leq d\nu \ , \quad \gamma \geq (2 - \eta)\nu \ , \quad \alpha + \beta(\delta + 1) \geq 2. \quad (1.33)$$

Scale-invariant theories satisfy these as equalities.

Having found the scaling relations, we now discuss how the correlator scaling (1.15) may be motivated from the **renormalization group** (RG). For simplicity, we only consider explicitly the two relevant variables t, h. A real-space renormalization group transformation performed on a spin system may be constructed as follows: the system, originally defined in terms of spin variables σ_i sitting on the sites i of a hypercubic lattice, is divided into cells, each of a volume $b \times b \times \ldots = b^d$, where $b \geq 1$ is some constant. In each cell, out of the individual spins a new **block spin** is formed. Then the interaction Hamiltonian $\mathcal{H}[\sigma; t, h]$ is rewritten in terms of the block spins, leading to a new Hamiltonian $\mathcal{H}'[\sigma'; t', h']$ where in general the couplings t, h are also transformed. For simplicity, we neglect here the creation of new interaction terms not yet contained in $\mathcal{H}$, which can be justified for example if t, h are the only two relevant scaling fields. Schematically, this change of variables leads to the following changes:

1. a block spin transformation which transforms the spin σ into a block spin σ';
2. a rescaling of distances according to $\boldsymbol{r} \to \boldsymbol{r}' = \boldsymbol{r}/b$;
3. the scaling fields are changed into $t \to t'$ and $h \to h'$.

such that the partition function $\mathcal{Z} = \mathcal{Z}(t, h) = \mathcal{Z}(t', h') = \mathcal{Z}'$ remains invariant. Now, if the system sits at a renormalization group **fixed point**, one has $t = t' = t^*$ and $h = h' = h^*$. Furthermore, the droplet picture suggests that the spin configurations are (on the average) self-similar under rescaling. Now, consider the special case $t = 0$ and assume that under a **dilatation** (scale transformation), we have

$$\boldsymbol{r} \to \boldsymbol{r}' = \boldsymbol{r}/b$$
$$h \to h' = h\, b^y. \quad (1.34)$$

Since h is conjugate to the spin σ, the mean value is

$$
\begin{aligned}
\langle \sigma(\boldsymbol{r}) \rangle &= \frac{1}{N\mathcal{Z}} \frac{\partial \mathcal{Z}}{\partial h(\boldsymbol{r})} \\
&= \frac{1}{N} \frac{1}{\mathcal{Z}'} \frac{\partial \mathcal{Z}'}{\partial h(\boldsymbol{r})} \\
&= \frac{1}{N'\, b^d} \frac{1}{\mathcal{Z}'} \frac{\partial h'(\boldsymbol{r}')}{\partial h(\boldsymbol{r})} \frac{\partial \mathcal{Z}'}{\partial h'(\boldsymbol{r}')} \\
&= b^{y-d} \langle \sigma'(\boldsymbol{r}') \rangle, \quad (1.35)
\end{aligned}
$$

where the factor b^y comes from the rescaling of h and the factor b^{-d} takes into account that to a single block spin σ' there correspond b^d individual spins before performing the dilatation. Now, defining the scaling dimension $x_\sigma = d - y$ of the spin σ, we have

$$\langle\sigma(\boldsymbol{r})\rangle_h = b^{-x_\sigma}\langle\sigma'(\boldsymbol{r'})\rangle_{h'}. \tag{1.36}$$

Higher correlators are treated along the same lines. For example, the spin–spin correlation function is

$$\bar{G}_\sigma(\boldsymbol{r}_1 - \boldsymbol{r}_2) = \langle\sigma(\boldsymbol{r}_1)\sigma(\boldsymbol{r}_2)\rangle = \frac{1}{N^2\,\mathcal{Z}}\frac{\partial^2\mathcal{Z}}{\partial h(\boldsymbol{r}_1)\partial h(\boldsymbol{r}_2)} \tag{1.37}$$

and we find its transformation under dilatations as before, thereby restoring the t dependence

$$\bar{G}_\sigma(\boldsymbol{r};t,h) = b^{-2x_\sigma}\bar{G}_\sigma(\boldsymbol{r}/b; tb^{y_t}, hb^{y_h}), \tag{1.38}$$

where $x_\sigma + y_h = d$. We thus recover the same scaling as (1.15), together with (1.22). Other correlators can be treated similarly.

The following notation is common and will be used throughout.

$$x_\phi = d - y_\phi \tag{1.39}$$

is called the **scaling dimension** of the **scaling operator** ϕ. The scaling operators are canonically conjugated to the scaling fields. In particular, y_t corresponds to the scaling field t and y_h corresponds to the scaling field h. The conjugated lattice operators are in first approximation the energy density ε for t

$$\varepsilon_i = \sum_{i'} J(i,i')\sigma_i\sigma_{i'} \tag{1.40}$$

and the magnetization density σ for h. Their scaling dimensions are related to the conventional exponents through

$$\boxed{x_\varepsilon := x_t = d - y_t = (1-\alpha)/\nu} \ , \ \boxed{x_\sigma := x_h = d - y_h = \beta/\nu} \ . \tag{1.41}$$

In the 2D Ising model, one has $x_\sigma = 1/8$ and $x_\varepsilon = 1$.

We have thus seen that scale invariance, as implemented by the renormalization group, yields relations between the critical exponents (or equivalently the relevant scaling dimensions), but without fixing their numerical value. In addition, the n-point correlation functions transform covariantly under a dilatation, at the critical point. In later chapters we shall use conformal invariance techniques to calculate exactly the scaling dimensions of *all* scaling operators (that is, relevant, marginal and irrelevant) for a large variety of systems. This amounts to give the full **operator content** of a spin system.

1.3 Some Simple Spin Systems

We now give a list of some well-known models, which we shall study later in much detail. In this sections, models will be conveniently defined on isotropic square lattices. In later chapters, we shall often consider anisotropic models where the couplings in the two different directions of the square lattice are different. That does not affect the universality class of any critical point in these models, however. Unless stated explicitly otherwise, the models to be defined now are all conformally invariant at their critical points.

1.3.1 Ising Model

Although the Ising model Hamiltonian was already defined in (1.1) and the phase diagram is displayed in Fig. 1.1, we begin our short list of statistical models with this classic example. Applications of the Ising model occur in very different settings, for example binary fluids or alloys or uniaxial magnets.

While the mapping of the Ising model magnet onto a lattice gas, a binary fluid[12] or a binary alloy is a standard textbook matter, it might be worth to recall some of the physical foundations behind the applicability of the Ising model Hamiltonian to magnetic systems, following standard texts, e.g. [34, 365]. The most naive interpretation of a simple spin system such as an Ising model would be that $\mathcal{H}$ represented the interaction energy of two magnetic dipole moments sitting at the lattice sites i and i' (and for the Ising model, being somehow constrained such that only certain orientations are possible). In any *realistic* magnet, such terms are present. However, in most systems these cannot account for the observed magnetic ordering, since a simple calculation shows that dipolar interactions are (i) long-ranged and (ii) have interaction energies which correspond to critical temperatures of around 2 K, which is far from the observed values of the critical temperatures in most magnets. Rather, the Ising model interaction is an *exchange interaction* which arises from the antisymmetry of the electron wave function under particle exchange, leading to much larger values of the coupling J and to a very rapid fall-off of the couplings with the spatial distance. While in most theoretical models, the dipolar terms are simply ignored, the interplay between exchange and dipolar interactions offers a few subtle points.

In principle, the presence of the long-range dipolar interactions may change the universality class of the system. This question can be addressed using renormalization group methods. Surprisingly, it matters here whether the magnetic ordering is ferromagnetic or antiferromagnetic [11, E21]. For antiferromagnetic order, the long-range nature of the interactions is irrelevant. Since most uniaxial magnetic systems order indeed antiferromagnetically, it is then completely legitimate to ignore the dipolar terms from the beginning. On the other hand, for ferromagnetic ordering, the dipolar terms are relevant. Then the upper critical dimension is $d^* = 3$, rather than $d^* = 4$ as for the short-ranged Ising model. This situation is observable in a few uniaxial systems without exchange interactions such as $LiTbF_4$ and $LiHoF_4$ [E12, E21], where indeed $T_c \simeq 1 - 3$ K.

However, since dipolar terms are present in any real magnet, one should expect that very close to the critical point, the dipolar terms may induce

---

[12] Historically, the first observation of critical behaviour seems to go back to Cagniard de la Tour in 1822 in experiments on the vaporisation of water and alcohol. The continuity of that transition was apparently first discussed by Herschel (1830) and Faraday (1844). The resulting confusion was finally cleared up and the notion of a *critical point* was introduced by Andrews (1869) following experiments on the critical opalescence in CO_2. See [542, 110] for more information.

a cross-over to a new renormalization group fixed point for ferromagnetic materials, even though that region will be very tiny for the conventional ferromagnets (in $d = 3$ dimensions, the critical behaviour induced by the dipolar terms should be mean-field like, up to logarithmic factors) and should escape attention at the temperature resolution currently available in experiments.

The reader may also wonder why we bother about 'classical' spins at all, while every textbook on quantum mechanics deals at length with the quantum nature of spin. Rather than considering spin variables $\sigma = \pm 1$, one should write down a quantum Hamiltonian, which in the simplest cases might take the form (in 1D)

$$H = -J \sum_n \left(t\sigma_n^x + \sigma_n^z \sigma_{n+1}^z + h\sigma_n^z \right), \tag{1.42}$$

where the $\sigma^{x,z}$ are Pauli matrices and t and h are couplings. This model is referred to as the **quantum Ising model** and we shall discuss it in detail later. Indeed, it can be shown through renormalization group arguments that for non-vanishing temperatures, the critical behaviour of the quantum Ising model reduces to the one of the classical Ising Hamiltonian (1.1), and in the *same* space dimension d. For vanishing temperature $T = 0$, however, quantum effects do become important and must be included in the analysis. We shall recall later in this chapter that to each model of classical spins defined by a classical spin Hamiltonian $\mathcal{H}$ in $d + 1$ space dimensions, one can associate a quantum spin model characterized by a quantum Hamiltonian H in d dimensions. The quantum spin model has at $T = 0$ a phase transition which is in the same universality class as the classical spin model defined through $\mathcal{H}$, where the **transverse field** t plays the role of a temperature. In Chap. 8 we shall show how to derive the quantum Hamiltonian H from a given classical Hamiltonian $\mathcal{H}$ and in particular, how to obtain (1.42) from the usual 2D Ising model.

The relationship between the temperature T in the classical spin system and the transverse field t in the associated quantum system at zero temperature has recently been confirmed experimentally in the dipolar-coupled Ising ferromagnet LiHoF$_4$ [E12].

1.3.2 Tricritical Ising Model

The first generalization of Ising criticality is obtained when considering an Ising *anti*ferromagnet with Hamiltonian

$$\mathcal{H} = +J \sum_{\langle i, i' \rangle} \sigma_i \sigma_{i'} - B \sum_i \sigma_i - B_s \sum_i (-1)^i \sigma_i, \tag{1.43}$$

where $J > 0$ is the exchange coupling between spins at the nearest neighbour sites i and i' and B is the magnetic field. The last term will be explained shortly, for the moment we put the **staggered magnetic field** $B_s = 0$. On a cubic lattice, one of the zero-field $(B = 0)$ ground states is shown in Fig. 1.4a. Here, the magnetic field B is *not* conjugate to the order parameter. Rather, we see that the system decomposes into two interprenetrating sublattices, one with spin $\langle \sigma \rangle = +1$ and the other one with $\langle \sigma \rangle = -1$. That may be described by writing $\sigma_j = (-1)^j M_s := M_s \exp(i r_j \cdot q)$, where $q = (\pi/a, \pi/a)$ for a cubic lattice with lattice constant a and M_s is called the **staggered magnetization**. That is the order parameter conjugate to the staggered magnetic field B_s. For $B = 0$, it is an easy exercise to rewrite the system as a ferromagnetic Ising model. One thus expects for $B = B_s = 0$ a second-order transition at the same critical temperature as for the ferromagnetic Ising model and in the Ising universality class.

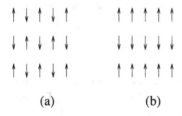

(a)   (b)

Fig. 1.4. (a) Antiferromagnetic and (b) metamagnetic ground states

The magnetic field B tends to align all spins and thus favours a paramagnetic ordering. That leads to a *competition* between the two ordered states and at zero temperature, there will be a first-order transition between them, which on a square lattice occurs at $B = 4J$.

A typical phase diagram[13] is sketched in Fig. 1.5, where the lines of first-order and second-order transitions meet in the **tricritical point**, located at $B = B_t$ and $T = T_t$. While at a normal critical point two phases become indistinguishable, at a tricritical point *three* physically distinct phases merge. For that reason, models such as (1.43) are referred to as **tricritical Ising model**. At the tricritical point, the system goes over into a new universality class, with values of the critical exponents[14] different from those of the simple ferromagnetic Ising model. For reviews on tricritical scaling, see [430, 408].

---

[13] In practice, one adds ferromagnetic next-nearest-neighbour interactions in (1.43), which stabilize the antiferromagnetic ground state and are needed in order to have a tricritical point with $T_t \neq 0$. More general situations are depicted in [12].

[14] At the tricritical point, there occur new tricritical exponents beyond those already introduced which may be related to the four x_i, see [430]. Let t and g be the scaling fields associated with the scaling operators ε and ε' and denote by h and h_3 the scaling field corresponding to σ and σ'. Then the free energy scaling (1.25) generalizes to $f_{\sin}(t, g, h, h_3) = |t|^{2-\alpha} W_{\pm}(gt^{-\phi}, ht^{-\Delta}, h_3 t^{-\Delta_3})$ or else $f_{\sin}(t, g, h, h_3) = |g|^{2-\alpha_t} W_{t,\pm}(gt^{-\phi_t}, ht^{-\Delta_t}, h_3 t^{-\Delta_{3t}})$ where the exponents indexed by a t are **tricritical exponents**. The index $\pm$ refers to $t, g > 0$ and $t, g < 0$, respectively. Scaling relations can used to relate the tricritical exponents $\alpha_t, \beta_t, \gamma_t, \delta_t, \nu_t, \eta_t$ to the scaling dimensions of the four relevant operators. In particular, (1.27) is readily rederived. Besides, the tricritical point can be described in terms of a supersymmetric field theory, see Chap. 12.

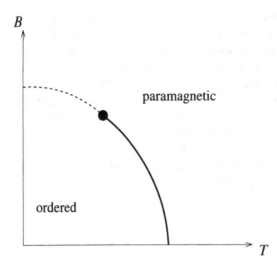

B

paramagnetic

ordered

T

Fig. 1.5. Phase diagram of the tricritical Ising model. The (broken) first-order line and the (full) line of Ising critical points meet at the tricritical point (B_t, T_t), shown as black dot

For the tricritical Ising model, there are four relevant scaling operators at the tricritical point, rather than two as for the normal Ising model. In 2D, the order parameter σ has scaling dimension $x_\sigma = 3/40$ and there is a subleading magnetic operator σ' with scaling dimension $x_{\sigma'} = 7/8$. For energy density ε, one has $x_\varepsilon = 1/5$ and there is a subleading energy-like operator ε' with $x_{\varepsilon'} = 6/5$. These results will be derived from conformal invariance in Chap. 12.

Phase diagrams with the topology of Fig. 1.5 are observed in 3D **meta-magnetic** systems like FeCl$_2$ [12]. These contain antiferromagnetic planes which themselves are coupled ferromagnetically, leading to ground states as in Fig. 1.4b. Alternatively, one might consider an Ising model with annealed vacancies, modeled by a vacancy variable $v_i = 0, 1$ and the Hamiltonian

$$\mathcal{H} = -J \sum_{\langle i,i' \rangle} \sigma_i \sigma_{i'} v_i v_{i'} - \mu \sum_i v_i - B \sum_i \sigma_i v_i. \qquad (1.44)$$

In the partition function, the sum is now over all configurations in *both* σ_i and v_i. Here, the extent to which the lattice is filled with spins is controlled through the chemical potential μ. Replacing $B \rightarrow B_s$ and $e^{-\mu} \rightarrow B$, one recovers the same phase diagram topology as in Fig. 1.5. Here, the scaling operator ε' can be interpreted as a *vacancy operator*. Finally, one may also define $s_i := \sigma_i v_i$, which takes the values $s_i = -1, 0, 1$. One then obtains the **Blume–Capel model** with Hamiltonian

$$\mathcal{H} = -J \sum_{\langle i,i' \rangle} s_i s_{i'} - B \sum_i s_i - \mu \sum_i s_i^2. \qquad (1.45)$$

1.3.3 q-States Potts Model

An important generalization of the Ising model was introduced by Potts in 1951 (see [625, 450] for reviews). On each site, there is attached a variable Θ, which is allowed to take q different values, for example $\Theta = 0, 1, \ldots, q - 1$. The Hamiltonian is

$$\mathcal{H} = -J \sum_{\langle i, i' \rangle} \delta(\Theta_i - \Theta_{i'}) \tag{1.46}$$

and the sum is over pairs of nearest neighbours, as usual. Here, we write $\delta(\Theta) = \delta_{\Theta, 0}$ for the Kronecker delta, which is unity for $\Theta = 0$ and zero otherwise. Obviously, the global symmetry of the model is the permutation group $\mathbb{S}_q$ of q distinct objects (see e.g. [293]).

The Ising model is recovered as the special case $q = 2$. Defining the Ising spin variable $\sigma = 2(\Theta - 1/2)$, the interaction term becomes $\delta(\Theta_i - \Theta_{i'}) = (\sigma_i \sigma_{i'} + 1)/2$. Thus, up to a multiplicative factor in the coupling J and a trivial additive constant, the Hamiltonian (1.1) results.

The q-states Potts model displays a disordered high-temperature phase and an ordered low-temperature phase. The transition between these two phases is of first order for $q > 4$ and of second order for $q \leq 4$, which are the cases we are interested in. The operator contents of the 3-state and 4-state Potts model are more complicated than for the Ising model and will be discussed in Chap. 12. Values for some of the conventional critical exponents are given in Table 1.1 below.

The two distinct phases in the q-states Potts model are related through a **duality transformation**. The presentation given here follows the one presented in [625] (see [169] for an alternative derivation). Let $K = J/T$ and consider the partition function

$$\begin{aligned} \mathcal{Z} &= \sum_{\{\Theta\}} \exp \left(K \sum_{\langle i, i' \rangle} \delta(\Theta_i - \Theta_{i'}) \right) \\ &= \sum_{\{\Theta\}} \prod_{\langle i, i' \rangle} \exp \left(K \delta(\Theta_i - \Theta_{i'}) \right) \\ &= \sum_{\{\Theta\}} \prod_{\langle i, i' \rangle} \left(1 + v \delta(\Theta_i - \Theta_{i'}) \right), \end{aligned} \tag{1.47}$$

where $v = e^K - 1$. For simplicity, one considers a square lattice with $\mathcal{N}$ sites. Now, multiply out out the terms in (1.47). The contribution of an individual spin configuration can then represented in a graphical way. Each site i of the lattice is occupied with a variable Θ_i. Draw a bond between neighbouring sites i and i' if $\Theta_i = \Theta_{i'}$. One obtains a graph G associated with that configuration. Let $b(G)$ be the number of bonds for the graph, see Fig. 1.6. Then the contribution to the partition function is $v^{b(G)}$. Furthermore, the sum over all configurations can be decomposed as $\sum_{\{\Theta\}} = \sum_G \sum_{\text{values}}$, where the second sum refers to a sum over all possible values of the Θ_i in a given cluster of G. If $n(G)$ is the number of disconnected components of the graph G, the second sum will simply produce a factor $q^{n(G)}$. The partition function reads

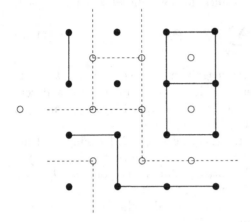

Fig. 1.6. Duality in the Potts model. Normal sites are full dots and dual sites are open dots. Full lines indicate bonds between sites, which make up the graph G and broken lines are the dual bonds of the dual graph D. The single dual site of D at the exterior of G is connected to all outgoing broken lines

$$\mathcal{Z} = \sum_G v^{b(G)} q^{n(G)}. \tag{1.48}$$

This form of the partition function may be viewed as a high-temperature (v small) expansion. The duality transformation now relates $\mathcal{Z}$ to the low-temperature behaviour.

To do so, one introduces a **dual lattice** and dual bonds [618, 625]. An example is given in Fig. 1.6. For the square lattice, **dual sites** correspond to plaquettes. Two dual sites are joined by a **dual bond**, provided that it does not intersect a bond of G. Let $c(G)$ be the number of loops (cycles) of the graph G. By construction, each loop of G encircles a dual site of the dual graph D and each dual loop of D encircles a site of G. Thus

$$n(D) = c(G) + 1 \;, \;\; n(G) = c(D) + 1. \tag{1.49}$$

The **Euler relation** states that for *any* graph $c(G) = b(G) + n(G) - \mathcal{N}$. Finally, the total number of bonds on the lattice is $\mathcal{B} = b(G) + b(D)$.

With these concepts, the partition function (1.48) can be rewritten in terms of dual graphs, where $\mathcal{N}^*$ is the number of dual sites of the dual lattice

$$\begin{aligned}
\mathcal{Z} &= \sum_D v^{\mathcal{B} - b(D)} q^{c(D)+1} \\
&= v^{\mathcal{B}} \sum_D v^{-b(D)} q^{b(D)+n(D)-\mathcal{N}^*+1} \\
&= v^{\mathcal{B}} q^{1-\mathcal{N}^*} \sum_D \left(\frac{q}{v}\right)^{b(D)} q^{n(D)}. \tag{1.50}
\end{aligned}$$

Define the **dual coupling** $K^* = J^*/T$ through $q/v = v^* = e^{K^*} - 1$. In the limit $\mathcal{N} \to \infty$ of the square lattice, the original and the dual lattice coincide and one has $\mathcal{N}^* = \mathcal{N}$ and $\mathcal{B} = 2\mathcal{N}$. Then the partition functions $\mathcal{Z}(K)$ and $\mathcal{Z}(K^*)$ are related through a **self-duality** relation

$$\lim_{N \to \infty} \left(\frac{1}{N} \ln \mathcal{Z}(K) \right) = \lim_{N \to \infty} \left(\frac{1}{N} \ln \mathcal{Z}(K^*) \right) + \ln \left(\left(e^K - 1 \right) / \left(e^{K^*} - 1 \right) \right),$$

$$(1.51)$$

where

$$\left(e^K - 1 \right) \left(e^{K^*} - 1 \right) = q. \tag{1.52}$$

Any singularity occurring in a thermodynamic quantity at a critical point $K = K_c$ is mapped through the duality transformation to another coupling $K = K_c^*$. Provided that there is a single phase transition in the q-state Potts model (and this can be checked by other means), it must be *at* the fixed point of the duality transformation. This fixes the critical point $K_c = J/T_c$ exactly at

$$K_c = \ln \left(\sqrt{q} + 1 \right). \tag{1.53}$$

We shall rederive this result by other methods in Chaps. 8 and 9. For the Ising model, it agrees of course with the classical result of Kramers and Wannier (1941).

Of course, one might also consider the Potts model on a lattice with annealed vacancies. As we had done before for the Ising model, one may thus arrive at the **tricritical Potts model**, at least for $q \leq 4$.

1.3.4 Vector Potts Model

There is a variant of the q-state Potts model, sometimes called **clock model**, but which is equivalent to the q-state Potts model only for $q = 2, 3$. The site variables are defined by $\Theta = (2\pi/q)r$, $r = 0, \ldots, q - 1$. The classical Hamiltonian is

$$\mathcal{H} = -J \sum_{\langle i, i' \rangle} \cos(\Theta_i - \Theta_{i'}) \tag{1.54}$$

and the model is self-dual. The global symmetry group is the cyclic group $\mathbb{Z}_q$, a true subgroup of the permutation group $\mathbb{S}_q$ [293]. The structure of the phase diagram is quite different from the one of the ordinary Potts model, however [218]. While there is a single second-order transition for $q \leq 4$, for $q > 4$, the model has *several* distinct critical points. These separate a high-temperature disordered phase, an intermediate phase and a low-temperature vortex phase. While the transition between the first two of these phases is a conventional second-order transition (for a discussion of the operator content see [538]), the other transition is in the universality class of the XY model to be discussed below.

An interesting variant of this is the **chiral Potts model**, which sometimes is also referred to as *asymmetric clock model* [488, 346, 343]. For recent reviews, see [35, 340]. The Hamiltonian is

$$\mathcal{H} = -J_x \sum_{i,j} \cos(\Theta_{(i+1,j)} - \Theta_{(i,j)} - \delta_x) - J_y \sum_{i,j} \cos(\Theta_{(i,j+1)} - \Theta_{(i,j)} - \delta_y),$$

$$(1.55)$$

where the pair (i, j) labels the points of a square lattice and $\delta_{x,y}$ are free parameters. The phase diagram is very rich and displays commensurate-incommensurate and melting transitions. The model is self-dual if $\delta_x = \pm i\delta_y$. The special case of this, when $\delta_x = \pi/6$ (for $q = 3$), is called the **super-integrable** chiral Potts model and can be solved through the Yang–Baxter equations [16, 65]. However, this model is somewhat unusual in that rotational symmetry is *not* recovered in the scaling limit (and therefore cannot be described by a conformal field theory). Nevertheless, the exponents α and β_j of the local order parameters found at the self-dual critical point appear to be the same as when the chirality is set to zero, i.e. $\delta_x = \delta_y = 0$ [343, 17, 310]. On the other hand, the correlation length scaling depends on the direction and for $q = 3$, one expects $\nu_\perp = 2/3$ in the incommensurate direction and $\nu_\parallel = 1$ in the commensurate direction. Similarly, one should have direction-dependent exponents $\eta_{\perp,\parallel}$ and some of the scaling relations must be generalized, where $\theta = \nu_\parallel/\nu_\perp$ is the **anisotropy exponent**

$$2 - \alpha = (d - 1)\nu_\perp + \nu_\parallel \ , \quad \gamma = (2 - \eta_\perp)\,\nu_\perp = (2/\theta - \eta_\parallel)\,\nu_\parallel \qquad (1.56)$$

(in d dimensions). We shall come back to this model in the context of relevant perturbations of conformally invariant models (Chap. 14).

We shall see later that conformal invariance yields a classification of the partition functions of critical systems. It is therefore interesting to compare with previous attempts to get an overwiew on the possible universality classes in two dimensions. In the context of the lattice gas realizations of the discrete spin systems, Landau theory arguments were used to propose [202, 555] that only four universality classes were possible for the critical melting transitions

Table 1.1. Possible melting transitions of adatoms. Some examples of materials with a certain adsorption site symmetry are also given (after [555, E27])

		Ising	Potts-3	Potts-4	XY	Examples
Adsorption Site Symmetry	α	0(log)	1/3	2/3	non-universal	
	β	1/8	1/9	1/12		
	γ	7/4	13/9	7/6		
	ν	1	5/6	2/3		
Skew (p1) or Rectangular (p2mm)		(2×1) (1×2) $c(2 \times 2)$				Ni(110) Au(110) W(112)
Centered Rect. (c2mm) or Square (p4mm)		$c(2 \times 2)$			(2×2) (1×2) (2×1)	Ni(100) W(100) W(110)
Triangular (p6mm)			$(\sqrt{3} \times \sqrt{3})$	(2×2)		graphite
Honeycomb (p6mm)		(1×1)		(2×2)		
Honeycomb in crystal field (p3m1)			$(\sqrt{3} \times \sqrt{3})$	(2×2)		Ni(111)

of adatoms adsorbed to a surface. This is illustrated in Table 1.1. The columns give the possible realizations together with the exactly known critical exponents and some experimental examples. The non-universal critical behaviour of the XY model (with extra cubic anisotropy terms) will be seen later in the example of the *Ashkin–Teller model* [202] (see Chaps. 11, 12). For the details of the classification scheme and the crystallographic notation used we refer to the literature. In fact, there are many more universality classes than the four considered by [202, 555] known in two dimensions. These can for example be defined in terms of restricted RSOS models which are introduced below.

1.3.5 XY Model

The classical XY model is defined by the Hamiltonian

$$\mathcal{H} = -J \sum_{\langle i, i' \rangle} \cos(\Theta_i - \Theta_{i'}), \tag{1.57}$$

where the spin continuous variables $\Theta_i \in [0, 2\pi]$. It can therefore be viewed as the $q \to \infty$ limit of the vector Potts model. The XY model has a global U(1) symmetry. The critical behaviour of a huge variety of systems can be described in terms of the XY model, including liquid helium,[15] superconductors,[16] ordering in liquid crystals or the roughening transition. In two dimensions, the Mermin-Wagner theorem [464] states that no *local* order parameter in a model with a continuous symmetry can take a non-zero value at a non-vanishing temperature. As a consequence, the transition from the disordered to the ordered phase is described in terms of free *vortices* in the high-temperature phase and bound vortices in the low temperature phase (Berezinskii [83], Kosterlitz and Thouless [415]). This unusual transition mechanism implies that most of the thermodynamic quantities do no longer show power-law singularities. Rather, their singular behaviour is expected to be described by essential singularities of the form, for $T > T_c$

$$\xi \sim \exp\left(bt^{-1/2}\right) \ , \quad \chi \sim \xi^{2-\eta} \tag{1.58}$$

(where b is a constant) while the specific heat is finite at T_c. The conventional critical exponents can no longer be defined. However, it is still possible to define scaling dimensions. At the critical point, the two-point correlation function has the long-distance behaviour

$$G(r) \sim r^{-1/4} (\ln r)^{1/8} \tag{1.59}$$

---

[15] Space shuttle experiments give $\alpha = -0.01285(38)$ in 3D [E52].
[16] For example the Meißner transition in a bulk type II superconductor without external magnetic field, e.g. [177, 485, E64].

which implies[17] $\eta = 1/4$ since logarithmic factors are suppressed in the critical exponent, as a consequence of (1.10). In the entire low-temperature phase, $G(r) \sim r^{-\eta}$, where the exponent $\eta = \eta(T)$ depends continuously on the temperature.

Alternatively, the XY model can be defined in terms of two-dimensional spin vectors $\boldsymbol{S}_i$ of absolute value $\boldsymbol{S}_i^2 = 1$

$$\mathcal{H} = -J \sum_{\langle i,i' \rangle} \boldsymbol{S}_i \cdot \boldsymbol{S}_{i'}. \tag{1.60}$$

If one were to consider n-dimensional vector spins instead, one would obtain the O(n) model. The familiar classical **Heisenberg model** corresponds to $n = 3$. The $n \to \infty$ limit of these is the exactly solvable **spherical model** [88, 575] and for $n = 1$ one is back to the Ising model again. There is a very rich body of results on the critical exponents of these models in more than two dimensions.[18] For example, in 3D the perturbative expansion of the underlying ϕ^4 field theory for the scaling dimensions has been carried out until seven loops [469], while in $4 - \epsilon$ dimensions the critical exponents are known to five loop order [275, 285]. Here, we merely list the lowest orders

$$\eta = \frac{n+2}{2(n+8)^2}\epsilon^2 + \mathcal{O}(\epsilon^3) \tag{1.61}$$

$$\gamma = 1 + \frac{n+2}{2(n+8)}\epsilon + \frac{(n+2)(n^2+22n+52)}{4(n+8)^3}\epsilon^2 + \mathcal{O}(\epsilon^3)$$

from which the other critical exponents can be found via the scaling relations. For further details and numerical estimates, see [G9, G20, 155, 285].[19] However, the Mermin-Wagner theorem shows that these systems (for $n > 2$) do not have a phase transition at a non-vanishing temperature in two dimensions. They are thus of little interest here. On the other hand, if $-2 \leq n \leq 2$, a critical point exists and the operator content can be found (see Chap. 11).

In general, however, there is no reason that in a given system there should be a complete isotropy in spin space, as written in (1.60). For example, for the case of the Heisenberg model ($n = 3$), one might write a Hamiltonian of the form

$$\mathcal{H} = -\sum_{\langle i,i' \rangle} [J_x S_i^x S_{i'}^x + J_y S_i^y S_{i'}^y + J_z S_i^z S_{i'}^z]. \tag{1.62}$$

If, for example, one has $J_z \gg J_x, J_y$, one is back to the Ising model and if $J_z \ll J_x = J_y$, the XY model is recovered.

---

[17] It is still surprisingly difficult to recover these results precisely from numerical studies of lattice systems, even when large lattices and correlation lengths of order 140 are considered [367].

[18] Numerical values for $K_c = J/T_c$ in 3D are given in Table 12.15.

[19] One may also add further terms into the Hamiltonian with a cubic symmetry, which leads to a rich phase diagram. The ϵ-expansion has been obtained to $\mathcal{O}(\epsilon^5)$ in this case as well [406]. Tetragonal interactions were added in [468].

The above remarks on the existence of phase transitions in $O(n)$ models apply to the bulk. For *thin films*, however, surface anisotropies can stabilize long-range order at finite temperatures. For example, for thin films with a cubic symmetry two situations can arise:

1. There exists an uniaxial anisotropy, either perpendicular or parallel to the film plane. Then the transition will be in the Ising universality class [44].
2. There is no uniaxial surface anisotropy, so that the magnetization is in the film plane. The critical behaviour may then depend on crystal fields within the plane. For four-fold fields, renormalization group arguments [375] indicate non-universal critical behaviour, as may be found in the XY model or else in the same universality class as the self-dual critical line in the Ashkin–Teller model (see Chap. 12). More recently, it has been argued [100, 101] that because of the power-law decay of correlations at low temperatures, finite-size effects may render the thermodynamic limit of the XY model inaccessible in practice. In it then concluded that instead of a Kosterlitz–Thouless transition the magnetization should be described by a conventional power law, with an effective exponent $\beta = 3\pi^2/128 \simeq 0.231$.

1.3.6 Yang–Lee Edge Singularity

The models defined so far are formulated in terms of spin models with real parameters. As we shall see later, conformal invariance also allows to discuss a large class of more exotic-looking systems, of which the Yang–Lee model is the simplest one. In spite of the unrealistic appearance of this model as it is usually defined, this kind of model does occur in very physical applications, as further discussed in Chap. 12. Reconsider the Ising model Hamiltonian

$$\mathcal{H} = -J \sum_{\langle i,i' \rangle} \sigma_i \sigma_{i'} - B \sum_i \sigma_i \qquad (1.63)$$

with nearest-neighbour interactions. The analyticity of the magnetization per site M was related in a remarkable theorem by Yang and Lee [628] to the distribution of the zeroes of the partition function $\mathcal{Z} = \mathcal{Z}(B)$ as a function of the *complex* magnetic field B, in the limit $\mathcal{N} \to \infty$. The Yang–Lee theorem [628] states that the zeroes of $\mathcal{Z}$ are all on the imaginary axis Re $B = 0$. Now consider how the distribution of the zeroes changes with temperature.[20] If $T < T_c$, there is an accumulation point of the zeroes at Im $B = 0$, related to the first-order phase transition with a jump discontinuity in M when the line Re $B = 0$ is crossed. However, if $T > T_c$, there is a T-dependent lower bound $|B_c|$ for the absolute values of the imaginary parts of the zeroes of the

---

[20] The density of the imaginary field Yang–Lee zeroes of the partition function has been measured in the 2D Ising ferromagnet $FeCl_2$ [E96].

partition function. The points are accumulation points for the zeroes of $\mathcal{Z}$. In their vicinity, the magnetization per spin scales as

$$M \sim (\mathrm{Im}\, B - B_c)^{\sigma_{YL}} \tag{1.64}$$

for $|\mathrm{Im}\, B| \gtrsim |B_c|$. Fisher [229] showed that the points $B = \pm i B_c$ are new critical points of the theory. They are called **Yang–Lee edge singularities**. Here, $\sigma = \sigma_{YL} = 1/\delta_{YL}$ is often referred to as Fisher exponent. It is related to the exponent η_{YL} which describes the scaling of the spin-spin correlation function $G(r) \sim r^{-(d-2+\eta_{YL})}$ in the usual way, viz. $\sigma_{YL} = (d-2+\eta_{YL})/(d+2-\eta_{YL})$. In one dimension, it is straightforward to establish that $\sigma_{YL} = -1/2$ and $\eta_{YL} = -1$.

The critical theories at $B = \pm i B_c$ are conformally invariant. We shall see in Chap. 12 that in two dimensions $\sigma_{YL} = -1/6$ and that the corresponding scaling operator has scaling dimension $x = -2/5$. Alternatively to the lattice model defined, the Yang–Lee singularity may also be described in terms of a non-unitary scalar field theory with the Landau–Ginzburg–Wilson action [229]

$$S = \int \mathrm{d}^d r \left(\frac{1}{2}(\partial\phi)^2 + i(\mathrm{Im}\, B - B_c)\phi + \frac{1}{3}i\lambda\phi^3 \right). \tag{1.65}$$

Landau–Ginzburg–Wilson actions will be discussed in more detail in Chap. 8. From this action, the critical exponents can be calculated in $6 - \epsilon$ dimensions with the result [229]

$$\sigma_{YL} = \frac{1}{2} - \frac{\epsilon}{12} + \mathcal{O}(\epsilon^2) \ , \quad \eta_{YL} = -\frac{\epsilon}{9}\left(1 + \frac{43}{81}\epsilon\right) + \mathcal{O}(\epsilon^3) \tag{1.66}$$

A two-point Padé approximant, which also uses the known results in 1D, allows to extrapolate this down to two dimensions. One finds to good accuracy $\sigma_{YL} \simeq -1/6$, in agreement with the result of conformal invariance, and consistent with series expansion studies, which yield $\sigma_{YL} = -0.155(10)$ in 2D and $\sigma_{YL} = 0.098(12)$ in 3D, respectively [229].

1.3.7 Percolation

The simplest example of **geometric phase transitions** is provided by *percolation*, see [193, 577, 428, 14] for detailed introductions and reviews. This model is defined as follows. Consider a hypercubic lattice and put on each site with a probability p a particle. Then all occupied nearest neighbor sites are joined to form clusters. One then asks for the probability $P(p)$ that in this way an infinitely large cluster can be formed. Indeed, it can be shown that $P(p) = 0$ for all values of p less than a critical value p_c (the value of which depends on the lattice) and it is non-vanishing for $p > p_c$. It transpires that $P(p)$ can be seen as the analogue of the order parameter of the magnetic systems introduced before.

The model just defined is called **site percolation**. There is a variant, called **bond percolation**. In that model, all sites are occupied but between nearest neighbor sites, bonds are present with probability p and absent with probability $1-p$. The rest of the model definition remains unchanged. Now, if n_s is the number of clusters with size s, consider the moments $M_k := \sum_s n_s s^k$. Close to p_c, it can then be shown that

$$M_2 \sim (p-p_c)^{-\gamma}$$
$$P \sim (p-p_c)^{\beta} \qquad (1.67)$$
$$M_0^{\sin} \sim (p-p_c)^{2-\alpha},$$

where $P(p) + M_1 = p$. While the values of p_c are in general different for site and bond percolation, the values of the exponents are the same for both models. In addition, one may define a correlation function $g(r)$ as the probability that two sites at a distance r belong to the same cluster, which allows to define exponents ν and η in a standard fashion. The analogy with the conventional critical phenomena is apparent, in particular, the *same* scaling relations as found for the magnetic systems hold for percolation as well.

Formally, the bond percolation problem can be related to the Potts model [625]. To see this, consider the q-state[21] Potts model Hamiltonian (1.46) with an additional external field $\mathcal{H} = -J\sum_{\langle i,i'\rangle} \delta(\Theta_i - \Theta_{i'}) - J'\sum_i \delta(\Theta_i)$. Generalizing the procedure explained above for the simple q-states Potts model, one easily rewrites $\mathcal{Z}$ as a sum over all possible graphs G, where $K = J/T$ and $K' = J'/T$

$$\mathcal{Z} = \sum_G (e^K - 1)^{b(G)} \prod_c \left(e^{K' s_c(G)} + q - 1\right), \qquad (1.68)$$

where $b(G)$ is the number of occupied bonds and $s_c(G)$ the number of occupied sites in a cluster. The sum runs over all subgraphs with these bonds and sites occupied and the product is over all connected clusters in G (including isolated sites). On the other hand, the generating function for bond percolation is defined as

$$Z = \sum_G p^{b(G)}(1-p)^{\mathcal{B}-b(G)} r^{s_c(G)}, \qquad (1.69)$$

where $\mathcal{B}$ is the total number of bonds. Taking logarithmic derivatives of Z with respect to r, one recovers $P(p)$ and $M_2(p)$ after setting $r = 1$. With the identification $p = 1 - e^{-K}$ and $r = e^{-K'}$, one obtains the relation

$$Z = \lim_{q \to 1} \frac{\partial}{\partial q} \ln \mathcal{Z}. \qquad (1.70)$$

This relation is often referred to by the abridged statement that *"percolation is the $q \to 1$ limit of the q-states Potts model"*. Again, this correspondence

---

[21] In later chapters this is sometimes referred to as p-states Potts model.

reinforces the analogy between the percolation critical point and the magnetic phase transitions described before. Similarly, correlators of the percolation model may be related to the $q \to 1$ limit of correlators in the q-states Potts model. From the known critical point (1.53) of the q-states Potts model one has $p_c = 1/2$ for the percolation threshold for bond percolation on the square lattice. Critical exponents may then be obtained through an analysis of the Potts model as well. In 2D, one finds $\alpha = -2/3$, $\beta = 5/36$, $\gamma = 43/18$ and $\nu = 4/3$ [625, 193, 577].

1.3.8 Linear Polymers

Another example of a non-thermal problem showing a behaviour analogous to criticality is provided by linear polymers. For a detailed introduction, see e.g. [182, 201]. A polymer consists of a chain of identical building blocks, called monomers. Ignoring the detailed chemical structure entirely, this might be modeled as follows. Imagine a particle effecting a **self-avoiding random walk** on a lattice. Starting from the initial site, join the sites which the particle visits along its walk. As for percolation above, one can now define a generating function for these walks. Furthermore, following the presentation given by [G6], these graphs can be related to the high-temperature expansion of an $O(n)$ symmetric spin model in a simple way. Rather than taking the usual Heisenberg Hamiltonian, consider

$$\mathcal{H} = -T \sum_{\langle i,i' \rangle} \ln(1 + \lambda \, \boldsymbol{S}_i \cdot \boldsymbol{S}_{i'}) \underset{\lambda \to 0}{\simeq} -\lambda T \sum_{\langle i,i' \rangle} \boldsymbol{S}_i \cdot \boldsymbol{S}_{i'}, \qquad (1.71)$$

where the $\boldsymbol{S}_i$ are n-component vectors of length unity and the sum extends over nearest neighbors only and λ is a function of the temperature T. The partition function, when expanded in powers of λ, becomes

$$\mathcal{Z} = \mathrm{Tr} \prod_{\langle i,i' \rangle} (1 + \lambda \, \boldsymbol{S}_i \cdot \boldsymbol{S}_{i'}) = \sum_{\text{loops}} n^c \lambda^b, \qquad (1.72)$$

where c is the number of loops (cycles) in a given configuration and b is the number of bonds. Spins were normalized such that $\mathrm{Tr} S_i^a S_{i'}^b = \delta^{a,b} \delta_{i,i'}$. Only closed loops survive because of the trace over all spin configurations. In the $n \to 0$ limit, two more simplifications occur. First, diagrams with four or more lines in a vertex are suppressed.[22] Second, from (1.72), only those configurations with a single closed loop remain. This makes the random walk a self-avoiding one. Similarly, one may now consider the spin-spin correlation function $G(\boldsymbol{r}_1 - \boldsymbol{r}_2; \lambda) = \langle S_1(\boldsymbol{r}_1) S_2(\boldsymbol{r}_2) \rangle$, assuming translation invariance for large distances. If $c_N(R)$ is the number of self-avoiding walks between two

---

[22] From the $O(n)$ symmetry, for four spins on the same site, one must have $\mathrm{Tr} S^a S^b S^c S^d = A(n)(\delta^{ab}\delta^{cd} + \delta^{ac}\delta^{bd} + \delta^{ad}\delta^{bc})$. Setting $a = b$ and $c = d$ and summing over a and c then gives $A(n) \simeq n/2$ in the $n \to 0$ limit.

points at a distance R, the same argument as before can be used to show that [G6]

$$\sum_N c_N(R)\lambda^N = \lim_{n \to 0} G(R; \lambda).$$ (1.73)

Now, consider the total number $c_N = \sum_R c_N(R)$ of self-avoiding walks (SAW) with N steps. Its generating function is $\sum_N c_N \lambda^N = \sum_R G(R, \lambda) = \chi(\lambda)$, the susceptibility of the associated O(n) model. Close to a critical point λ_c, it will display a singularity of the form $\chi \sim (\lambda_c - \lambda)^{-\gamma}$. Thus, one expects for $N \to \infty$ that

$$c_N \sim N^{\gamma-1}\mu^N,$$ (1.74)

where $\mu = 1/\lambda_c$. Also, it becomes apparent that the limit $N \to \infty$ of the number of monomers is formally analogous to the approach $T \to T_c$ towards a critical point [182]. In other words, the physics of long polymers is automatically close to the critical point of some associated spin model. By universality, one would then expect that finer points of the model, such as the initial lattice formulation, should be irrelevant for the value of exponents like γ. Another quantity of interest is the end-to-end distance, given by $\langle R^2 \rangle = \sum_R R^2 c_N(R)/c_N$. Using again the link (1.73) to the spin model, one expects for $N \to \infty$ that

$$\langle R^2 \rangle \sim N^{2\nu}.$$ (1.75)

A similar scaling law, with the same exponent ν, is expected for the **radius of gyration** $R_G^2 := (1/2N^2)\sum_{r_1,r_2}(r_1 - r_2)^2 \sim N^{2\nu}$, where the sum is over pairs of points on the polymer.

So far, we considered a single polymer chain. It is possible to generalize this and allow for interactions between different monomers which are not directly linked by the chain itself. Formally, these may be introduced in a simple way by including an extra fugacity term $\lambda_{contact}^k$ into the polymer generating function, where k is the number of contacts (see Chap. 12). It turns out that these interactions may cause a collapse of the polymer into a compact structure. The transition between the compact and the self-avoiding phase is called a **theta point**. In the associated spin model, this will correspond to a *tricritical* point [182]. Although at the Θ-point the relation with the simple spin Hamiltonian (1.71) will not hold, the phenomenological relations (1.74,1.75) remain valid, see e.g. [579] for a recent review.

In 2D, numerical values for the exponents are $\nu = 3/4$, $\gamma = 43/32$ and $\eta = 5/24$ for the self-avoiding walk [476] and $\nu = 4/7$, $\gamma = 8/7$ and $\eta = 0$ for the Θ-point [211, 579]. Obviously, in the compact phase $\nu = 1/d$, where d is the number of space dimensions.

1.3.9 Restricted Solid-On-Solid Models

Finally, we present a class of spin systems whose definition may at first sight appear artificial. However, as we shall see later, these **RSOS models** contain most of the universality classes defined so far as special cases and furthermore, there are deep connections with integrable systems. To define them, one assigns to each site i of the square lattice a **height variable** $\hat{l}_i$ which may have only a restricted set of integer values $l_i = 1, \ldots, m$. Moreover, the heights $\hat{l}_{i,j}$ of nearest neighbour sites i, j are *constrained* to satisfy $\hat{l}_i - \hat{l}_j = \pm 1$. The interactions depend on the heights of the four sites (i, j, k, l) of a given square plaquette $P(i, j, k, l)$

$$\mathcal{H} = -\sum_{P(i,j,k,l)} \epsilon(i,j,k,l) = -\sum_{P(i,j,k,l)} \begin{array}{cc} l & k \\ \boxed{} \\ i & j \end{array}, \tag{1.76}$$

where the site i is in the lower left corner and the other labels are distributed in an anticlockwise manner. The partition function becomes the product of the **Boltzmann weights** $W(\hat{l}_i, \hat{l}_j, \hat{l}_k, \hat{l}_l) = \exp(\epsilon(i,j,k,l)/T)$ associated to each plaquette

$$\mathcal{Z} = \sum \prod_{P(i,j,k,l)} W(\hat{l}_i, \hat{l}_j, \hat{l}_k, \hat{l}_l). \tag{1.77}$$

These models are also often called **IRF models** (Interaction Round a Face) [G19, 32]. Because of the constraint on the $\hat{l}_i$, only the weights

$$W(i, i+1, i, i-1), \quad W(i, i-1, i, i+1), \quad W(i+1, i, i-1, i),$$
$$W(i-1, i, i+1, i), \quad W(i+1, i, i+1, i), \quad W(i-1, i, i-1, i) \tag{1.78}$$

are non-vanishing. This model might be viewed as a rough, discrete analogon of a gently fluctuating surface of a liquid, where neighbouring points cannot have heights which differ much from each other. The local energy density ϵ is given by the surface energy. The phase structure is quite complicated, see [347, 32].

However, there is a critical point in the model which is of interest to us. In fact, there is a connection with the **Dynkin diagrams** of certain Lie algebras. Dynkin diagrams are a graphic representation of the Cartan matrix $\hat{A}$ of a Lie algebra [270, 377, G12, G18]. In Fig. 1.7, we recall the Dynkin diagrams of the simply laced finite-dimensional and affine simple Lie algebras.[23] The connection to the RSOS model is made through the **adjacency matrix** $\hat{G} := 2 - \hat{A}$ obtained from the Cartan matrix. Consider the m allowed values of the heights $\hat{l}_{i,j}$ in the RSOS model on nearest neighbour sites. If $\hat{l}_i - \hat{l}_j = \pm 1$, the constraint is satisfied and the corresponding matrix element $G_{\hat{l}_i, \hat{l}_j} = 1$. Otherwise, $G_{l,l'} = 0$. Clearly, the constraint in the RSOS model defined above is encoded in the Dynkin diagram of the Lie algebra A_m.[24] Therefore, the model (1.76) is called the A_m-**RSOS model**.

---

[23] Affine Lie algebras are briefly discussed in Chap. 7. For the moment, a reader not familiar with them should simply accept that the Dynkin diagrams may be used to define an algebra. For detailed treatments, see [270, 377, 173, G18].

[24] A more familiar notation of these algebras may be the unitary Lie algebras SU($m+1$).

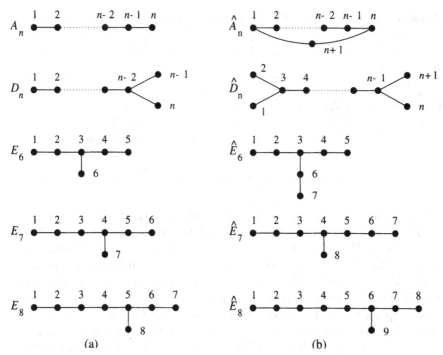

Fig. 1.7. Dynkin diagrams of the simply-laced (**a**) finite-dimensional and (**b**) affine Lie algebras

The classification of the simple Lie algebras through their Dynkin diagram is of major importance for the classification of 2D universality classes, see Chap. 11. In fact, RSOS models can be defined for any simply-laced Lie algebra. The point of these definitions are deep connections with integrable models [32]. For reference, we collect in Table 1.2 the **Coxeter number** and the **Coxeter exponents**[25] of these Lie algebras [270, 377, G18].

Let us make contact with known spin models. As examples, we take A_3, D_4 and $\hat{D}_4$. The RSOS constraint is expressed through the Dynkin diagrams, respectively,

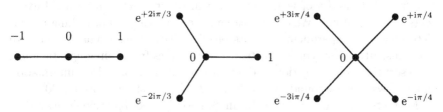

---

[25] For simply-laced Lie algebras, the Coxeter number is related to the quadratic Casimir operator in the adjoint representation, while the eigenvalues of $\hat{G}$ are $2\cos(n_i\pi/h)$, where the n_i are the Coxeter exponents. $n_i + 1$ are the degrees of the Casimir operators of the algebra.

Table 1.2. Coxeter number $\hat{h}$ and exponents of the simply-laced Lie algebras

finite-dimensional			affine		
	$\hat{h}$			$\hat{h}$	
A_n	$n+1$	$1, 2, \ldots, n$	$\hat{A}_n$	$n+1$	$0, 2, 4, \ldots, 2n$
D_n	$2n-2$	$1, 3, 5, \ldots, n-1$	$\hat{D}_n$	$2n-4$	$0, 2, 4, 6, \ldots, n-2, n-2$
E_6	12	$1, 4, 5, 7, 8, 11$	$\hat{E}_6$	6	$0, 2, 2, 3, 4, 4, 5$
E_7	18	$1, 5, 7, 9, 11, 13, 17$	$\hat{E}_7$	12	$0, 3, 4, 6, 6, 8, 9, 12$
E_8	30	$1, 7, 11, 13, 17, 19,$ $23, 29$	$\hat{E}_8$	30	$0, 6, 10, 12, 15, 18, 20,$ $24, 30$

where the height variables are also indicated. The A_3 RSOS model is equivalent to the Ising model. The three height values may be taken to be $\hat{l} = -1, 0, 1$. If at the site i, one has $\hat{l}_i = \pm 1$, at each neighbour site i' one must have $\hat{l}_{i'} = 0$. Thus the lattice decomposes into two interpenetrating sublattices, on one of which $\hat{l} = 0$ and $\hat{l} = \pm 1$ on the other. For each plaquette, two diagonally opposite corners have height zero and the interaction reduces to a coupling of two Ising spins along the diagonals of the plaquettes. The three-state Potts model corresponds to the D_4-RSOS model. Again, there are two sublattices, one with $\hat{l} = 0$ while the other one has three possible states with a natural $\mathbb{Z}_3$ symmetry, characteristic for the critical three-state Potts model. The correspondence of the four-state Potts model with the affine $\hat{D}_4$ Lie algebra is seen similarly.[26]

Integrable RSOS models are discussed in Chaps. 11 and 14. An important application is the exact solution of the Ising model in a magnetic field, but at $T = T_c$.

1.4 Some Experimental Examples

For a reader who might have got the impression that the discussion of two-dimensional critical phenomena were a purely theoretical issue, we shall quote here from the results of work performed in order to *measure* critical exponents in real systems. We do not attempt here anything approaching a complete review of experimental results on 2D critical phenomena. Rather, we shall illustrate a few typical effects, using examples from thin magnetic films, chemisorption, physisorption and surface reconstruction. More information and reviews may be found in [E4, E21, E24, E27, E69, E87, E88, E90].

Magnetic systems behaving in an essentially two-dimensional way may be produced from the molecular structure of certain compounds. A classic example is the K_2NiF_4 family, with a tetragonal structure such that the magnetic ions form a square lattice in two dimensions [E21]. Adsorbed atoms

---

[26] RSOS models are also closely related to the theory of knots and links [626].

on crystalline surfaces provide further examples of experimentally accessible two-dimensional systems [E27, E87]. One should distinguish between physisorption and chemisorption. In **physisorption**, the adatoms are typically bound to the surface by a van der Waals bond, with energies of the order 10^{-2} eV, are relatively free to move across the surface and to desorb at temperatures near to T_c. Experiments are thus performed in equilibrium with the gas phase.[27] In **chemisorption**, the adatom is chemically bound to the substrate, with energies of the order 10^0 eV, and at a well-defined place. Near T_c there should be little desorption and one has a strictly 2D system with constant coverage. In recent years, considerable progress has been in achieved in the technology required for the *epitaxial growth* of magnetic materials on a substrate, see [E68, E7]. While the adatom may or may not be magnetic in the three-dimensional bulk, the interactions with the substrate should strongly affect its electronic structure and thus its magnetic properties. The change from a 3D to a 2D material is then accompanied by a change of the symmetries of the interaction Hamiltonian as well, which might induce a change of the universality class. We shall not enter into this discussion, but rather limit ourselves to a phenomenological discussion of some experimental results.

As a first example, consider the results for the susceptibility exponent γ for a layer of Gd(0001) on W(110) of varying thickness as measured by electron-spin resonance [E30], listed in Table 1.3. It can be seen that the value of T_c does depend on the layer thickness and the ferromagnetic transition does occur below[28] the bulk critical temperature. We shall discuss the finite-size behaviour of T_c systematically in Chap. 3. Furthermore, the value of the effective critical exponent γ also depends on the layer thickness. Two

Table 1.3. Curie temperatures and susceptibility exponents for different layer thicknesses (given in Å or monolayer coverages θ_A, respectively and atomic monolayers normalized to $\theta_A = 0.8$) of Gd(0001) on W(110). The errors are estimated to be $\Delta T_c = 2K$ and $\Delta \gamma = 0.2$ (after [E30])

Layer	# atoms	T_c (K)	γ
bulk	1935	292.5	1.25
80 Å	35	288	1.25
$\theta_A = 1.6$	2	281	1.74
$\theta_A = 0.8$	1	271	1.90

---

[27] In fact, the first examples of 2D criticality appear to have been observed in this setting, although for a long time, were dismissed as uninteresting due to ubiquitous inhomogeneities! The first clear example of a 2D phase diagram with a 'normal' triple point and a critical point was found by Thomy and Duval (1970) for Xe on graphite [E86]. See [E88] for a historical account.

[28] This should be expected, since close to the surface of a thin layer not all the bonds between spins can be saturated. It should thus become easier for the thermal fluctuations to destroy the long-range order.

Table 1.4. Critical points T_c (in K) and magnetization exponent β for ultrathin Fe or Ni layers on non-magnetic substrates as a function of the layer thickness n, given im monolayers (ML). The numbers in brackets give the uncertainty in the last digit. Data for Fe(110)/Ag(111) are from [E70], data for Fe/Au(100) are from [E23] and data for Ni(111)/W(110) are from [E51]

Fe(110)/Ag(111)			Fe/Au(100)			Ni(111)/W(110)		
n	T_c	β	n	T_c	β	n	T_c	β
1.8	338.1	0.139(6)	1.0	315	0.22	2	325	0.13(6)
1.9	450.5	0.139(4)	1.5	-	0.21	3	371	0.13(6)
2.0	466.4	0.130(3)	2.0	-	0.22	4	435	0.13(6)
			2.5	521	0.21			

regimes can be recognised in Table 1.3. For the bulk system down to moderately thin layers, γ is close to the theoretical prediction[29] of the 3D Ising model $\gamma_{3D} \simeq 1.24$. For Gd thicknesses of just a few[30] monolayers, the critical behaviour goes over to a two-dimensional one, with exponent value close to the theoretically expected value $\gamma_{2D} = 7/4$ for the Ising model. The crossover between 2D and 3D critical behaviour occurs at layer thicknesses of about $d_\times \sim 3 - 5$ monolayers. Apart from the cross-over between 2D and 3D criticality, the critical exponents are independent of the layer thickness d [E30, E24, E70, E51], in agreement with the theoretically expected universality. We illustrate this further in Table 1.4, which lists the critical points and the exponent β as obtained for thin layers of Fe(110) on Ag(111) through the Kerr effect [E70], for Fe/Au(100) through spin-polarized low-energy electron diffraction [E23] and for N(111)/W(110) through magnetic resonance [E51]. It is clearly seen that, while the Curie temperature T_c of the film is greatly reduced with respect to the bulk value[31] and furthermore varies strongly with the thickness of the Fe/Ni layer, the critical exponent β remains constant within the experimental accuracy. This observation confirms the expected universality of the critical exponents. For larger thicknesses of the Fe/Ni layer, β crosses over rapidly to values of $\beta \simeq 0.33 - 0.36$ typical for 3D systems, see [E51] for a particularly clear example.

---

[29] In fact, the magnetic properties of Gd are unusual and it has proved difficult to describe it in terms of the simple spins models defined. This is attributed to the presence of anisotropic dipolar interactions, see [E25, E62] for details. This does not affect the distinction between 2D and 3D criticality through the quoted effective exponents, however.

A 2D Heisenberg model with *both* exchange and dipolar interactions was studied in [553]. For sufficiently large lattices, an ordered circular phase is found (experimentally, impurities may act to bound the maximal correlation length and thus the effective size of the system). Rough estimates for exponents are $\alpha \simeq -0.4$, $\beta \simeq 0.23$ and $\nu \simeq 1.20$.

[30] Of course, the film must be thick enough to sustain long-range order. For Fe/Au(100), for example, this minimum thickness is about $0.6 - 0.9$ monolayers [E23].

[31] $T_{c,bulk} \simeq 1043$ K for iron and $T_{c,bulk} \simeq 627$ K for nickel [34].

A remarkable confirmation of universality was obtained recently using an epitaxial Fe film on a non-magnetic W(110) substrate [E9], see also [576]. For film thicknesses between one and two atomic layers, scanning tunneling microscopy shows that a complete atomic layer of Fe is deposited, followed by a two-dimensional network of irregular patches. This makes apparent that this system is certainly not well modeled by a lattice with perfect translational symmetry underlying the spin Hamiltonians discussed above. Also, Fe does not have localized spins, which rather imitate an insulating ferromagnet with localized moments than a broad-band metallic ferromagnet like Fe [E9]. Moreover, the strength of the various magnetic anisotropies is quite small, of the order 1K as compared to $T_c \simeq 300$K [E9]. Nevertheless, data for the magnetization (measured through the Kerr effect) as a function of both T and B collapse nicely when the scaled variables $BM^{-\delta}$ and $tM^{-1/\beta}$ of the equation of state are used and moreover agree with the theoretical prediction of the equation of state obtained for the 2D Ising model [253]. For this, the growing of the lateral correlation length to values ($\sim 10^3$nm [E9]) much larger than the film thickness is essential. The agreement between the experimentally measured scaling function and the theoretical one extends over 18 orders of magnitude of the scaling variable $x = tM^{-1/\beta}$. This remarkable agreement should be regarded as a stringent test of the universality hypothesis in the critical region.[32]

In Tables 1.5, 1.6 and 1.7 we list some results for the critical exponents of two-dimensional systems. In compiling the data, we used previous collections [E21, E27, E90] of results for critical exponents, but we do not distinguish data obtained above T_c from those found below the critical point. For detailed reviews, further results and a discussion of the experimental techniques, we refer to [E21, E24, E27, E69, E87, E90, E94]. In particular, a closer inspection of the real systems presented here may in some cases suggest that their behaviour close to the critical point might be more complex than can be described by a single critical exponent. Furthermore, since many experiments involve one or two decades of data, the exponents finally quoted should often be regarded as effective exponents, whose interpretation might not always be straightforward.

Nevertheless, if the quoted values represent indeed the true critical exponents, the comparison of the experimentally measured values with the theoretical predictions, for example from Table 1.1, shows an overall agreement. In particular, since several systems have been studied by different groups, the scatter of the results gives a fair indication to what extent critical exponents of two-dimensional systems can be reliably measured. In addition, data have been obtained for several closely related systems, which allows for experimental checks of universality through the numerical values of the exponents. On the other hand, it becomes also clear that the numerical values of the

---

[32] In fact, the temperature dependence of the magnetization far away from the critical point deviates strongly from the 2D Ising behaviour [E8].

Table 1.5. Experimentally measured critical exponents of some two-dimensional systems, part I. The numbers in brackets give the estimated uncertainties. KT indicates a Kosterlitz–Thouless behaviour, see (1.58). XYcf stands for the universality class of the XY model with a crystal field. (NTDF = $Nb/(Tb_{0.27}Dy_{0.73})_{0.32}Fe_{0.68}$, LSMO = $La_{1.2}Sr_{1.8}Mn_2O_7$)

	β	γ	ν	Class	source	notes
Antiferromagnetic systems						
Rb_2CoF_4	0.119(8)	1.34(22)	0.89(10)	Ising	[E77, E21]	$\eta = 0.2(1)$
	0.115(16)	1.67(9)	0.99(4)		[E43, E21]	
	0.114(4)				[E39, E21]	
K_2CoF_4	0.123(8)	1.71(4)	0.97(4)	Ising	[E42, E21]	
		1.73(5)	1.02(5)		[E22, E21]	
Ba_2FeF_6	0.135(3)			Ising	[E14]	
Epitaxial magnetic films						
Co/Cu(001)	$\sim 0.14^*$	2.395(70)		Percolation	[E78, E79]	*from scaling relation
V/Ag(100)	0.128(10)			Ising	[E73]	
Gd/W(110)	$\sim 1/8$	1.8		?	[E30, E31]	cf Table 1.3
Ni/W(110)	0.13(6)			Ising	[E51]	
Fe/W(110)	0.13(2)	1.74(5)		Ising	[E9]	$\delta = 14(5)$
	0.123		0.93(14)		[E28]	cf Chap. 3
Fe/Pd(100)	0.125(10)			Ising	[E74]	
	0.127(4)				[E53]	
Fe/Ag(111)	0.137(8)			Ising	[E70]	
Fe/Ag(100)	0.124(2)			Ising	[E71]	
NTDF	0.126(20)	1.75(3)		Ising	[E60]	$\delta = 15.1(10)$ amorphous film
Fe/Au(100)	0.22(5)			XYcf	[E23, E24]	see text
Fe/W(100)	0.22(3)	~ 5?		XYcf	[E29]	see text
LSMO	0.25(2)			XYcf	[E66]	2 transitions
$BaNi_2(PO_4)_2$	0.23		KT	XY	[E76]	

critical exponents of the 2D universality classes are often relatively close to each other (see Table 1.1) so that it is difficult to distinguish different universality classes on the basis of the measured values of critical exponents. This is particularly so if the order parameter exponent β is the only quantity extracted (an extensive list of further examples of this kind is compiled in [E13]). In fact, as discussed in more detail in [E27], the exclusive reliance on measurements of the long-range order might be risky,[33] in particular near weak first-order transitions where very small values of the effective exponent

---

[33] For example, in the Kr/graphite system, early measurements gave a value of β consistent with the three-states Potts model but the transition is in fact of first order [E83, E27]. For the O/N(111) system, the long-disputed transition [E75, E90] also appears to be weakly first-order [E92].

Table 1.6. Experimentally measured critical exponents of some two-dimensional systems, part II. $0(\log)$ indicates a logarithmic dependence on t, viz. $C \sim \log|t|$. The chiral Potts model displays anisotropic scaling with $\nu_\perp \neq \nu_\parallel$, both of which are quoted. For the S/Ru(0001) system, the indices I,II refer to the $p(2 \times 2)$ and $(\sqrt{3} \times \sqrt{3})R30°$ superstructures, respectively

	β	γ	ν	α	η	Class	source
Liquid-vapor transition							
CH$_4$/graphite	0.127(20)			$0(\log)$		Ising	[E45, E27]
^{4}He/graphite				0.28		Potts-3	[E84, E27]
			0.36(3)				[E15]
^{3}He/graphite			0.37(2)			Potts-3	[E15]
H$_2$/graphite			0.36(5)			Potts-3	[E61]
			0.33(3)				[E34, E94]
HD/graphite			0.33(3)			Potts-3	[E94]
D$_2$/graphite			0.31(2)			Potts-3	[E35, E94]
Surface reconstruction, Reorientation and Chemisorption							
CO/graphite		$\sim 7/4$		$0(\log)$		Ising	[E33, E95]
C$_2$F$_6$/graph.				$0(\log)$		Ising	[E6]
Au(110)	0.13(2)	1.75(3)	1.02(2)	0.00(5)		Ising	[E18, E90]
O/W(112)	0.13(1)	1.79(14)	1.09(11)			Ising	[E93, E90]
O/Rh(100)	0.120(8)	1.70(20)	1.01(12)	0.04(12)		Ising	[E11]
Br/Ag(001)	0.125(40)					Ising	[E63]
Cl/Ag(001)	0.12(4)					Ising	[E63]
O/Ru(0001)	0.085($^{+15}_{-5}$)	1.08(7)	0.68(3)	0.60(4)		Potts-4	[E65]
H/Ni(111)	0.11(2)	1.23(10)	0.68(5)	0.68(7)	0.27(10)	Potts-4	[E16, E92]
S/Ru(0001),I	0.11(2)	1.04(8)	0.66(6)	0.66(7)	0.32(10)	Potts-4	[E82, E92]
S/Ru(0001),II	0.14(3)	1.18(14)	0.81(9)	0.40(5)	0.30(8)	Potts-3	[E82, E92]
Si(113)	0.11(2)	1.56(13)	0.65(7)			chiral	[E1]
			1.06(7)			Potts-3	
				$\sim 1/3$			[E89]

Table 1.7. Experimentally measured critical exponents of some two-dimensional systems, part III. α_t and β_t are tricritical exponents as defined in the text

	β	ν	α	Class	source
Polymer chains					
PVAc		0.79(1)		SAW	[E91]
PMMA		0.56(1)		Θ-point	[E91]
Tricritical point of liquid-vapor transitions					
	β_t	α_t			
HD/graphite		-1.57(10)		tricritical Potts-3	[E94]
D$_2$/graphite		-1.48(10)		tricritical Potts-3	[E94]
N$_2$/graphite	0.55(2)			tricritical Potts-3	[E59, E27]

β_{eff} may arise. The majority of systems investigated so far apparently fall into the four universality classes listed in Table 1.1.

An intriguing problem is the origin of the rather large values of β for a number of systems, three of which are quoted in Table 1.5 (Another interesting example of this class is provided by the CuO_2 planes in high-T_c superconductors like $YBa_2Cu_3O_{6+x}$, see [E76]. More examples are quoted in [E13]). These systems are known to have the magnetization in the film plane, see [E71, E29]. In order to have a non-vanishing magnetization at all, a n-fold crystal field within the plane to break the continuous symmetry will be needed. The effect of this has been investigated using renormalization group arguments [375]. For $n < 4$ the field should become a relevant variable at a temperature greater than the critical point of the XY model. For $n = 2, 3$ one expects to recover the Ising and three-states Potts universality classes, respectively. On the other hand, for $n > 4$ the crystal field should be irrelevant and the system remains in the XY universality class. Finally, for $n = 4$ the field is marginal and the exponent β should in principle depend continuously on the crystal field. At first sight, one would expect a wide distribution of possible values of β, depending on the individual value of the crystal field, but in contrast to observation [E13, 101].[34] It was argued, however, that finite-size effects in this type of system are so large that the cross-over between the regime of an effective power law (with an exponent $\beta = 3\pi^2/128 \simeq 0.23$) and the thermodynamic limit (with true XY behaviour) is outside the range of the macroscopic length scales one can expect to find in experiments [100, 101].[35]

All in all, the quoted results give a fair impression to what extent theoretical predictions for critical exponents can presently be controlled experimentally.[36]

Finally, we mention that because close to a critical point material properties can be varied to a large extent by only slight changes in control parameters such as temperature, the applications of supercritical fluids (in the temperature-pressure range $1 < T/T_c < 1.1$ and $1 < P/P_c < 2$) as sol-

---

[34] On a purely phenomenological basis, one might argue that for a $n = 4$ crystal field, the transition might be described by the universality class of the self-dual line of the Ashkin–Teller model, where exponents vary continuously with a model parameter, and which is terminated by a Kosterlitz–Thouless transition point, see Chap. 12. A value $\beta \simeq 0.23$ would then imply that the system were close to that point. Furthermore, rather large values of the other exponents should be expected, for example $\gamma \simeq 3.2$. Interestingly, recent results for the susceptibility are equally compatible with an exponential law or a power law with a huge exponent $\gamma \sim 5$ [E29].

[35] The system Se/Ni(100) was proposed as a candidate for the realization of the Ashkin–Teller model [E27].

[36] Throughout this book, we shall limit ourselves to *pure* systems. Disorder may change the critical behaviour, however. Experimentally, this has been checked for example in the H/Ni(111) system, where impurities were created through the adsorption of oxygen atoms. Then the critical exponents change from the Potts-4 universality class to $\beta = 0.15(2)$, $\gamma = 1.68(15)$ and $\nu = 1.03(8)$ [E16].

vents for chemical and materials processing is currently being investigated, see [E26] for a review.

1.5 Correspondence Between Statistical Systems and Field Theory

As a preparation for the field-theoretic considerations to follow in the next chapters, we now recall the formal correspondence between statistical systems and quantum field theory. We have already seen that the important aspect to be emphasized in a second-order transition is the role of the fluctuations, as illustrated by the droplet picture (see Fig. 1.3). Due to universality, the *presence* of strong fluctuations is important, but for a description of the critical behaviour, many of their detailed properties are irrelevant in the renormalization group sense and can be neglected. This is the physical idea behind the formalism which formally relates the thermal fluctuations in statistical mechanics and the quantum fluctuations of field theory.

Consider a quantum mechanical system. We shall use the non-relativistic notation for simplicity, though the results extend to the relativistic case. The time evolution is described by the time-evolution operator

$$\mathcal{U}(t',t) = \exp\left(-\frac{i}{\hbar}H\left(t'-t\right)\right), \tag{1.79}$$

where H is the Hamiltonian, which we have assumed to be time-independent. $\mathcal{U}$ gives the time evolution of a wave function $\psi(t)$ via $\psi(t') = \mathcal{U}(t',t)\psi(t)$. Now take the time difference $t'-t$ to be infinitesimal. We can write the transition amplitude $\mathcal{Z}$ between two space-time points (t_a, x_a) and (t_b, x_b) via a Feynman path integral

$$\mathcal{Z} = \int \prod_i \mathcal{U}(t_{i+1}, t_i)\mathrm{d}x_i = \int \mathcal{D}x \exp\left(-\frac{i}{\hbar}S\right), \tag{1.80}$$

where the integral represents the sums over all possible paths (see Fig. 1.8) and $S(t_1, t_2) = \int_{t_1}^{t_2} dt\mathcal{L}$ is the classical action. We also note that a matrix element of $\mathcal{U}$ takes the form

$$\mathcal{U}(t_{i+1}, t_i) = \exp\left(-\frac{i}{\hbar}S(t_{i+1}, t_i)\right) \tag{1.81}$$

We need not describe how the functional integral can be precisely defined, since for our purposes it is sufficient to show a formal correspondence with analogous expressions arising in the context of statistical mechanics. Now, consider a statistical system defined on a hypercubic lattice (see Fig. 1.9), e.g. an Ising model. Single out one of the d directions and call it *"time"* τ. The remaining $d-1$ other directions will be referred to as *"space"* r.

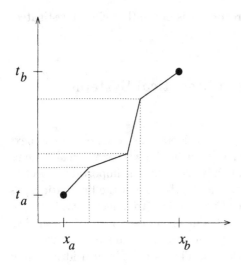

Fig. 1.8. Classical path joining space-time points (x_a, t_a) and (x_b, t_b)

Consider two times τ_1, τ_2 and their τ-dependent configurations of (Ising) spins $\{\sigma\}(\tau_1)$, $\{\sigma\}(\tau_2)$. The **transfer matrix** T links these configurations and has the elements

$$\langle\{\sigma\}(\tau_1)|T|\{\sigma\}(\tau_2)\rangle = \exp\left(-\beta\tilde{\mathcal{H}}(\tau_1, \tau_2)\right), \tag{1.82}$$

where $\mathcal{H} = \sum_\tau \tilde{\mathcal{H}}(\tau, \tau+1)$ is the classical Hamiltonian of the spin system and $\tilde{\mathcal{H}}$ includes the sum over the interacting pairs in the space directions. The partition function $\mathcal{Z}$ is

$$\begin{aligned}
\mathcal{Z} &= \sum_{\{\sigma\}} e^{-\beta\mathcal{H}} \\
&= \sum \langle\{\sigma\}(\tau_a)|T|\{\sigma\}(\tau_1)\rangle\langle\{\sigma\}(\tau_1)|T|\{\sigma\}(\tau_2)\rangle\cdots\langle\{\sigma\}(\tau_T)|T|\{\sigma\}(\tau_b)\rangle \\
&= \operatorname{tr} T^M, \tag{1.83}
\end{aligned}$$

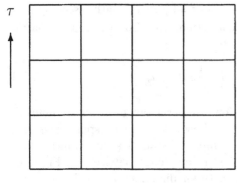

Fig. 1.9. Hypercubic lattice with "time" and "space" directions (after [169])

where M is the number of sites in the time direction. Periodic boundary conditions $|\{\sigma\}(\tau_a)\rangle = |\{\sigma\}(\tau_b)\rangle$ in the time direction were assumed.

Comparing these expressions with the ones for the quantum mechanical system, we note a few formal analogies, which are collected in Table 1.8. In particular, $\mathcal{T}$ may be interpreted as an Euclidean evolution operator

$$\mathcal{T} = \exp\left(-\tau H\right). \tag{1.84}$$

This defines the **quantum Hamiltonian** H. Therefore we obtain a map of a d-dimensional problem with classical, commuting variables onto a $(d-1)$-dimensional problem with quantum, non-commuting variables. In particular, we stress that $1/\hbar$ and $\beta = 1/k_B T$ parametrize the quantum and thermal fluctuations in a similar way. Absence of fluctuations corresponds to the classical $(\hbar \to 0)$ or the zero-temperature $(T \to 0)$ limits, respectively.

Table 1.8. Analogies between quantum theory and statistical mechanics

quantum theory		statistical mechanics	
real time	t	distance in 'time'	$\tau = it$
Planck's constant	$1/\hbar$	coupling	$\beta = 1/k_B T$
generating function	$\mathcal{Z}$	partition function	$\mathcal{Z}$
time evolution operator	$\mathcal{U}$	transfer matrix	$\mathcal{T}$

Later on, we shall rely heavily on the quantum Hamiltonian when applying conformal invariance techniques to statistical systems and most of the results will be phrased such that they apply immediately to H. Using (1.84), it is however a straightforward matter to reformulate everything in terms of the transfer matrix $\mathcal{T}$ and its eigenvalues.

1.6 Correspondence of Physical Quantities

We now establish the explicit dictionary between quantities calculated in the context of statistical mechanics with those obtained within quantum theory. The results of this section are summarized in Table 1.9.

Table 1.9. Correspondence of physical quantities of statistical and quantum systems

Statistical systems	Quantum systems
classical Hamiltonian, $\beta\mathcal{H}$	classical (Euclidean) action, $\frac{1}{\hbar}S$
infinitesimal transfer matrix, $1 - \mathcal{T}$	quantum Hamiltonian, H
Equilibrium state	ground state
Ensemble averages	ground state expectation values
Correlation functions	propagators
free energy density, f	ground state energy density, ω_0
inverse correlation lengths, ξ_i^{-1}	energy gaps, $E_i - E_0$

1.6.1 Free Energy Density

Since the classical Hamiltonian is real, the transfer matrix is hermitian and has a real eigenvalue spectrum $e^{-E_i \tau}$ and an eigenvector basis $|i\rangle$. One can arrange the eigenvalues to be ordered, i.e. $E_0 \leq E_1 \leq E_2 \leq \dots$. Then, using the spectral representation $\mathcal{T} = \sum_i |i\rangle e^{-E_i \tau} \langle i|$, one has[37]

$$\mathcal{T}^N = \sum_i |i\rangle e^{-E_i N \tau} \langle i|. \tag{1.85}$$

For a system with a finite number of degrees of freedom per site and a finite number of sites, $\mathcal{T}$ is a finite-dimensional matrix. If the Boltzmann weights $e^{-\beta \mathcal{H}}$ are positive, the Perron-Frobenius theorem (e.g. [G5]) states that the largest eigenvalue of $\mathcal{T}$, corresponding to E_0, is unique. Taking the limit $N \to \infty$, only the contribution from the eigenstate $|0\rangle$ survives, with exponentially small corrections[38]

$$\mathcal{T}^N \to |0\rangle e^{-E_0 M} \langle 0| \ , \quad \text{with } M = N\tau \tag{1.86}$$

and the partition function $\mathcal{Z} = \exp{-E_0 M} = \exp{-F}$, where F is the free energy (into which we have absorbed a factor $k_B T$). F is extensive in both T and the "spatial" volume V, while E_0 is extensive in V

$$F = fMV \ , \quad E_0 = \omega_0 V \tag{1.87}$$

and we have for the free energy density $f = \omega_0$.

From (1.86) we see that the lattice extent M in the temporal direction plays the role of an inverse temperature. This relationship is sometimes used in field theory calculations (see Chap. 3).

1.6.2 Correlation Functions

The field theory propagator is defined by a vacuum expectation value

$$\Delta(t, \boldsymbol{r}) := \langle 0|\phi(t, \boldsymbol{r}), \phi(0, \boldsymbol{0})|0\rangle. \tag{1.88}$$

The field operator $\phi(t, \boldsymbol{r})$, given in the Heisenberg picture, can be rewritten in the Schrödinger picture

$$\phi(t, \boldsymbol{r}) = e^{i/\hbar H t} \phi(\boldsymbol{r}) \, e^{-i/\hbar H t} \tag{1.89}$$

$$\Delta(t, \boldsymbol{r}) = \langle 0|\phi(\boldsymbol{r}) \, e^{-i/\hbar H t} \phi(\boldsymbol{0})|0\rangle e^{i/\hbar E_0 t}. \tag{1.90}$$

---

[37] We have implicitly added a constant to $\mathcal{H}$ to make the eigenvalues of $\mathcal{T}$ positive. This corresponds to adding a smooth background term which does not influence the critical behaviour.

[38] As it stands, (1.86) is correct above the critical point. Below T_c, the ground state is degenerate. However, for a finite ground state degeneracy, the correspondence $f = \omega_0$ remains true in the $N \to \infty$ limit.

The statistical mechanics correlation function is defined by

$$G(n_0, \boldsymbol{n}) = \frac{1}{\mathcal{Z}} \sum_{\phi(n'_0, n')} \phi(n_0, \boldsymbol{n})\phi(0, \boldsymbol{0}) \, e^{-\beta \mathcal{H}}$$

$$= \frac{\mathrm{tr} \left(T^A \phi(\boldsymbol{n}) T^{n_0} \phi(\boldsymbol{0}) T^B \right)}{\mathrm{tr} \, T^M}, \tag{1.91}$$

where n_0 measures the distance in time-direction and $\boldsymbol{n}$ in space direction. In writing this, we assume that $\langle \phi \rangle = 0$, which can be always achieved by redefining $\phi \to \phi - \langle \phi \rangle$. Periodic boundary conditions in the τ-direction are implied and $A + n_0 + B = M$. Taking the limit $A, B, n_0 \to \infty$, one has from the spectral representation (τ is the lattice constant in "time")

$$G(n_0, \boldsymbol{n}) \simeq \langle 0 | \phi(\boldsymbol{n}) e^{-n_0 H \tau} \phi(\boldsymbol{0}) | 0 \rangle \, e^{n_0 E_0 \tau} \tag{1.92}$$

and comparison yields as the final result (see Table 1.8)

$$G(n_0, \boldsymbol{n}) = \Delta \left(\frac{\hbar}{i} n_0 \tau, \boldsymbol{n} \right) \tag{1.93}$$

which illustrates again the analytic continuation between the time in quantum mechanics and the Euclidean time direction of statistical mechanics.

1.6.3 Correlation Lengths

Correlation lengths can be defined by the exponential decay of the two-point function

$$G(n_0, \boldsymbol{0}) \sim \exp\left(-|n_0|/\xi\right) \quad , \quad \text{if } |n_0| \gg \xi. \tag{1.94}$$

On the other hand, we have from equation (1.92)

$$G(n_0, \boldsymbol{0}) = \sum_i \exp\left[-(E_i - E_0)n_0\tau\right] |\langle 0 | \phi(\boldsymbol{0}) | i \rangle|^2. \tag{1.95}$$

Consider all the (scaling) operators contained in a theory. These operators can be given a number i such that i corresponds to the first eigenstate $|i\rangle$ for which $\langle 0 | \phi(\boldsymbol{0}) | i \rangle \neq 0$. One obtains an ordered sequence $\phi_0 < \phi_1 < \phi_2 \ldots$ of operators. If necessary, one has to redefine operators, e.g. $\phi_0 = 1, \phi_1$, $\phi'_2 = \phi_2 - a_{2,1}\phi_1 - a_{2,0}\phi_0$, $\phi'_3 = \phi_3 - a_{3,2}\phi_2 - a_{3,1}\phi_1 - a_{3,0}\phi_0, \ldots$ such that $\langle 0 | \phi_1(\boldsymbol{0}) | 1 \rangle \neq 0$, $\langle 0 | \phi_2(\boldsymbol{0}) | 1 \rangle = 0$, $\langle 0 | \phi_2(\boldsymbol{0}) | 2 \rangle \neq 0, \ldots$. For the Ising model, ϕ_1 would be the order parameter, ϕ_2 the energy density (then $a_{2,0} = 0$) and so on. Then $G_{\phi_i}(n_0, \boldsymbol{0}) \sim \exp(-(E_i - E_0)n_0\tau)$ and we thus have for the correlation length ξ_i

$$\xi_i^{-1} = E_i - E_0, \tag{1.96}$$

where the lattice constant in time direction τ is put to unity.

In writing this, we have implicitly assumed *isotropic* critical behaviour. More precisely, we have obtained a time correlation length $\xi_{\parallel}$, which in principle could be different from a space correlation length $\xi_{\perp}$. Further, the choice of the lattice units is arbitrary. This leads to an ambiguous, non-universal prefactor for ξ. It is not important for the definition of the critical exponents (see (1.10)), but does become important for the calculation of critical amplitudes. There are techniques to fix its value uniquely (Chaps. 9, 10).

Exercises

1. Consider the 1D three-states Potts model. Determine the transfer matrix T and find the free energy as a function of the temperature T.

2. Consider the (connected) two-point functions for spin-spin and energy-energy correlations

$$G_\sigma(\mathbf{r}_1 - \mathbf{r}_2) = \langle \sigma(\mathbf{r}_1)\sigma(\mathbf{r}_2)\rangle - \langle\sigma(\mathbf{r}_1)\rangle\langle\sigma(\mathbf{r}_2)\rangle$$
$$G_\epsilon(\mathbf{r}_1 - \mathbf{r}_2) = \langle \epsilon(\mathbf{r}_1)\epsilon(\mathbf{r}_2)\rangle - \langle\epsilon(\mathbf{r}_1)\rangle\langle\epsilon(\mathbf{r}_2)\rangle. \tag{1.97}$$

Derive the fluctuation-dissipation theorems relating G_σ and G_ϵ to the susceptibility χ and the specific heat C, respectively. At the critical point $t = h = 0$, show that they transform covariantly under a scale transformation $\mathbf{r} \to \mathbf{r}/b$ and satisfy relations of the same kind as the 'unconnected' two-point function $\bar{G}_\sigma$ discussed in the text. From this, rederive the scaling relations of the exponents and discuss hyperscaling.

3. Consider a (spin-spin) two point function $G(r_\perp, r_\parallel)$ in d dimensions which depends on two coordinates $r_{\perp,\parallel}$, which may be labelled as 'space' and 'time'. At criticality, assume that (d_* is the lower critical dimension)

$$G(r_\perp, 0) \sim r_\perp^{-[d - d_* + \eta_\perp]} \quad , \quad G(0, r_\parallel) \sim r_\parallel^{-[(d - d_*)/\theta + \eta_\parallel]}.$$

Out of criticality, $G(r_\perp, r_\parallel)$ decays exponentially, with correlation lengths, respectively, $\xi_{\perp,\parallel} \sim (T - T_c)^{-\nu_{\perp,\parallel}}$ where $\theta = \nu_\parallel/\nu_\perp$ and scales as

$$G(r_\perp, r_\parallel; t, h) = b^{-2x/\theta}G(r_\perp/b, r_\parallel/b^\theta; tb^{y_t}, hb^{y_h}).$$

Invoke a fluctuation-dissipation relation to derive (1.56).

4. For a spin system at a tricritical point, derive the relations between the tricritical exponents and the scaling dimensions of the relevant scaling operators [430]

$$\alpha_t = -\frac{x_\epsilon}{d - x_\epsilon} \quad , \quad \beta_t = \frac{x_\sigma}{d - x_\epsilon} \quad , \quad \gamma_t = \frac{d - 2x_\sigma}{d - x_\epsilon} \quad , \quad \delta_t = \frac{d - x_\sigma}{x_\sigma}$$
$$\nu_t = \frac{1}{d - x_\epsilon} \quad , \quad \eta_t = 2 - d + 2x_\sigma \quad , \quad \phi_t = \frac{d - x_{\epsilon'}}{d - x_\epsilon}. \tag{1.98}$$

Find their values for the tricritical Ising model.

5. What universality class do you expect for the critical A_4 RSOS model?

6. To what extent can the scaling relations for the critical exponents be experimentally confirmed?

2. Conformal Invariance

2.1 From Scale Invariance to Conformal Invariance

In the first chapter, we have seen that critical theories may be characterized through the covariance of the correlators under scale transformations,
or dilatations, of the length scales by a constant factor b, for example for a
two-point function

$$G(\boldsymbol{r}) = b^{-2x} G(\boldsymbol{r}/b), \qquad (2.1)$$

where we took $t = h = 0$. From now on, this will always be assumed to be
the case and we shall omit these variables altogether. For a field-theoretic
interpretation of this, it appears, at least heuristically, natural to assume the
same sort of covariance under dilatations for the scaling operators ϕ_i from
which multipoint correlators are built, viz.

$$\phi_i(\boldsymbol{r}) = b^{-x_i} \phi_i(\boldsymbol{r}/b), \qquad (2.2)$$

where x_i is the scaling dimension of ϕ_i.

At a critical point, the long-range properties of the system become independent of the details of the underlying physical system, for example the
precise lattice structure. This may lead to the invariance of the system under
a larger group than explicitly specified when writing down a lattice Hamil-
tonian. We now ask: are other coordinate transformations $\boldsymbol{r} \to \boldsymbol{r}' = b(\boldsymbol{r})\boldsymbol{r}$,
with a *space-dependent* rescaling $b(\boldsymbol{r})$, compatible with dilatation covariance
and such that the scaling operators (or at least a sufficiently large subset of
them) transform covariantly with respect to them?

It appears natural to require that the scaling operators should transform
as

$$\phi_i(\boldsymbol{r}) \to \phi_i'(\boldsymbol{r}) = (J(\boldsymbol{r}))^{x_i/d} \phi_i(\boldsymbol{r}/b(\boldsymbol{r})) \; ; \; J(\boldsymbol{r}) = \left| \frac{D\boldsymbol{r}'}{D\boldsymbol{r}} \right|, \qquad (2.3)$$

in d dimensions, where J is the Jacobian. For a dilatation, $\boldsymbol{r}' = \boldsymbol{r}/b$,
$J(\boldsymbol{r}) = b^{-d}$ and (2.2) is recovered. There are obvious transformations which
will leave a correlator G invariant, namely translations $\boldsymbol{r}' = \boldsymbol{r} + \boldsymbol{a}$ and rotations $\boldsymbol{r}' = \Lambda \boldsymbol{r}$ when the ϕ_i are scalars under rotation (otherwise, they
will transform according to the representations of the rotation group $SO(d)$).
These invariances are simply a consequence of the irrelevance of the detailed

lattice structure for critical behaviour. Is there a systematic way to find out whether the ϕ_i are covariant under other coordinate transformations?

To outline an answer, it may be helpful to consider infinitesimal transformations $\boldsymbol{r}' = \boldsymbol{r} + \boldsymbol{\epsilon}(\boldsymbol{r})$, with ϵ small in some sense. The covariance condition (2.3) then gives the change $\phi_i(\boldsymbol{r}) \rightarrow \phi_i(\boldsymbol{r}) + \delta\phi_i(\boldsymbol{r})$ as

$$\delta\phi_i(\boldsymbol{r}) = \left(\frac{x_i}{d}\boldsymbol{\nabla} \cdot \boldsymbol{\epsilon} + \boldsymbol{\epsilon} \cdot \boldsymbol{\nabla}\right)\phi_i(\boldsymbol{r}). \tag{2.4}$$

Later, we shall refer to scaling operators which transform in this way as *(quasi)primary*. Up to this point, it might appear that only the local properties of the scaling operators were important. In general, however, coordinate transformations will also change the global geometry of the system, which in turn will affect the local covariance properties of the scaling operators. We shall show that this can be taken into account through the conformal *Ward identity* for the n-point correlators

$$\sum_{p=1}^{n}\langle\phi_1(\boldsymbol{r}_1)\ldots\left(\frac{x_i}{d}\boldsymbol{\nabla} \cdot \boldsymbol{\epsilon} + \boldsymbol{\epsilon} \cdot \boldsymbol{\nabla}\right)\phi_p(\boldsymbol{r}_p)\ldots\phi_n(\boldsymbol{r}_n)\rangle$$

$$+\frac{1}{S_d}\int d^d\boldsymbol{r}\,\langle\phi_1(\boldsymbol{r}_1)\ldots\phi_n(\boldsymbol{r}_n)T_{\mu\nu}(\boldsymbol{r})\rangle\partial^\mu\epsilon^\nu = 0, \tag{2.5}$$

where $T_{\mu\nu}$ is the energy–momentum tensor, $\boldsymbol{\nabla}$ acts on the operator ϕ_p and S_d is the area of the sphere in d dimensions ($S_2 = 2\pi$). Now, rotation and scale invariance require that T is symmetric and traceless. Since this already implies the covariance of scaling operators under *conformal* transformations, it becomes apparent that *rather than being an extra requirement, conformal invariance is the logical extension of scale invariance*, provided of course that a conformal Ward identity can be written down. That is already enough to fix the form of the two- and three-point functions in d dimensions. In 2D, the conformal group is infinite-dimensional. However, from the same Ward identity, it follows that quantum fluctuations do break 2D conformal invariance in a subtle way, leading to the *conformal anomaly*, parametrized by the *central charge c*. In later chapters, we shall see that different values of the central charge lead to different universality classes.

Having thus motivated the advent of conformal symmetry, we shall now turn to a more formal presentation of the subject, largely following [169]. The basic ideas presented in this chapter are due to Belavin, Polyakov and A. Zamolodchikov [68].

2.2 Conformal Transformations in d Dimensions

Let us begin by defining conformal transformation in d dimensions (for a mathematically fully rigorous discussion, see [G27]). Consider a line element in a space with metric $g_{\mu\nu}$

$$ds^2 = g_{\mu\nu}dr^\mu dr^\nu. \tag{2.6}$$

Under an arbitrary local change of coordinates $r \to r'$ the metric changes covariantly

$$g'_{\mu\nu}(r') = \frac{\partial r^\alpha}{\partial r'^\mu}\frac{\partial r^\beta}{\partial r'^\nu}g_{\alpha\beta}(r). \tag{2.7}$$

Now, **conformal transformations** are defined as those special coordinate transformations which keeps the metric invariant, up to a local scale factor

$$g'_{\mu\nu}(r') = \Omega(r)g_{\mu\nu}(r). \tag{2.8}$$

Geometrically, this means that the angle $\cos\theta = r \cdot r'/(r^2 r'^2)^{1/2}$ between the two vectors r and r' is kept invariant, with $r \cdot r' = g_{\mu\nu}r^\mu r'^\nu$. This is illustrated in Fig. 2.1. The set of conformal transformations forms a group,

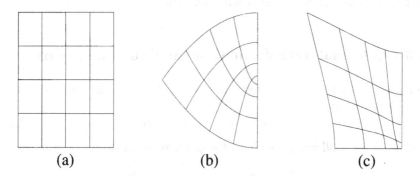

(a)   (b)   (c)

Fig. 2.1. Coordinate transformations. The transformation from the square lattice in (**a**) onto the lattice in (**b**) is conformal, while the transformation onto the lattice in (**c**) is not

the **conformal group**. In $d > 2$ dimensions, the conformal group is finite-dimensional. It is generated by global rotations $r \to r' = \Lambda r$, global translations $r \to r' = r + a$, global **dilatations** $r \to r' = br$, and the so-called **special conformal transformations**

$$r \to r' = \frac{r + ar^2}{1 + 2a \cdot r + a^2 r^2}. \tag{2.9}$$

The geometric meaning of the special transformation becomes clear when rewriting it as

$$\frac{r'}{r'^2} = \frac{r}{r^2} + a, \tag{2.10}$$

which is the combination of an inversion $r \to r' = r/r^2$ followed by a translation a and again an inversion.

Let us find all conformal transformations in d dimensions. Under an infinitesimal transformation $r^\mu \to r^\mu + \epsilon^\mu(\boldsymbol{r})$ the line element transforms as

$$ds^2 \longrightarrow ds^2 + (\partial_\mu \epsilon_\nu + \partial_\nu \epsilon_\mu) dr^\mu dr^\nu. \qquad (2.11)$$

Because of (2.8), the second term must be proportional to $g_{\mu\nu}$. Let $g^{\mu\nu}$ be the inverse matrix of $g_{\mu\nu}$. Taking the trace $g^{\mu\nu}(\partial_\mu \epsilon_\nu + \partial_\nu \epsilon_\mu)$ leads to the constraint

$$\partial_\mu \epsilon_\nu + \partial_\nu \epsilon_\mu = \frac{2}{d}(\partial \cdot \epsilon) g_{\mu\nu}. \qquad (2.12)$$

It is easy to check that translations, rotations and dilations and, finally, the special transformations are solutions of this equation, of order zero, one and two, respectively. In addition, it can be shown from the constraint (2.12) that for $d > 2$, the only infinitesimal conformal transformations besides translations, rotations and dilatations are those obtained from the special conformal transformations (2.10).[1] For $d > 2$, therefore, the given examples already exhaust the possibilities for conformal transformations.

2.3 Conformal Transformations in Two Dimensions

For $d = 2$, equations (2.12) become the Cauchy-Riemann equations in an Euclidean metric

$$\partial_1 \epsilon_1 = \partial_2 \epsilon_2 \qquad \partial_1 \epsilon_2 = -\partial_2 \epsilon_1. \qquad (2.13)$$

Naturally, we shall use throughout the complex coordinates

$$z = r_1 + i r_2 \quad , \quad \bar{z} = r_1 - i r_2$$
$$\epsilon(z) = \epsilon_1 + i \epsilon_2 \quad , \quad \bar{\epsilon}(\bar{z}) = \epsilon_1 - i \epsilon_2. \qquad (2.14)$$

The conformal transformations introduced in the previous section are easily rewritten. A translation is $z \to z' = z + a$. A rotation by the angle α is $z \to z' = e^{i\alpha} z$ and a dilatation becomes $z \to z' = bz$, where α and b are both real. For the special transformations, let $a = a_1 - i a_2$. Then, from (2.9)

$$z' = \frac{z + \bar{a} z \bar{z}}{1 + \bar{a}\bar{z} + az + a\bar{a}z\bar{z}} = \frac{z(1 + \bar{a}\bar{z})}{(1 + az)(1 + \bar{a}\bar{z})} = \frac{z}{1 + az}. \qquad (2.15)$$

Similar expressions apply to the transformation of $\bar{z}$.

From (2.13), the group of conformal transformations in two dimensions is isomorphic to the (infinite-dimensional) group of arbitrary analytic coordinate transformations $z \to w(z)$ and $\bar{z} \to \bar{w}(\bar{z})$. For illustration, consider

---

[1] Straightforward but lengthy calculations show that $(1 - d) \, \Box \, \partial^\lambda \epsilon_\lambda = 0$, which implies that conformal invariance is trivial for $d = 1$. If $d \neq 1$, if follows further that $(d - 2)\partial_\mu \partial_\nu \partial^\lambda \epsilon_\lambda = 0$. For $d \neq 2$, ϵ_μ can be at most quadratic in the r^μ and from (2.12) it then follows that the only conformal transformations are those already listed (see exercises).

again Fig. 2.1, where the conformal transformation $w_{a \to b}(z) = z^2$ is clearly analytic while the non-conformal transformation $w_{a \to c}(z) = z|z|$ is not. A line element transforms under a conformal transformation as

$$\mathrm{d}s^2 \longrightarrow \left(\frac{\partial w}{\partial z}\right)\left(\frac{\partial \bar{w}}{\partial \bar{z}}\right)\mathrm{d}s^2. \tag{2.16}$$

Writing $w(z) = z + \epsilon(z), \bar{w}(\bar{z}) = \bar{z} + \bar{\epsilon}(\bar{z})$, the infinitesimal transformations are given by

$$\epsilon_n = -\epsilon z^{n+1} \quad , \quad \bar{\epsilon}_n = -\bar{\epsilon}\bar{z}^{n+1}. \tag{2.17}$$

Their Lie algebra generators are

$$l_n = -z^{n+1}\partial_z \quad , \quad \bar{l}_n = -\bar{z}^{n+1}\partial_{\bar{z}} \tag{2.18}$$

which satisfy the following commutation relations

$$\begin{aligned} [l_n, l_m] &= (n - m)\, l_{m+n} \\ [\bar{l}_n, \bar{l}_m] &= (n - m)\, \bar{l}_{m+n} \\ [l_n, \bar{l}_m] &= 0. \end{aligned} \tag{2.19}$$

This algebra is called the **loop algebra**. The third commutator shows that this algebra decomposes into the direct sum of two algebras, one generated by $\{l\}$ and the other by $\{\bar{l}\}$. Because of this simple structure, it is often enough to consider the merely the z-dependence of correlators. Unless global and analyticity properties of the correlators are explicitly needed (see Chap. 5), one may treat z and $\bar{z}$ formally as separate, independent, coordinates.

Some properties of $\{l_n\}$ (and $\{\bar{l}_n\}$) are important. Obviously, l_n is regular at $z = 0$ for $n \geq -1$. Performing the transformation $z = 1/v$, however, one finds $l_n \to -v^{-n+1}\partial_v$ singular at $v = 0$ (i.e. $z \to \infty$) for $n > 1$. Correspondingly l_n is singular at $z = 0$ for $n < -1$ but regular at infinity. Therefore only the projective subset $\{l_{-1}, l_0, l_1\}$ is defined everywhere and induces analytic transformations of the complex plane $\mathbb{C}$ onto itself. The other transformations are therefore *not* analytic everywhere and do not possess a global inverse. Their existence is peculiar to two dimensions.

It is easy to go back from the infinitesimal generators $l_n = -z^{n+1}\partial_z$ to the corresponding finite conformal transformations. Formally, this would be the transformation $\exp(-\epsilon z^{n+1}\partial_z)$. To calculate this, define the function $f(\epsilon, z) := \exp(-\epsilon z^{n+1}\partial_z)z$. Taking the formal derivative with respect to ϵ, one arrives at the differential equation

$$\frac{\partial f(\epsilon, z)}{\partial \epsilon} + z^{n+1}\frac{\partial f(\epsilon, z)}{\partial z} = 0 \quad , \quad f(0, z) = z \tag{2.20}$$

because for $\epsilon = 0$, one should come back to the trivial transformation. For $n \neq 0$, the general solution is $f(\epsilon, z) = F(\epsilon + (nz^n)^{-1})$, where $F(u)$ is an arbitrary function (see [383] for a compilation of partial differential equations and their solutions). The functional equation $F(z^{-n}/n) = z$ is solved by inspection and the conformal transformation finally becomes $z \to w(z) = f(\epsilon, z)$, with

$$f(\epsilon, z) = \frac{z}{(1 + \epsilon n z^n)^{1/n}} \,, \quad n \neq 0. \tag{2.21}$$

The case $n = 0$ must be treated separately and is left as an exercise.

The meaning of *global* conformal transformations is straightforward. Translations are generated by l_{-1} and $\bar{l}_{-1}$. A dilatation is a combination $l_0 + \bar{l}_0 = r^\mu \partial_\mu$, and $\mathrm{i}(l_0 - \bar{l}_0) = \epsilon^{\mu\nu} r_\mu \partial_\nu$ (where $\epsilon^{\mu\nu}$ is the totally antisymmetric tensor) generates a rotation. The last generators l_1 and $\bar{l}_1$ correspond to the special conformal transformations. The set $\{l_{-1}, l_0, l_1\}$ generates the **projective conformal transformations** (sometimes also referred to as *Möbius transformations*)

$$z \longrightarrow w_p(z) = \frac{az + b}{cz + d} \tag{2.22}$$

with a, b, c, d complex such that $ad - bc = 1$. There is a relation of the $w_p(z)$ to the matrix group $\mathrm{SL}(2, \mathbb{C})/\mathbb{Z}_2$. To see this, consider matrices of the form, associated to $w_p(z)$

$$\begin{pmatrix} a & b \\ c & d \end{pmatrix}. \tag{2.23}$$

Successive application of projective conformal transformations corresponds to matrix multiplication. Translations, rotations, dilatations, and special conformal transformations are expressed by

$$\text{Translation} \Leftrightarrow \begin{pmatrix} 1 & b \\ 0 & 1 \end{pmatrix}, \qquad \text{Rotation} \Leftrightarrow \begin{pmatrix} e^{\mathrm{i}\theta/2} & 0 \\ 0 & e^{-\mathrm{i}\theta/2} \end{pmatrix},$$

$$\text{Dilatation} \Leftrightarrow \begin{pmatrix} \sqrt{a} & 0 \\ 0 & \sqrt{1/a} \end{pmatrix}, \qquad \text{Special} \Leftrightarrow \begin{pmatrix} 1 & 0 \\ c & 1 \end{pmatrix}. \tag{2.24}$$

The subalgebra generated by $\{l_0, \bar{l}_0\}$ is the *Cartan subalgebra*. As explaining in the next section, physical operators will be eigenstates of this subalgebra with eigenvalues $\{\Delta, \bar{\Delta}\}$, called **conformal weights**. There is no need to have $\bar{\Delta}$ being equal to Δ. Since $l_0 + \bar{l}_0$ and $\mathrm{i}(l_0 - \bar{l}_0)$ generate dilatations and rotations respectively, the **scaling dimension** x and the **spin** s are given by

$$\boxed{x = \Delta + \bar{\Delta}} \qquad \text{and} \qquad \boxed{s = \Delta - \bar{\Delta}} \,. \tag{2.25}$$

The spin s must not be confused with the spin operator encountered in the previous sections.[2]

---

[2] s is referred to as spin since it is the eigenvalue of the rotation operator $\mathrm{i}(l_0 - \bar{l}_0)$. In distinction to the situation in higher dimensions, s is not restricted to integer or half-integer values.

2.4 Conformal Invariance in Two Dimensions

Having defined the conformal group and written down its Lie algebra, we now
have to implement the requirement of conformal invariance in a statistical
system. To do so, we have to define how scaling operators should 'transform'
under the action of the conformal group.[3] Of course, this must be done in a
way that covariance under global dilatations is recovered. Since not all scaling operators transform in the same way under conformal transformations,
a precise terminology is required, as introduced in [68]. One distinguishes
between so-called *primary* operators, which transform covariantly (examples
are the order parameter density σ or the energy density ε in the Ising model),
and the *secondary* ones, whose transformation law is more complicated. We
choose a formal way and state the required properties as postulates before
explaining how to apply them.

A **conformal theory** is described by the correlation functions of a large
set of scaling operators $A(z, \bar{z})$, with the following properties [68]. Operators
in this set are called **local**.

1. If $A(z, \bar{z})$ is a local operator, all derivatives of A are local as well. Thus
   the set of local operators $\{A(z, \bar{z})\}$ is in general infinite.
2. There is a subset $\{\Phi(z, \bar{z})\} \subset \{A(z, \bar{z})\}$ of local operators, called **quasi-
   primary**, which 'transform covariantly' under projective[4] conformal
   transformations w_p (2.22)

$$\Phi(z, \bar{z}) \longrightarrow \left(\frac{\partial w_p}{\partial z}\right)^{\Delta} \left(\frac{\partial \bar{w}_p}{\partial \bar{z}}\right)^{\overline{\Delta}} \Phi(w_p(z), \bar{w}_p(\bar{z})) \qquad (2.26)$$

3. Any local operator can be written as a linear combination of the quasi-
   primary operators and their derivatives.
4. The *vacuum* is invariant under projective conformal transformations.

Property 2 means that correlation functions of quasi-primary operators transform covariantly under projective conformal transformations

$$\langle \Phi_1(z_1, \bar{z}_1) \cdots \Phi_n(z_n, \bar{z}_n) \rangle = \left(\frac{\partial w_p}{\partial z_1}\right)^{\Delta_1} \left(\frac{\partial \bar{w}_p}{\partial \bar{z}_1}\right)^{\overline{\Delta}_1} \cdots \left(\frac{\partial w_p}{\partial z_n}\right)^{\Delta_n} \left(\frac{\partial \bar{w}_p}{\partial \bar{z}_n}\right)^{\overline{\Delta}_n}$$
$$\cdot \langle \Phi_1(w_p(z_1), \bar{w}_p(\bar{z}_1)) \cdots \Phi_n(w_p(z_n), \bar{w}_p(\bar{z}_n)) \rangle. \qquad (2.27)$$

As an exercise, the reader may check by considering a dilatation $z \to bz, \bar{z} \to b\bar{z}$ and a rotation $z \to e^{i\alpha}z, \bar{z} \to e^{-i\alpha}\bar{z}$ that (2.25) is recovered. It is also
easy to check that a derivative $\partial\Phi$ of a quasi-primary operator cannot be
itself be quasi-primary, unless $\Phi = \mathbf{1}$. The meaning of property 4 will become clear in Chap. 4 when studying the representation theory of the conformal algebra. Projective transformations are special in that they map the

---

[3] Writing a 'transformation' of a scaling operator is a purely formal device used for
heuristic purposes which only acquires a meaning when inserted into a correlator.
[4] These are sometimes called global conformal transformations.

entire complex plane onto itself and do not change the global geometry. This should correspond to the invariance of the physical ground state under these transformations. That is no longer so, however, for more general conformal mappings. geometry of the system.[5] An important example is the conformal map $w = \ln z$, which maps the entire plane onto a strip of finite width.

Quasi-primary operators are the most fundamental operators for $d > 2$. Unfortunately, in general there exists an infinite number of them. In two dimensions, we can take advantage of the special features of the conformal algebra and impose the much stronger condition of covariant transformation under *arbitrary* conformal transformations w, rather than just the projective ones w_p. Thus, and only in 2D, one may identify a new subset of operators [68].

5. The subset $\{\phi(z, \bar{z})\} \subset \{\Phi(z, \bar{z})\}$ of **primary** operators contains those quasi-primary operators which 'transform' under *any* conformal transformation $w(z)$ covariantly

$$\phi(z, \bar{z}) \longrightarrow \left(\frac{\partial w}{\partial z}\right)^{\Delta} \left(\frac{\partial \bar{w}}{\partial \bar{z}}\right)^{\overline{\Delta}} \phi(w(z), \bar{w}(\bar{z})). \qquad (2.28)$$

Correlation functions of primary operators transform similarly to (2.27) but now for *any* conformal transformation w

$$\langle \phi_1(z_1, \bar{z}_1) \cdots \phi_n(z_n, \bar{z}_n) \rangle = \left(\frac{\partial w}{\partial z_1}\right)^{\Delta_1} \left(\frac{\partial \bar{w}}{\partial \bar{z}_1}\right)^{\overline{\Delta}_1} \cdots \left(\frac{\partial w}{\partial z_n}\right)^{\Delta_n} \left(\frac{\partial \bar{w}}{\partial \bar{z}_n}\right)^{\overline{\Delta}_n}$$
$$\cdot \langle \phi_1(w(z_1), \bar{w}(\bar{z}_1)) \cdots \phi_n(w(z_n), \bar{w}(\bar{z}_n)) \rangle. \qquad (2.29)$$

The average brackets $\langle \rangle$ of this last equation have to be taken with caution. As mentioned above, they evaluate functionals of operators in the same geometry only if w is projective. Examples will be given below.

Obviously, every primary operator is also quasi-primary, but a quasi-primary operator may or may not be primary. Primary operators transform in a simpler manner than the remaining operators, called **secondary**. A given secondary operator may or may not be quasi-primary. We shall see examples of both in the following sections. The important point of the distinction made between primary and secondary operators is, as we shall show, that all properties of any secondary operator can be entirely deduced from the knowledge of the primary operators. Moreover, many physical systems only have a finite number of primary operators. For example, the only two primary operators of the Ising model are just the order parameter density and the energy density.

As a first application, we now show how some simple correlation functions of quasi-primary operators can be completely determined.

---

[5] Many practical examples are nicely presented in [363].

2.5 Correlation Functions of Quasi-primary Operators

The projective conformal group is sufficient to determine the functional form of a few simple correlation functions. Although written here in a 2D form, the results of this section are readily generalized to any dimension $d \geq 2$. Under infinitesimal conformal transformations $z \rightarrow z + \epsilon(z)$, $\bar{z} \rightarrow \bar{z} + \bar{\epsilon}(\bar{z})$, primary operators transform as

$$\delta_\epsilon \phi(z, \bar{z}) = \left((\Delta \partial_z \epsilon + \epsilon \partial_z) + (\overline{\Delta} \partial_{\bar{z}} \bar{\epsilon} + \bar{\epsilon} \partial_{\bar{z}}) \right) \phi(z, \bar{z}). \qquad (2.30)$$

The same holds for quasi-primary operators Φ if infinitesimal projective transformations (written ϵ_p) are considered. Such transformations do not change the correlation functions of any (quasi-)primary operator. We have therefore

$$\sum_{i=1}^{n} \left[(\Delta_i \partial_{z_i} \epsilon_p(z_i) + \epsilon_p(z_i) \partial_{z_i}) + (\overline{\Delta}_i \partial_{\bar{z}_i} \bar{\epsilon}_p(\bar{z}_i) + \bar{\epsilon}_p(\bar{z}_i) \partial_{\bar{z}_i}) \right] \langle \phi_1 \cdots \phi_n \rangle$$
$$= \delta_{\epsilon_p} \langle \phi_1 \cdots \phi_n \rangle = 0. \qquad (2.31)$$

Let us investigate the consequences for the two-point function [507]. We first consider the translation $\epsilon_p = 1, \bar{\epsilon}_p = 0$. Equation (2.31) becomes the differential equation

$$(\partial_{z_1} + \partial_{z_2}) \langle \phi_1 \phi_2 \rangle = 0 \qquad (2.32)$$

which proves that the two-point function only depends on $z_{12} = z_1 - z_2$, and $\bar{z}_{12} = \bar{z}_1 - \bar{z}_2$. We take now $\epsilon_p = z, \bar{\epsilon}_p = 0$.

$$(\Delta_1 + \Delta_2 + z_1 \partial_{z_1} + z_2 \partial_{z_2}) \langle \phi_1 \phi_2 \rangle = 0. \qquad (2.33)$$

Taking translation invariance into account, we conclude that

$$\langle \phi_1 \phi_2 \rangle \sim z_{12}^{-\Delta_1 - \Delta_2} \bar{z}_{12}^{-\overline{\Delta}_1 - \overline{\Delta}_2}. \qquad (2.34)$$

Finally, with $\epsilon_p = z^2, \bar{\epsilon}_p = 0$, we have to solve

$$(2z_1 \Delta_1 + 2z_2 \Delta_2 + z_1^2 \partial_{z_1} + z_2^2 \partial_{z_2}) \frac{1}{z_{12}^{\Delta_1 + \Delta_2}} = 0. \qquad (2.35)$$

This constrains the exponents to be equal $\Delta_1 = \Delta_2 = \Delta$ and, similarly, $\overline{\Delta}_1 = \overline{\Delta}_2 = \overline{\Delta}$. Any two-point function of primary or quasi-primary operators must take the form

$$\langle \phi_1 \phi_2 \rangle = C_{12} z_{12}^{-2\Delta} \bar{z}_{12}^{-2\overline{\Delta}}. \qquad (2.36)$$

The scaling dimension is given by $x = \Delta + \overline{\Delta}$ and the spin $s = \Delta - \overline{\Delta}$. The coefficient C_{12} can be arbitrarily chosen. Actually, we can always find a basis of primary operators, such that $C_{12} = 0$ if $\phi_1 \neq \phi_2$ for real operators or $\phi_1 \neq \phi_2^*$ for complex operators. Therefore, C_{12} is not a free parameter left undetermined by conformal invariance but simply a normalization factor.

We can repeat these calculations for any n-point function. The covariance conditions are called **projective Ward identities**, written here for the z-dependence only. These express the covariance of the theory under translations, dilatations and special conformal transformations, respectively[6]

$$\sum_i \partial_{z_i} \langle \phi_1 \cdots \phi_n \rangle = 0$$
$$\sum_i (z_i \partial_{z_i} + \Delta_i) \langle \phi_1 \cdots \phi_n \rangle = 0 \qquad (2.37)$$
$$\sum_i (z_i^2 \partial_{z_i} + 2\Delta_i z_i) \langle \phi_1 \cdots \phi_n \rangle = 0 .$$

Although these do not fix all possible correlation functions, they neverthe-less reduce the number of degrees of freedom. For example the three-point function $\langle \phi_1 \phi_2 \phi_3 \rangle$ must have the form [507]

$$\langle \phi_1 \phi_2 \phi_3 \rangle = C_{123} \, z_{12}^{-(\Delta_1+\Delta_2-\Delta_3)} \, z_{23}^{-(\Delta_2+\Delta_3-\Delta_1)} \, z_{13}^{-(\Delta_1+\Delta_3-\Delta_2)}$$
$$\times \bar{z}_{12}^{-(\bar{\Delta}_1+\bar{\Delta}_2-\bar{\Delta}_3)} \, \bar{z}_{23}^{-(\bar{\Delta}_2+\bar{\Delta}_3-\bar{\Delta}_1)} \, \bar{z}_{13}^{-(\bar{\Delta}_1+\bar{\Delta}_3-\bar{\Delta}_2)} . \qquad (2.38)$$

We point out that the value of the coefficient C_{123} is *not* arbitrary.[7] We shall return to the important problem of its determination in Chaps. 5–7.[8]

The four-point functions are not completely determined. From the pro-jective transformation (2.22), one can fix a, b, c, d so that the four points (z_1, z_2, z_3, z_4) are transformed into $(\eta, 0, 1, \infty)$ respectively with

$$\eta = \frac{z_{12} z_{34}}{z_{32} z_{14}}. \qquad (2.39)$$

The parameter η is invariant under projective transformations. The projective Ward identities determine the form of a correlation function up to projectively invariant terms. Therefore [507]

$$\langle \phi_1 \phi_2 \phi_3 \phi_4 \rangle = \prod_{i<j} z_{ij}^{-\gamma_{ij}} \, \bar{z}_{ij}^{-\bar{\gamma}_{ij}} \, F(\eta, \bar{\eta}), \qquad (2.40)$$

where $\gamma_{ij}, \bar{\gamma}_{ij}$ are parameters which satisfy the conditions

$$\sum_i \gamma_{ij} = 2\Delta_j \, , \quad \sum_i \bar{\gamma}_{ij} = 2\bar{\Delta}_j \qquad (2.41)$$

---

[6] In terms of the multi-body generators l_n, these conditions on $C = \langle \phi_1 \ldots \phi_n \rangle$ might also be written as $l_{-1}C = 0$, $l_0 C = \sum_i \Delta_i C$ and $l_1 C = 2 \sum_i \Delta_i z_i C$. Here, the l_n only contain the effects of the variable change $r \to r'$ and *not* the transformation of ϕ itself.

[7] With the normalization $C_{ij} = \delta_{ij}$ in (2.36), the C_{ijk} are universal. In Chap. 5, we shall see that $C_{ijk} = \mathbf{C}_{ijk}$, the operator product expansion coefficient.

[8] The C_{ijk} also play an important role in the renormalization group. Consider a Hamiltonian $\mathcal{H} = \mathcal{H}^* + \sum_k g_k \phi_k$ perturbed away from the fixed point Hamiltonian $\mathcal{H}^*$. Under a length rescaling by $1 + \delta\ell$, it can be shown that $dg_k/d\ell = y_k g_k - \frac{1}{2} S_d \sum_{i,j} C_{ijk} g_i g_j + \ldots$ [G6], where y_k are the renormaliza-tion group eigenvalues and S_d is the surface of the hypersphere in d dimensions. Thus the C_{ijk} determine how RG trajectories start from the fixed points g_k^*, characterized by $dg_k^*/d\ell = 0$.

and $F = F(\eta, \bar{\eta})$ is an arbitrary function.[9] Consequently, covariance under projective transformations only determines the correlators up to functions depending merely on projectively invariant terms like $\eta, \bar{\eta}$. For a n-point function, one may build $n - 3$ independent invariants (twice as much if we count separately the $\bar{\eta}$), because there is always a projective transformation which maps three points onto $0, 1, \infty$.

More information will be needed to find four-point (and higher order) correlators. That information is only available in 2D and will be developed in the next few chapters. As a first step towards this goal, we need to analyse the energy–momentum tensor.

2.6 The Energy–Momentum Tensor

In the previous section, we studied simple correlation functions of quasi-primary operators. Stronger results can only be obtained by considering the more restricted set of primary operators. However, primary operators transform covariantly under any 2D conformal transformation (2.29). According to Liouville's theorem [13, 363], an analytic function cannot be bounded in the entire complex plane, unless it is a constant. This implies that non-trivial analytic transformations (2.17) must have singularities and they cannot be small everywhere. To cope with this problem, we consider a local coordinate transformation $\boldsymbol{r}' = \boldsymbol{r} + \hat{\boldsymbol{\epsilon}}(\boldsymbol{r})$ with $\hat{\boldsymbol{\epsilon}}(\boldsymbol{r})$ small everywhere, which is analytic inside the region D_1 bounded by the curve C, and is a decreasing differentiable function in the external region D_2 and at the boundary C. We suppose that all points of the correlation function are in the inner region D_1, indicated by shading in Fig. 2.2.

Since $\hat{\boldsymbol{\epsilon}}$ is not analytic in the external region D_2, this is a non-conformal transformation. Therefore, its effect is to drive the system away from criticality. The fixed point Hamiltonian is modified by a term which couples to

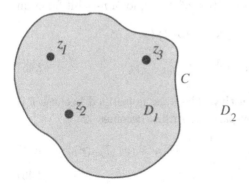

Fig. 2.2. Domains for bounded co-ordinate transformations

---

[9] Attempts to formally treat a covariance of $\langle \phi_1 \ldots \phi_4 \rangle$ under the action of the $l_n, n \neq \pm 1, 0$ in the same way as in (2.37) lead to $F = $ const., see [383, p. 187].

the small perturbing parameter $\hat{\epsilon}$, or rather to derivatives of $\hat{\epsilon}$, because the translations $\hat{\epsilon}(r) = $ cst are conformal. By definition, the scaling operator coupled to the lowest order in $\partial\hat{\epsilon}$ is the **energy–momentum tensor** $T_{\mu\nu}$.

$$\delta_{\hat{\epsilon}}\mathcal{H} = -\frac{1}{2\pi}\int_{D_2} d^2r\, \partial^\mu \hat{\epsilon}^\nu\, T_{\mu\nu}(r). \tag{2.42}$$

Sometimes, by analogy with elasticity theory, T is referred to as stress-energy tensor.[10] Certainly, the presence of relevant long-range interactions will invalidate the above expansion in $\hat{\epsilon}$.

An important result now follows. Invariance under rotations $\hat{\epsilon}^1 = \epsilon r^2$, $\hat{\epsilon}^2 = -\epsilon r^1$ implies that $T_{\mu\nu} = T_{\nu\mu}$ is symmetric. Global scale invariance $\hat{\epsilon}^\mu = \epsilon r^\mu$ implies that the trace $T^\mu_\mu = 0$. Now, for special conformal transformations, we have $\hat{\epsilon}^\mu = \epsilon^\mu r^2 - 2r^\mu \epsilon \cdot r$. Thus

$$T_{\mu\nu}\partial^\mu\hat{\epsilon}^\nu = 2T_{\mu\nu}\left(r^\mu\epsilon^\nu - \epsilon^\mu r^\nu\right) - 2T^\mu_\mu\,\epsilon \cdot r = 0 \tag{2.43}$$

and we see that the invariance under special conformal transformations does not add any further condition to the energy–momentum tensor. Summarizing, we have seen that, see [126]

$$\left.\begin{array}{r}\text{translation invariance}\\ \text{rotation invariance}\\ \text{scale invariance}\\ \text{short-ranged interactions}\end{array}\right\} \Longrightarrow \text{conformal invariance.} \tag{2.44}$$

If one of the above conditions is not met, full conformal invariance will not hold. However, sometimes it is possible to recover a sufficiently large subgroup of the full conformal group, such that at least some of the conformal techniques can still be used. This situation occurs for example close to free surfaces and will be discussed in detail in Chap. 15.

The consequences of an infinitesimal transformation $\hat{\epsilon}$ on the correlation function $\langle\phi_1\cdots\phi_n\rangle$ of primary operators can be perturbatively computed, taking into account that $\mathcal{H}^* \to \mathcal{H}^* + \delta_{\hat{\epsilon}}\mathcal{H}$ after the transformation. Formally expanding in $\hat{\epsilon}$ then yields the generalization of (2.31), for a n-point function

$$\sum_i \left((\Delta_i\partial_{z_i}\hat{\epsilon} + \hat{\epsilon}\partial_{z_i}) + (\overline{\Delta}_i\partial_{\bar{z}_i}\bar{\hat{\epsilon}} + \bar{\hat{\epsilon}}\partial_{\bar{z}_i})\right)\langle\phi_1\cdots\phi_n\rangle$$

$$= -\frac{1}{2\pi}\int_{D_2} d^2r\, \partial^\mu\hat{\epsilon}^\nu(r)\,\langle T_{\mu\nu}(r)\phi_1\cdots\phi_n\rangle \tag{2.45}$$

which is the **conformal Ward identity**. The l.h.s. is justified because $\hat{\epsilon}$ is conformal within D_1. Integrating by parts, the r.h.s. becomes

$$\frac{1}{2\pi}\int_{D_2} d^2r\, \hat{\epsilon}^\nu(r)\,\langle \partial^\mu T_{\mu\nu}(r)\phi_1\cdots\phi_n\rangle - \frac{1}{2\pi}\int_C d\Sigma\, n^\mu\hat{\epsilon}^\nu(r)\,\langle T_{\mu\nu}(r)\phi_1\cdots\phi_n\rangle,$$

$$\tag{2.46}$$

---

[10] Recall that $T_{\mu\nu}$ is the *improved* energy–momentum tensor, see [126], which is not always obtained from the Lagrangian in the canonical way.

where n^μ is orthonormal to the surface C. Since $\hat{\epsilon}$ is arbitrary, one may choose it with compact support in the interior of D_2. It follows that the first term of (2.46) must vanish for any such $\hat{\epsilon}$, yielding the *conservation law*

$$\partial^\mu T_{\mu\nu} = 0. \tag{2.47}$$

Since $T_{\mu\nu}$ is symmetric and traceless, one can define the two independent components of the energy–momentum tensor as

$$T(z) = T_{11} - T_{22} - 2\mathrm{i}T_{12} \ , \ \ \overline{T}(\bar{z}) = T_{11} - T_{22} + 2\mathrm{i}T_{12} \tag{2.48}$$

which satisfy, because of (2.47), the formal analyticity conditions

$$\partial_{\bar{z}}T(z) = \partial_z\overline{T}(\bar{z}) = 0. \tag{2.49}$$

Now, the dependence on z and $\bar{z}$ can be separated completely. To see this, one may use the Cauchy theorem for the l.h.s. of (2.45) and one obtains

$$\frac{1}{2\pi\mathrm{i}} \int_C \mathrm{d}z\, \hat{\epsilon}(z) \langle T(z)\phi_1 \cdots \phi_n \rangle - \frac{1}{2\pi\mathrm{i}} \int_C \mathrm{d}\bar{z}\, \overline{\hat{\epsilon}(z)} \langle \overline{T}(\bar{z})\phi_1 \cdots \phi_n \rangle$$
$$= \frac{1}{2\pi\mathrm{i}} \int_C \mathrm{d}z\, \hat{\epsilon}(z) \sum_i \left(\frac{\Delta_i}{(z - z_i)^2} + \frac{1}{z - z_i}\partial_{z_i} \right) \langle \phi_1 \cdots \phi_n \rangle$$
$$- \frac{1}{2\pi\mathrm{i}} \int_C \mathrm{d}\bar{z}\, \overline{\hat{\epsilon}(z)} \sum_i \left(\frac{\overline{\Delta}_i}{(\bar{z} - \bar{z}_i)^2} + \frac{1}{\bar{z} - \bar{z}_i}\partial_{\bar{z}_i} \right) \langle \phi_1 \cdots \phi_n \rangle. \tag{2.50}$$

Since the r.h.s. will vanish for any infinitesimal projective transformation (i.e. for $\hat{\epsilon} = 1, z, z^2$), a consistency condition for the vanishing of the l.h.s. is that $T(z)$ behaves for large z as z^{-4}. Moreover, being valid for any $\hat{\epsilon}$, we conclude that we can write the above in a form called local **conformal Ward identity**

$$\langle T(z)\phi_1(z_1, \bar{z}_1) \cdots \phi_n(z_n, \bar{z}_n) \rangle$$
$$= \sum_i \left(\frac{\Delta_i}{(z - z_i)^2} + \frac{1}{z - z_i}\partial_{z_i} \right) \langle \phi_1(z_1, \bar{z}_1) \cdots \phi_n(z_n, \bar{z}_n) \rangle \tag{2.51}$$

with a similar expression for $\overline{T}$. This is nothing but the formal expression of the covariance of the primary fields ϕ_i under an arbitrary analytic transformation. An insertion of the holomorphic part T of the energy–momentum tensor into a correlation function of primary operators amounts to apply a differential operator of dimension 2 to the correlation function without T. The scaling dimension of T (and $\overline{T}$) is therefore canonically 2. Its spin, however, is $+2$ for T and -2 for $\overline{T}$.[11]

---

[11] The singularities which arise when $z \to z_i$ indicate that results such as energy–momentum conservation $\partial^\mu T_{\mu\nu} = 0$ only have a precise meaning when inserted into a correlation function and when the point z is kept distinct from the z_i. Since in a more rigorous setting, $T(z)$ and the ϕ_i become operator-valued distributions, it is clear that the formal treatment presented here is somewhat heuristic. Since the results agree with those of a full treatment, we carry on and refer to texts on quantum field theory for the detailed justifications.

The energy–momentum tensor is more than the generator of conformal transformations. It is the *Noether current* which generates infinitesimal coordinate transformations in the sense that for any operator $A(z_1, \bar{z}_1)$ and any $\hat{\epsilon}$ (conformal or not)

$$\delta_{\hat{\epsilon}} A(z_1, \bar{z}_1) = \frac{1}{2\pi i} \oint_{C_1} d\Sigma \, n^\mu \hat{\epsilon}^\nu(\boldsymbol{r}) \, T_{\mu\nu}(\boldsymbol{r}) \, A(z_1, \bar{z}_1), \tag{2.52}$$

where the contour is chosen to surround the point z_1. This definition is true even if $\hat{\epsilon}$ is a general coordinate transformation because the definition of T is independent of the size of D_1 or the correlation function we want to study. We have to choose D_1 to contain the physical points only to derive the Ward identities (2.51) for specific primary operators. More generally, for any functional $X(z_1, \ldots, z_n)$ (in our notation the $\bar{z}$-dependence is suppressed)

$$\delta_{\hat{\epsilon}} \langle X(z_1, \ldots, z_n) \rangle = \frac{1}{2\pi i} \oint_{C_1 \cup \cdots \cup C_n} d\Sigma \, n^\mu \hat{\epsilon}^\nu(\boldsymbol{r}) \, \langle T_{\mu\nu}(\boldsymbol{r}) X(z_1, \ldots, z_n) \rangle. \tag{2.53}$$

Integrating over a large curve surrounding all points amounts to sum over smaller curves surrounding each point. We assume that all points are distinct.

To extract the algebraic information hidden in the structure of T, we have to consider $\delta_\epsilon \delta_\epsilon \langle X \rangle$ where T will appear twice. First notice that $\langle T \rangle = 0$ by translation invariance. However, the two-point function has no reason to vanish but, because of scale invariance, must have the form

$$\langle T(z_1) T(z_2) \rangle = \frac{c/2}{(z_1 - z_2)^4}. \tag{2.54}$$

This defines the **central charge** c, which plays a major role in the following. Below, we shall see that it describes the analogue of the Casimir effect and in Chap. 3, we shall relate it to the finite-size scaling of the free energy density at the critical point. The value of c is related to the universality class of the model at hand. For the Ising model, $c = 1/2$. Of course holomorphic and antiholomorphic parts of $T_{\mu\nu}$ do not couple and $\langle T\bar{T} \rangle = 0$. A consequence of the non-vanishing two-point function is that for ϵ analytic

$$\delta_\epsilon \langle T(z) \rangle = \langle \delta_\epsilon T(z) \rangle = \frac{1}{2\pi i} \oint_{C_z} dz' \, \epsilon(z') \, \langle T(z') T(z) \rangle = \frac{c}{12} \partial_z^3 \epsilon(z). \tag{2.55}$$

$\langle \delta T \rangle$ vanishes only if the third derivative is zero, i.e. for the infinitesimal generators of the projective group. This is normal because general conformal transformation will change the geometry of the plane. Nevertheless, the innocent-looking equation (2.54) has important consequences. Our starting point had been the recognition that the Ward identity implies conformal invariance as a natural ingredient at criticality, under very mild conditions. We had then extended in 2D the projective conformal group to the infinite-dimensional group of analytic transformations, expecting that the system would still be invariant under all these. The very same Ward identity tells us now that this is not possible, rather, fluctuation effects break conformal invariance, which is only present at the kinematical level. That type of phenomenon is common (and unavoidable) in quantum field theories and is called an *anomaly*. In particular, the conformal algebra is *not* given by the generators l_n (apart from the projective subalgebra) and we shall have to rederive it, taking care of the central charge.

It is well known that renormalization group transformations, here conformal coordinate transformations, do not change the canonical dimension of a Noether current. T must essentially transform like $(\partial_z)^2$ (i.e. like a rank 2 tensor) and δT (of dimension 3) must be expressed in terms of conserved operators (i.e. $\partial_{\bar{z}} A(z) = 0$). We only dispose of T, $\partial_z T$ (higher derivatives are ruled out for dimensional reasons), and of the identity operator. Therefore, under an infinitesimal conformal transformation

$$\delta_\epsilon T(z) = (2\partial_z \epsilon + \epsilon \partial_z) T(z) + \frac{c}{12} \partial_z^3 \epsilon. \tag{2.56}$$

The first two terms mean that T is a quasi-primary operator and the factor 2 is fixed by the dimension of T. Dimensional arguments do not permit other terms and the transformation rule for T is the most general we can write. If we compute now $\delta \langle T\phi_1 \cdots \phi_n \rangle$ with equations (2.53), (2.56) and (2.30), and following the same lines as before, using Cauchy's theorem, we obtain the local form of the Ward identity

$$\langle T(z)T(z')\phi_1 \cdots \phi_n \rangle = \frac{c/2}{(z-z')^4} \langle \phi_1 \cdots \phi_n \rangle$$

$$+ \left(\frac{2}{(z-z')^2} + \frac{1}{z-z'} \partial_{z'} + \sum_i \left(\frac{\Delta_i}{(z-z_i)^2} + \frac{1}{z-z_i} \partial_{z_i} \right) \right) \langle T(z')\phi_1 \cdots \phi_n \rangle. \tag{2.57}$$

T behaves almost like a primary operator, up to a c-term which picks up the two-point singularity of T. We have seen that the last term is present as a consequence of the global properties of the conformal transformations. It describes the **anomaly** due to the fluctuations on *all* length scales for critical statistical systems. It numerical value will therefore contain important information on the nature of the fluctuating critical system and it will be one of our main tasks to decode this in terms of universality classes.

Ward identities can be derived for any operator once we know how it transforms. We have seen that T acts onto other operators through singular terms $1/(z-z_i)^k$ for some k. k is smaller than or equal to 2 for primary operators and has the value $k = 4$ for T. Higher values of k are possible for general operators. For example, for the secondary operator $\partial\phi$ one has $k = 3$ (see exercises). A derivative of order n of a primary operator gives rise to a pole of order $k = n + 2$. It makes therefore sense to be interested in the Laurent coefficients of this development. This defines the **Virasoro generators** L_n acting on any operator $A(z_1, \bar{z}_1)$

$$L_n A(z_1, \bar{z}_1) = \frac{1}{2\pi i} \oint_{C_{z_1}} dz\, (z - z_1)^{n+1}\, T(z) A(z_1, \bar{z}_1). \qquad (2.58)$$

It is understood that these expressions must be inserted into a correlation function in order to make sense. The delicate problem how to define this in an operator sense will not be discussed here. To apply an operator such as L_n several times, we need a convention on the choice of the contours. This is given from the situation where two Virasoro generators are applied to $A(z_1, \bar{z}_1)$.

$$L_n L_m A(z_1, \bar{z}_1)$$
$$= \left(\frac{1}{2\pi i}\right)^2 \oint_{C'_{z_1,z}} dz' \oint_{C_{z_1}} dz\, (z' - z_1)^{n+1}(z - z_1)^{m+1} T(z') T(z) A(z_1, \bar{z}_1). \qquad (2.59)$$

The contour C_{z_1} surrounds z_1 and is itself surrounded by $C'_{z_1,z}$, see Fig. 2.3a.

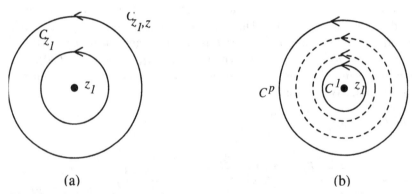

(a)   (b)

Fig. 2.3. Contours in the definition of the Virasoro generators [169]

If we apply p times L to the left of A, i.e. $L_{n_p} \cdots L_{n_1} A(z_1, \bar{z}_1)$, we have to choose p non-intersecting contours surrounding z_1, with C^p corresponding to the upmost left operator L_{n_p} surrounding all others, and C^1 corresponding to L_{n_1} placed inside all the others (Fig. 2.3b). Permuting the order of two L amounts to interchange contours. For example in the case of (2.59), commuting L_n and L_m means that the contour $C'_{z_1,z}$ must move inside C_{z_1}. The difference, i.e. the commutator $[L_n, L_m]$, is given by the residue of the integration over z' at the pole $z' \to z$.

$$[L_n, L_m] A(z_1, \bar{z}_1)$$
$$= \left(\frac{1}{2\pi i}\right)^2 \oint_{C_z} dz' \oint_{C_{z_1}} dz\, (z' - z_1)^{n+1}(z - z_1)^{m+1} T(z') T(z) A(z_1, \bar{z}_1). \qquad (2.60)$$

The contour C_z surrounds the point z only as illustrated in Fig. 2.4.

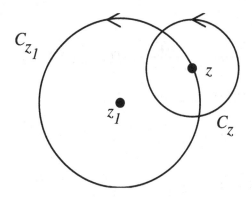

Fig. 2.4. Commutation contours for the Virasoro algebra [169]

We know that the insertion of $T(z')$ induces an infinitesimal transformation. Its action on $T(z)$ is given by (2.56) which can be written in an integral form and gives rise to the pole structure

$$T(z')\,T(z) = \frac{c/2}{(z'-z)^4} + \frac{2T(z)}{(z'-z)^2} + \frac{\partial_z T(z)}{z'-z} + \text{regular terms} \qquad (2.61)$$

as can be read from (2.57). This is the first example of an **operator product expansion** which we shall encounter several times in the sequel. The "regular terms" stand for the infinitesimal transformation of A and all other operators which may be in a general correlation function. For general operators A, the regular terms will be much more complicated than in (2.57) but they do not contribute to the integration in the variable z' along C_z. In fact, the universal long-distance behaviour of the theory is already contained in the singular term spelled out in (2.61). We compute now the three contributions to the commutator $[L_n, L_m]$ coming from the three poles in $(z'-z)$ in (2.61). The first term is given by

$$\frac{c}{2}\left(\frac{1}{2\pi i}\right)^2 \oint_{C_z} dz' \oint_{C_{z_1}} dz\, \frac{(z'-z_1)^{n+1}(z-z_1)^{m+1}}{(z'-z)^4} A(z_1, \bar z_1)$$

$$= \frac{cn(n^2-1)}{12}\left(\frac{1}{2\pi i}\right) \oint_{C_{z_1}} dz\,(z-z_1)^{n+m-1} A(z_1, \bar z_1)$$

$$= \frac{cn(n^2-1)}{12}\delta_{n+m,0}\, A(z_1, \bar z_1). \qquad (2.62)$$

This term gives the central extension of the loop algebra (2.19). The second and third contributions will couple to L_{n+m}. The second gives

$$2\left(\frac{1}{2\pi i}\right)^2 \oint_{C_z} dz' \oint_{C_{z_1}} dz\, \frac{(z'-z_1)^{n+1}(z-z_1)^{m+1}}{(z'-z)^2} T(z) A(z_1, \bar z_1)$$

$$= 2\,(n+1)\left(\frac{1}{2\pi i}\right) \oint_{C_{z_1}} dz\,(z-z_1)^{n+m+1} T(z) A(z_1, \bar z_1)$$

$$= 2\,(n+1)\, L_{n+m} A(z_1, \bar z_1) \qquad (2.63)$$

and the last one is

$$\left(\frac{1}{2\pi i}\right)^2 \oint_{C_z} dz' \oint_{C_{z_1}} dz \frac{(z'-z_1)^{n+1}(z-z_1)^{m+1}}{(z'-z)} \partial_z T(z) A(z_1, \bar{z}_1)$$

$$= \left(\frac{1}{2\pi i}\right) \oint_{C_{z_1}} dz\, (z-z_1)^{n+m+2} \partial_z T(z) A(z_1, \bar{z}_1)$$

$$= -\left(\frac{1}{2\pi i}\right) \oint_{C_{z_1}} dz\, \partial_z (z-z_1)^{n+m+2} T(z) A(z_1, \bar{z}_1)$$

$$= -(n+m+2)\, L_{n+m} A(z_1, \bar{z}_1). \tag{2.64}$$

We add the three contributions and repeat the same operations for the conjugate algebra $\{\bar{L}_n\}$ generated by $\bar{T}$. We thus arrive at the desired double **Virasoro algebra** [604]

$$\boxed{\begin{array}{rcl} [L_n, L_m] & = & (n-m)\, L_{n+m} + \frac{1}{12}\, c\, n(n^2-1)\, \delta_{n+m,0} \\ [L_n, \bar{L}_m] & = & 0 \\ [\bar{L}_n, \bar{L}_m] & = & (n-m)\, \bar{L}_{n+m} + \frac{1}{12}\, c\, n(n^2-1)\, \delta_{n+m,0} \end{array}} \tag{2.65}$$

The **central extension** proportional to c disappears for $n = -1, 0, 1$ where we recover the loop algebra (2.19) for the projective transformations. The two Lie algebras (2.19) and (2.65) only differ by this central term. Summarising, we have reduced the treatment of a complicated operator like the energy–momentum tensor to a problem involving an infinite-dimensional Lie algebra. Important results will come from its representation theory, see Chap. 4.

Finally, reconsider the energy–momentum tensor T. We have only shown how it transforms under infinitesimal transformations. In several applications, we need to know how it transforms under a general conformal transformation. This is given by

$$T_{\text{plane}}(z) = T_{\text{new}}(z')\, (\partial_z z')^2 + \frac{c}{12}\{z', z\}, \tag{2.66}$$

where we introduce the subscripts "plane" and "new" to emphasize that the geometry is not the same. The last term introduces the **Schwarzian derivative** defined by

$$\{w, z\} = \frac{\partial_z^3 w}{\partial_z w} - \frac{3}{2}\left(\frac{\partial_z^2 w}{\partial_z w}\right)^2. \tag{2.67}$$

Schwarz introduced such a derivative while working on the problem of mapping circular polygons conformally [332, 333]. Such a derivative appears in the study of second order differential equations $y'' + Q(z)y = 0$. It obeys several important properties. Given a projective transformation, we have

$$\left\{\frac{aw+b}{cw+d}, z\right\} = \{w, z\} \quad \text{and} \quad \left\{w, \frac{az+b}{cz+d}\right\} = \{w, z\}\, (cz+d)^4. \tag{2.68}$$

Under a sequence of conformal maps $z \to w \to \tilde{w}$, we have the equality

$$\{\tilde{w}, z\} = \{\tilde{w}, w\} (\partial_z w)^2 + \{w, z\}. \tag{2.69}$$

The two last equations reflect the consistency condition for (2.66) to transform correctly under successive analytic transformations.

The particular case of the **strip geometry** will become very important in the sequel. It is defined by $z' = w(z) = (L/2\pi) \ln(z)$ which maps the entire plane onto a strip of width L. The negative part of the real axis corresponds to both edges of the strip, which has therefore the topology of a cylinder. Under this particular transformation, (2.66) can be written

$$T_{\text{strip}}(z') = \left(\frac{2\pi}{L}\right)^2 \left(T_{\text{plane}}(z)z^2 - \frac{c}{24}\right). \tag{2.70}$$

Consequently, the expectation value of T in the strip is different from zero

$$\langle T_{\text{strip}}(z') \rangle_{\text{strip}} = -\frac{c}{24} \left(\frac{2\pi}{L}\right)^2 \tag{2.71}$$

even if $\langle T_{\text{plane}} \rangle = 0$. This is the reason why the central charge is often called **conformal anomaly number**. Recalling the definition of the stress tensor as the variation of the Hamiltonian under a coordinate transformation, we can see that the free energy is now modified

$$\delta_{\hat{\epsilon}} \ln \mathcal{Z} = \langle \delta_{\hat{\epsilon}} \mathcal{H} \rangle = \tag{2.72}$$
$$-\frac{1}{2\pi} \int d^2 z' \left(\langle T_{\text{strip}}(z') \rangle_{\text{strip}} \partial_{\bar{z}'} \hat{\epsilon} + \langle \overline{T}_{\text{strip}}(\bar{z}') \rangle_{\text{strip}} \partial_{z'} \bar{\hat{\epsilon}} \right).$$

The parameter c measures therefore the variation of the free energy due to finite size effects. Recalling from Chap. 1 that the free energy of two-dimensional systems corresponds to the ground state (or vacuum) energy of a one-dimensional quantum system, we realize the analogy with the **Casimir effect** in quantum electrodynamics (with c instead of $\hbar$). The equations (2.71) and (2.72) will be used to discuss the finite-size scaling of the free energy in the next chapter.

Summarising, we have introduced two of the main ingredients of 2D conformal field theory, namely the covariance of primary operators under any conformal transformation and the central charge. The third and last ingredient will be the operator product algebra, to be discussed in Chap. 5.

Exercises

1. Consider an infinitesimal conformal transformation in d dimensions, of the form $r^\mu \to r^\mu + \epsilon^\mu$. Show that those transformations listed in the text exhaust the possible conformal transformations, unless $d = 1, 2$ (see also [G18, 277]).

Hints: Let $f(\boldsymbol{r}) := (2/d)\partial_\lambda \epsilon^\lambda$. For simplicity, assume also that the transformation starts from flat space, with metric $g_{\mu\nu} = \delta_{\mu\nu}$. From (2.12), show first that

$$2\partial_\mu\partial_\nu\epsilon_\lambda = (\delta_{\mu\lambda}\partial_\nu + \delta_{\nu\lambda}\partial_\mu - \delta_{\mu\nu}\partial_\lambda) f$$

and then deduce the following relations, one after the other

$$2\partial^2\epsilon_\mu = (2-d)\partial_\mu f \; , \quad (2-d)\partial_\mu\partial_\nu f = \delta_{\mu\nu}\partial^2 f \; , \quad (d-1)\partial^2 f = 0.$$

It follows for $d \neq 1,2$ that $\partial_\mu\partial_\nu f = 0$. Discuss the general solution of this. A solution quadratic in r^μ is $\epsilon^\mu = c^\mu{}_{\nu\lambda}r^\nu r^\lambda$. Show, using (2.12), that

$$c_{\mu\nu\lambda} = \delta_{\mu\lambda}b_\nu + \delta_{\mu\nu}b_\lambda - \delta_{\nu\lambda}b_\mu \; , \quad b_\mu := \frac{1}{d}c^\lambda{}_{\lambda\mu}.$$

Deduce that these correspond to the special conformal transformations.

2. Let t, r be two real coordinates and consider the generator

$$X_n = -t^{n+1}\partial_t - \frac{1}{2}(n+1)t^n r\partial_r.$$

Find the commutator $[X_n, X_m]$ and the finite transformations $t \to t'$, $r \to r'$ generated by X_n, at least for $n = -1, 0, 1$ (see Chap. 16).

3. Show that the massless free Dirac equation is conformally invariant (Pauli 1940) [494].

4. In d dimensions, the Jacobian of the special transformation (2.9) is

$$J(\boldsymbol{r}) = (1 + 2\,\boldsymbol{a}\cdot\boldsymbol{r} + a^2 r^2)^{-d}.$$

From (2.3), derive the two- and three-point functions, see (2.36,2.38).

5. For the nonprimary scaling operator $\partial\phi$, verify that it transforms as

$$\delta_\epsilon(\partial_z\phi) = ((\Delta+1)\partial_z\epsilon + \epsilon\partial_z)(\partial_z\phi) + \Delta(\partial_z^2\epsilon)\phi$$

when ϕ is primary. Then show that

$$\langle T(z)(\partial\phi)(z_0)\,\phi_1\ldots\phi_n\rangle = \frac{2\Delta}{(z-z_0)^3}\langle\phi(z_0)\,\phi_1\ldots\phi_n\rangle$$
$$+ \left(\frac{\Delta+1}{(z-z_0)^2} + \frac{1}{z-z_0}\frac{\partial}{\partial z_0}\right)\langle(\partial\phi)(z_0)\,\phi_1\ldots\phi_n\rangle + \cdots,$$

where the suppressed terms act only on the scaling operators $\phi_1 \ldots \phi_n$.

6. Prove the identity (2.57).

3. Finite-Size Scaling

In the last chapter, conformal invariance was used to constrain the multipoint correlation functions of isotropic critical two-dimensional systems. Before following the field-theoretic developments further, we shall describe important applications to the study of finite-size effects. Besides being of interest in their own right, these results provide highly efficient computational tools for the practical calculations of central charges and scaling dimensions.

Finite-size effects are conventionally described by **finite-size scaling**. In this chapter, we shall give a presentation of the most basic facts useful for the later discussion in connection with conformal invariance. From a phenomenological description of the behaviour of observables of a statistical model like a specific heat, a magnetic susceptibility or a correlation length due to the influences of the presence of a finite geometry, we shall pass to recall the notions of finite-size scaling variables and finite-size scaling functions. We describe the phenomenological renormalization and discuss universality. Although usually presented here for the two-dimensional case, the results are in fact valid for all dimensions d up to the upper critical dimension, $d < d^*$. These considerations will set the frame into which the stronger results following from two-dimensional conformal invariance will be inserted.

For detailed reviews of finite-size scaling, we refer to [48, 521] and [G10].

3.1 Statistical Systems in Finite Geometries

Consider a statistical system confined to a finite domain of volume V, surface A, etc. The free energy $F(T, V, \mathcal{N})$ of the system can be written as

$$F(T, V, \mathcal{N}) = V f_b(T, \rho) + A f_s(T, \rho) + \ldots \qquad (3.1)$$

where T is the temperature, $\mathcal{N}$ the number of particles and we have suppressed other thermodynamic variables. Writing this, a limit $V \to \infty$, $\mathcal{N} \to \infty$ such that $\rho = \mathcal{N}/V$ is kept fixed is implied, that is, the **bulk free energy density** $f_b(T, \rho)$ is defined by

$$f_b(T, \rho) = \lim_{\mathcal{N} \to \infty, V \to \infty} F(T, V, \mathcal{N})/V \quad \text{with } \rho = \mathcal{N}/V \text{ fixed.} \qquad (3.2)$$

A similar limit is implied for the definition of the **surface free energy density** $f_s(T, \rho)$. Further finite-size terms can (and will) be defined if necessary.

We shall mainly use the following boundary conditions. Let $\Phi(r)$ be a space-dependent observable with r constrained to a finite volume, e.g. $0 \leq r \leq L$. We shall use **periodic** boundary conditions $\Phi(r + L) = \Phi(r)$ and also **free** boundary conditions $\Phi(r) = 0$ if $r < 0$ or $r > L$. For an example consider the 1D Ising model with Hamiltonian

$$\mathcal{H} = -\sum_{i=1}^{N} J_i \sigma_i \sigma_{i+1}. \tag{3.3}$$

Periodic boundary conditions correspond to $\sigma_1 = \sigma_{N+1}$ and free boundary conditions correspond to $J_N = 0$ (alternatively, one often sets $\sigma_{N+1} = 0$).

The bulk free energy density f_b is independent of the boundary conditions, while the surface free energy density f_s depends on the boundary conditions. For periodic boundary conditions and for values of T far enough from the critical point T_c, the finite-size corrections to f_b are exponentially small.

In what follows we shall restrict our attention to the case where $\bar{d}$ dimensions of the lattice are infinite and in d' dimensions we take a hypercubic lattice of linear extent L. We shall for brevity refer to L as the *size of the system*. Then $d = \bar{d} + d'$ and we have for the free energy density

$$f(T, \rho) = f_b(T, \rho) + \frac{1}{L} f_s(T, \rho) + \frac{1}{L^2} f_2(T, \rho) + \ldots \tag{3.4}$$

In the following sections, we shall mainly consider the case $d = 2$, $\bar{d} = d' = 1$. The quantity f_2 will be shown to play an important role in the following.

3.2 Finite-Size Scaling Hypothesis

The decomposition of the free energy density as written in (3.4) is only valid when the temperature T is far enough from the critical point T_c. Close to T_c, singularities will appear, invalidating the simple relationship written so far.

As an example, consider the specific heat of some lattice system as shown in Fig. 3.1 for two lattices of finite sizes $L > L'$. Although the specific heat of the *infinite* system shows a divergence $C_\infty(T) \sim (T - T_c)^{-\alpha}$, this singularity is rounded when the system size L is finite. We now define a few quantities which are useful to describe these new finite-size features, following phenomenological ideas first proposed by Fisher and Barber [227].

The curves show a maximum at some value of T which is shifted away from T_c. One can take the temperature of the finite-size maximum to be a definition of a finite-size estimate $T_c(L)$ of the true critical temperature $T_c(\infty) = T_c$. Then the finite-size shift in T_c is

$$\left(T_c(L) - T_c(\infty) \right) / T_c(\infty) \sim L^{-\lambda} \text{ if } L \to \infty. \tag{3.5}$$

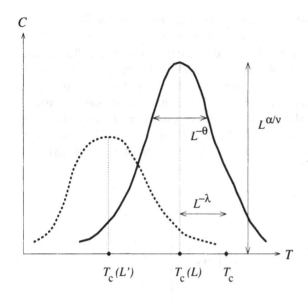

Fig. 3.1. Specific heat as a function of temperature for two lattices of sizes L (full) and $L' < L$ (broken). The finite-size scaling of the shift, the rounding and the maximal height is also indicated

This defines the **shift exponent** λ. Another useful quantity is the **rounding temperature** $T^*(L)$. This is defined such that if $|T - T_c|/T_c \geq |T^* - T_c|/T_c$, then the finite-size specific heat is close to the bulk value: $C_L(T) \simeq C_\infty(T)$. We use this to define a **rounding exponent** θ via

$$(T^*(L) - T_c(\infty))/T_c(\infty) \sim L^{-\theta} \text{ if } L \to \infty. \tag{3.6}$$

Asymptotically, θ characterizes the size of the finite-size scaling region.

Note that one has implicitly assumed here that there is no phase transition in the system for L finite. It is straightforward to generalize to this case as well. For the later applications of conformal invariance, we shall always have $\bar{d} = 1$ and no phase transition for $L < \infty$ can occur.

Now we state the basic hypothesis of finite-size scaling, namely that *there should be only one relevant length describing the rounding and shifting of the thermodynamic singularities.* For the correlation length, that means

$$\xi(T^*(L)) \sim L, \tag{3.7}$$

which, using $\xi_\infty \sim (T - T_c)^{-\nu}$ yields for the rounding exponent

$$\theta = 1/\nu. \tag{3.8}$$

As we shall see later, this result is confirmed by explicit calculation in a large variety of systems. For the shift exponent, one expects, if there is only one length describing the finite-size effects

$$\lambda = 1/\nu. \tag{3.9}$$

This result is also found to hold true for many systems, but exceptions occur in the spherical model and the ideal Bose gas, see [48]. Since these models do not have a phase transition at a non-vanishing temperature for $d = 2$, we shall use both of the relations (3.8) and (3.9) in the sequel.

Equations (3.5, 3.9) are very often used for lattice models. For illustration, consider a system consisting of a set of n chains of a spin $\frac{1}{2}$ Ising model, coupled to n chains of a spin 1 Ising model. The spin 1 subsystem orders at higher temperatures than the spin $\frac{1}{2}$ subsystem. Critical point estimates $T_{\max}$ were obtained from the maximum of the 'specific heat'. Since only the spin 1 subsystem orders, the model effectively behaves as if free boundary conditions had been imposed. In Fig. 3.2, the variation of $T_{\max}(n)$ with n is

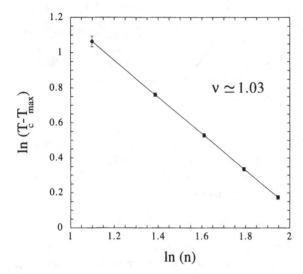

$\nu \simeq 1.03$

Fig. 3.2. Finite-size scaling of the critical point shift $|T_{\max}(n) - T_c|$ as a function of the number of chains n for a mixed spin $\frac{1}{2}$-spin 1 Ising model (after [391])

displayed, using values of $n = 3, \ldots, 7$. Clearly, a power law is obtained and the value of ν extracted from the slope is in agreement with the theoretical value $\nu = 1$ for the 2D Ising model. It is remarkable that (3.5), which in principle was derived in the $n \to \infty$ limit, holds down to very small n and, consequently, for quits large shift of $T_{\max}$ away from the true critical point $T_c = \lim_{n \to \infty} T_c(n)$.

We now proceed to state the finite-size scaling hypothesis for the scaling function. Introduce the finite-size scaling variable

$$\tilde{z} = L/\xi_\infty(T) \tag{3.10}$$

and distinguish three cases

1. $\tilde{z} \ll 1$: the correlation length is much larger than the system size. Finite-size and boundary effects will be very important.
2. $\tilde{z} \simeq 1$: this defines the rounding temperature $T^*(L)$.
3. $\tilde{z} \gg 1$: the correlation length is much smaller than the system size. One expects a cross-over to the bulk behaviour. Here we assume that the correlation length is still much larger than some microscopic scale.

We thus see that the **finite-size scaling region** is characterized by those values of T and L such that the finite-size scaling variable $\tilde{z}$ remains finite.

For $L \to \infty$, the specific heat behaves close to T_c like

$$C_\infty(T) \simeq \frac{A_\pm}{\alpha} \left(T - T_c\right)^{-\alpha},$$ (3.11)

where $A_\pm$ refers to $T - T_c \gtrless 0$, respectively.[1] One now postulates [227] that in the case of the **finite-size scaling limit**

$$T \to T_c, \quad L \to \infty, \quad \text{but} \quad \tilde{z} = L/\xi(T) \text{ fixed},$$ (3.12)

one can write C_L in the form

$$C_L(T) \sim L^{\bar{\omega}} \tilde{Q}(\tilde{z}) \sim L^{\bar{\omega}} Q\left(L^{1/\nu} t\right),$$ (3.13)

where the reduced temperature is

$$t := (T - T_c)/T_c.$$ (3.14)

This is the basic finite-size scaling form of a thermodynamic quantity. Using $\theta = \lambda = 1/\nu$, we can replace $T_c = T_c(\infty)$ by some finite-size estimate $T_c(L)$. This does not affect the finite-size scaling form as specified.

In order to reproduce for $\tilde{z} \gg 1$ the bulk behaviour, one makes the ansatz

$$Q(\tilde{x}) \sim q_\infty^\pm \tilde{x}^{-\rho} \quad \text{if } \tilde{x} \to \pm\infty$$ (3.15)

and finds by matching expressions $\bar{\omega} = \rho/\nu, \rho = \alpha$ and $q_\infty^\pm = A^\pm/\alpha$. Thus

$$C_L(T^*) \sim L^{\alpha/\nu} \tilde{Q}(\tilde{z}).$$ (3.16)

Again, with $\theta = \lambda = 1/\nu$ and assuming that $Q(\tilde{z})$ is finite and non-vanishing between $0 \leq \tilde{z} \leq 1$, one has

$$C_L(T_c(L)) \sim L^{\alpha/\nu}.$$ (3.17)

This result will be very important for a practical calculation of critical exponents from finite-size data.

To summarize, we have seen how to relate the finite-size rounding of the singularities of thermodynamic quantities close to a second order phase transition to the exponents of the bulk system. The discussion, presented here for the specific heat, immediately generalizes to any other quantity and to other thermodynamic variables than the temperature. In Table 3.1, we give a few examples. We also recall that the free energy density or the specific heat must be decomposed into a regular and a singular part: $f = f^{\text{reg}} + f^{\text{sin}}$ for example. While the regular part f^{reg} is an analytic function of T close to T_c, non-analyticities appear in the singular part f^{sin} and its derivatives. The background term f^{reg} (and C^{reg}) must therefore be subtracted before

---

[1] For self-dual systems, the specific heat amplitudes satisfy $A_+ = A_-$.

Table 3.1. Bulk scaling and finite-size scaling close to T_c with $\tilde{z} = 0$

		bulk	finite-size
specific heat	C^{sin}	$t^{-\alpha}$	$L^{\alpha/\nu}$
magnetic susceptibility	χ	$t^{-\gamma}$	$L^{\gamma/\nu}$
correlation length	ξ	$t^{-\nu}$	L
free energy density	f^{sin}	$t^{2-\alpha}$	L^{-d}
order parameter[a]	M	t^{β}	$L^{-\beta/\nu}$
latent heat[a]	ℓ_h	$t^{1-\alpha}$	$L^{(\alpha-1)/\nu}$

[a] Special care is needed to define non-vanishing lattice expressions for M and ℓ_h

the critical behaviour of either f or C can be analyzed. Background terms are absent for ξ, M or χ.

We note that the finite-size scaling behaviour of both ξ and f^{sin} at $\tilde{z} = 0$ does not depend on any exponent. This will be very useful in the context of phenomenological renormalization and for determining $T_c(L)$.

3.3 Universality

The critical behaviour of a statistical system is determined by its renormalization group (RG) properties. In particular, it follows from the renormalization group that the critical exponents only depend on very few characteristics of the model, like lattice dimension, number of components of the order parameter and the global symmetry, but are independent of the details of the lattice structure or the precise form of the interactions. This is referred to by saying that critical exponents are **universal**. In fact, similar universality properties also hold for the scaling functions. The presentation given follows the original work of Privman and Fisher [516].

For a simple ferromagnet, the singular bulk free energy density becomes

$$f^{\text{sin}}_{\infty} \sim A_1 |t|^{2-\alpha} W^{\pm}(A_2 h |t|^{-\beta-\gamma}), \qquad (3.18)$$

where $W^{\pm}$ is a scaling function, and the index $\pm$ refers to $t \gtrless 0$. It is one of the results of the renormalization group that all information on the "details" of the system can be absorbed into the two non-universal **metric factors** A_1, A_2. Therefore, $W^{\pm}$ is a universal function. From this, similar expressions are easily obtained for the derivatives of f^{sin}_{∞}. In writing this, one has implicitly assumed that the scaling function has neither zeroes nor poles close to criticality. This requires the absence of so-called **dangerous irrelevant variables** (which is the case if $d < d^*$ [G1]), and of marginal variables.

Equation (3.18) is readily generalized to include finite-size effects

$$f^{\text{sin}}(t, h; L) \sim A_1 |t|^{2-\alpha} \tilde{W}^{\pm}(A_2 h |t|^{-\beta-\gamma}; L/\xi_{\infty}), \qquad (3.19)$$

where $\tilde{W}^\pm$ is universal and $\xi_\infty \sim \xi_0 t^{-\nu}$ (as $t \to +0$) is the bulk correlation length. Note that there is no extra metric factor associated with L/ξ_∞. This is true if the dimension d is below the upper critical dimension d^*, but holds no longer for $d \geq d^*$ due to the presence of dangerous irrelevant variables, as can be seen explicitly in spherical model calculations [106, 440]. On the other hand, the non-universal constant ξ_0 does appear implicitly. It must also be present in the correlation function

$$G(\boldsymbol{r};t,h) := \langle s_0 s_{\boldsymbol{r}} \rangle - \langle s_0 \rangle \langle s_{\boldsymbol{r}} \rangle \tag{3.20}$$
$$\sim D_1 r^{2-d-\eta} X^\pm(\boldsymbol{r}/\xi_\infty, D_2 h|t|^{-\beta-\gamma}),$$

where $X^\pm$ is universal and D_1, D_2 are metric factors. The magnetic susceptibility is obtained as

$$\chi_\infty(t,h) = \int d\boldsymbol{r} \; G(\boldsymbol{r};t,h) \sim D_1 \xi_\infty^{2-\eta} \tilde{X}^\pm(D_2 h|t|^{-\beta-\gamma}). \tag{3.21}$$

The same scaling behaviour is also expected for the total correlation function

$$\Gamma(\boldsymbol{r};t,h) := \langle s_0 s_{\boldsymbol{r}} \rangle \sim D_1 r^{2-d-\eta} Z^\pm(\boldsymbol{r}/\xi_\infty, D_2 h|t|^{-\beta-\gamma}) \tag{3.22}$$

with a new universal scaling function $Z^\pm$. In the limit $r \to \infty$, $\Gamma(\boldsymbol{r}) \to m_\infty^2$, where m_∞ is the magnetization per spin. Thus

$$m_\infty^2(t,h) \sim D_1 \xi_\infty^{2-d-\eta} \tilde{Z}^\pm(D_2 h|t|^{-\beta-\gamma}). \tag{3.23}$$

Now, from (3.18), one has by taking derivatives

$$m_\infty(t,h) \sim A_1 A_2 |t|^\beta W_1^\pm(A_2 h|t|^{-\beta-\gamma})$$
$$\chi_\infty(t,h) \sim A_1 A_2^2 |t|^{-\gamma} W_2^\pm(A_2 h|t|^{-\beta-\gamma}), \tag{3.24}$$

where $W_1^\pm(x) = \partial_x W^\pm(x)$ and $W_2^\pm(x) = \partial_x W_1^\pm(x)$. Comparing coefficients, one finds for the susceptibility

$$A_1 A_2^2 |t|^{-\gamma} W_2^\pm(A_2 h|t|^{-\beta-\gamma}) = D_1 \xi_0^{2-\eta} |t|^{-(2-\eta)\nu} \tilde{X}^\pm(D_2 h|t|^{-\beta-\gamma}) \tag{3.25}$$

which returns the well-known scaling relation $\gamma = (2-\eta)\nu$, and for the magnetization density

$$A_1^2 A_2^2 |t|^{2\beta} W_1^{\pm\,2}(A_2 h|t|^{-\beta-\gamma}) = D_1 \xi_0^{2-\eta-d} |t|^{-(2-\eta-d)\nu} \tilde{Z}^\pm(D_2 h|t|^{-\beta-\gamma}) \tag{3.26}$$

which implies the hyperscaling relation $2\beta = (d-2+\eta)\nu$. Furthermore, the metric factors become related and one identifies the universal combinations

$$A_1 \xi_0^d =: Q_1$$
$$A_2/D_2 =: Q_2 \tag{3.27}$$
$$D_1 A_1^{-\psi} A_2^{-2} =: Q_3 \quad \text{if } \psi = 1 + \gamma/(d\nu)$$

whose universality follows from the universality of the scaling functions [516]. In particular, we have only two non-universal metric factors which describe, to leading order in the scaling behaviour, the non-universal details of the system under consideration. This universality goes a little beyond the conventional universality of ratios of some scaling amplitudes like A^+/A^- ((3.11)) since it does involve hyperscaling. In other words, we have obtained a **hyperuniversality relation** [516]

$$\lim_{t \to \pm 0} f^{\rm sin}(t)\xi^d(t) = u^\pm = \text{universal.} \tag{3.28}$$

This is known as *two-scale factor universality* or *Privman-Fisher hypothesis*.

Consequently, one can rewrite the singular free energy density with only two metric factors C_1, C_2

$$f^{\rm sin}(t, h) = L^{-d}Y(C_1 t L^{1/\nu}, C_2 h L^{(\beta+\gamma)/\nu}), \tag{3.29}$$

where $Y(x_1, x_2)$ is a universal finite-size scaling function. Similarly, one writes for the (inverse) correlation length

$$\xi^{-1}(t, h) = L^{-1}S(C_1 t L^{1/\nu}, C_2 h L^{(\beta+\gamma)/\nu}) \tag{3.30}$$

with the *same* metric factors C_1, C_2 and a universal finite-size scaling function $S(x_1, x_2)$. The new feature of hyperuniversality is the absence of a non-universal prefactor in the finite-size scaling for both $f^{\rm sin}$ and ξ. For later use, we define the finite-size scaling variables

$$z := tL^{y_t} \ , \quad \mu := hL^{y_h}, \tag{3.31}$$

where $y_t = 1/\nu$ and $y_h = (\beta+\gamma)/\nu$ are the renormalization group eigenvalues. In particular, we have the following universal amplitudes

$$Y(0,0), \quad S(0,0), \quad \lim_{z \to \pm\infty} Y(z,0)S^{-d}(z,0), \quad \lim_{\mu \to \pm\infty} Y(0,\mu)S^{-d}(0,\mu) \tag{3.32}$$

which is the main result of this section. Note that scaling considerations alone do not imply a relationship between the amplitudes evaluated in the limits $z(\mu) \to \pm\infty$. From the derivation of these results, it is clear that a confirmation of universality of the amplitudes in (3.32) is equivalent to a confirmation of hyperscaling. It will be one of the objectives of conformal invariance to relate these amplitudes to critical exponents.

Generically, a universality class is characterized by its set of critical exponents along with its set of universal amplitudes or amplitude ratios.

It follows that for each relevant scaling variable in the system there is an associated non-universal metric factor. For conventional *bulk* critical points, there are two independent amplitudes (or amplitude ratios). Additional amplitudes arise for tricritical points or surface critical phenomena.

We summarize the assumptions implicitly contained in the discussion [520]

1. There are no dangerous irrelevant variables. This means that there are no singularities in neither $f^{\rm sin}$ nor ξ^{-1} as the irrelevant couplings go to zero. This requires $d < d^*$. For many models, the upper critical dimension $d^* = 4$. Then hyperscaling is implied.

2. There is only a single length scale ξ_∞. Strong anisotropies are excluded.
3. There are no "accidental" zeroes in Y or S. (If they would occur, they could affect the apparent exponents. For a 2D example, see [311].)
4. The exponent $\alpha \neq 0, -1, -2, \ldots$. Otherwise, logarithmic factors appear [140, 519].
5. If further relevant fields different from t, h are present, additional metric factors will enter. This does happen, for example, for tricritical points, surface anisotropies, etc. as we shall see later.

The finite-size scaling form can be used to locate the critical point T_c. Its position is given by the maximum of the specific heat, i.e.

$$\frac{\partial^3 Y(z,\mu)}{\partial z^3}\bigg|_{z=z_0,\mu=0} = 0, \tag{3.33}$$

where z_0 is universal. The shift in the critical temperature is $T_c(L) - T_c(\infty) = C_1^{-1} z_0 L^{-1/\nu}$, returning the expected shift exponent $\lambda = 1/\nu$, see (3.9).

Although the finite-size scaling functions $Y(z,\mu)$ and $S(z,\mu)$ are universal, they do depend on the shape of the finite domain the system is confined to and also on the boundary conditions. This will become explicit from the study of examples below. Only when the bulk limit $z \to \infty$ or $\mu \to \infty$ is taken, these dependencies can disappear.

So far only the leading finite-size term was considered. On any lattice of finite extent L, there are corrections which may come from two sources.

1. **Non-linear scaling fields.** The scaling form written in (3.19) only includes the linear part of the scaling fields corresponding to t and h. While this is sufficient close enough to the critical point, it is sometimes necessary to include the full nonlinear scaling fields

$$\begin{aligned} u_t &= C_1 t L^{1/\nu} + \mathcal{O}(t,h) \\ u_h &= C_2 h L^{(\beta+\gamma)/\nu} + \mathcal{O}(t,h). \end{aligned} \tag{3.34}$$

This will lead to analytic correction terms. Conformal invariance shows that these are related to the secondary operators in the conformal tower.

2. **Irrelevant scaling fields.** These have a *negative* RG eigenvalue $y_{ir} = -\theta_{ir}/\nu < 0$ and do not change the critical point, but they may affect how the critical point is approached. For the correlation length, for example,

$$\begin{aligned} \xi^{-1} &\sim L^{-1} \bar{S}(C_1 t L^{1/\nu}, C_2 h L^{(\beta+\gamma)/\nu}, C_{ir} g L^{-\theta_{ir}/\nu}) \tag{3.35} \\ &\simeq L^{-1} S(C_1 t L^{1/\nu}, C_2 h L^{(\beta+\gamma)/\nu}) \\ &\quad + C_{ir} g L^{-1-\theta_{ir}/\nu} S_{ir}(C_1 t L^{1/\nu}, C_2 h L^{(\beta+\gamma)/\nu}), \end{aligned}$$

where the second term contains the non-universal amplitude $C_{ir} g$ and $S_{ir}(z,\mu) = \frac{\partial}{\partial g} \bar{S}(z,\mu,g)|_{g=0}$ is universal. This type of corrections gives rise to non-analytic corrections, of whom the leading one may be described by Wegner's **correction exponent** $\omega = \Delta_1/\nu$ [610] (see exercises).[2] From the point of view of conformal invariance, these are related to the presence of irrelevant primary operators, see Chap. 13.

3.4 Phenomenological Renormalization

We now describe a practical way to find the critical point and the critical exponents from given finite-lattice data. Consider the correlation length and take $h = 0$ for simplicity. According to finite-size scaling it has the form, to leading order in $1/L$

$$\xi^{-1}(T; L) \sim L^{-1} S(z, 0). \tag{3.36}$$

Consider two lattices of sizes L, L' and with temperatures T, T' such that the following condition is satisfied

$$\xi(T; L)/L = \xi(T'; L')/L'. \tag{3.37}$$

We can take this as describing a relationship between T and T' via

$$T' - T_c = (T - T_c) \left(\frac{L}{L'} \right)^\theta, \tag{3.38}$$

where $\theta = 1/\nu$ is the rounding exponent. Nightingale [478, 480] proposed to reinterpret (3.37) as giving a renormalization group mapping from a temperature T to T'

$$T \rightarrow T' = \mathcal{R}_{L,b}(T), \tag{3.39}$$

where $b = L/L'$ is the rescaling factor. This can be done exactly only in the limit $L, L' \rightarrow \infty$, but the definition of $\mathcal{R}$ is considered to make sense for L finite at least to a good approximation. Explicit model calculations show this to be normally the case.

Now, find a fixed point $T^*(L, L')$ of the RG-mapping $\mathcal{R}$, which is given by

$$\xi(T^*; L)/L = \xi(T^*; L')/L' \tag{3.40}$$

and one expects that $T^*(L, L') \rightarrow T_c(\infty)$ as $L, L' \rightarrow \infty$. The solution of this equation is sketched in Fig. 3.3. This yields a sequence of estimates $T^*(L, L')$ expected to converge to some limit T_c. Indeed, this is the case in exactly solvable models like the 2D Ising model [190]. The explicit calculation of the finite-size corrections also suggests that the convergence of the sequence of the T^*'s is fastest if $|L/L' - 1|$ is made as small as possible, e.g. $L' = L + a$, where a is the lattice constant.

Alternatively, one can use the results of the last section. Since $L/\xi(T; L) = S(0,0)$ is universal, one can define an estimate T^* by using a single lattice

---

[2] In general, Δ_1 (not to be confused with a conformal weight) describes the leading corrections to scaling of *any* observable $\mathcal{X}(t)$, viz. $\mathcal{X}(t) \sim |t|^{-x}(1 + a_\mathcal{X}|t|^{\Delta_1} + \ldots)$. In [155, 285] numerical values for the $O(n)$ vector model are compiled.

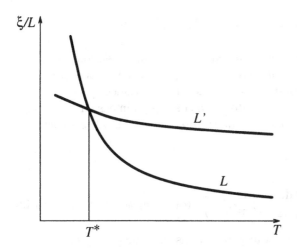

Fig. 3.3. Solving the phenomenological renormalization condition to find T^* [169]

size. This may be advantageous if the value of $S(0,0)$ is known. Then, one can show the convergence to be of order $\sim L^{-(1/\nu+|y_{ir}|)}$, while one expects a convergence of order $\sim L^{-1/\nu}$ otherwise. Since only information from one lattice is used, we obtain for a maximal size $L_{\max}/a$ a sequence with $L_{\max}/a - 2$ entries (the lattices with one or two sites in one direction do not contain much information about the $L \to \infty$ behaviour) while the conventional method yields a sequence with $L_{\max}/a - 3$ entries. This might become important if only very few distinct lattices are available.

From the correlation length scaling the exponent ν can be estimated

$$\left(\frac{L'}{L}\right)^{1+1/\nu} = \frac{\partial \xi(T^*; L)}{\partial T} \Big/ \frac{\partial \xi(T^*; L')}{\partial T} \tag{3.41}$$

and one obtains a sequence of estimates $\nu(T^*; L, L')$ converging to the exponent ν. It may seem that the final result for ν depends on the value chosen for T^*. However, since all that is needed is the finite-size scaling form, which is the same inside the entire finite-size scaling region, one can replace T^* by any sequence of estimates converging towards T_c. This can be done in such a way as to improve the apparent convergence of the sequence. In particular, the final estimate for ν does not depend on the final estimate for T_c (unless, of course, the estimates $\nu(T_c(\infty); L, L')$ are used).

The calculation of other critical exponents like α or γ follows the same lines, using the results given in Table 3.1. In Chap. 9, we discuss numerical algorithms for the extrapolation towards the limit $L \to \infty$ of the finite-size sequences.

3.5 Consequences of Conformal Invariance

The finite-size scaling results presented are valid for all dimensions below the upper critical dimension, $d < d^*$. We now turn exclusively to the case $d = 2$ and impose the stronger constraints of conformal invariance.

Consider a scalar primary operator ϕ with scaling dimension x. It transforms covariantly under a analytic transformation $z \to w(z)$, $\bar{z} \to \bar{w}(\bar{z})$ as shown in (2.29)

$$\langle \phi(z_1, \bar{z}_1)\phi(z_2, \bar{z}_2)\rangle_z$$

$$= \left(\frac{dw}{dz}(z_1)\frac{dw}{dz}(z_2)\right)^\Delta \left(\frac{d\bar{w}}{d\bar{z}}(\bar{z}_1)\frac{d\bar{w}}{d\bar{z}}(\bar{z}_2)\right)^{\overline{\Delta}} \langle \phi(w_1, \bar{w}_1)\phi(w_2, \bar{w}_2)\rangle_w. \quad (3.42)$$

The indices of the expectation values remind us that the conformal transformation $w(z)$ changes in general the geometry of the space.

Here, we consider the effect of the **logarithmic transformation** (Fig.3.4)

$$w = \frac{L}{2\pi}\ln z. \quad (3.43)$$

Under this transformation, the infinite complex plane parametrized by z is transformed into an infinitely long strip of finite width L. Writing $w = u + iv$ and $z = \rho e^{i\phi}$, one sees that v corresponds to the polar angle $v = L\phi/2\pi$, while u measures the radius via $u \sim \ln \rho$. In particular, since the operator ϕ is continuous in the infinite z-plane, it should also be so in the w-strip. This implies that one has *periodic boundary conditions* in the strip geometry.

Scale invariance alone is enough to fix the two-point correlation function of ϕ in the infinite z-plane (see (2.36))

$$\langle \phi(z_1, \bar{z}_1)\phi(z_2, \bar{z}_2)\rangle_z = (z_1 - z_2)^{-2\Delta}(\bar{z}_1 - \bar{z}_2)^{-2\overline{\Delta}}. \quad (3.44)$$

Let us calculate the corresponding correlation function in the strip geometry. Using $z = \exp(2\pi L^{-1}w) = \exp(2\pi L^{-1}(u + iv))$, we have

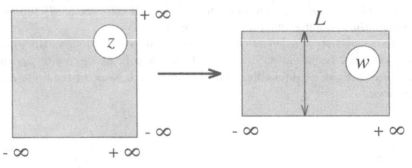

Fig. 3.4. Logarithmic transformation (3.43) of the complex plane onto the strip

$$\langle \phi(w_1, \bar{w}_1)\phi(w_2, \bar{w}_2)\rangle_w = \left(\frac{2\pi}{L}\right)^{2\Delta+2\overline{\Delta}} \left(\frac{z_1^{1/2} z_2^{1/2}}{z_1 - z_2}\right)^{2\Delta} \left(\frac{\bar{z}_1^{1/2} \bar{z}_2^{1/2}}{\bar{z}_1 - \bar{z}_2}\right)^{2\overline{\Delta}}$$

$$= \left(\frac{2\pi}{L} \frac{\exp[\frac{\pi}{L}(w_1 + w_2)]}{\exp(\frac{2\pi}{L}w_1) - \exp(\frac{2\pi}{L}w_2)}\right)^{2\Delta} \cdot \left(\frac{2\pi}{L} \frac{\exp[\frac{\pi}{L}(\bar{w}_1 + \bar{w}_2)]}{\exp(\frac{2\pi}{L}\bar{w}_1) - \exp(\frac{2\pi}{L}\bar{w}_2)}\right)^{2\overline{\Delta}}$$

$$= \left(\frac{\pi}{L} \frac{1}{\sinh[\frac{\pi}{L}(w_1 - w_2)]}\right)^{2\Delta} \cdot \left(\frac{\pi}{L} \frac{1}{\sinh[\frac{\pi}{L}(\bar{w}_1 - \bar{w}_2)]}\right)^{2\overline{\Delta}}, \tag{3.45}$$

where $w_1 - w_2 = (u_1 - u_2) + i(v_1 - v_2)$. This gives the exact two-point function of a scalar primary operator in the strip geometry with periodic boundary conditions. To understand the result, consider two limit cases.

1. $u_1 - u_2 \ll L$. By a rotation of the coordinate system in the plane, one can always arrange for $v_1 = v_2$. Then one recovers the form valid for the infinite plane.
2. $u_1 - u_2 \gg L$. Then one finds an asymptotic exponential decay

$$\langle \phi(u_1, v_1), \phi(u_2, v_2)\rangle_{\text{strip}} \simeq \left(\frac{2\pi}{L}\right)^{2x}$$

$$\times \exp\left[-\frac{2\pi}{L}(\Delta + \overline{\Delta})(u_1 - u_2) - i\frac{2\pi}{L}(\Delta - \overline{\Delta})(v_1 - v_2)\right] \tag{3.46}$$

which is the usual way to define a correlation length ξ via $\langle \phi(u, v)\phi(0, 0)\rangle \sim \exp(-u/\xi)$. Since $x = \Delta + \overline{\Delta}$, comparison yields the result (Cardy [131])

$$\boxed{\xi = L/(2\pi x)} . \tag{3.47}$$

It is not exaggerated to call (3.47) one of the most important results for the application of conformal invariance to critical phenomena.[3]

Let us make the connection with the finite-size scaling functions. Consider a set of scaling operators ϕ_i (like magnetization density, energy density, ...) which have RG eigenvalues y_i. Take the y_i to be ordered, e.g. $y_1 \geq y_2 \geq y_3 \geq y_4 \geq \dots$. The corresponding scaling dimensions x_i are $x_i + y_i = d$ ($d = 2$ here). To each scaling operator one has a correlation length ξ_i. Then finite-size scaling theory had shown that

$$\xi_i^{-1} = L^{-1}S_i(C_1 t L^{y_t}, C_2 h L^{y_h}) \tag{3.48}$$

with $y_t = y_2$ and $y_h = y_1$ for a conventional ferromagnet. Comparison with (3.47) shows that, for an infinitely long strip of finite width L and with periodic boundary conditions

$$\boxed{S_i(0, 0) = 2\pi x_i} . \tag{3.49}$$

A few comments are in order:

---

[3] See Chaps. 8 and 9 for the application of (3.46) to quantum Hamiltonians, e.g. (9.29)

1. Conformal invariance relates universal scaling amplitudes explicitly to universal critical exponents. This is an *a posteriori* confirmation of the finite-size scaling hypothesis. One could also say that the renormalization group tends to emphasize *qualitative* aspects in identifying universal quantities, while conformal invariance emphasizes the *quantitative* aspect by giving relations between amplitudes and critical exponents.

2. Equation (3.49) gives a simple means to calculate the exponents of many non-leading operators, since correlation lengths are easily obtained from the eigenvalues of the transfer matrix. From the point of view of conformal invariance there is a clear interest in calculating more than one eigenvalue of the transfer matrix, in distinction to the usual approach, which is satisfied by knowing the largest eigenvalue of the transfer matrix, related to the free energy.

3. As it stands, (3.47) only applies to fully isotropic lattice systems. This restriction can be removed, if only the RG fixed point corresponds to an isotropic system. We remark that the result (3.47) can be rederived from conformally mapping strips of different widths onto each other [517].

4. Some care is needed with the correct normalization of S_i and Y. While transfer matrix data obtained on a square lattice are already correctly normalized, this is not so for triangular lattices [516, 359] or for data obtained from the quantum Hamiltonian, see Chap. 8. Anisotropic couplings are discussed in [479, 359].

5. Historically, it is interesting that (3.47) was conjectured from model calculations quite some time before the proof could be given [504, 190, 439, 479, 499, 255]. Also, explicit model results suggest that the validity of (3.47) apparently extends beyond the cases where the given derivation from conformal invariance applies, see Chaps. 12 and 15.

6. In the derivation, it was assumed that the scaling operators involved are primary. Nevertheless, from model examples it transpires that the validity of (3.47) is apparently independent of this condition.

7. The derivation assumed the isotropy in space as described by the complex coordinates z and $\bar{z}$. It may occur, however, that the (in a terminology borrowed from string theory) left-moving (corresponding to z) and right-moving sectors of the theory are normalized by $\zeta_L \neq \zeta_R$, where $\zeta_{L,R}$ are the respective speeds of sound in the lattice model (their determination will be described in Chaps. 8 and 9). The above result can be generalized to this situation, see [18] for details.

A similar result can be obtained for the singular free energy density. We know from the last chapter how the partition function $\mathcal{Z}$ transforms under conformal transformations, in particular from the complex plane to the strip geometry ((2.71) and (2.72)). Consider now a strip of width L and a new coordinate transformation, which changes the width of the strip $L \rightarrow L + \delta\epsilon$. Then the derivatives in (2.72) reduce to $\delta\epsilon$. Since for an infinite strip $\ln \mathcal{Z}$ is infinite, we consider the well-defined free energy per unit longitudinal length

$$\mathcal{F}(L) := -\lim_{M \to \infty} \frac{1}{M} \ln \mathcal{Z}(L, M). \tag{3.50}$$

Combining with (2.71) and (2.72), one finds

$$\delta \mathcal{F}(L) = \delta \epsilon L \frac{d\mathcal{F}(L)}{dL} = \frac{c}{12} \frac{2\pi}{L} \delta \epsilon \tag{3.51}$$

yielding [6, 93]

$$\mathcal{F}(L) = -\frac{\pi c}{6} \frac{1}{L}. \tag{3.52}$$

This result for $\mathcal{F}$ does not contain a bulk contribution. Bulk terms were already subtracted in (2.71) since for its derivation the infinite plane expectation value

$$\langle T_{\mathrm{plane}}(z) \rangle_{\mathrm{plane}} = 0. \tag{3.53}$$

We can rephrase the result in terms of the finite-size scaling function $Y(x_1, x_2)$ of the singular free energy density $f^{\mathrm{sin}} = L^{-1}\mathcal{F}$, for an infinitely long strip of width L and periodic boundary conditions

$$\boxed{Y(0, 0) = -\pi c/6} \ . \tag{3.54}$$

This result defines an algorithm to compute the central charge c from a lattice Hamiltonian. The comments made in connection with the correlation length also apply here.

To conclude, we have seen how conformal invariance can be used to make quantitative statements on universal scaling amplitudes. The results obtained so far give rise to two questions. First, one would like to use (3.49) and (3.54) as practical tools for the calculation of both c and the x_i in a given lattice system. Second, what can be said about the finite-size scaling functions $Y(z, \mu)$ and $S_i(z, \mu)$? We will return frequently to these questions in the following chapters (in particular Chaps. 10, 12–15).

We close this section with an application of (3.52,3.54) to quantum chains. From Chap. 1, we recall the relationship of the free energy of a two-dimensional classical statistical system on a strip of infinite length and finite width L and the ground state energy of a one-dimensional quantum system as temperature $T \sim (k_B L)^{-1}$. For such a quantum system, conformal invariance is applicable provided the energy spectrum obeys a *linear* dispersion relation $E_k = v_s |k|$, where k is the wave vector and v_s the speed of sound[4] (and in general non-universal). If that is the case, (3.54) gives the amplitude for the low-temperature behaviour of the free energy per unit length $f/(k_B T)$ of the quantum system. The specific heat is (with all constants restored) [6, 93]

$$C = \frac{\pi c}{3} \frac{k_B^2}{\hbar v_s} T, \tag{3.55}$$

---

[4] In relativity, this corresponds to ultrarelativistic particles and v_s becomes the speed of light. We always use units such that $v_s = 1$.

in the $T \to 0$ regime. Thus a measurement of the coefficient of the linear variation of C with temperature allows to extract the central charge c.

To illustrate this, consider an ideal ultrarelativistic Bose gas. We work with units such that in the dispersion relation $E_p = |p|$, where p is the particle momentum, thus $v_s = 1$. The total energy is $E = \sum_p E_p/(\exp(\beta E_p) - 1)$. In the continuum limit

$$E = \frac{1}{\pi} \int_0^\infty dp\, p \left(e^{\beta p} - 1\right)^{-1} = \frac{k_B^2 T^2}{\pi} \sum_{n=0}^\infty \int_0^\infty du\, u\, e^{-(n+1)u} = \frac{1}{6}\pi k_B^2 T^2,$$

(3.56)

where $\beta = 1/(k_B T)$. The specific heat is then

$$C = \frac{\partial E}{\partial T} = \frac{\pi}{3} k_B^2 T$$

(3.57)

and we read off $c = 1$. Similarly, for n distinct types of free bosons, one finds $c = n$.[5] The case of a free fermion (ideal Fermi-Dirac gas) is left as an exercise. The *quantum* chain two-point function for $T \neq 0$ is given in (13.55).

3.6 Comparison with Experiments

We close this chapter with a brief phenomenological discussion on some experimental tests of finite-size scaling. Most examples studied so far deal with finite-size scaling in 3D, but a 2D example has already been found as well. Presently, data are only available for the shift $T_c - T_c(L)$.

The natural setting for these experiments are thin epitaxial layers of a magnetic substance on a non-magnetic substrate. One may either consider just a single magnetic layer or else build up magnetic multilayers, provided there is no magnetic coupling between the individual layers. In any case, one ends up with a geometry of thin *films* with free boundary conditions in the finite direction.

In Chap. 1, we have already seen that as a function of layer thickness, there is a rapid cross-over for between 2D and 3D criticality at thickness of more than just a few monolayers. So these experiments consider thin, but not ultrathin films. Also, one expects that the shift exponent λ should be related through (3.9) to the critical exponents of the 3D universality classes. Empirically, data may agree better with power laws if (3.5) is replaced by

$$\Delta T := (T_c(L) - T_c(\infty))/T_c(L) \sim (L - L_0)^{-\lambda},$$

(3.58)

which is supposed to take into account that in some systems the magnetic order is completely suppressed for a finite layer thickness L_0 (dead layer).

---

[5] This may suggest that in some sense c counts the number of critical degrees of freedom.

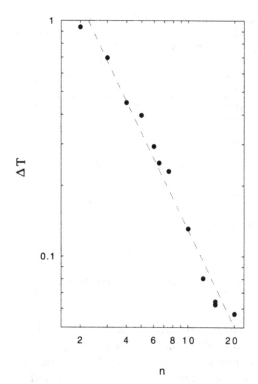

Fig. 3.5. Finite-size scaling of the shift ΔT in Ni(111) films on W(110). Layer thicknesses are given in monolayers. The slope of the straight line is $\lambda \simeq 1.4$. The data are from [E51]

We have already seen in Fig. 3.2 an example that the asymptotic shift of the critical point $T_c(L)$ with L remains valid down to quite small values of L in many theoretical models. The same apparently also holds for real systems. To illustrate this, we show in Fig. 3.5 the shift ΔT (with $L_0 = 0$) as a function of the layer thickness n for Ni(111) films on W(110). The data [E51] are clearly seen to be in agreement with a power law down to very small values of n. Remarkably, this phenomenological relation appears to be valid even for those values of n where the exponent β has crossed over from 3D to 2D values, which occurs below $n \simeq 4 - 6$ [E51].

In Table 3.2 we collect experimental results. FeF_2 and CoO are antiferromagnets with well localized magnetic moments. Estimates for $T_c(L)$ were obtained from thermal expansion (related to the specific heat) and the susceptibility, respectively. For the Ni/Cu system, $T_c(L)$ was obtained from the magnetization measured through Kerr effect and for Ni/W, through the magnetic resonance linewidths. Data for Cr come from neutron diffraction. Data for liquid ^{4}He films come from specific heat measurements. For Gd/W again the magnetic properties were measured, while the Gd/glass data come from electrical resistivity measurements. Finite-size shifts were also found in SF_6 in pores [E85].

It can be seen that the values of λ are close to the theoretical values, at least for those systems for which the bulk critical behaviour is well under-

Table 3.2. Experimental results for the shift exponent λ in films of magnetic systems and liquid helium. The theoretical values are $\lambda_I \simeq 1.588$, $\lambda_X \simeq 1.495$ and $\lambda_H \simeq 1.417$, where I, X and H refer to the 3D Ising, XY and Heisenberg universality classes. A * indicates that ν was quoted in the source and $\lambda = 1/\nu$

Class	Substance	λ	source	comments
I	FeF_2/ZnF_2	$1.56(10)^*$	[E49]	$\nu = 0.64(4)$
	CoO/SiO_2	$1.55(5)$	[E2]	
X	^4He	1.5^*	[E58]	$\nu \sim 0.67$
?	Gd/W	1.6^*	[E32]	$\nu \sim 0.63$, see [E62]
	Gd/W	$1.5(1)$	[E44]	
	Gd/glass	$0.91(2)$	[E17]	
H	Ni/Cu	$1.44(20)$	[E10]	
	Ni/W	$1.4(1)$	[E50, E51]	
	Ni/Cu	$1.32(14)^*$	[E41]	$\nu = 0.76(8)$
				(mean value of Ni
				and Co_1Ni_9 films)
	Fe/Cr	$1.4(3)$	[E36]	data for Cr Néel transition

stood. Error bars are just small enough to be able to distinguish the Ising from the Heisenberg universality classes. Since the same exponents are found for physically different systems, universality is confirmed. For the experiments performed on ^4He films, confined between two silicon wafers and of thicknesses between 0.1 to $0.7\mu m$, both the bulk and the finite-size scaling is considered [E58]. A very clear collapse of the data, obtained for different film thicknesses, is found for the theoretically predicted values of the critical exponents (for $T > T_c$, over five decades in the variable $tL^{1/\nu}$). The form of the finite-size scaling function is in qualitative agreement with previous field-theoretical calculations. Given the more complicated nature of the magnetism in Gd, see [E25, E62], its critical properties cannot be straightforwardly described in terms of the simplest universality classes. The very low value of λ obtained in [E17], if correct, looks more like a finite-size effect in a 2D bulk system. It might be worth recalling that the data presented here had been obtained by varying the layer thickness by at most one order of magnitude.

In a few cases, experiments yield effective values of λ which may be about a factor of two larger than the values expected from the usual relation $\lambda = 1/\nu$. This may be explained by invoking finite-size corrections to the leading finite-size scaling behaviour. For example, in the Fe/Ir(100) system, the critical point for a layer of thickness L is determined from the condition that the magnetization vanishes [E40]. The magnetization should scale as

$$M_L \sim L^{-\beta/\nu} Z\left(tL^{1/\nu}, uL^{y_3}\right), \qquad (3.59)$$

Table 3.3. Effective values of the shift exponent λ in 3D of the O(n) model ($n = 1, 2, 3$) and comparison of the finite-size scaling prediction λ_{th} from (3.60) with some experimental results

model	n	Δ_1	$1/\nu$	λ_{th}	λ_{exp}	system	Ref.
Ising	1	0.508(7)	1.588(3)	3.20(3)	3.4(3)	CoO/SiO$_2$	[E3]
XY	2	0.533(8)	1.495(4)	3.09(3)			
Heisenberg	3	0.556(10)	1.417(7)	2.99(3)	3.15(15)	Fe/Ir(100)	[E40]

where u is the scaling field associated with the leading irrelevant scaling operator and $y_3 < 0$ is its renormalization group eigenvalue.[6] Now, a finite-size estimate for the critical point $T_c(L)$, and thus t_L, comes from the condition $Z(t_L L^{1/\nu}, u L^{y_3}) = 0$. In the same way, t_{L-1} for a layer of thickness $L - 1$ is determined. Assuming phenomenologically that $t_L \sim L^{-\lambda}$, it follows by expanding in $1/L$ and matching exponents that $\lambda = 1/\nu - y_3$ [515, E40]. Now, the systems under consideration are expected to be described theoretically by the O(n) Heisenberg model. In that case, $|y_3| = \omega = \Delta_1/\nu$ should be related to the Wegner correction exponent. However, since the corresponding scaling operator is even under spin reversal while the magnetization is odd, the first-order term in the above expansion vanishes and the second-order terms will determine the effective value of λ, leading to [E40]

$$\lambda_{\text{eff}} = (1 + 2\Delta_1)/\nu. \tag{3.60}$$

In 3D, values for Δ_1 and ν have been determined numerically [285]. In Table 3.3, we collect values for ν and Δ_1 for the O(n) model in 3D for $n = 1, 2, 3$ and compare the resulting predictions with measured results of λ_{eff}. Within the errors, an agreement is observed, which offers an experimental test on the Wegner correction exponent.

Finally, for *submonolayer* magnetism, a finite-size scaling behaviour corresponding to the cross-over between a 2D magnet and a 1D system can be observed [E28]. This has been carried out for submonolayer films of Fe(110) on a W(110) substrate. Through scanning tunneling microscopy, it was established that when the Fe film is grown at sufficiently high temperatures[7] ($T_p \simeq 600$K), one obtains continuous Fe stripes along the steps of the substrate for all values of the coverage ($0.11 \leq \theta \leq 0.8$) considered. In other words, the coverage θ is a measure of the in-plane width of the iron film. Well-defined films of a definite width can be obtained for monolayer coverages up to $\theta \leq 0.8$ [E28]. Spin polarized low energy electron diffraction was then used to measure the order parameter and the critical temperature $T_c(\theta)$

---

[6] If finite-size corrections are small, u can be put to zero and the usual result $\lambda = 1/\nu$ is recovered. The following discussion should apply to situations where finite-size corrections are sizable, which should be the case when $T_c(L) \ll T_c(\infty)$.

[7] If the film is grown at room temperature, one observes a growth in form of small islands of Fe for small θ, which join to form a magnetic system only for $\theta \geq 0.6$, in a form quite analogous to a percolation transition [E28, E78].

is obtained. A finite-size (or rather finite-width) scaling analysis using (3.5) then yielded the shift exponent $\lambda = 1.03(14)$ [E28], in good agreement with the expected value of unity for the 2D Ising model. Ising critical behaviour should have been expected, since the magnetization is in-plane and subjected to a strong uniaxial surface type in-plane anisotropy. This agrees with the measured value $\beta \simeq 0.123$, obtained in the critical region for $\theta = 0.8$ [E28].

Exercises

1. From a huge simulation (or an elaborate experiment) precise data for the magnetic susceptibility $\chi_L(T)$ as a function of T on a hypercube (length of one side L) in d dimensions were obtained for several values of L. Assuming that no other information is available, show how the critical point $T_c(\infty)$ and the exponent γ/ν can both be found through a finite-size scaling analysis (see e.g. [313, 82]).

2. Consider the finite-size scaling form (3.35) with $h = 0$ for the correlation length. Assuming that the first-order finite-size corrections are not negligible and dominant, show that the shift exponent $\lambda = (1 + \theta_{ir})/\nu$ [515]. Show that for the leading term, $\theta_{ir} = \Delta_1 = \omega\nu$.

3. Consider the (mean) **spherical model**, which is defined in terms of continuous spins s_i which can take on any real value, and has the Hamiltonian

$$\mathcal{H} = -J \sum_{\langle i,i' \rangle} s_i s_{i'} - \lambda \sum_i s_i^2,$$

where the Lagrange multiplier λ is determined from the (mean) spherical constraint $\langle \sum_i s_i^2 \rangle = \mathcal{N}$, where $\mathcal{N}$ is the total number of sites.
It is easy to see that $\mathcal{H}$ can be diagonalized by going over from the real-space spins s_i to spin variables in Fourier space s_q and that the spherical constraint can be obtained by considering the derivative $\partial \ln Z / \partial \lambda$. Consider the spherical model in a slab geometry, where $d - 1$ directions are of infinite extent, while the last one only has L layers of spins. Work out the finite-size scaling function of the free energy (or any other observable you prefer) and compare with the predictions of finite-size scaling theory, e.g. [54, 106, 440, 570]. For simplicity, use periodic boundary conditions.

4. Consider a *scalar* scaling operator $\phi(z, \bar{z})$ with scaling dimension x. Writing $w = u + iv$ for the complex coordinates on the strip, show that [131]

$$\langle \phi_1 \phi_2 \rangle_w = \left(\frac{2\pi}{L} \right)^{2x} \left(2 \cosh \frac{2\pi}{L} (u_1 - u_2) - 2 \cos \frac{2\pi}{L} (v_1 - v_2) \right)^{-x},$$

where $\phi_i = \phi(w_i, \bar{w}_i)$, $i = 1, 2$. Derive the two-point function in any other geometry you like, e.g. [42, 404, 405], see also Chap. 15.

5. Show that an ideal Fermi-Dirac gas of ultrarelativistic spinless particles has central charge $c = 1/2$ in one dimension (for generalizations to parastatistics, see [558]).

4. Representation Theory of the Virasoro Algebra

In this and the following chapters, we shall develop the necessary techniques for the explicit calculation of four-point correlation functions. The tools needed are the representation theory of the Virasoro algebra and the operator product algebra. These will be presented in the next two chapters. The first application of this will be to the Ising model in Chap. 6.

The major difference between a critical two-dimensional system and any other physical system lies in the presence of the enlarged symmetry algebra given by the two infinite dimensional Virasoro algebras (2.65). To understand its importance, its representation theory must be discussed. For the purposes of this chapter, it is enough to discuss a *single* Virasoro algebra generated by $\{L_n\}$

$$[L_n, L_m] = (n - m) L_{n+m} + \frac{1}{12} c\, n(n^2 - 1)\delta_{n+m,0}. \tag{4.1}$$

Any results for L_n can be formally taken over to $\bar{L}_n$ immediately, since the two Virasoro algebras commute. The full conformal algebra is of course generated by the tensor product of the Virasoro algebras $\{L_n\} \otimes \{\bar{L}_m\}$ and must be taken into account when studying correlation functions or the partition function. In the next chapter, we shall see that these must satisfy certain analyticity conditions.

The physical motivation for this exercise in representation theory will be provided in Chap. 8, where it will be shown that the critical point quantum Hamiltonian H, obtained from the transfer matrix T (1.84) and defined on a chain of N sites with periodic boundary conditions, is related to the Virasoro generators through

$$\boxed{H = \frac{2\pi}{N} \left(L_0 + \bar{L}_0\right) - \frac{\pi c}{6} \frac{1}{N}}. \tag{4.2}$$

In other words, the Virasoro algebra acts as a **dynamical symmetry**.

The results described in this chapter were derived in the seminal work by Belavin, Polyakov and A. Zamolodchikov [68] and, for the last two sections, by Kac [376, 377, 378], Feigin and Fuks [223] and Rocha–Caridi [539]. We follow here essentially the presentation given in [169].

4.1 Verma Module

The representations of the Virasoro algebra describe how the energy–momentum tensor acts on the physical scaling operators. For a primary operator ϕ, the essential information is contained in the conformal Ward identities (2.51) and involves only the analytic component $T = T(z)$. Indeed, we only need the singular part of this, which may be written down as the formal operator product expansion

$$T(z)\phi(z_1, \bar{z}_1) = \frac{\Delta}{(z - z_1)^2}\phi(z_1, \bar{z}_1) + \frac{1}{z - z_1}\partial_{z_1}\phi(z_1, \bar{z}_1) + \text{regular terms.} \tag{4.3}$$

The "regular terms" include the parts of the correlation function where T couples with the other operators. On the other hand, the definition (2.58) for the Virasoro generators L_n may be rewritten as

$$T(z)A(z_1, \bar{z}_1) = \sum_{n=-\infty}^{\infty} (z - z_1)^{-n-2} L_n A(z_1, \bar{z}_1), \tag{4.4}$$

where the operator $A(z_1, \bar{z}_1)$ can be either a primary or a secondary operator. Equation (4.4) is indeed consistent with the definition (2.58) of the generators L_n, since

$$\frac{1}{2\pi i} \oint_{C_{z_1}} dz\, (z - z_1)^{m+1} T(z) A(z_1, \bar{z}_1)$$

$$= \sum_{n=-\infty}^{\infty} \frac{L_n}{2\pi i} \oint_{C_{z_1}} dz\, (z - z_1)^{m-n-1} A(z_1, \bar{z}_1) = L_m A(z_1, \bar{z}_1). \tag{4.5}$$

Now, let $A(z_1, \bar{z}_1)$ be a primary operator $\phi(z_1, \bar{z}_1)$. Compare the leading terms in (4.3) and in (4.4). It follows that an *algebraic definition* of a primary operator can be formulated as

$$\begin{aligned} L_n\phi &= 0 \quad \text{if } n \geq 1 \\ L_0\phi &= \Delta\phi. \end{aligned} \tag{4.6}$$

In fact, for ϕ to be primary, it is necessary and sufficient that

$$L_0\phi = \Delta\phi \ , \quad L_1\phi = 0 \ , \quad L_2\phi = 0. \tag{4.7}$$

To see this, note that the other constraints $L_n\phi = 0$ for $n \geq 3$ can be deduced by recursion from (4.7). From (4.1) we have $[L_n, L_1] = (n - 1)L_{n+1}$. Now, by induction over n

$$L_{n+1}\phi = \frac{1}{n - 1}(L_n L_1\phi - L_1 L_n\phi) = 0 \tag{4.8}$$

follows for all $n \geq 3$, if it is valid for $n = 1, 2$. This elementary observation will save a lot of work later on.

L_0 measures the holomorphic part of the scaling dimension of the operator. This holds for primary as well as for secondary operators. Notice also that L_{-1} generates the translations. Therefore we have the identification

$$L_{-1} = \partial_z. \tag{4.9}$$

A primary operator defines the **highest weight** of a representation of the conformal symmetry.[1] An infinite set of secondary descendant operators are defined by the negatively indexed operators L_{-n}. They build up the **Verma module** of ϕ whose elements are given by

$$\phi^{(-n_k,\ldots,-n_1)} = L_{-n_k} \cdots L_{-n_1}\phi. \tag{4.10}$$

Often, the set of secondary operators obtained from ϕ is also referred to as the **conformal tower** of ϕ. The integer $n := n_1 + n_2 + \ldots n_k$ is called **level** of the operator $\phi^{(-n_k,\ldots,-n_1)}$. The commutation relations (4.1) lead to linear dependencies between different descendant operators. A simple way to take care of at least some of these relations is to impose some ordering on the indices of $\phi^{(-n_k,\ldots,-n_1)}$. For example, one may impose $n_1 \leq n_2 \leq \ldots \leq n_k$.

The secondary operators are all eigenvectors of L_0 with eigenvalue $\Delta + \sum_i n_i$, since

$$
\begin{aligned}
L_0\phi^{(-n_k,\ldots,-n_1)} &= L_0 L_{-n_k} \cdots L_{-n_1}\phi \\
&= n_k L_{-n_k} \cdots L_{-n_1}\phi + L_{-n_k} L_0 \cdots L_{-n_1}\phi \\
&= (n_k + \ldots + n_1 + \Delta)\,\phi^{(-n_k,\ldots,-n_1)}.
\end{aligned} \tag{4.11}
$$

The primary operator ϕ is the only element of the Verma module with conformal weight Δ and $L_{-1}\phi$ the only one with conformal weight $\Delta + 1$. At conformal weight $\Delta + 2$, we have two elements, $L_{-2}\phi$ and $(L_{-1})^2\phi$.

The argument leading to the algebraic conditions (4.6) can be extended to non-primary operators as well. Consider for example the Ward identity (2.57) together with the expansion (4.4)

$$
\begin{aligned}
\langle T(z)T(z_1)\cdots\rangle &= \sum_{n=-\infty}^{\infty} \frac{L_n}{(z-z_1)^{n+2}}\langle T(z_1)\cdots\rangle \\
&= \left(\frac{2}{(z-z_1)^2} + \frac{1}{z-z_1}\frac{\partial}{\partial z_1} + \cdots\right)\langle T(z_1)\cdots\rangle + \frac{c/2}{(z-z_1)^4}\langle 1\cdots\rangle,
\end{aligned} \tag{4.12}
$$

where the $\cdots$ refer to other scaling operators.[2] Comparing the coefficients, we read off, besides recovering (4.9)

---

[1] The standard example of highest weight representations occurs in the algebraic theory of angular momentum, described in any text on elementary quantum mechanics. The main difference with the case at hand is that the representations of angular momentum usually considered are finite-dimensional, while the Verma module is infinite-dimensional.

[2] Using the operator product expansion (2.61) avoids this cumbersome notation.

$$L_0 T = 2T(z_1) \ , \ \ L_1 T = 0 \ , \ \ L_2 T = \frac{c}{2} \ , \ \ L_n T = 0 \ ; \ \text{ for } n \geq 3. \quad (4.13)$$

Now insert, as an important special case of a primary operator, the identity operator $\mathbf{1}$ into (2.58) (or (4.4)). Because of the regularity of the integral involved for $n \geq -1$, the identity operator is primary with $L_0 \mathbf{1} = 0$ and $L_{-1} \mathbf{1} = 0$. However, $L_{-2} \mathbf{1}$ is non-vanishing

$$L_{-2} \mathbf{1} = T(z). \quad (4.14)$$

The energy–momentum tensor T is therefore a secondary operator of the identity. We say that it belongs to the **conformal tower** of the identity. However, T is quasi-primary, since $L_1 T = L_1 L_{-2} \mathbf{1} = 3L_{-1} \mathbf{1} + L_{-2} L_1 \mathbf{1} = 0$, see also (4.13). We recall that a quasi-primary operator Φ transforms like a primary one under the projective group only. Consequently, for any quasi-primary Φ

$$L_1 \Phi = 0. \quad (4.15)$$

This is the *defining algebraic condition* for a quasi-primary operator. In general, $L_n \Phi = 0$ holds for some positive n. In the case of T, we have $L_2 T = L_2 L_{-2} \mathbf{1} = 4L_0 \mathbf{1} + (c/2)\mathbf{1} = (c/2)\mathbf{1}$ and $L_n T = 0$ for $n \geq 3$.

We will now show that *any quasi-primary operator is the secondary operator of a primary one*. This implies that the primary operators are the basic building blocks of a conformal theory, which by themselves fix the structure of the theory completely. The proposition follows because the operator L_0 should be bounded from below. Namely, if L_0 were unbounded, there would exist operators with conformal weights $\Delta = -\infty$, which should have an everywhere infinite two-point function. Such an operator would be difficult to interpret physically and therefore we can suppose L_0 bounded from below by $\Delta_{\min}$, in any physically reasonable theory. Any quasi-primary operator Φ with conformal weight Δ_Φ at least verifies $L_{n_1} \ldots L_{n_k} \Phi = 0$ for any k and any positive $n_1, \ldots, n_k$ such that $\Delta_\Phi - n_1 - \cdots - n_k < \Delta_{\min}$. There is therefore by construction at least one operator[3] ϕ with conformal weight $\Delta_\Phi \geq \Delta_\phi \geq \Delta_{\min}$ which is primary and contains Φ in its Verma module.

To summarize, *an infinite-dimensional representation of the Virasoro algebra, called Verma module, has a primary operator as its highest weight*. Descendant operators of higher conformal weights are either quasi-primary operators or derivatives of the primary or quasi-primary operators (we shall show below that some descendant quasi-primary operators may also be primary.).

We now give some examples of quasi-primary operators. The highest weight ϕ is of course quasi-primary. At level 2, the most general operator in the conformal tower of ϕ is of the form $(aL_{-2} + bL_{-1}^2)\phi$, where a, b are constants. It is easy to verify that

---

[3] There could be many operators if degeneracies occur. Φ can be a linear combination of quasi-primary operators which are secondary to primary operators with conformal weights equal or differing by integers.

$$\Phi^{(2)} := \left(L_{-2} - \frac{3}{2(2\Delta + 1)}L_{-1}L_{-1}\right)\phi \qquad (4.16)$$

satisfies $L_1\Phi^{(2)} = 0$ (and is unique up to normalization). At level 3, there are two derivative operators $L_{-1}L_{-1}L_{-1}\phi$, $L_{-1}\Phi^{(2)}$, and a new quasi-primary operator given is by

$$\Phi^{(3)} := \left(L_{-3} - \frac{2}{\Delta + 2}L_{-1}L_{-2} + \frac{1}{(\Delta + 2)(\Delta + 1)}L_{-1}L_{-1}L_{-1}\right)\phi. \qquad (4.17)$$

Two quasi-primary operators appear at level 4, one containing $L_{-4}\phi$, the other with $L_{-2}L_{-2}\phi$. The number of quasi-operators obviously increases with the level n (see (14.41,14.42) for explicit expressions).

Up to now, we have not specified the possible values for Δ and c. A conformal field theory contains Verma modules of primary operators with conformal weights Δ. In a first step, we assume that the infinite dimensional representation, made out of the descendant operators $L_{-n_k} \cdots L_{-n_1}\phi$ with $n_k \geq \cdots \geq n_1 > 0$, is irreducible. This means that all descendant operators at a given **level** $n = n_k + \cdots + n_1$ are linearly independent. We shall see in the next sections that the most interesting representations are not irreducible but it is simpler to show the structure of the Verma modules starting from the generic case. In this case and at a given level n, the number of independent operators is the number of partitions $p(n)$ of n. We define the **generic character** (for $\Delta \neq 0$) of a Verma module by

$$\chi(\Delta, c) = q^{\Delta - \frac{c}{24}} \sum_{n=0}^{\infty} p(n)q^n = q^{\Delta - \frac{c}{24}} \prod_{i=1}^{\infty} \frac{1}{1 - q^i} \; ; \; \Delta \neq 0. \qquad (4.18)$$

The character encodes the information on the structure of the Verma module. The prefactor $q^{\Delta - c/24}$ leads to simple modular properties of the character, as will be shown later. We also remark that the product in (4.18) is up to a factor $q^{1/24}$ the inverse Dedekind's function $\eta(\tau)$ with $q = \exp(2\pi i\tau)$. The formula (4.18) does not hold for $\Delta = 0$ because $p(1) = 1$. In this case $L_{-1}1 = 0$ and the number of operators at level 1 is zero. This is also true for any operator $L_{-n_k} \ldots L_{-n_1}L_{-1}1 = 0$. Therefore, the generic character of the identity operator is

$$\chi(0, c) = (1 - q)\, q^{-\frac{c}{24}} \prod_{i=1}^{\infty} \frac{1}{1 - q^i} = q^{-\frac{c}{24}} \prod_{i=2}^{\infty} \frac{1}{1 - q^i}. \qquad (4.19)$$

The factor $(1 - q)$ suppresses the counting of all vanishing operators. We say that $L_{-1}1$ is a **null operator** and the Verma module is called **degenerate**.

For future applications, we notice how to count the quasi-primary operators. Suppose that we have p operators $A_1, \ldots, A_p$ at level n, then at level $n + 1$ we will have p derivative operators $L_{-1}A_1, \ldots, L_{-1}A_p$. The remaining ones at level $n + 1$ are the new quasi-primary operators. Thus, the difference in the number of operators between two successive levels counts the number of quasi-primary operators. In the character formula (4.18), two successive levels are distinguished by a factor q. Therefore, a function which counts the number of quasi-primary operators at each level is given by

$$\chi_{QP}(\Delta, c) = (1 - q)\chi(\Delta, c) + q\delta_{\Delta,0}. \tag{4.20}$$

The last term of (4.20) is best checked through explicit expansion of the first few terms.

4.2 Hilbert Space Structure

In order to define the representations, we have to specify on which states they should act. We shall do so now in a simple way [169] which allows to proceed quickly to the Kac formula, which the goal of these considerations. Most of the results of the previous section are independent of the coordinates z_1 of the operators. It is therefore natural to define the **vacuum state** as the highest weight of the identity operator evaluated at $z = 0$.

$$|0\rangle := \mathbf{1}(z = 0). \tag{4.21}$$

This state is defined by $L_n|0\rangle = 0$ $(n \geq -1)$. The true physical vacuum, which recombines the z and $\bar{z}$ dependence, is the tensor product $|0\rangle \otimes \overline{|0\rangle}$. A primary operator can now be viewed as a mapping between two vector spaces such that the vacuum state is mapped onto a highest weight state defined by

$$|\Delta\rangle := \phi(0)|0\rangle \tag{4.22}$$

with the properties $L_n|\Delta\rangle = 0$ $(n > 0)$ and $L_0|\Delta\rangle = \Delta|\Delta\rangle$.[4] We can connect this state at $z_1 = 0$ to any value of z_1 with the Taylor series

$$\phi(z_1)|0\rangle = e^{z_1 L_{-1}}|\Delta\rangle. \tag{4.23}$$

We need to define a *dual state* to the vacuum state to obtain an Hilbert space structure. The hermitian conjugate operator L_n^+ of L_n must verify

$$\langle 0|L_n^+ = 0 \qquad \text{if } n \geq 0. \tag{4.24}$$

In our case, a natural dual transformation is given by the projective map $z' = -1/z$ for which we can verify

$$T(z)\, z^2 = T(z')\, z'^2. \tag{4.25}$$

---

[4] If we were to follow this mathematical development further, ϕ could be properly defined as an operator, justifying the terminology we had introduced earlier.

Though interchanging the points at zero and at infinity, such a map transforms the complex plane into itself and does not modify the theory. Equation (4.4) taken for example with the identity operator at $z_1 = 0$ can be written

$$z^2 T(z)\mathbf{1} = \sum_{n=-\infty}^{\infty} z^{-n} L_n \mathbf{1} = z'^2 T(z')\mathbf{1} = \sum_{n=-\infty}^{\infty} z'^{+n} L_{-n}^+ \mathbf{1}. \qquad (4.26)$$

The consequence for the generators of the Virasoro algebra is expressed by

$$\boxed{L_n^+ = L_{-n}} \ . \qquad (4.27)$$

This is natural because the commutator $[L_n, L_{-n}]$ belongs to the Cartan subalgebra spanned by $(L_0, \{c\})$. This means that $L_0^+ = L_0$ and the spectrum of conformal weights is real as expected in a physical theory. Notice that for an infinite-dimensional algebra, the hermiticity of the algebra does *not* imply the unitarity of the representations (i.e. the positivity of the states). Unitarity will be discussed in the next section.

We now show how hermitian conjugation acts on primary operators. The map $z' = -1/z$ modifies the primary operators as follows

$$\phi(z') = z^{2\Delta} \phi(z). \qquad (4.28)$$

By duality, the state $|\Delta\rangle$ is mapped to a dual state located at infinity[5]

$$\langle \Delta | = \lim_{z_1 \to \infty} \langle 0 | \phi(z_1) (z_1)^{2\Delta} \qquad (4.29)$$

which verifies the orthogonality property $\langle \Delta | \Delta' \rangle = \delta_{\Delta, \Delta'}$. It is interesting to see that everything is consistent with the two-point function. The Taylor expansion at infinity gives the dual state at finite value of z_1

$$\langle 0 | \phi(z_1) = \langle \Delta | z_1^{-2L_0} e^{(1/z_1)L_1}. \qquad (4.30)$$

To check that these definitions are sensible, combine this dual state with (4.23), and prove the formula

$$\langle \Delta | (L_1)^n (L_{-1})^m | \Delta \rangle = \delta_{n,m}\, n! \frac{(2\Delta - 1 + n)!}{(2\Delta - 1)!}. \qquad (4.31)$$

---

[5] For a heuristic motivation of these definitions, recall a scattering process. An incoming state Φ_{in}, defined at times $t \to -\infty$, is scattered into an outgoing state $\Psi_{\text{out}} = S\Phi_{\text{in}}$ for times $t \to \infty$, where the scattering operator is $S \sim \exp(-iHt)$ and transition probabilities are matrix elements $\langle \Psi_{\text{out}} | S | \Phi_{\text{in}} \rangle$. Through the transformation $z = e^w = e^{u+iv}$ which maps the plane onto the strip and where u plays the role of the time, the dual state is defined analogously, viz. $\langle \Delta | \sim \lim_{u \to -\infty} \langle 0 | \phi\, e^{2uL_0}$, where L_0 is related to the quantum Hamiltonian H. It is possible to make this analogy quite precise and to define correlators in terms of time-ordered ground state expectation values of quantum operators defined on the strip. This is referred to as **radial quantization**.

Then, it is easy to verify the two point function

$$\langle 0|\phi(z_1)\phi(z_2)|0\rangle = \frac{1}{(z_1 - z_2)^{2\Delta}} = \langle \phi(z_1)\phi(z_2)\rangle. \qquad (4.32)$$

The algebraic language encodes the complete information on the operators even if the z dependence is not explicit. For example in the two-point function, the explicit z dependence does not really matter. What really matters is on one side the conformal weight Δ given as eigenvalue of L_0 and on the other side the condition that the two-point function does not vanish. This last condition is encoded in the hermitian product $\langle \Delta|\Delta'\rangle$.

Though not explicitly mentioned in the beginning of this section, there is more in the definition of (4.21): while introducing the hermitian product, we included an information on the two-point function. Later, we will investigate tensor product representations and show how n-point functions can be obtained. Notice that the brackets $\langle\rangle$ have now taken another meaning as the vacuum state and its dual in a quantum field theory.

4.3 Null Vectors

When constructing the unitary irreducible representations of an infinite-dimensional Lie algebra such as the Virasoro algebra, we must be careful. In contrast to the habit the reader might have acquired from quantum mechanics, the hermiticity of the generators is *not* enough to guarantee the unitarity of the irreducible representations. It is therefore necessary to investigate under which conditions unitary representations of the Virasoro algebra generators satisfying (4.27) can be constructed.[6]

The first models we are going to discuss are the **minimal models** which, unitary or not, are characterized by the fact that the number of distinct Verma modules is finite. **Unitarity** is characterized by the absence of negative-norm states in the theory. States at level 0 do not provide any information, since without restriction of generality, we can normalize $\langle \Delta|\Delta\rangle = 1$. At level one we have the state $L_{-1}|\Delta\rangle$ with norm

$$\langle \Delta|L_1 L_{-1}|\Delta\rangle = 2\Delta. \qquad (4.33)$$

Therefore, a necessary condition for unitarity is $\Delta \geq 0$. Another simple condition is given by the two-point function of the energy–momentum tensor. The norm is given by

$$\langle 0|L_2 L_{-2}|0\rangle = \frac{c}{2}. \qquad (4.34)$$

---

[6] The requirement of unitarity arises from the need to formulate conformally invariant theories as consistent quantum field theories. In statistical mechanics applications, however, unitarity is not a basic axiom. Although many systems do indeed satisfy unitarity, there are also well-defined models which do not.

Therefore $c \geq 0$ is necessary to ensure unitarity. So far, we have

$$\text{unitarity} \quad \Longrightarrow \quad c \geq 0 \ , \ \Delta \geq 0. \tag{4.35}$$

To gain new information, we have to investigate higher levels in the the Verma module. At level 2, we have two independent vectors $L_{-2}|\Delta\rangle$ and $(L_{-1})^2|\Delta\rangle$. The unitarity condition requires that the determinant

$$
\det{}_2(c, \Delta) := \begin{vmatrix} \langle \Delta | L_2 L_{-2} | \Delta \rangle & \langle \Delta | L_2 (L_{-1})^2 | \Delta \rangle \\ \langle \Delta | (L_1)^2 L_{-2} | \Delta \rangle & \langle \Delta | (L_1)^2 (L_{-1})^2 | \Delta \rangle \end{vmatrix}
$$

$$
= \begin{vmatrix} 4\Delta + \frac{1}{2}c & 6\Delta \\ 6\Delta & 4\Delta(2\Delta + 1) \end{vmatrix}
$$

$$
= 2\Delta \left(16\Delta^2 + 2(c - 5)\Delta + c \right) \tag{4.36}
$$

is positive definite. This is equivalent to the requirement that the matrix in (4.36) is positive definite, which is the case if and only if its eigenvalues are positive. An arbitrary choice of the conformal weight Δ and the central charge c may result in negative eigenvalues at level 2, therefore breaking unitarity. That is so because the determinant

$$\det{}_2(c, \Delta) = 32\Delta \left(\Delta - \Delta_+ \right) \left(\Delta - \Delta_- \right), \tag{4.37}$$

where

$$\Delta_\pm(c) := \frac{5 - c \pm \sqrt{(1 - c)(25 - c)}}{16} \tag{4.38}$$

changes sign when passing in the (c, Δ) plane one of the curves $\Delta = \Delta_\pm(c)$. In Fig. 4.1, we display these two curves. They are called **vanishing curves** because $\det{}_2(c, \Delta)$ vanishes along them. From the discussion of the states at level 1 it is already clear that we need only consider the quadrant $c \geq 0$ and

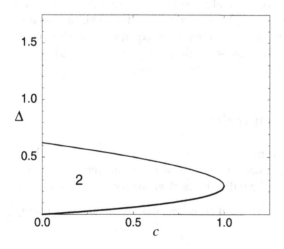

Fig. 4.1. Vanishing curves for level 2. In the region labeled 2, the Kac determinant $\det{}_2(c, \Delta)$ is negative

$\Delta \geq 0$. In addition, from the definition of the $\Delta_{\pm}(c)$ it can be explicitly seen that constraints from unitarity are *only* imposed for theories with $c < 1$.[7] Finally, since for a sufficiently small Δ and a large enough c, $\det_2(c, \Delta)$ is seen from (4.36) to be positive, it follows that $\det_2(c, \Delta) < 0$ in the region between the two curves $\Delta = \Delta_{+}(c)$ and $\Delta = \Delta_{-}(c)$ which is labeled 2 in Fig. 4.1. Consequently, no unitary conformal theory is possible in that region.

An interesting case occurs if Δ takes one of the values $\Delta = \Delta_{\pm}(c)$ because then one of the eigenvalues will be zero and the other positive. It follows that there is a linear combination of $L_{-2}|\Delta\rangle$ and $L_{-1}^2|\Delta\rangle$ with zero norm. This combination must be a quasi-primary operator due to the presence of L_{-2}. Notice that the norm of the derivative of a quasi-primary operator is always proportional to the norm of the quasi-primary operator itself, which is located at a lower level. The vanishing norm of a derivative operator always implies that the norm of its quasi-primary ascendant is zero. At level 2 the only quasi-primary state is $|\Phi^{(2)}\rangle$, defined in (4.16), which norm is now zero. It is easy to verify that, using (4.38)

$$L_2|\Phi^{(2)}\rangle = 0. \tag{4.39}$$

The state $|\Phi^{(2)}\rangle$ with $\Delta = \Delta_{\pm}(c)$ is primary. Besides having zero norm, it is *orthogonal to any other operator in the Verma module* (even its own derivatives)

$$\langle\Delta|L_{n_1}\ldots L_{n_k}|\Phi^{(2)}\rangle = 0 \tag{4.40}$$

with $n_1, \ldots, n_k$ positive. A scaling operator with this property is called a **null operator** (or *null vector* when considering the state $|\Phi^{(2)}\rangle = \Phi^{(2)}(0)|0\rangle$). The spectrum for Δ and c is restricted if we want to have semi-positivity, i.e. at most null-vectors in the Verma module. Such a Verma module is called **degenerate**. Null operators can appear at any level. They are always primary and their submodule is orthogonal with themselves and with any other operator.

These results are not simply abstract remarks for the mere benefit of representation theory. In Chap. 5, we shall come back to the special case where Δ and c satisfy (4.38) with $|\Phi^{(2)}\rangle$ as a null-vector. The presence of null-vectors will permit to derive extra differential equations for the higher n-point correlation functions. In a sense, null-vectors may be considered as *the ingredients which make it possible to solve a model*.

4.4 Kac Formula and Unitarity

Having discussed the null vectors and vanishing curves which occur at level 2 in the degenerate Verma module, we now adress the same question for any level. The general discussion of the null-vectors at arbitrary level was done by

---

[7] For $c > 25$, the $\Delta_{\pm}(c)$ are negative and therefore play no role in the unitarity discussion.

Kac [376]. Since the mathematical proofs [223] are quite complicated (more details may be found in [G12, 378]), we merely quote some important results.

The determinant computed in (4.36) at level 2 can be generalized to any level. At level n, Kac found the following formula

$$\det_n(c, \Delta) = a_n \prod_{r,s=1;\, 1 \le rs \le n}^{n} (\Delta - \Delta_{r,s})^{p(n-rs)}, \qquad (4.41)$$

where a_n is a positive constant. A particular parametrization was introduced to write this formula. The central charge is defined as

$$c := 1 - \frac{6}{m(m+1)}, \qquad (4.42)$$

where m is *a priori* real or complex. A model is called **minimal** if m is a rational number. The zeroes of the determinant take the values[8]

$$\Delta_{r,s} = \Delta_{m-r,m+1-s} = \frac{[r(m+1) - sm]^2 - 1}{4m(m+1)}, \qquad (4.43)$$

where r and s are positive integers and $p(k)$ is the number of partitions of the integer k. For r, s given, the first null-vector appears at level rs and propagates with increasing multiplicity at all levels larger than rs. For example $\det_2(c, \Delta)$ (see (4.36)) is proportional to Δ, and $\det_1(c, \Delta) = \Delta$ because $\Delta_{1,1} = 0$. This reflects the fact that $\det_n(c, \Delta)$ always divides $\det_{n+1}(c, \Delta)$. The proof of the Kac formula was given by Feigin and Fuks [223].

Here, we shall limit ourselves to illustrate its consequences. In Fig. 4.2, we show the curves $\Delta = \Delta_{r,s}$ for the levels $n = 2, 3, 4$. The discussion is

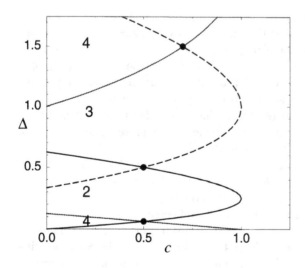

Fig. 4.2. Vanishing curves $\Delta = \Delta_{r,s}$ for the first four levels. The lines correspond to level 2 (full), level 3 (dashed) and level 4 (dotted). In the regions labeled 2,3,4, the determinant $\det_n(c, \Delta)$ with $n = 2, 3, 4$ is negative. The points mark the first intersections at $c = 1/2$ and $\Delta = 1/16, 1/2$ and $c = 7/10$ and $\Delta = 3/2$

---

[8] It is easily checked that $\Delta_{1,2} = \Delta_-(c)$ and $\Delta_{2,1} = \Delta_+(c)$ as defined in (4.38).

now largely parallel to the one done for the case of level 2 above. For a given level n, the determinant $\det_n(c, \Delta)$ changes sign across the *vanishing curves* $\Delta = \Delta_{r,s}$ with $rs = n$. One identifies the regions labeled $2, 3, 4$ in Fig. 4.2 where a conformal theory is manifestly non-unitary, since $\det_n(c, \Delta) < 0$ in them. As before, for $c \geq 1$, no further constraints follow from the unitarity condition. This can be seen from the explicit form of the $\Delta_{r,s}$.

Of particular interest are the points where two vanishing curves cross. These points are called **first intersections**. At a first intersection, from a given primary operator ϕ one can construct at least two null operators of different levels. We recognize from Fig. 4.2 that up to level 4, first intersections occur for $c = 1/2$ at $\Delta = 1/16$ and $\Delta = 1/2$ and for $c = 7/10$ at $\Delta = 3/2$. In these cases, we see from (4.42) that $m = 3, 4$, respectively.

Of course, one has now to investigate systematically in what regions in the (c, Δ) plane *unitary irreducible representations of the Virasoro algebra* (where all vectors have semi-positive norms) are possible. This problem was solved by Friedan, Qiu and Shenker [246] who showed that unitarity is only possible in one of the following two cases.

$c \geq 1$: unitarity only implies $\Delta \geq 0$. However, the number of primary operators is infinite.

$c < 1$: unitarity can be enforced for $c = 1 - 6/m(m+1)$ with $m \in \mathbb{N}$ starting from $m = 2$. In this case, we have a *finite* number of primary operators with conformal weights $\Delta_{r,s}$ with $1 \leq r \leq m - 1$ and $1 \leq s \leq m$. The corresponding theory is called a **unitary minimal model**. Notice that each operator is counted twice here because of the symmetry $\Delta_{r,s} = \Delta_{m-r,m+1-s}$. It is remarkable that only the points which correspond to first intersections can correspond to primary operators in an unitary theory at all.

The infinite sequence of unitary conformal models which corresponds to the values $m = 2, 3, 4, 5, \ldots$ is called the **unitary minimal series**. For each value of m, the table of possible conformal weights $\Delta_{r,s}$ with $r = 1, \ldots, m - 1$ and $s = 1, \ldots, m$ is called the **Kac table**. Let us discuss the first few examples. With $m = 2$, one has $c = 0$ and the only unitary representation is the vacuum one with conformal weight $\Delta_{1,1} = \Delta_{1,2} = 0$. However, one sees from computing $\det_2(0, 0)$ that there is a double zero. This is equivalent to say that in this case only the trivial one-dimensional representation $|0\rangle$ exists. That means $T = 0$, not a very exciting theory.

More interesting is the case $m = 3$, where $c = 1/2$, shown in Table 4.1. Here, one has the vacuum representation and two non-trivial representations. There are $|\Delta_{1,2}\rangle$ with $\Delta_{1,2} = \Delta_{2,2} = 1/16$ and $|\Delta_{1,3}\rangle$ with $\Delta_{1,3} = \Delta_{2,1} = 1/2$. The observed equalities between the $\Delta_{r,s}$ are reflected in Fig. 4.2. We see that the first intersection at $c = 1/2$, $\Delta = 1/16$ comes from the intersection of the vanishing curves of level 2 and of level 4. This implies that the corresponding primary operator ϕ must satisfy *two* null vector conditions, one of level 2 (which we have shown above to be $\Phi^{(2)} = 0$ with $\Phi^{(2)}$ defined in (4.16)) and

Table 4.1. Kac table for the Ising model ($m = 3$)

r

2	1/2	1/16	0
1	0	1/16	1/2
	1	2	3	s

a similar one of level 4. In the same way, the first intersection at $c = 1/2$, $\Delta = 1/2$ comes from the intersection of the vanishing curves of level 2 and of level 3. This implies that the corresponding primary operator ϕ, which serves as the highest weight state from which the unitary representation is built, must satisfy two null vector conditions, one of level 2 and one of level 3. Similar remarks apply to any first intersection. Finally, in all models of the unitary minimal series, $\Delta_{1,1} = 0$, which suggests the identification $\phi_{1,1} = 1$ throughout.

We may tentatively identify the $m = 3$ unitary conformal model with the *Ising model* where $|\Delta_{1,2}\rangle$ corresponds to the holomorphic part of the spin-density operator (recall that $x_\sigma = 1/8 = 2\Delta_{1,2}$) and $|\Delta_{1,3}\rangle$ represents the holomorphic part of the energy-density operator (since $x_\epsilon = 1 = 2\Delta_{1,3}$). That amounts to the identifications [68]

$$\sigma = \phi_{1,2} = \phi_{2,2} \ , \quad \epsilon = \phi_{2,1} = \phi_{1,3} \ ; \quad \text{(for } m = 3\text{)}. \tag{4.44}$$

Because of the relations (1.41), these results are in agreement with the values of the conventional critical exponents $\alpha = 0, \beta = 1/8$ and $\nu = 1$ known from the exact solution of the Ising model. In Chap. 3, we have already shown how to calculate the central charge from the finite-size scaling of the free energy density of a lattice model and we shall present the complete calculation of c for the Ising model in Chap. 10, thereby confirming that indeed $c = 1/2$.

The *three-state Potts model* is obtained for $m = 5$ ($c = 4/5$), together with the A_5-RSOS model. The respective Kac table is given in Table 4.2. Here, it turns out that only part of the Kac table is connected to the physical scaling operators which actually occur in the three-states Potts model, see Chap. 12.[9] Experimenting with the known critical exponents of the three-states Potts model (see Chap. 1) suggests the following tentative identification of the order parameter σ and the energy density ϵ, assumed to be scalars under rotation

$$\sigma = \phi_{3,2} = \phi_{3,3} \ , \quad \epsilon = \phi_{2,1} = \phi_{3,5} \ ; \quad \text{(for } m = 5, \text{ three-states Potts model)}. \tag{4.45}$$

---

[9] This can be also understood from the operator product algebra discussed in Chaps. 5–7. It can be shown that the primary scaling operators highlighted in Table 4.2 for $m = 5$ form a subalgebra, see especially (7.51).

Table 4.2. Kac table for the A_5 RSOS model ($m = 5$). The boxes indicate the sub-table corresponding to the three-state Potts model

r

4	3	13/8	2/3	1/8	0
3	7/5	21/40	1/15	1/40	2/5
2	2/5	1/40	1/15	21/40	7/5
1	0	1/8	2/3	13/8	3
	1	2	3	4	5 s

From this, one would deduce $x_\sigma = 2/15$ and $x_\epsilon = 4/5$ and through (1.41), the conventional critical exponents $\alpha = 1/3$, $\beta = 1/9$ and $\nu = 5/6$ are correctly reproduced.

The full operator content of the $m = 5$ case is realized for the A_5 RSOS model. Here, the most relevant scaling operator is $\phi_{2,2} = \phi_{3,4}$ which should be identified with the order parameter σ, leading to $x_\sigma = 1/40$.[10] This example shows that the value of the central charge c alone does not yet specify the universality class completely, but that extra information on the operator content is needed.

We close this discussion by remarking that the four-state Potts model is reached in the limit $m \to \infty$. In fact, all critical A_m-RSOS models yield examples of unitary conformal theories for all values of m permitted by the Kac formula, see Chap. 11.

Notice that the Kac formula itself applies for *any* value of m. The interesting case when m is *rational* defines the **minimal models**. They are always non-unitary, unless m is an integer ≥ 2, but are nevertheless characterized by a *finite* number of primary operators. Suppose $m = p/(p' - p)$, p and p' coprime, then the irreducible highest weight representations are given in the *Kac table* where $\Delta_{r,s} = \Delta_{p-r,p'-s}$ and $1 \leq r \leq p-1$, $1 \leq s \leq p'-1$. For example, the non-unitary Yang–Lee model corresponds to $p = 2$, $p' = 5$. Therefore $c = -22/5$ and the two distinct primary operators have highest weights $\Delta_{1,1} = 0$ and $\Delta_{1,2} = -1/5$. We shall denote the minimal models by $\mathcal{M}_{p,p'}$. The unitary minimal series corresponds to $\mathcal{M}_{m,m+1}$.

Although we did not discuss the second Verma module generated by $\bar{L}$, it is clear that $\bar{c} = c$ because the two components of the energy–momentum

---

[10] A physical realization is provided through *tetracritical* points in diluted Ising models.

tensor must behave like the complex conjugates of each other in a physical theory. However, physical operators can be a mixture of non-equivalent holomorphic and anti-holomorphic representations. For example, this is the case for $m = 5$ where we have two different physical models: the RSOS model and the three-state Potts model. The determination of physical operators is related to the *monodromy problem* to be discussed in Chap. 5.

In conclusion, we have seen how the requirement of unitarity restricts the possible values of the central charge c and of the conformal weights through the Kac table. Several different physical models may be realized for a given value of the central charge. The conformal weights of all unitary minimal models must belong to the set defined by the Kac formula (4.43). So far, this has only been a *necessary* criterion for unitarity. Conversely, unitary conformal field theories can be constructed for each integer value $m \geq 3$ via the so-called *coset construction*, as shown by Goddard, Kent and Olive [271]. Lattice realizations of these are provided by the A_m and D_m RSOS models introduced in Chap. 1. That means that the Kac formula gives *a complete and exhaustive classification of all unitary conformal field theories* with central charge $c < 1$.

4.5 Minimal Characters

The characters of the degenerate representations give the content in secondary operators, level by level. They were explicitly obtained by Rocha–Caridi [539]. The Kac determinant formula (4.41) determines more than the values of Δ for the highest weight representations. It also provides the structure of null-vectors. Begin with a primary operator of conformal weight $\Delta_{r,s}$. We know there is a null-vector at level rs with conformal weight $\Delta_{r,s} + rs = \Delta_{r+m,m+1-s}$. Because of the symmetry $\Delta_{r,s} = \Delta_{m-r,m+1-s}$, there is a second null-vector[11] at level $(m - r)(m + 1 - s)$ with conformal weight $\Delta_{r,s} + (m - r)(m + 1 - s) = \Delta_{2m-r,s}$. Both null-vectors are primary and their Kac determinants can be computed with the Kac formula. Up to now, we only discussed the consequences of one null-vector. The existence of an intricated pattern of null-vectors increases the complexity of the Verma module. However, the number of non-trivial vectors is reduced and we have to analyse the Kac formula to determine what is left from the generic Verma module.

A careful study [539] shows that the corresponding null-modules of the two null-vectors intersect and therefore they have some common descendants. The intersection can be decomposed into two new Verma null-submodules the primary highest weights of which are of course new primary null-vectors. These latter two null-vectors have once again an intersection given by new

---

[11] For the first few examples, this is illustrated in Fig. 4.2, since the allowed conformal weights correspond to first intersections of two vanishing curves of different levels.

null-submodules. The same pattern takes place step by step. The sequence of inclusions is depicted in Fig. 4.3 [539]. At the top, we indicate by r, s the highest weight primary vector with conformal weight $\Delta_{r,s}$. It has two null-vectors indexed by $2m - r, s$ and $r + m, m + 1 - s$. Both null-vectors intersect at the null-vectors $3m - r, m + 1 - s$ and $r + 2m, s$. Once again these new null-vectors intersect at new null-vectors. The never-ending structure of null submodules of $|\Delta_{r,s}\rangle$ can be computed recursively with the Kac formula.

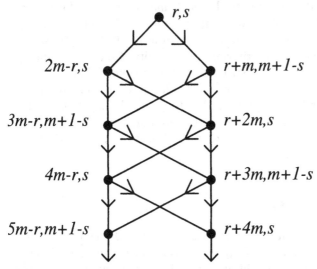

Fig. 4.3. Sequence of Verma submodules with conformal weights $\Delta_{a,b}$ where a, b are listed at the sides of the graph ([169] after [539])

The character of the degenerate Verma module tells us how many non-trivial vectors there are at a given level. The generic formula (4.18) gives the content of a non-degenerate Verma module. Each null-module has basically the structure of a non-degenerate module. The fact that they intersect in a complicated way has to be taken into account.

From the generic character $\chi(\Delta_{r,s}, c)$ computed in (4.18), one must subtract $\chi(\Delta_{2m-r,s}, c)$ and $\chi(\Delta_{r+m,m+1-s}, c)$ to discard in the counting of independent vectors the contributions from the null-vectors. However, while doing this, we subtracted twice the content of the intersection of the submodules with highest weights $\Delta_{3m-r,m+1-s}$ and $\Delta_{r+2m,s}$. To compensate, we have to add $\chi(\Delta_{3m-r,m+1-s}, c)$ and $\chi(\Delta_{r+2m,s}, c)$. So far, one has

$$\chi(\Delta_{r,s}, c) - \chi(\Delta_{2m-r,s}, c) - \chi(\Delta_{r+m,m+1-s}, c) \qquad (4.46)$$
$$+ \chi(\Delta_{3m-r,m+1-s}, c) + \chi(\Delta_{r+2m,s}, c).$$

From Fig. 4.3, one may read off where the two last submodules considered will intersect and this means that this intersection has now been counted twice. The submodules describing this intersection have to be subtracted.

This procedure must be repeated infinitely many times. The character of the degenerate Verma module becomes an infinite series of generic characters. In Fig. 4.3, one recognizes four kinds of series of indices and one arrives at the **Rocha–Caridi formula** for the character [539]

$$\chi_{r,s}^{(m)} = \chi(\Delta_{r,s},c) - \sum_{k=1}^{\infty} \left[\chi(\Delta_{2km-r,s},c) + \chi(\Delta_{r+m(2k-1),m+1-s},c) \right.$$

$$\left. - \chi(\Delta_{(2k+1)m-r,m+1-s},c) - \chi(\Delta_{r+2km,s},c) \right]$$

$$= \sum_{k=-\infty}^{+\infty} \left[\chi(\Delta_{r+2km,s},c) - \chi(\Delta_{r+m(2k-1),m+1-s},c) \right]. \tag{4.47}$$

Notice how positive and negative contributions to the character were written in the last line of (4.47). It is only a question of practical purpose whether to choose the expression with a summation over positive values of k or the more compact expression with k running from $-\infty$ to $+\infty$. All generic characters essentially differ by their leading terms $q^{\Delta - c/24}$, which can be factorized out. One may rewrite (4.47) in the form

$$\chi_{r,s}^{(m)} = q^{\Delta_{r,s} - \frac{c(m)}{24}} \prod_{i=1}^{\infty} \frac{1}{1-q^i}$$

$$\times \left(1 - \sum_{k=1}^{\infty} \left[q^{(km-r)(k(m+1)-s)} + q^{((k-1)m+r)((k-1)(m+1)+s)} \right. \right.$$

$$\left. \left. - q^{k(km(m+1)-(m+1)r+ms)} - q^{k(km(m+1)+(m+1)r-ms)} \right] \right)$$

$$= q^{\Delta_{r,s} - \frac{c(m)}{24}} \prod_{i=1}^{\infty} \frac{1}{1-q^i}$$

$$\times \sum_{k=-\infty}^{+\infty} \left[q^{k(km(m+1)+(m+1)r-ms)} - q^{((k-1)m+r)((k-1)(m+1)+s)} \right]. \tag{4.48}$$

The summation integer k is the same as in (4.47). The dependence of the character of the central charge becomes important when studying the modular properties of the characters. In Chap. 11, we shall describe how the different characters of a minimal model build up a representation of the modular group. They are therefore strongly related to each other. For the practitioner more important is perhaps how to expand $\chi_{r,s}$ in powers of q in order to obtain the number of secondary states at a certain level. A simple scheme to do this will be presented in Chap. 10.

The character formulas for the general minimal models $\mathcal{M}_{p,p'}$ are very similar to the unitary characters. It is sufficient to systematically replace the values m and $m+1$ by p and p' respectively.

Exercises

1. Starting from the Ward identity for the scaling operator $(\partial\phi)(z)$ (see Chap. 2), where ϕ is primary, find $L_n(z_1)(\partial\phi)(z)$. Check against the Virasoro commutator.

2. Show that the operators (4.16,4.17) are the only primary operators at levels two and three, respectively. How many independent primary operators exist at level four?

3. Find the Kac table for $m = 4$. Assuming that all scaling operators are scalars, try to identify a physical system described by it.

4. Considering only excited states up to level 3, verify the Kac formula directly.

5. Correlators, Null Vectors and Operator Algebra

In the previous chapter, we have seen that null vectors play an essential role in the characterization of minimal conformal models. However, they also lead to more tangible consequences. In particular, one may derive from them linear partial differential equations the solution of which determines the four-point functions. Physical operators, however, have holomorphic and antiholomorphic contributions. These parts are connected because the coordinates z and $\bar{z}$ are complex conjugates of each other. Physically sensible correlation functions of scalar operators ($\Delta = \overline{\Delta}$) must be single-valued functions of z. Correlation functions of operators with non-trivial spins $s = \Delta - \overline{\Delta}$ may pick up a phase under the exchange of two operators. However such a change must be consistent all over the possible physical n-point functions. This constitutes the monodromy problem for correlation functions and is the the subject of this chapter. Here, a particular property of two-dimensional conformal theories will become very helpful, namely the **operator product algebra** [619]. This is an extension of the short-distance expansion of the product of two operators to conformal theories. The particular features of the latter theories results in the existence of finite algebras of operators whose correlation functions can be exactly computed.

The technique was formulated in [68]. The presentation given here is at a more elementary level, following [169].

5.1 Null Vectors and Correlation Functions

We have seen that because of orthogonality, the Verma module is reduced by the null submodule which is orthogonal to all other vectors. We consider now the impact on the correlation functions. It is clear that a correlation function $\langle (\phi_1 \dots \phi_k)\phi \rangle$ will be computed from the projection of $|\Delta_1\rangle \otimes \cdots \otimes |\Delta_n\rangle$ onto the vector $|\Delta\rangle$ related to the primary operator ϕ. This can be seen through a formal tensor product of primary operators. Consider a null primary operator $\Phi^{(2)}$ of level two. Certainly

$$\langle \phi_1(z_1) \dots \phi_k(z_k)\Phi^{(2)}(z) \rangle = 0. \tag{5.1}$$

This introduces a constraint which can be expressed as a differential operator acting on $\langle \phi_1 \dots \phi_n \phi \rangle$ which can be solved. To do so, we first have to relate a correlation function containing a secondary operator like $L_{-p}\phi(z)$ to a correlator with primary operators only. By definition

$$\langle\phi_1(z_1)\ldots\phi_k(z_k)L_{-p}\phi(z)\rangle$$
$$= \frac{1}{2\pi i}\oint_{C_z} dz'(z'-z)^{-p+1}\langle\phi_1(z_1)\ldots\phi_k(z_k)T(z')\phi(z)\rangle. \qquad (5.2)$$

The contour C_z is analytically deformed into the combination of a large contour C surrounding all the points and k smaller ones C_i surrounding each point z_i ($i = 1\ldots k$), as shown in Fig. 5.1. Let us first discuss the integration

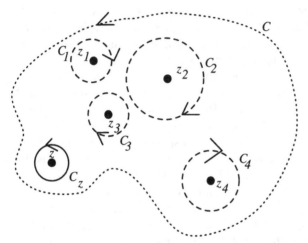

Fig. 5.1. Deformation of an oriented contour: $\oint_{C_z} = \oint_C - \sum_i \oint_{C_i}$ [169]

over the large contour C. This integral is very similar to the one in (2.50) from which we obtained the Ward identities. The Ward identities hold for any infinitesimal deformation $z \to z + \epsilon$. However, we did not discuss further the actual value of the integral. Suppose $\epsilon(z')$ behaves at large distance as $(z')^{-p+1}$ like in (5.2). We know that $T(z')$ behaves like $(z')^{-4}$ asymptotically. Therefore the integrand in (5.2) will behave as $(z')^{-p-3}$ at large distances. The integrand is therefore regular at infinity if $p \geq -1$ for which the integral over C vanishes. This is the case we are interested in because null-vectors are built up with L_{-p} with $p > 0$. Therefore the integral along C is zero and the residue of the integral along C_z is minus the sum of the integrals along each C_i. These other residues are found by introducing the Ward identity (2.51) in (5.2). Then

$$\langle\phi_1(z_1)\ldots\phi_k(z_k)L_{-p}\phi(z)\rangle = -\frac{1}{2\pi i}\sum_{i=1}^{k}\oint_{C_i} dz'(z'-z)^{-p+1}$$
$$\times \left(\frac{\Delta_i}{(z'-z_i)^2} + \frac{1}{z'-z_i}\partial_{z_i}\right)\langle\phi_1(z_1)\ldots\phi_k(z_k)\phi(z)\rangle. \qquad (5.3)$$

This can be integrated out and we define the differential operator $\mathcal{L}_{-p}$ by

$$\langle \phi_1(z_1)\ldots\phi_k(z_k)L_{-p}\phi(z)\rangle = \mathcal{L}_{-p}\langle\phi_1(z_1)\ldots\phi_k(z_k)\phi(z)\rangle \qquad (5.4)$$

with

$$\mathcal{L}_{-p} = \sum_{i=1}^{k}\left(\frac{(p-1)\Delta_i}{(z_i-z)^p} - \frac{1}{(z_i-z)^{p-1}}\partial_{z_i}\right). \qquad (5.5)$$

The expression $\langle\phi_1\ldots\phi_k L_{-p_r}\ldots L_{-p_1}\phi(z)\rangle$ is treated recursively in the same way. We first deform the contour corresponding to L_{-p_r}, i.e. the one surrounding all the others. The poles between the different energy–momentum tensors do not contribute. We successively deform the contours and compute the residues until we have done this for the most interior one related to L_{-p_1}. The result is

$$\langle\phi_1\ldots\phi_k L_{-p_r}\ldots L_{-p_1}\phi(z)\rangle = \mathcal{L}_{-p_1}\cdots\mathcal{L}_{-p_r}\langle\phi_1\ldots\phi_k\phi(z)\rangle. \qquad (5.6)$$

Each $\mathcal{L}$ acts on everything placed to its right. The central charge does not appear because it comes from poles which are not considered with the new set of contours $\{C_i\}$. The c-dependence is hidden in the correlation function of the primary operators. Notice that it is sometimes interesting to further work out $\mathcal{L}$. For example

$$\mathcal{L}_{-1} = -(\partial_{z_1} + \cdots + \partial_{z_k}) = \partial_z \qquad (5.7)$$

as a consequence of the projective Ward identities (2.37).

We are now ready to return to the null-vector problem (5.1). We have proven that the correlation function of a descendant operator with several primary ones can be expressed as a differential operator. Consequently, a correlation function with a null vector like (5.1) can be expressed as a differential equation acting on the correlation function of the primary operators. In this case, we have

$$\left(\mathcal{L}_{-2} - \frac{3}{2(2\Delta+1)}\partial_z^2\right)\langle\phi_1(z_1)\ldots\phi_k(z_k)\phi(z)\rangle =$$

$$\left[\sum_{i=1}^{k}\left(\frac{1}{z-z_i}\partial_{z_i} + \frac{\Delta_i}{(z-z_i)^2}\right) - \frac{3}{2(2\Delta+1)}\partial_z^2\right]\langle\phi_1(z_1)\ldots\phi_k(z_k)\phi(z)\rangle = 0. \qquad (5.8)$$

It is easy to verify that this differential equation is compatible with the two and three-point functions. More important is the fact that this equation provides solutions to the four-point function. We introduce the projective invariant $\eta = (z-z_1)(z_2-z_3)/(z-z_3)(z_2-z_1)$ and write

$$\langle\phi_1(z_1)\phi_2(z_2)\phi_3(z_3)\phi(z)\rangle = \prod_{i=1}^{3}(z-z_i)^{-\gamma_{0i}}\prod_{i<j}(z_i-z_j)^{-\gamma_{ij}}F(\eta). \qquad (5.9)$$

In Chap. 2, the general constraint on the exponents γ was given by equation (2.41). Now, we choose

$$\begin{aligned}
\gamma_{01} &= \gamma_{02} = 0, & \gamma_{03} &= 2\Delta, & \gamma_{12} &= +\Delta + \Delta_1 + \Delta_2 - \Delta_3, \\
\gamma_{13} &= -\Delta + \Delta_1 - \Delta_2 + \Delta_3, & \gamma_{23} &= -\Delta - \Delta_1 + \Delta_2 + \Delta_3.
\end{aligned}$$
(5.10)

Then $F(\eta)$ can be defined through the limit

$$F(\eta) := \lim_{\xi \to \infty} \xi^{2\Delta_3} \langle \phi_1(0)\phi_2(1)\phi_3(\xi)\phi(\eta) \rangle$$
(5.11)

and it follows that $F(\eta)$ must be a solution of the differential equation

$$\left[-\frac{3}{2(2\Delta+1)}\partial_\eta^2 - \left(\frac{1}{\eta} - \frac{1}{1-\eta}\right)\partial_\eta \right.$$
$$\left. + \frac{\Delta_1}{\eta^2} + \frac{\Delta + \Delta_1 + \Delta_2 - \Delta_3}{\eta(1-\eta)} + \frac{\Delta_2}{(1-\eta)^2} \right] F(\eta) = 0.$$
(5.12)

As shown for a level-two null vector, any null-vector of degenerate Verma module implies solvable differential equations for the correlation functions. A differential equation like (5.8) will have several distinct solutions (for (5.12) one expects two, i.e. F_1 and F_2). On the other hand, the primary operator $\phi(z, \bar{z})$ also has a null vector which constrains the $\bar{z}$-dependence. The most general solution for the full four-point function is therefore given by

$$\langle \phi_1(z_1, \bar{z}_1)\phi_2(z_2, \bar{z}_2)\phi_3(z_3, \bar{z}_3)\phi(z, \bar{z}) \rangle =$$
$$\prod_{i=1}^{3} (z - z_i)^{-\gamma_{0i}}(\bar{z} - \bar{z}_i)^{-\bar{\gamma}_{0i}} \prod_{i<j} (z_i - z_j)^{-\gamma_{ij}}(\bar{z}_i - \bar{z}_j)^{-\bar{\gamma}_{ij}} \sum_{k,l} \mathcal{N}_{kl} F_k(\eta)\bar{F}_l(\bar{\eta}).$$
(5.13)

The physical four-point function is of course evaluated for Euclidean coordinates where $\bar{z}$ is the complex conjugate of z. Invariance under physical constraints, such as rotation invariance, implies strong constraints on the coefficients $\mathcal{N}_{kl}$. The problem of solving these constraints constitutes the *monodromy problem* which will be discussed in the rest of this chapter.

5.2 Operator Algebra and Associativity

We have seen how the two Virasoro algebras, which reflect the z and $\bar{z}$ behaviour of the theory, can be studied independently. Physical primary operators formally decompose as

$$\phi(z, \bar{z}) \approx \phi(z) \otimes \bar{\phi}(\bar{z}).$$
(5.14)

The tensor product of highest weight states of each Virasoro algebra is exactly the heighest weight state of a physical primary operators

$$|\Delta, \overline{\Delta}\rangle = |\Delta\rangle \otimes |\overline{\Delta}\rangle. \tag{5.15}$$

The decomposition of physical operators (5.14) is more rigorous for the (secondary) energy–momentum tensor where we can actually write $T(z) = T(z) \otimes \overline{\mathbf{1}}$ and $\overline{T}(\bar{z}) = \mathbf{1} \otimes \overline{T}(\bar{z})$. However, the energy–momentum tensor is not the only operator which generates an algebra. In fact, a new ingredient of conformal field theories is needed:

Operator Product Algebra Hypothesis [68]: *The set of operators of the theory is complete in the sense that the product of any pair of operators can be expressed as the convergent infinite sum*

$$A_i(z, \bar{z}) A_j(0, 0) = \sum_k C_{i,j}^k(z, \bar{z}) A_k(0, 0). \tag{5.16}$$

The operator product coefficients $C_{i,j}^k(z, \bar{z})$ must be single-valued functions to ensure well-defined local properties of correlation functions

$$\langle A_i(z, \bar{z}) A_j(0, 0) X \rangle = \sum_k C_{i,j}^k(z, \bar{z}) \langle A_k(0, 0) X \rangle, \tag{5.17}$$

where X stands for any functional of local operators $\{A_s(z_s, \bar{z}_s)\}$ of the theory. This is the last important ingredient needed for the construction of conformal field theories.

The operator algebra hypothesis extend Wilson's [619] **Operator Product Expansion** (OPE) by adding stronger properties. It is assumed that such a sum is convergent for a certain range of the local coordinates. Using translation invariance, we can reduce a general OPE, where two arbitrary points appear, to (5.16). All operators of conformal field theory are primary operators or descendants of them. The OPE of primary operators is

$$\phi_i(z, \bar{z}) \phi_j(0, 0) = \sum_k \sum_{(n_1, \dots, n_p)} \sum_{(\bar{n}_1, \dots, \bar{n}_p)} \mathbf{C}_{i,j}^{k;\,(n_1, \dots, n_p; \bar{n}_1, \dots, \bar{n}_p)}$$

$$\times z^{\Delta_k - \Delta_i - \Delta_j + \sum_s n_s} \bar{z}^{\bar{\Delta}_k - \bar{\Delta}_i - \bar{\Delta}_j + \sum_s \bar{n}_s} \phi_k^{(-n_1, \dots, -n_p; -\bar{n}_1, \dots, -\bar{n}_p)}(0, 0). \tag{5.18}$$

The operators on the r.h.s. of (5.18) are descendant operators of the primary operator ϕ_k.

$$\phi_k^{(-n_1, \dots, -n_p; -\bar{n}_1, \dots, -\bar{n}_p)}(0, 0) = L_{-n_1} \cdots L_{-n_p} \overline{L}_{-\bar{n}_1} \cdots \overline{L}_{-\bar{n}_p} |\Delta_k, \overline{\Delta}_k\rangle. \tag{5.19}$$

The powers in z or $\bar{z}$ result from dimensional considerations. The operator product coefficients $\mathbf{C}_{i,j}^{k;\,(n_1, \dots, n_p; \bar{n}_1, \dots, \bar{n}_p)}$ are now real or complex numbers. It may seem untractable to compute this infinite number of coefficients. This is fortunately not the case. Actually, without any restriction of generality

$$\mathbf{C}_{i,j}^{k;\,(n_1, \dots, n_p; \bar{n}_1, \dots, \bar{n}_p)} = \mathbf{C}_{i,j}^k \, \omega_{i,j}^{k;\,(n_1, \dots, n_p)} \, \bar{\omega}_{i,j}^{k;\,(\bar{n}_1, \dots, \bar{n}_p)}, \tag{5.20}$$

where $\mathbf{C}_{i,j}^k$ is the coupling to the primary operator ϕ_k. The corrections to this value due to descendant operators factorizes in the two ω-terms. This is the consequence of the full decomposition of the energy–momentum tensor into holomorphic and anti-holomorphic parts. The coefficients ω and $\bar{\omega}$, though tricky to compute, are fixed by conformal invariance. To show this, notice that the coefficient $\mathbf{C}_{i,j}^k$ appears in the three-point function in the limit (comparison with (2.38) implies that $\mathcal{C}_{ijk} = \mathbf{C}_{i,j}^k$, check it)

$$\lim_{\xi,\bar{\xi}\to\infty} \xi^{2\Delta_k}\bar{\xi}^{2\overline{\Delta}_k}\langle\phi_k(\xi,\bar{\xi})\phi_i(1,1)\phi_j(0,0)\rangle = \langle\Delta_k,\overline{\Delta}_k|\phi_i(1,1)|\Delta_j,\overline{\Delta}_j\rangle = \mathbf{C}_{i,j}^k.$$
(5.21)

Suppose that we take a similar limit with a descendant operator of ϕ_k. We recall the differential equation (5.6) to relate such a three-point function to (5.21). Obviously the result will be proportional to $\mathbf{C}_{i,j}^k$. The factorization of the z and $\bar{z}$-dependent differential operators $\mathcal{L}_{-n}$ and $\overline{\mathcal{L}}_{-\bar{n}}$ implies the factorization of the other terms, proportional to ω, $\bar{\omega}$, and to the norm of the secondary operator. This argument is sufficient to prove that $\{\omega,\bar{\omega}\}$ are fixed by conformal invariance. A practical way to calculate them explicitly can be found in Appendix B of [68]. However, it is in most cases not necessary to compute them.

The OPE of a primary operator with itself couples to the identity. This is the same as writing

$$\langle\Delta_i,\overline{\Delta}_i|\Delta_j,\overline{\Delta}_j\rangle = \mathbf{C}_{i,j}^0 = \delta_{i,j}.$$
(5.22)

We assume here that we deal with real operators. A generalization to operators which appear in complex conjugate pairs is obvious. The same care must be applied to $\mathbf{C}_{i,j}^k$ which is completely symmetric only for real operators. Besides the central charge and the critical exponents, the expansion coefficient $\mathbf{C}_{i,j}^k$ are the last fundamental parameters. They are not determined by conformal invariance only. A more popular way to write the OPE is the formal expression

$$\phi_i \cdot \phi_j = N_{i,j}^k \phi_k,$$
(5.23)

where $N_{i,j}^k$ is zero if $\mathbf{C}_{i,j}^k$ is zero and an integer if $\mathbf{C}_{i,j}^k \neq 0$. This integer is in general one for minimal models. However, there are models where an operator is degenerate and $N_{i,j}^k$ is related to this degeneracy. In particular in models with a continuous symmetry, where each operator carries a representation of the symmetry algebra, the OPE is related to the irreducible decomposition of the tensor product of representations and $N_{i,j}^k$ counts the multiplicity of a representation in the tensor product.

We have seen in Chap. 2 some examples of OPE. The Ward identities state what appears in the OPE of the secondary operator $T(z)$ with a primary operator or with itself (see (2.51) and (2.61)). These examples can be summarized by

$$\mathbf{1}\cdot\phi = \phi, \qquad \mathbf{1}\cdot\mathbf{1} = \mathbf{1}.$$
(5.24)

The determination of the OPE coefficients is fundamental for solving all correlation functions. Two and three-point functions are unsufficient to determine them. The required information is given by the solutions to the four-point functions. The condition that four-point functions must be analytically well-defined means that we have to discuss their properties under analytic continuations in their variables z and $\bar{z}$. For the sake of simplicity and to avoid redundant variables, it is better to study the analytical properties of four-point functions expressed with the projectively invariant quantities $\eta = (z_j - z_i)(z_k - z_l)/(z_k - z_i)(z_j - z_l)$ and $\bar{\eta} = (\bar{z}_j - \bar{z}_i)(\bar{z}_k - \bar{z}_l)/(\bar{z}_k - \bar{z}_i)(\bar{z}_j - \bar{z}_l)$. We make use of (5.11) to define

$$G_{i,j}^{l,k}(\eta, \bar{\eta}) := \lim_{\xi, \bar{\xi} \to \infty} \xi^{2\Delta_l} \bar{\xi}^{2\overline{\Delta}_l} \langle \phi_l(\xi, \bar{\xi}) \phi_k(1,1) \phi_j(\eta, \bar{\eta}) \phi_i(0,0) \rangle$$

$$= \langle \Delta_l, \overline{\Delta}_l | \phi_k(1,1) \phi_j(\eta, \bar{\eta}) | \Delta_i, \overline{\Delta}_i \rangle. \tag{5.25}$$

This function will be very useful to discuss the neighbourhood of $\eta = \bar{\eta} = 0$. The choice made in defining $G_{i,j}^{l,k}$ is of course arbitrary. It is useful to discuss other choices and to relate them to each other. For example we can introduce $\eta' = (z_j - z_k)(z_i - z_l)/(z_i - z_k)(z_j - z_l) = 1 - \eta$ and its conjugate $\bar{\eta}' = 1 - \bar{\eta}$. This is the same as evaluating ϕ_k at 0 and ϕ_i at 1 or permuting both. The limit taken in (5.25) defines in a similar manner a new function

$$G_{k,j}^{l,i}(1 - \eta, 1 - \bar{\eta}) = \langle \Delta_l, \overline{\Delta}_l | \phi_i(1,1) \phi_j(1 - \eta, 1 - \bar{\eta}) | \Delta_k, \overline{\Delta}_k \rangle \tag{5.26}$$

related to the previous one by

$$G_{k,j}^{l,i}(1 - \eta, 1 - \bar{\eta}) = G_{i,j}^{l,k}(\eta, \bar{\eta}). \tag{5.27}$$

This second choice modifies the choice of the asymptotic states $|\Delta, \overline{\Delta}\rangle$. We obviously have a third choice with the projective transformation which permutes 0 and ∞. It is used to define a third function of the projective invariant parameters $(1/\eta, 1/\bar{\eta})$

$$G_{l,j}^{i,k}\left(\frac{1}{\eta}, \frac{1}{\bar{\eta}}\right) = \langle \Delta_i, \overline{\Delta}_i | \phi_k(1,1) \phi_j(1/\eta, 1/\bar{\eta}) | \Delta_l, \overline{\Delta}_l \rangle. \tag{5.28}$$

This last function G is related to the first one by

$$\eta^{-2\Delta_j} \bar{\eta}^{-2\overline{\Delta}_j} G_{l,j}^{i,k}\left(\frac{1}{\eta}, \frac{1}{\bar{\eta}}\right) = G_{i,j}^{l,k}(\eta, \bar{\eta}). \tag{5.29}$$

The two formulas (5.27) and (5.29) are called **crossing conditions**. The prefactor on the l.h.s. of (5.29) is due to the particular definition of the asymptotic state at infinity (see (4.30)). Each of the three functions G is completely equivalent to the same four-point function. However, each of them will be most useful to discuss the specific singular behaviours as two of the four points get closer. This occurs in the formalism with the variables $(\eta, \bar{\eta})$ as $\eta = \bar{\eta} \to 0, 1$ and ∞.

We can now replace $\phi_j(\eta, \bar{\eta})|\Delta_i, \overline{\Delta}_i\rangle$ in $G_{i,j}^{l,k}(\eta, \bar{\eta})$ by the OPE $\phi_j \cdot \phi_i = N_{i,j}^s \phi_s$. We obtain the infinite sum over three-point functions which can be written

$$G_{i,j}^{l,k}(\eta, \bar{\eta}) = \sum_s \mathbf{C}_{k,s}^l \mathbf{C}_{j,i}^s \mathcal{A}_{i,j}^{l,k}(s; \eta, \bar{\eta}). \qquad (5.30)$$

We define here the functions $\mathcal{A}_{i,j}^{l,k}(s; \eta, \bar{\eta})$ which collect the contributions from the descendant operators of ϕ_s in the OPE $\phi_j \cdot \phi_i$. Due to the decoupling between T and $\overline{T}$, we can decompose $\mathcal{A}_{i,j}^{l,k}$ into holomorphic and anti-homomorphic parts $\mathcal{F}$ and $\overline{\mathcal{F}}$ called **conformal blocks**

$$\mathcal{A}_{i,j}^{l,k}(s; \eta, \bar{\eta}) = \mathcal{F}_{i,j}^{l,k}(s; \eta) \overline{\mathcal{F}}_{i,j}^{l,k}(s; \bar{\eta}). \qquad (5.31)$$

The same dimensional reasons as in (5.18) permit to write

$$\mathcal{F}_{i,j}^{l,k}(s; \eta) = \eta^{\Delta_s - \Delta_i - \Delta_j} F_{i,j}^{l,k}(s; \eta). \qquad (5.32)$$

where $F_{i,j}^{l,k}(s; \eta)$ is an infinite power series in η which takes into account the contributions of the holomorphic part of the descendant operators. The anti-holomorphic behaviour is similarly.

This reduction of a four-point function into a sum of three-point functions can be done with $G_{k,j}^{l,i}(1 - \eta, 1 - \bar{\eta})$ and the OPE $\phi_j \cdot \phi_k = N_{i,k}^t \phi_t$ for $\phi_j(1 - \eta, 1 - \bar{\eta})|\Delta_k, \overline{\Delta}_k\rangle$. We obtain a new decomposition

$$G_{k,j}^{l,i}(1 - \eta, 1 - \bar{\eta}) = \sum_t \mathbf{C}_{i,t}^l \mathbf{C}_{j,k}^t \mathcal{A}_{k,j}^{l,i}(t; 1 - \eta, 1 - \bar{\eta}). \qquad (5.33)$$

where

$$\mathcal{A}_{k,j}^{l,i}(t; 1 - \eta, 1 - \bar{\eta}) = \mathcal{F}_{k,j}^{l,i}(t; 1 - \eta) \overline{\mathcal{F}}_{k,j}^{l,i}(t; 1 - \bar{\eta}) \qquad (5.34)$$

and we can extract the asymptotic behaviour from the conformal blocks with

$$\mathcal{F}_{k,j}^{l,i}(t; 1 - \eta) = (1 - \eta)^{\Delta_t - \Delta_k - \Delta_j} F_{k,j}^{l,i}(t; 1 - \eta). \qquad (5.35)$$

The crossing condition (5.27) implies the equality

$$\sum_s \mathbf{C}_{k,s}^l \mathbf{C}_{j,i}^s \mathcal{A}_{i,j}^{l,k}(s; \eta, \bar{\eta}) = \sum_t \mathbf{C}_{i,t}^l \mathbf{C}_{j,k}^t \mathcal{A}_{k,j}^{l,i}(t; 1 - \eta, 1 - \bar{\eta}). \qquad (5.36)$$

It is on purpose that we chose the indices s and t for the functions $\mathcal{A}$. This is just the s and t-channel decomposition of scattering theory. Equation (5.36) is the *crossing symmetry condition* graphically depicted in Fig. 5.2. Three-point functions are described by the vertices and the two-point functions by propagators. An equivalent statement to the crossing symmetry condition is the requirement of **associativity** for the formal OPE

$$\sum_s N_{k,s}^l N_{j,i}^s = \sum_t N_{i,t}^l N_{j,k}^t. \qquad (5.37)$$

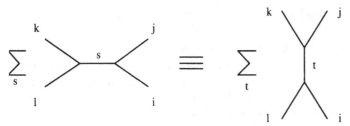

Fig. 5.2. Crossing symmetry condition for the four-point functions ([169] after [68])

We can introduce the OPE in any n-point function and replace it by a infinite sum over $(n-1)$-point functions. Introducing step by step OPEs into correlation functions results in multiple infinite sums which are graphically represented in Fig. 5.3. The choice of OPEs is arbitrary and therefore crossing symmetry is analogous to the duality condition in string theory. Figure 5.3 only shows one possible computation of a n-point function.

The implementation of the crossing condition for the four-point functions implies the duality of any n-point function. It is thus fundamental but sufficient to consistently solve all four-point functions of all primary operators.

Finally, one can also discuss the u-channel and introduce the OPE $\phi_j \cdot \phi_l = N_{j,l}^u \phi_u$ into $G_{l,j}^{i,k}(1/\eta, 1/\bar{\eta})$, leading the last decomposition

$$G_{l,j}^{i,k}\left(\frac{1}{\eta}, \frac{1}{\bar{\eta}}\right) = \sum_u \mathbf{C}_{k,u}^i \mathbf{C}_{j,l}^u \, \mathcal{A}_{l,j}^{i,k}\left(u; \frac{1}{\eta}, \frac{1}{\bar{\eta}}\right). \tag{5.38}$$

As before, we can decompose $\mathcal{A}_{l,j}^{i,k}$ into its conformal blocks

$$\mathcal{A}_{l,j}^{i,k}\left(u; \frac{1}{\eta}, \frac{1}{\bar{\eta}}\right) = \mathcal{F}_{l,j}^{i,k}\left(u; \frac{1}{\eta}\right) \overline{\mathcal{F}}_{l,j}^{i,k}\left(u; \frac{1}{\bar{\eta}}\right). \tag{5.39}$$

However, from the crossing condition (5.29), one has the particular asymptotic behaviour

$$\mathcal{F}_{l,j}^{i,k}\left(u; \frac{1}{\eta}\right) = \left(\frac{1}{\eta}\right)^{\Delta_u - \Delta_l + \Delta_j} F_{l,j}^{i,k}\left(u; \frac{1}{\eta}\right). \tag{5.40}$$

The possible crossing conditions can be straightforwardly derived.

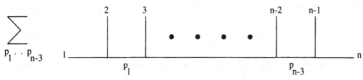

Fig. 5.3. A possible decomposition of a n-point function ([169] after [68])

5.3 Analyticity and the Monodromy Problem

The decomposition of a four-point function into its conformal blocks gives the necessary insight into the analytic structure of the correlation function. The conformal blocks have always a particularly simple behaviour under specific analytic continuations in their parameter η. Recall for example the holomorphic conformal blocks in the s-channel which have the structure

$$\mathcal{F}_{i,j}^{l,k}(s;\eta) = \eta^{\Delta_s - \Delta_i - \Delta_j}\, F_{i,j}^{l,k}(s;\eta). \tag{5.41}$$

The power series $F_{i,j}^{l,k}$ provides an explicit expansion of the correlation function in the neighbourhood of zero. The analytic continuation of $\mathcal{F}_{i,j}^{l,k}(s;\eta)$ along a curve surrounding the point zero (Fig. 5.4) modifies the conformal block by a phase $\Omega_{s,s}^{(0)} = e^{2\pi i(\Delta_s - \Delta_i - \Delta_j)}$. This is called a **monodromy transformation** [332, 333]. All conformal blocks $\{\mathcal{F}_{i,j}^{l,k}(s;\eta)\}_s$ in the s-channel are eigenvectors of the **monodromy matrix** associated with the monodromy transformation $\Omega^{(0)}$.

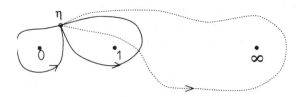

Fig. 5.4. Monodromy transformations: η moves around 0 and 1 (solid lines), and ∞ (dotted line) [169]

As a result, such an analytic continuation modifies $\mathcal{A}_{i,j}^{l,k}(s;\eta,\bar{\eta})$ by a phase term.

$$\mathcal{A}_{i,j}^{l,k}(s;\eta,\bar{\eta}) \xrightarrow{\Omega^{(0)}} e^{2\pi i\,(\Delta_s - \Delta_i - \Delta_j - \overline{\Delta}_s + \overline{\Delta}_i + \overline{\Delta}_j)}\mathcal{A}_{i,j}^{l,k}(s;\eta,\bar{\eta}). \tag{5.42}$$

The contributions from $\overline{\mathcal{F}}$ have a negative sign with respect to the ones from $\mathcal{F}$ because the corresponding complex conjugate monodromy transformations winds up around zero in the opposite direction. Different phase terms cause the correlation function to be ill-defined. A well-defined correlation function requires the s-dependent phases of $\mathcal{A}_{i,j}^{l,k}$ to be all identical (up to integer multiples of $2\pi i$).

Single-valued two-, three-, and four-point correlation functions need a trivial phase. This is the case if the spin of all primary operators are integer.

$$|\Delta_a - \overline{\Delta}_a| \in \mathbb{N}. \tag{5.43}$$

Physical theories where all primary operators are scalars or have integer spins are called **local**, as all their correlation functions are locally well-defined.

However, it is possible to relax somehow this statement because we know that statistical systems with particular boundary conditions have correlators which are multi-valued and gain a phase term under analytic continuation. This means that it is physically possible that a global s-independent phase term appears. However, such phase terms must be consistent throughout all possible correlation functions. This situation occurs in parafermionic sectors of the theory where fractional spins appear. It is however reasonable to investigate first the strict monodromy invariance because any physical model possesses a local sector which already contains the essence of the model.

Moreover, if the analytic continuation around zero are simple, the analytic continuation $\Omega^{(1)}$ of $\mathcal{F}_{i,j}^{l,k}(s;\eta)$ along a curve surrounding the point one (Fig. 5.4) is more complicated. This transformation is not diagonal but becomes the linear combination

$$\mathcal{F}_{i,j}^{l,k}(s;\eta) \xrightarrow{\Omega^{(1)}} \sum_{s'} \Omega_{s,s'}^{(1)} \mathcal{F}_{i,j}^{l,k}(s';\eta) \tag{5.44}$$

defining the monodromy matrix $\Omega^{(1)}$. The s-channel basis $\{\mathcal{F}_{i,j}^{l,k}(s;\eta)\}_s$ is not the appropriate basis to discuss this monodromy transformation. The t-channel basis of conformal block is more appropriate because it provides a natural expansion in the parameter $(1 - \eta)$ as shown in (5.35). This basis diagonalizes the monodromy transformation given by the monodromy matrix $\tilde{\Omega}^{(1)}$ and we have

$$\mathcal{F}_{k,j}^{l,i}(t;1-\eta) \xrightarrow{\Omega^{(1)}} \sum_{t'} \tilde{\Omega}_{t,t'}^{(1)} \mathcal{F}_{k,j}^{l,i}(t';1-\eta) = \delta_{t,t'}\, e^{2\pi i(\Delta_t - \Delta_k - \Delta_j)} \mathcal{F}_{k,j}^{l,i}(t';1-\eta). \tag{5.45}$$

The eigenvalues of $\tilde{\Omega}^{(1)}$ and $\Omega^{(1)}$ are the same because the two sets of functions $\{\mathcal{F}_{i,j}^{l,k}(s;\eta)\}_s$ and $\{\mathcal{F}_{k,j}^{l,i}(t;1-\eta)\}_t$ are linearly dependent and describe the different solutions to the same differential operator (due to a specific null vector).

The conformal blocks in a given channel are the eigenfunctions of a given monodromy matrix. The monodromy group contains $\Omega^{(0)}$, $\Omega^{(1)}$ and of course $\Omega^{(\infty)}$ which is diagonal in the u-channel (see (5.40)). However the latter transformation is not independent of the others because the points zero, one and infinity are the only singular points of the correlation functions. Performing successively the three monodromy transformations is equivalent to analytically continue the conformal blocks along a curve surrounding a portion of the space empty of singular points. The conformal blocks are invariant under such a continuation and it follows that

$$\left(\Omega^{(\infty)}\Omega^{(1)}\Omega^{(0)}\right)_{s,s'} = \delta_{s,s'}. \tag{5.46}$$

We know that both sides of (5.36) describe the same correlation functions and we know that a necessary condition to implement locality is the integer-spin condition (5.43). However such a condition is not sufficient to ensure single-valued correlation functions because the other monodromy transformations will mix up the different conformal blocks and modify the pairings between holomorphic and anti-holomorphic parts.

The monodromy problem may now be precisely stated: *The four-point functions must be invariant under any analytic continuation in η along any curves surrounding the points zero, one and infinity.* This guarantees locality and satisfies the crossing conditions. The solution to this comes form the determination of the operator product expansion coefficients $\mathbf{C}_{i,j}^k$. This is still a complicated task which gives rise to connections with *quantum groups*. We shall determine the OPE coefficients for minimal models in the next two chapters.

The condition of strict monodromy invariance may be relaxed to let correlation functions change by a phase factor under the exchange of two operators. This amounts to the invariance under a specific subgroup of the monodromy group. Such a discussion can be done once the strict monodromy invariance problem has been fully solved. We shall illustrate this for the critical Ising model in the next chapter.

5.4 Riemann's Method

We conclude this chapter by a beautiful discussion, due to Riemann, which enlightens the conditions under which the monodromy problem may be solvable, see [13]. The technique was extensively used in [164, 165], to which we refer for details, see also [169].

A four-point correlation function is given by OPE coefficients $\mathbf{C}_{i,j}^k$ and a set of intricated functions $\{\mathcal{F}_{i,j}^{l,k}(s;\eta)\}$ for which we know the leading behaviour as η goes to zero. Due to the crossing conditions we can use equivalent other families $\{\mathcal{F}_{k,j}^{l,i}(t;1-\eta)\}$ and $\{\mathcal{F}_{l,j}^{i,k}(u;1/\eta)\}$ where we know their asymptotic behaviour as η goes respectively to one and infinity. The elements of these families are the different solutions of the differential equation $D_N \mathcal{F} = 0$ of order N which constrains the correlation function. N is the level of the corresponding null vector. Therefore the number of elements in the three possible families must the same and equal to N. The details of these functions or of the differential equation are very complicated. However, Riemann suggested that such a complicated family is described by its singular behaviour, complemented by some general properties. In our four-point function, we know the singular behaviour at zero, one and infinity given respectively by η^{α_s}, $(1 - \eta)^{\beta_t}$, $(1/\eta)^{\gamma_u}$ where the **singular exponents** are related to the conformal weights by

$$\alpha_s = \Delta_s - \Delta_i - \Delta_j, \qquad \beta_t = \Delta_t - \Delta_k - \Delta_j, \qquad \gamma_u = \Delta_u - \Delta_l + \Delta_j \qquad (5.47)$$

for $1 \leq s, t, u \leq N$. We assume that the singular exponents in each channel are distinct and do not differ by integers. This is important to avoid logarithmic behaviour like $\eta^\alpha \ln \eta$.

Lemma: *The vector space generated by any basis of functions (we know one in each channel) must transform into itself under any analytic continuation along paths which do not cross one of the singular points 0, 1, and ∞. Therefore the choice of the singular exponents α, β, and γ is restricted by the condition of closure under analytic continuations and must satisfy the sum rule*

$$\boxed{\sum_{r=1}^{N} (\alpha_r + \beta_r + \gamma_r) = \frac{N(N-1)}{2} - \sum_a R_a} \,, \tag{5.48}$$

where $\{R_a\}$ forms a finite set of positive integers. Moreover, none of the numbers $\alpha_r - \alpha_{r'}$, $\beta_r - \beta_{r'}$, $\gamma_r - \gamma_{r'}$ may be an integer.[1]

This result is obviously necessary to build correlation functions which preserve their values but also their interpretations under monodromy transformations. A generalization of the proposition to a n-point function, related to differential operators having more singular points $(n-1)$ is straightforward. To simplify the notation, we rename

$$\begin{aligned}
f_s^{(0)}(\eta) &= \mathcal{F}_{i,j}^{l,k}(s;\eta) \\
f_t^{(1)}(\eta) &= \mathcal{F}_{k,j}^{l,i}(t;1-\eta) \\
f_u^{(\infty)}(\eta) &= \mathcal{F}_{l,j}^{i,k}(u;1/\eta).
\end{aligned} \tag{5.49}$$

The three bases of functions $f^{(\ell)}$, $\ell = 0, 1, \infty$, describe the same family. A major information comes from the study of the Wronski determinant defined in any basis by

$$W(\eta) = \begin{vmatrix} f_1^{(\ell)} & \cdot & \cdot & f_N^{(\ell)} \\ \partial_\eta f_1^{(\ell)} & \cdot & \cdot & \partial_\eta f_N^{(\ell)} \\ \cdot & & & \cdot \\ \cdot & \cdot & \cdot & \cdot \\ \partial_\eta^{N-1} f_1^{(\ell)} & \cdot & \cdot & \partial_\eta^{N-1} f_N^{(\ell)} \end{vmatrix}. \tag{5.50}$$

Notice that $W(\eta)$ is independent of ℓ because all three bases of functions are linearly related. The logarithmic derivative $\partial_\eta \ln W(\eta)$ is a meromorphic function with single poles located at zero, one, and infinity because of the corresponding singular behaviour of the family. Moreover, the Wronski determinant may vanish for some regular values $\eta = \eta_a$ which contribute to the pole structure. There is no way to specify more closely the values of the parameters η_a which are arbitrary and have no physical interpretation. The residues are given by the leading behaviour of the elements of the family. At finite values of η we can compute

$$\partial_\eta \ln W(\eta) = \frac{(\sum_r \alpha_r) - \frac{N(N-1)}{2}}{\eta} + \frac{(\sum_r \beta_r) - \frac{N(N-1)}{2}}{\eta - 1} + \sum_a \frac{R_a}{\eta - \eta_a}. \tag{5.51}$$

---

[1] If $\alpha_r - \alpha_{r'}$ etc. happens to be an integer, one should consider logarithmic behaviour of the type $\eta^\alpha \ln \eta$. The singularities must be regular singular (see [332, 333]).

The coefficients R_a may be quite complicated. They are positive integers because the zeroes are regular. We can compare the limit $\eta \to \infty$ of (5.51) with the asymptotic behaviour given by $\{f_s^{(\infty)}\}$ which gives

$$\partial_\eta \ln W(\eta) \underset{\eta \to \infty}{\sim} - \frac{(\sum_r \gamma_r) + \frac{N(N-1)}{2}}{\eta} + \mathcal{O}\left(\frac{1}{\eta^2}\right). \qquad (5.52)$$

The result is the equality given by the proposition (5.48)

$$\sum_{r=1}^{N} (\alpha_r + \beta_r + \gamma_r) = \frac{N(N-1)}{2} - \sum_a R_a. \qquad (5.53)$$

We can introduce the definition (5.47) to express the last equation as a consistency **sum rule** for the conformal weights

$$\sum_{s=1}^{N} \Delta_s + \sum_{t=1}^{N} \Delta_t + \sum_{u=1}^{N} \Delta_u = N(\Delta_i + \Delta_j + \Delta_k + \Delta_l) + \frac{N(N-1)}{2} - \sum_a R_a.$$
$$(5.54)$$

This equation is very powerful because it restricts the conformal weights of the possible operators which contribute to a non-trivial correlation function. The restriction actually constrains the possible operator content of the OPE of two primary operators.

Notice that the differential operator D_N is partially determined by the singular behaviour. Any solution $f(\eta)$ is a linear combination of the basis elements $f^{(\ell)}$, $\ell = 1, \ldots, N$. Therefore the following determinant vanishes

$$\begin{vmatrix} f & f^{(1)} & \cdot & \cdot & f^{(N)} \\ \partial_\eta f & \partial_\eta f^{(1)} & \cdot & \cdot & \partial_\eta f^{(N)} \\ \cdot & \cdot & & \cdot & \\ \partial_\eta^N f & \partial_\eta^N f^{(1)} & \cdot & \cdot & \partial_\eta^N f^{(N)} \end{vmatrix} = 0. \qquad (5.55)$$

This can be expressed as the differential equation, where we have normalized the coefficient in $\partial_\eta^N f(\eta)$ to one

$$\partial_\eta^N f(\eta) + \partial_\eta \ln W(\eta) \partial_\eta^{N-1} f(\eta) + \cdots + \widehat{W}_0(\eta) f(\eta) = 0. \qquad (5.56)$$

The coefficient in front of $\partial_\eta^{N-1} f(\eta)$ is given in (5.51). The other coefficients $\widehat{W}_j$ are not completely determined by the singular behaviour. The points $\eta = 0$, 1 and ∞ are called **singular regular points** if the singular exponents in each channel do not differ by integers. The other points with residues R_a are **regular points** which do not affect the monodromy properties. Since the null-vector related differential operators are constructed from (5.4) and (5.5), we know that they are always free of such regular singularities. The differential operator D_N is called **Fuchs differential operator**. In the particular situation where $N = 2$ and there are no regular points, the differential equation is completely determined by the singular exponents, which satisfy $\alpha_1 + \alpha_2 + \beta_1 + \beta_2 + \gamma_1 + \gamma_2 = 1$. We have

$$D_2 f = \left[\partial_\eta^2 - \left(\frac{\alpha_1 + \alpha_2 - 1}{\eta} - \frac{\beta_1 + \beta_2 - 1}{1 - \eta} \right) \partial_\eta \right.$$
$$\left. + \frac{\alpha_1 \alpha_2}{\eta^2} - \frac{\gamma_1 \gamma_2 - \alpha_1 \alpha_2 - \beta_1 \beta_2}{\eta(1 - \eta)} + \frac{\beta_1 \beta_2}{(1 - \eta)^2} \right] f(\eta) = 0. \quad (5.57)$$

After the transformation

$$f(\eta) = \eta^{\alpha_1} (1 - \eta)^{\beta_1} g(\eta) \quad (5.58)$$

we recognize the hypergeometric differential equation for $g(\eta)$.

$$\left[\partial_\eta^2 - \left(\frac{\alpha_2 - \alpha_1 - 1}{\eta} - \frac{\beta_2 - \beta_1 - 1}{1 - \eta} \right) \partial_\eta \right.$$
$$\left. - \frac{(\gamma_1 + \alpha_1 + \beta_1)(\gamma_2 + \alpha_1 + \beta_1)}{\eta(1 - \eta)} \right] g(\eta) = 0. \quad (5.59)$$

We write the two solutions which diagonalize the monodromy matrix $\Omega^{(0)}$ in the usual notation

$$g(\eta) = \begin{cases} {}_2F_1(\gamma_1 + \alpha_1 + \beta_1, \gamma_2 + \alpha_1 + \beta_1, 1 + \alpha_1 - \alpha_2; \eta) \\ \eta^{\alpha_2 - \alpha_1} {}_2F_1(\gamma_1 + \alpha_2 + \beta_1, \gamma_2 + \alpha_2 + \beta_1, 1 + \alpha_2 - \alpha_1; \eta) \end{cases}. \quad (5.60)$$

This shows that any four-point function containing the primary operator $\phi_{(1,2)}$ or $\phi_{(2,1)}$ with conformal weights $\Delta_{1,2}$ and $\Delta_{2,1}$ is up to a prefactor made out of hypergeometric functions ${}_2F_1$.

It is not possible to tell much more about D_N for $N > 2$. However, the sum rule (5.54) remains a very powerful tool. Equation (5.54) can detect possible analytically closed subfamilies. We know from Chap. 4 that a unitary primary operators ϕ has many null-vectors. Each of them implies a differential equation of the same order as its Verma module level. Obviously the null-vector at the lowest possible level, say N, and its differential operator, say D_N, must completely determine the correlation functions containing ϕ, and therefore the OPEs of ϕ with any other primary operator. To be consistent with each other, another differential operator $D_{N'}$ of order $N' > N$ (related to a higher level null-vector in the same Verma module) must admit a N-dimensional subspace of solutions which overlap with the N-dimensional space of solutions of D_N. The following decomposition is therefore necessary

$$D_{N'} = D_{N'-N} \cdot D_N. \quad (5.61)$$

Such a decomposition is detectable in the N'-dimensional family of $D_{N'}$ with the sum rule (5.54) which will show if a N-dimensional subfamily may satisfy the sum rule and therefore be analytically closed. In this case, both $D_{N'-N}$ and D_N will be free of regular singularities.

In the next chapter, we will discuss the Ising model and we will show that there are decompositions of differential operators into a product of differential operators with this time regular singularities which have nothing to do with the null-vector structure. Such decomposition are related to the fact that a subset of the conformal blocks may form a analytically closed subfamily which can be detected by the sum rule (5.54). The corresponding decomposition of differential operators cannot be foreseen in the study of the Verma module and the Riemann point of view provides a very simple tool to gain more information on the possible operator content and OPEs of a physical system.

Exercises

1. We recall a classical theorem of Fuchs for solving differential equations of the form $y''(x) + f(x)y'(x) + g(x)y(x) = 0$ through a Frobénius series of the form $y(x) = \sum_{k=0}^{\infty} a_k x^{k+s}$, where the a_k and s must be determined and $a_0 \neq 0$. *Iff both $xf(x)$ and $x^2 g(x)$ can be expanded around $x = 0$ in converging power series, then at least one solution of the above differential equation is a Frobénius series.* Substituting the above ansatz into the differential equation, the most singular terms in x then yields a quadratic equation for s and the a_k are found from a recurrence relation. If the two solutions for s do not differ by an integer, both solutions of the differential equation take the Frobénius form. Use this technique to derive the solutions (5.60) of the differential equation (5.59).

2. Consider the operator $\sigma = \phi_{2,2}$ of a unitary minimal model. Determine the associated null vector $|\Phi^{(4)}\rangle$ and derive the differential equation

$$
\left[\mathcal{L}_{-4} - \frac{4\Delta}{9} \mathcal{L}_{-2}^2 - \frac{4\Delta + 15}{6\Delta + 18} \mathcal{L}_{-1} \mathcal{L}_{-3} \right.
$$
$$
\left. + \frac{2\Delta + 3}{3\Delta + 9} \mathcal{L}_{-1}^2 \mathcal{L}_{-2} - \frac{1}{4\Delta + 12} \mathcal{L}_{-1}^4 \right] \langle \sigma \sigma \sigma \sigma \rangle = 0.
$$

Write the four solutions in terms of the hypergeometric function $_2F_1$ [250] (see Chap. 14 for explicit expressions of the higher level null vectors).

6. Ising Model Correlators

We shall now apply the tools developed so far to the benchmark example of the critical Ising model. The (primary) operator content is read from the Kac table (Table 4.1). One has the identity operator $\mathbf{1} = \phi_{(1,1)}$ (conformal weights $\Delta_{1,1} = \overline{\Delta}_{1,1} = 0$) which contains the energy–momentum tensor in its conformal tower, the spin-density $\sigma = \phi_{(1,2)} = \phi_{(2,2)}$ ($\Delta_{1,2} = \overline{\Delta}_{1,2} = 1/16$), and the energy-density $\varepsilon = \phi_{(2,1)} = \phi_{(1,3)}$ ($\Delta_{2,1} = \overline{\Delta}_{2,1} = 1/2$). The central charge is $c = 1/2$ and the model is labeled by $m = 3$ in the Kac notation (see (4.42) and (4.43)). We shall calculate the most important four-point functions and determine exactly the monodromy group and the operator product expansion coefficients. All this will be found without making use of the standard free fermion techniques, that is, without appealing to specific properties of the Ising model which render it soluble. Consequently, the conformal methods also work when free fermions are not available. The Ising model fermionic operators will appear when the constraint of locality is relaxed to invariance under a subgroup of the monodromy group. The presentation follows [169].

6.1 Spin-Density Four-Point Function

We first discuss the correlation functions of the spin-density operator σ. We can now collect the results of the previous chapters. Holomorphic and anti-holomorphic parts are both constrained by the null-vectors at level 2, explicitly given by (4.16). Along the same line as in the previous chapter, we discuss the four-point function reduced to the projectively invariant form.

$$G_\sigma(\eta, \bar{\eta}) = \langle 1/16, 1/16 | \sigma(1, 1)\sigma(\eta, \bar{\eta}) | 1/16, 1/16 \rangle. \tag{6.1}$$

Let us first discuss the holomorphic behaviour. We know that (5.12) expresses the consequence of the level 2 null vector of $\sigma(\eta, \bar{\eta})$ and we have to solve

$$\left[-\frac{4}{3}\partial_\eta^2 - \left(\frac{1}{\eta} - \frac{1}{1-\eta} \right) \partial_\eta + \frac{(1/16)}{\eta^2} + \frac{(1/8)}{\eta(1-\eta)} + \frac{(1/16)}{(1-\eta)^2} \right] G_\sigma(\eta) = 0. \tag{6.2}$$

We can discuss the asymptotic behaviour of the solutions as η becomes close to 0, 1, or ∞. It is straightforward to determine the leading powers of the conformal blocks $\eta^\alpha F(\eta)$. According to (5.32), we have $\alpha = \Delta - 2\Delta_{1,2}$ where Δ stands for the conformal weights of the operators which enter into the OPE $\sigma \cdot \sigma$. We now look at the leading terms close to $\eta = 0$. Since the conformal blocks are holomorphic in η, $F(\eta)$ may be expanded as a power series in η. Inserting the solution $\eta^\alpha F(\eta)$ into the differential equation (6.2), the most singular term is of order $\mathcal{O}(\eta^{\alpha-2})$. Its amplitude prefactor has to vanish because there is no other term, coming from the higher orders in the expansion of $F(\eta)$, in (6.2) to compensate it. This leads to the condition

$$-\frac{4}{3}\alpha(\alpha - 1) - \alpha + \frac{1}{16} = 0. \tag{6.3}$$

Simple calculation gives the two solutions, and, since here $\Delta_{1,2} = \Delta_\sigma = 1/16$

$$\Delta = 0 = \Delta_{1,1}, \quad \text{and} \quad \Delta = \frac{1}{2} = \Delta_{2,1} \tag{6.4}$$

and we obtain the OPE

$$\sigma \cdot \sigma = 1 + \varepsilon. \tag{6.5}$$

This result is obviously confirmed by the study of the asymptotic behaviour as $\eta \to 1$, because of the manifest symmetry of (6.2) under the exchange $\eta \leftrightarrow 1 - \eta$. The conformal blocks for $\eta \to \infty$ have slightly different leading powers given in (5.40), but as expected in agreement with the OPE (6.5). Notice that the sum rule (5.54) is verified with no additive term R_a.

Notice now that (6.2) becomes the well-known hypergeometric differential equation for the function $g_\sigma(\eta)$ defined by

$$G_\sigma(\eta) = \left(\frac{1}{\eta(1 - \eta)}\right)^{1/8} g_\sigma(\eta). \tag{6.6}$$

The hypergeometric differential equation for $g_\sigma(\eta)$ reads

$$\left[\partial_\eta^2 + \left(\frac{1/2}{\eta} - \frac{1/2}{1 - \eta}\right)\partial_\eta + \frac{1/16}{\eta(1 - \eta)}\right] g_\sigma(\eta) = 0. \tag{6.7}$$

This confirms the result obtained with the Riemann point of view of Chap. 5. Notice that the second-order differential equations related to the level 2 null-vectors of $\phi_{(1,2)}$ and $\phi_{(2,1)}$ of *any* minimal model will always be transformable into hypergeometric differential operators, see equations (5.57) to (5.60).

The two solutions to this equation and the monodromy properties can be found in any table of integrals, see [1, 276]. They are normally expressed as power series or as Euler contour integrals. Fortunately the Ising model has the exceptional property that all correlation functions are algebraic and it is simple to verify the following pair of solutions of (6.7)

$$g_\sigma^\pm(\eta) = \sqrt{1 \pm \sqrt{1 - \eta}}. \tag{6.8}$$

Up to the term factorized in (6.6) and a normalization, the two solutions $g_\sigma^\pm(\eta)$ are the two conformal blocks of the s-channel.

Notice that the invariance of the differential operator (6.7) under the exchange $\eta \leftrightarrow 1 - \eta$ implies the following second basis of solutions

$$g_\sigma^+(\eta) \pm g_\sigma^-(\eta) = \sqrt{2}\, g_\sigma^\pm(1-\eta) = \sqrt{2}\,\sqrt{1 \pm \sqrt{\eta}}. \tag{6.9}$$

Both solutions $g_\sigma^\pm(\eta)$ are mapped onto themselves under the analytic continuations along a curve surrounding the origin, generated by the monodromy transformation $\Omega^{(0)}$. To see this, consider their leading behaviour in the $\eta \to 0$ limit, which is

$$g_\sigma^+(\eta) = \sqrt{2}\,(1 + \mathcal{O}(\eta))$$
$$g_\sigma^-(\eta) = \frac{1}{\sqrt{2}}\, \eta^{1/2}\,(1 + \mathcal{O}(\eta)). \tag{6.10}$$

The associated monodromy matrix $\Omega^{(0)}$ is therefore (see exercises)

$$\Omega^{(0)} = \begin{pmatrix} 1 & 0 \\ 0 & -1 \end{pmatrix}. \tag{6.11}$$

The transformations of the two solutions is more complicated under a monodromy transformation surrounding the point 1. We expand the solutions in $(1 - \eta)$ to derive

$$g_\sigma^\pm(\eta) = \left(1 \pm \frac{1}{2}(1-\eta)^{1/2}\right)\left(1 + \mathcal{O}(1-\eta)\right). \tag{6.12}$$

The analytic continuation along a curve surrounding the point one exchanges the two functions and the monodromy matrix is

$$\Omega^{(1)} = \begin{pmatrix} 0 & 1 \\ 1 & 0 \end{pmatrix}. \tag{6.13}$$

The second basis of solutions (6.9) has therefore a particular interpretation. The symmetric combination $(g_\sigma^+ + g_\sigma^-)$ is regular at $\eta = 1$ while the other, $(g_\sigma^+ - g_\sigma^-)$, behaves like $(1 - \eta)^{1/2}$ as $\eta \to 1$. These two combinations diagonalize $\Omega^{(1)}$ and are proportional to the two conformal blocks defined in (5.35) for the t-channel. It is finally interesting to derive the two leading behaviours as $\eta \to \infty$

$$g_\sigma^\pm(\eta) = \sqrt{1 \pm i\eta^{1/2}}\sqrt{1 - \frac{1}{\eta}} = e^{\pm i\frac{\pi}{4}}\eta^{1/4}\sqrt{1 - \frac{1}{2\eta} \mp \frac{i}{\eta^{1/2}} + \mathcal{O}\left(\frac{1}{\eta^2}\right)}$$
$$= e^{\pm i\frac{\pi}{4}}\eta^{1/4}\left(1 \mp \frac{i}{2\eta^{1/2}}\right)\left(1 + \mathcal{O}\left(\frac{1}{\eta}\right)\right) \tag{6.14}$$
$$= e^{\pm i\frac{\pi}{4}}\left(\frac{1}{\eta}\right)^{-1/4} + \frac{1}{2}e^{\mp i\frac{\pi}{4}}\left(\frac{1}{\eta}\right)^{1/4} + \dots.$$

The monodromy matrix $\Omega^{(\infty)}$ is therefore given by

$$\Omega^{(\infty)} = \begin{pmatrix} 0 & 1 \\ -1 & 0 \end{pmatrix}. \tag{6.15}$$

The linear combinations $g_\sigma^+ \pm i g_\sigma^-$ diagonalize $\Omega^{(\infty)}$ and are proportional to the u-channel conformal blocks given in (5.40). As expected, the three monodromy matrices satisfy (5.46).

We are now in position to restore the anti-holomorphic part. The physical monodromy invariant four-point function is logically the linear combination

$$G_\sigma(\eta, \bar{\eta}) = \frac{1}{2} \left(\frac{1}{\eta\bar{\eta}(1-\eta)(1-\bar{\eta})} \right)^{1/8} (g_\sigma^+(\eta)g_\sigma^+(\bar{\eta}) + g_\sigma^-(\eta)g_\sigma^-(\bar{\eta})). \tag{6.16}$$

It is easy to verify the monodromy invariance of $G_\sigma(\eta, \bar{\eta})$, using the explicit monodromy matrices. Alternatively, monodromy invariance can be proven by checking that this solution satisfies the crossing conditions (5.27,5.29).

The amplitude $1/2$ in front of (6.16) is fixed by the normalization of the two point functions given by $\mathbf{C}_{\sigma,\sigma}^1 = 1$. The square of this coefficient[1] appears in (5.30) for the contribution of the identity operator to the OPE $\sigma \cdot \sigma$. This contribution is given in (6.16) by the leading behaviour in $1/(\eta\bar{\eta})^{1/8}$ coming from the term $g_\sigma^+(\eta)g_\sigma^+(\bar{\eta})$. We deduce from the expansion (6.10) the amplitude $1/2$.

Next, the expansion of (6.16) for small values of η and $\bar{\eta}$ gives the leading terms of the full decomposition into conformal blocks given in (5.30)

$$G_\sigma(\eta, \bar{\eta}) = \left(\frac{1}{\eta\bar{\eta}(1-\eta)(1-\bar{\eta})} \right)^{1/8} \left(1 + \frac{1}{4}(\eta\bar{\eta})^{1/2} \right) \left(1 + \mathcal{O}(\eta, \bar{\eta}) \right). \tag{6.17}$$

We can read off the OPE coefficients. The contribution from ϵ to the OPE $\sigma \cdot \sigma$ appears in the $g_\sigma^-(\eta)g_\sigma^-(\bar{\eta})$ term. The coefficient in front of the corresponding leading term $\eta^{1/2-1/8}\bar{\eta}^{1/2-1/8}$ is equal to $(\mathbf{C}_{\sigma,\sigma}^\epsilon)^2$. We obtain therefore, for the primary operators

$$\mathbf{C}_{\sigma,\sigma}^1 = 1 \qquad \mathbf{C}_{\sigma,\sigma}^\epsilon = \frac{1}{2}. \tag{6.18}$$

Our solution is now complete. Notice that $\mathbf{C}_{\sigma,\sigma}^\sigma = 0$ because the contribution of σ to $\sigma \cdot \sigma$ would imply a singular behaviour $\eta^{-1/16}$ in the correlation function as $\eta \to 0$. Such a term is obviously absent.

We can restore the z_1, z_2, z_3, z_4 and $\bar{z}_1, \bar{z}_2, \bar{z}_3, \bar{z}_4$-dependence to write the correlation function $\langle \sigma(z_4, \bar{z}_4)\sigma(z_3, \bar{z}_3)\sigma(z_1, \bar{z}_1)\sigma(z_2, \bar{z}_2) \rangle$. The relation was given by (5.9) and (5.10) and we can write

---

[1] Recall that the OPE coefficient $\mathbf{C}_{123} = \mathcal{C}_{123}$, defined in the three-point function (2.38).

$$\langle \sigma(z_4, \bar{z}_4) \sigma(z_3, \bar{z}_3) \sigma(z_1, \bar{z}_1) \sigma(z_2, \bar{z}_2) \rangle$$

$$= \frac{1}{|z_1 - z_4|^{1/4}} \frac{1}{|z_3 - z_2|^{1/4}} G_\sigma \left(\frac{(z_1 - z_2)(z_3 - z_4)}{(z_3 - z_2)(z_1 - z_4)}, \frac{(\bar{z}_1 - \bar{z}_2)(\bar{z}_3 - \bar{z}_4)}{(\bar{z}_3 - \bar{z}_2)(\bar{z}_1 - \bar{z}_4)} \right)$$

$$(6.19)$$

in agreement with the exact result in the scaling limit of the Ising model [627, G22].

6.2 Energy-Density Four-Point Function

We proceed along the same lines as in the last section to discuss the four-point function

$$G_\varepsilon(\eta, \bar{\eta}) = \langle 1/2, 1/2 | \varepsilon(1, 1) \varepsilon(\eta, \bar{\eta}) | 1/2, 1/2 \rangle. \tag{6.20}$$

Due to the level 2 null-vector of $\phi_{(2,1)}$, we have again to solve second-order differential equations (given by (5.12)) which holomorphic part is

$$\left[-\frac{3}{4} \partial_\eta^2 - \left(\frac{1}{\eta} - \frac{1}{1-\eta} \right) \partial_\eta + \frac{(1/2)}{\eta^2} + \frac{1}{\eta(1-\eta)} + \frac{(1/2)}{(1-\eta)^2} \right] G_\varepsilon(\eta) = 0. \tag{6.21}$$

We can again determine the $\eta \to 0$ asymptotic behaviour of the s-channel conformal blocks $\eta^\alpha F(\eta)$. From this follows the OPE $\varepsilon \cdot \varepsilon$. With $\alpha = \Delta - 2\Delta_{2,1}$, $\Delta_{2,1} = 1/2$, we obtain

$$\Delta = 0 = \Delta_{1,1}, \qquad \text{and} \qquad \Delta = \frac{5}{3} = ?. \tag{6.22}$$

The second solution does not correspond to an operator of the Kac table (Table 4.1). In Chap. 13, we shall describe a failed attempt [94] to establish its presence in an antiferromagnetic Ising model. However we can examine the one-dimensional subfamily from the Riemann point of view. Taking only the first solution amounts to have the OPE

$$\varepsilon \cdot \varepsilon = 1 \tag{6.23}$$

which is of course expected, due to the well-known free fermionic representation of the critical Ising model [G22, G19]. The sum rule (5.54) for the $N = 1$ subfamily can be written

$$0 = 2 - \sum_a R_a \tag{6.24}$$

which is perfectly solvable, R_a being positive integers. Therefore, we expect the differential operator (6.21) to decompose into a product $D'_1 D_1$ of two first order differential operators. The two zeroes[2] of the Wronski determinant (5.50) which give rise to the additive terms R_a appear in the first order differential operators a simple poles. After some computations, we obtain

---

[2] Or the single zero of order two.

$$D'_1 = -\frac{1}{4}\left[3\partial_\eta + \left(\frac{1}{\eta} - \frac{1}{1-\eta}\right) + 3\left(\frac{1}{\eta - \eta_1} + \frac{1}{\eta - \eta_1^*}\right)\right]$$

$$D_1 = \partial_\eta + \left(\frac{1}{\eta} - \frac{1}{1-\eta}\right) - \left(\frac{1}{\eta - \eta_1} + \frac{1}{\eta - \eta_1^*}\right),\qquad(6.25)$$

where (η_1, η_1^*) are the complex conjugate solutions of

$$\eta^2 - \eta + 1 = 0.\qquad(6.26)$$

This is the first example of the regular points discussed in Chap. 5. They appear as expected in a decomposition of a differential operator which cannot be explained from the knowledge of the Verma module of ε only. The common solution of (6.21) and of $D_1 G_\varepsilon(\eta) = 0$ is given by

$$G_\varepsilon(\eta) = \left(\frac{1}{\eta} + \frac{1}{1-\eta} - 1\right) = \frac{\eta^2 - \eta + 1}{\eta(1-\eta)}.\qquad(6.27)$$

Due to the one-dimensionality of the solution space, $G_\varepsilon(\eta) = W(\eta)$ is the Wronski determinant (5.50) the numerator of which has regular zeroes. The physical local four-point function is finally

$$G_\varepsilon(\eta, \bar\eta) = \left(\frac{1}{\eta} + \frac{1}{1-\eta} - 1\right)\left(\frac{1}{\bar\eta} + \frac{1}{1-\bar\eta} - 1\right).\qquad(6.28)$$

The expression is easier to recognize if we restore general coordinates. Taking (5.9) and (5.10) into account results in the well-known expression

$$\langle\varepsilon(z_4, \bar z_4)\varepsilon(z_3, \bar z_3)\varepsilon(z_1, \bar z_1)\varepsilon(z_2, \bar z_2)\rangle$$

$$= \left(\frac{1}{(z_4 - z_3)(z_1 - z_2)} - \frac{1}{(z_4 - z_1)(z_3 - z_2)} + \frac{1}{(z_4 - z_2)(z_3 - z_1)}\right)$$

$$\times \left(\frac{1}{(\bar z_4 - \bar z_3)(\bar z_1 - \bar z_2)} - \frac{1}{(\bar z_4 - \bar z_1)(\bar z_3 - \bar z_2)} + \frac{1}{(\bar z_4 - \bar z_2)(\bar z_3 - \bar z_1)}\right).$$

$$(6.29)$$

We recognize the Wick theorem for the free massless fermions with which we can decompose the energy-density operator

$$\varepsilon(z, \bar z) = i\psi(z)\psi(\bar z).\qquad(6.30)$$

The factor i takes into account that $\psi(z)$ and $\psi(\bar z)$ anticommute and that ε is a real scalar operator.

This is a simple example of a decomposition of a differential operator which cannot be predicted from an elementary overview of the structure of the Verma module of ε.

6.3 Mixed Four-Point Functions

We deduce from the OPEs (6.5) and (6.23) that the last possible non-zero four-point function involves two operators σ and two ε. We define

$$G_{\sigma,\varepsilon}(\eta, \bar{\eta}) = \langle 1/16, 1/16 | \varepsilon(1,1)\varepsilon(\eta,\bar{\eta}) | 1/16, 1/16 \rangle. \qquad (6.31)$$

The differential equation for the holomorphic part is again given by the level 2 null-vector of $\varepsilon(\eta)$.

$$\left[-\frac{3}{4}\partial_\eta^2 - \left(\frac{1}{\eta} - \frac{1}{1-\eta} \right) \partial_\eta + \frac{(1/16)}{\eta^2} + \frac{1}{\eta(1-\eta)} + \frac{(1/2)}{(1-\eta)^2} \right] G_{\sigma,\varepsilon}(\eta) = 0. \qquad (6.32)$$

We can proceed this time with another method. We know from the OPEs (6.5) and (6.23) the operator content of $\varepsilon \cdot \sigma$ which is given by

$$\varepsilon \cdot \sigma = \sigma. \qquad (6.33)$$

This means that only one conformal block enters in the description of each channel of $G_{\sigma,\varepsilon}$. The asymptotic behaviours as $\eta \to 0, 1$ and ∞ are read from (5.32), (5.35), and (5.40).

$$\begin{aligned}
\eta \to 0: & \qquad \eta^{\Delta\sigma - \Delta\varepsilon - \Delta\sigma} = \eta^{-\frac{1}{2}} \\
\eta \to 1: & \qquad (1-\eta)^{\Delta_1 - 2\Delta\varepsilon} = (1-\eta)^{-1} \\
\eta \to \infty: & \qquad \left(\frac{1}{\eta} \right)^{\Delta\sigma + \Delta\varepsilon - \Delta\sigma} = \left(\frac{1}{\eta} \right)^{\frac{1}{2}}.
\end{aligned} \qquad (6.34)$$

The one dimensional family having such asymptotic singularities satisfies the sum rule (5.54) with one zero with residue $R_1 = 1$. It is easy to verify that this asymptotic behaviour is compatible with the differential equation (6.32), whose second possible solution does not correspond to a contribution in the OPE from any operator in the Kac table (Table 4.1). The differential operator (6.32) must once again decompose into the product of two first-order differential operators $D'_1 D_1$. Simple calculations show that

$$D'_1 = -\frac{1}{4} \left[3\partial_\eta + \left(\frac{5}{2\eta} - \frac{1}{1-\eta} + \frac{3}{\eta+1} \right) \right]$$

$$D_1 = \partial_\eta + \left(\frac{1}{2\eta} - \frac{1}{1-\eta} - \frac{1}{\eta+1} \right) \qquad (6.35)$$

and the holomorphic solution to $D_1 G_{\sigma,\varepsilon} = 0$ is

$$G_{\sigma,\varepsilon}(\eta) = \frac{\eta+1}{\eta^{1/2}(1-\eta)}. \qquad (6.36)$$

The properly normalized crossing symmetric four-point function is finally

$$G_{\sigma,\varepsilon}(\eta,\bar\eta) = \frac{1}{4}\left|\frac{\eta+1}{\eta^{1/2}(1-\eta)}\right|^2. \qquad (6.37)$$

The factor $1/4$ corresponds to $(\mathbf{C}^\sigma_{\varepsilon,\sigma})^2$ (see (6.18))in the asymptotic limits $\eta \to 0$ or ∞. As $\eta \to 1$, we have $1+\eta \to 2$ and the asymptotic behaviour is normalized by $\mathbf{C}^1_{\varepsilon,\varepsilon}\mathbf{C}^1_{\sigma,\sigma} = 1$, which is the correct result. The four-point function expressed in terms of general coordinates is

$$\langle\sigma(z_4,\bar z_4)\varepsilon(z_3,\bar z_3)\varepsilon(z_1,\bar z_1)\sigma(z_2,\bar z_2)\rangle =$$

$$\frac{1}{4}\frac{1}{|z_1-z_3|^2|z_2-z_4|^{1/4}}\left|\left(\frac{(z_1-z_2)(z_3-z_4)}{(z_3-z_2)(z_1-z_4)}\right)^{\frac12} + \left(\frac{(z_3-z_2)(z_1-z_4)}{(z_1-z_2)(z_3-z_4)}\right)^{\frac12}\right|^2$$

$$(6.38)$$

in agreement with the crossing conditions (5.27) and (5.29).

To summarize, we have computed all possible four-point functions of primary operators and all expansion coefficients. The OPEs are given by

$$\boxed{\sigma\cdot\sigma = 1+\varepsilon \ , \ \ \varepsilon\cdot\varepsilon = 1 \ , \ \ \sigma\cdot\varepsilon = \sigma} \ . \qquad (6.39)$$

The remaining n-point functions, which are all algebraic, can be computed along the same lines.

6.4 Semi-Local Four-Point Functions

The critical algebra $\{1,\sigma,\varepsilon\}$ is obtained from the high-temperature phase where the spin-density operator σ plays the role of order operator. In the low-temperature phase, the disorder operator μ, dual to the order operator, becomes a distinct operator at the critical temperature with conformal weights $\Delta_\mu = \bar\Delta_\mu = 1/16$. The critical theory contains therefore another critical algebra $\{1,\mu,\varepsilon\}$ which is completely isomorphic to the former one. OPEs and correlation functions within this new algebra are totally identical to the ones computed in the previous sections. However, it is well-known that order and disorder operators are non-local with respect to each other. This means that the exchange of such operators results in a phase factor in their correlation functions. Consider the following four-point function:

$$G_{\mu,\sigma}(\eta,\bar\eta) = \langle\mu|\sigma(1,1)\sigma(\eta,\bar\eta)|\mu\rangle. \qquad (6.40)$$

The existence of null-vector implies a differential equations, which is the same as in (6.2). The conformal blocks discussed in that section are the same. However, the correlation function must gain a $(-)$ sign when analytically continued in η along curves surrounding the points 0 and ∞ where disorder operators μ are located. Correlation functions are therefore invariant under the subgroup of the monodromy group generated by $(\Omega^{(0)})^2$, $\Omega^{(1)}$, and $(\Omega^{(\infty)})^2$. The only possible correlation function is the combination

$$G_{\mu,\sigma}(\eta,\bar{\eta}) = \frac{1}{2}\left(\frac{1}{\eta\bar{\eta}(1-\eta)(1-\bar{\eta})}\right)^{1/8}\left(g_\sigma^+(\eta)g_\sigma^-(\bar{\eta}) + g_\sigma^-(\eta)g_\sigma^+(\bar{\eta})\right) \quad (6.41)$$

which has eigenvalue -1 under $\Omega^{(0)}$ and $\Omega^{(\infty)}$. The normalization factor $1/2$ is determined by the asymptotic limit $\eta \to 1$ which should be normalized to $\mathbf{C}^1_{\sigma,\sigma}\mathbf{C}^1_{\mu,\mu} = 1$. The expansion (6.12) gives the necessary information to expand (6.41), up to factors $1 + \mathcal{O}(1-\eta, 1-\bar{\eta})$

$$G_{\mu,\sigma}(\eta,\bar{\eta}) \simeq \left(\frac{1}{\eta\bar{\eta}(1-\eta)(1-\bar{\eta})}\right)^{1/8}\left(1 - \frac{1}{4}\sqrt{(1-\eta)(1-\bar{\eta})}\right). \quad (6.42)$$

The factor $-1/4$ is equal to $\mathbf{C}^\epsilon_{\sigma,\sigma}\mathbf{C}^\epsilon_{\mu,\mu}$, giving therefore $\mathbf{C}^\epsilon_{\mu,\mu} = -1/2$. In the previous sections of this chapter, we computed the square of the expansion coefficients and we arbitrarily chose positive operator product coefficients, leaving a phase freedom undiscussed. The reason is that, as in quantum mechanics, exact phases are in general not important, while relative phases obey very strong constraints. This last calculation shows the importance of relative phases.

We can now discuss the new operators which appear in the limit as $\eta \to 0$ where we have

$$G_{\mu,\sigma}(\eta,\bar{\eta}) = \frac{1}{2}\left(\frac{1}{\eta\bar{\eta}}\right)^{\frac{1}{8}}\left(\bar{\eta}^{\frac{1}{2}} + \eta^{\frac{1}{2}} + \ldots\right). \quad (6.43)$$

The non-local singularities are due to the presence of fermionic operators in the OPE $\sigma \cdot \mu$ which can be written

$$\sigma \cdot \mu = \bar{\psi} + \psi, \quad (6.44)$$

where $\psi(z)$ and $\bar{\psi}(\bar{z})$ are the free fermions which build up the energy-density operator. They imply the new expansion coefficients $\mathbf{C}^\psi_{\mu,\sigma} = \mathbf{C}^{\bar{\psi}}_{\mu,\sigma} = 1/\sqrt{2}$. The fermionic operators couple locally with the identity and with themselves. There is therefore a last consistent local operator algebra $\{\mathbf{1}, \psi, \bar{\psi}, \epsilon\}$ which is nothing but the free fermion theory. The OPEs are:

$$\begin{array}{rcl} \psi \cdot \psi &=& \mathbf{1} \\ \bar{\psi} \cdot \bar{\psi} &=& \mathbf{1} \\ \psi \cdot \bar{\psi} &=& \epsilon \\ \psi \cdot \epsilon &=& \bar{\psi} \\ \bar{\psi} \cdot \epsilon &=& \psi \end{array} \quad . \quad (6.45)$$

By construction, $\mathbf{C}^\epsilon_{\psi,\bar{\psi}} = -\mathbf{C}^\epsilon_{\bar{\psi},\psi} = i$. Such an operator spectrum can be found in the Ising model on the torus with mixed periodic and anti-periodic boundary conditions along the two homological cycles, as will be shown in Chap. 10. Notice that the largest possible semi-local algebra is naturally given by the operators $\{\mathbf{1}, \psi, \bar{\psi}, \epsilon, \sigma, \mu\}$. The remaining correlation functions can easily be deduced from the OPEs and from the correlation functions discussed so far.

Exercises

1. Why does one expand in (6.42) around $\eta = 1$, $\bar{\eta} = 1$ to find the OPE coefficient $\mathbf{C}_{\mu\mu}^{\epsilon}$?
2. Check the operator product expansions given in the text. Verify in particular the given values of the operator product expansion coefficients.
3. In Chap. 8 it will be shown that the Ising model may alternatively be described in terms of a ϕ^4 field theory. The classical equation of motion may be written in the form

$$\partial^2 \phi(x) + \mu^2 \phi(x) = g\phi^3(x)$$

and the field ϕ provides a continuum description of the order parameter σ. In a field theory setting, composite objects such as ϕ^3 must be defined through a point-splitting procedure

$$\phi^3(x) \to \Phi(x, a) := \lim_{a \to 0} \phi(x + a)\phi(x)\phi(x - a),$$

where the limit implies that first an average over all directions of a is taken and then a is allowed to become "small". Use the operator product of the conformal operator σ to show that $\Phi(x, a) \simeq a^{-4\Delta_\sigma}[\alpha_1 + \alpha_2 a^2 \partial^2 + \ldots]\phi(x)$ and determine the coefficients $\alpha_{1,2}$. Compare with the classical equation of motion of the ϕ^4 theory [508].

4. Consider a complex function $F(z, \bar{z})$. The monodromy transformations $\Omega^{(0,1)}$ around the points $z_0 = 0$ and $z_1 = 1$ are defined by

$$F(z, \bar{z}) \xrightarrow[\Omega^{(0)}]{} \lim_{\tau \to 1-} F\left(z e^{2\pi i \tau}, \bar{z} e^{-2\pi i \tau}\right)$$

$$F(z, \bar{z}) \xrightarrow[\Omega^{(1)}]{} \lim_{\tau \to 1-} F\left(1 + (z - 1)e^{2\pi i \tau}, 1 + (\bar{z} - 1)e^{-2\pi i \tau}\right).$$

Now consider a pair $\{F_1, F_2\}$ of functions. If this set is closed under a monodromy transformation, one may associate a **monodromy matrix** with the map $\Omega^{(0)}$

$$F_1 \xrightarrow[\Omega^{(0)}]{} \sum_{i=1}^{2}\left(\Omega^{(0)}\right)_{1i} F_i \ , \quad F_2 \xrightarrow[\Omega^{(0)}]{} \sum_{i=1}^{2}\left(\Omega^{(0)}\right)_{2i} F_i$$

and similarly for $\Omega^{(1)}$. Find the monodromy matrices $\Omega^{(0,1)}$ for the pair of functions

$$F_1(z) := \int_1^\infty dw \, w^a (w - 1)^b (w - z)^c \ , \quad F_2(z) := \int_0^z dw \, w^a (1 - w)^b (z - w)^c,$$

where a, b, c are constants. Write $F_{1,2}$ as hypergeometric functions and use their transformation properties [1] to get the matrix $\Omega^{(1)}$. If $F(z, \bar{z}) = \sum_{i,j=1}^{2} \lambda_{ij} F_i(z, \bar{z}) \bar{F}_j(z, \bar{z})$ is monodromy invariant, show that invariance under $\Omega^{(0)}$ implies that $\lambda_{ij} = \delta_{ij} \lambda_i$. Invariance under $\Omega^{(1)}$ then yields $\lambda_1/\lambda_2 = [\sin(\pi b)\sin(\pi(a + b + c))]/[\sin(\pi a)\sin(\pi c)]$.

7. Coulomb Gas Realization

Having looked at the Ising model in some detail, it would of course be possible to try to solve other models in the same way. However, often it is helpful to rely on the very general and efficient techniques coming from the intimate connections of 2D critical systems with the Coulomb gas in two dimensions, as developed by Vl. Dotsenko and Fateev [205, 206]. In general, any minimal model may be reformulated in terms of a Coulomb gas with a charge at infinity. Furthermore, this provides useful computational tools. As an introduction to the subject, we shall concentrate on some formal calculations but do not discuss the justification of these which come from quite deep quantum field theory results. Our aim is to present a technique which allows to compute any correlation function of a minimal model. We largely follow here [169].

7.1 The Free Bosonic Scalar Field

Consider a free massless bosonic scalar field φ and the classical action

$$S \sim \int dz\, d\bar{z}\, \partial_z \varphi \partial_{\bar{z}} \varphi. \tag{7.1}$$

Although that model might at first sight appear to be quite trivial, its consistent definition as a quantum field theory is far from obvious. Infrared and ultraviolet problems can be formally resolved by the introduction of a infrared cut-off a and a ultraviolet cut-off R, which of course introduce new (length) scales in the model, which in turn break scale invariance. The equation of motion for the propagator is

$$\partial_z \partial_{\bar{z}} \langle \varphi(z, \bar{z})\, \varphi(0, 0) \rangle \sim \delta^{(2)}(z, \bar{z}). \tag{7.2}$$

When the cut-off's are removed, the propagator is not defined and breaks the positivity of the two-point function. Here, we choose the normalization

$$\langle \varphi(z, \bar{z})\, \varphi(0, 0) \rangle = -4\ln(|z|/R) = -2\ln(z/R) - 2\ln(\bar{z}/R). \tag{7.3}$$

Of course, this is Coulomb's law in two dimensions, explaining the name **Coulomb gas** for this model. The theory is plagued by inconsistencies, because of the ill-defined propagator. This becomes apparent when one tries to compute correlation functions for any functional of the field φ. For the purposes of this chapter, however, we may restrict ourselves to a subsector of the model which is well-defined in the asymptotic limit. In this sector, correct results can be obtained through rather formal manipulations. The simplest operators which allow such a formal treatment, free from renormalization problems, are

$$J = \left(\frac{i}{2}\right) \partial_z \varphi , \quad \bar{J} = -\left(\frac{i}{2}\right) \partial_{\bar{z}} \varphi. \tag{7.4}$$

Their two-point functions, formally obtained from (7.3), are

$$\langle J(z_1, \bar{z}_1) J(z_2, \bar{z}_2) \rangle = \frac{1/2}{(z_1 - z_2)^2}$$

$$\langle \bar{J}(z_1, \bar{z}_1) \bar{J}(z_2, \bar{z}_2) \rangle = \frac{1/2}{(\bar{z}_1 - \bar{z}_2)^2} \tag{7.5}$$

and $\langle J\bar{J} \rangle = 0$. Since the free boson theory obeys the Wick theorem for the n-point functions, their calculation reduces to the calculation of the two-point functions. In addition, since the model is massless and because of (7.3) one may decompose φ into its holomorphic and anti-holomorphic components

$$\varphi(z, \bar{z}) = \varphi(z) + \overline{\varphi}(\bar{z}). \tag{7.6}$$

The two operators J and $\bar{J}$ are, respectively, holomorphic or antiholomorphic operators. Their scaling dimensions are unity and their spin is ± 1. Both are conserved currents ($\partial_{\bar{z}} J = \partial_z \bar{J} = 0$) which generate U(1) current algebras.

It is often enough to consider merely the holomorphic operator $J(z)$. We now show how the techniques of Chap. 2 can be adapted to construct the energy–momentum tensor T and derive the Virasoro algebra. First, define the moments of the current J, in analogy with (2.58)

$$J_n A(z_1, \bar{z}_1) = \frac{1}{2\pi i} \oint_{C_{z_1}} dz \, (z - z_1)^n J(z) A(z_1, \bar{z}_1), \tag{7.7}$$

where C_{z_1} is a closed curve surrounding the point z_1 and A is a scaling operator. From (7.5), we have formally the OPE

$$J(z') J(z) = \frac{1/2}{(z' - z)^2} + \text{regular terms}. \tag{7.8}$$

As in Chap. 2, one can write the commutator as

$$[J_n, J_m] A(z_1, \bar{z}_1) = \frac{1}{(2\pi i)^2} \oint_{C_z} dz' \oint_{C_{z_1}} dz \, (z' - z_1)^n (z - z_1)^m$$
$$\times J(z') J(z) A(z_1, \bar{z}_1)$$
$$= \frac{1}{2(2\pi i)^2} \oint_{C_z} dz' \oint_{C_{z_1}} dz \, (z' - z_1)^n (z - z_1)^m \frac{1}{(z' - z)^2} A(z_1, \bar{z}_1)$$

$$= \frac{1}{4\pi i} \oint_{C_{z_1}} dz \, (z - z_1)^{n+m-1} n A(z_1, \bar{z}_1)$$

$$= \frac{n}{2} \delta_{n+m,0} A(z_1, \bar{z}_1) \tag{7.9}$$

and we also see that the regular terms (in $z' - z$) not detailed in (7.8) do not contribute to the commutator. We have thus found the commutation relations of the so-called $\hat{U}(1)$ **Kac–Moody algebra**.

$$[J_n, J_m] = \frac{n}{2} \delta_{n+m,0}. \tag{7.10}$$

Generally, for a finite-dimensional Lie algebra **g** with commutation relations $[J^a, J^b] = i f^{ab}{}_c J^c$, there is an **affine extension** $\hat{g}$ of level k or Kac–Moody algebra [270, 377]

$$[J_n^a, J_m^b] = i f^{ab}{}_c J_{n+m}^c + k n \, \delta^{a,b} \delta_{n+m,0}, \tag{7.11}$$

where n, m are integers. Kac–Moody (or affine Lie) algebras are infinite-dimensional extensions of the familiar finite-dimensional simple Lie algebras. Lie algebras are characterized through their commutator relations, which may be compactly presented in the Chevalley basis and then written in terms of the Cartan matrix. The Cartan matrix of a Kac–Moody algebra has to be positive semidefinite, rather than strictly positive definite as required for the finite-dimensional simple Lie algebras. The possible Cartan matrices for simple Kac–Moody algebras can be classified. A simple way of graphically presenting this classification is through the Dynkin diagrams. In Fig. 1.7, we had listed the Dynkin diagrams of the simply-laced finite-dimensional and affine Lie algebras, which is the only case we shall need here and in the sequel. Kac–Moody algebras have since a long time played an important role in particle physics, under the name of current algebras. For detailed introductions, see [270, 377, 173, G18].

The conserved currents $J(z)$ can be used to express the energy–momentum tensor which has the so-called **Sugawara–Sommerfield** form [581, 573]

$$T(z) = :JJ: (z) = -\frac{1}{4} :\partial_z \varphi \partial_z \varphi: (z) = \lim_{\epsilon \to 0} \left(J(z + \epsilon) J(z - \epsilon) - \frac{1}{8\epsilon^2} \right). \tag{7.12}$$

The **normal ordering** $::$ subtracts the self-interaction of the field φ with itself. This amounts to subtract the singular part in the OPE of J with itself and leads to $\langle T(z) \rangle = 0$, as expected from translation invariance. Next, we have to find the OPE of T with J

$$T(z') J(z) = \lim_{\epsilon \to 0} \left(J(z' + \epsilon) J(z' - \epsilon) J(z) - \frac{1}{8\epsilon^2} J(z) \right)$$

$$= \lim_{\epsilon \to 0} \left(\frac{1}{2(z' - z + \epsilon)^2} J(z' - \epsilon) + \frac{1}{2(z' - z - \epsilon)^2} J(z' + \epsilon) - \frac{1}{8\epsilon^2} J(z) \right)$$

$$= \frac{1}{(z' - z)^2} J(z) + \frac{1}{z' - z} \partial_z J(z) + \text{regular terms} \tag{7.13}$$

since $J(z') = J(z) + (z' - z)\partial_z J(z) + \ldots$. The terms not explicitly spelled out are regular in the limit $z' - z \to 0$. Equation (7.13) shows that J is primary. In the same way as before, the commutation relations between the Virasoro generators L_n with the $\hat{U}(1)$ generators J_n can be found from (7.13)

$$[L_n, J_m] = -m\, J_{n+m}. \tag{7.14}$$

Again, the regular terms do not contribute. So far, the normal ordering prescription had no perceptible influence on the operator product expansions. That is different for the OPE of T with itself, which defines the central charge of the model

$$T(z')T(z) = \lim_{\epsilon \to 0}\left\{ T(z')\left(J(z+\epsilon)J(z-\epsilon) - \frac{1}{8\epsilon^2} \right)\right\} \tag{7.15}$$

$$= \lim_{\epsilon \to 0}\left\{ \frac{1}{(z'-z-\epsilon)^2}\left[J(z+\epsilon)J(z-\epsilon) - \frac{1}{8\epsilon^2} + \frac{1}{8\epsilon^2} \right] \right.$$

$$+ \frac{1}{(z'-z+\epsilon)^2}\left[J(z+\epsilon)J(z-\epsilon) - \frac{1}{8\epsilon^2} + \frac{1}{8\epsilon^2} \right]$$

$$+ \frac{1}{z'-z-\epsilon}\left[(\partial_z J(z+\epsilon))\, J(z-\epsilon) + J(z+\epsilon)\, (\partial_z J(z-\epsilon)) \right]$$

$$\left. + \left[\frac{1}{z'-z+\epsilon} - \frac{1}{z'-z-\epsilon} \right] J(z+\epsilon)\, (\partial_z J(z-\epsilon)) - \frac{1}{8\epsilon^2} T(z') \right\}. \tag{7.16}$$

Because of

$$\langle J(z_1,\bar{z}_1)\partial_{z_2} J(z_2,\bar{z}_2)\rangle = \frac{1}{(z_1 - z_2)^3} \tag{7.17}$$

the last line in the above expansion is potentially singular in the $\epsilon \to 0$ limit and must be treated carefully. We find, using again the normal ordering prescription (7.12)

$$T(z')T(z) = \lim_{\epsilon \to 0}\left\{ \frac{2}{(z'-z)^2}T(z) + \frac{1}{4\epsilon^2}\frac{1}{(z'-z)^2} + \frac{3}{4}\frac{1}{(z'-z)^4} \right. \tag{7.18}$$

$$\left. + \frac{1}{z'-z}\partial_z T(z) - \frac{\epsilon}{4\epsilon^3}\frac{1}{(z'-z)^2} - \frac{1}{4}\frac{1}{(z'-z)^4} \right\} + \text{regular terms}$$

$$= \frac{1}{2}\frac{1}{(z'-z)^4} + \frac{2}{(z'-z)^2}T(z) + \frac{1}{z'-z}\partial_z T(z) + \text{regular terms} \tag{7.19}$$

as expected. The last two terms in (7.18) come from the last line in (7.16). The terms which are singular for $\epsilon \to 0$ as well as for $z' - z \to 0$ cancel out, as it should be for a well-defined theory. We read off the value of the central charge $c = 1$, in agreement with the finite-lattice calculation of Chap. 3. This example also shows explicitly that non-vanishing central charges result from normal ordering, which is necessary for the proper definition of quantum field theories. In purely classical theories, there is no normal ordering and central charges vanish. In addition, one may show that the singular part of the operator product expansions (7.13,7.19) are independent of the precise form of the normal ordering (see exercises). This illustrates once more the universality of the field theory underlying critical phenomena.

The sector of the free massless boson theory we are dealing with may be built from the components of the $\hat{U}(1)$ current $J(z)$, which is a scaling operator. For $c = 1$, there are infinitely many primary operators with respect to the Virasoro algebra. One may use the conserved current $J(z)$ and its associated Kac–Moody algebra to sharpen the notion of a primary scaling operator in a way consistent with the model at hand. We say that a **primary operator** ϕ must be a primary with respect to the Virasoro algebra and in addition also primary with respect to the Kac–Moody algebra, that is

$$J_n \phi = 0 \quad \text{if} \quad n > 0. \tag{7.20}$$

That means that several irreducible representations of the $c = 1$ Virasoro algebra alone are combined into a single irreducible representation of the Kac–Moody-Virasoro algebra spanned by the set $\{L_n, J_n\}$. The maximal symmetry of a given model is found when the entire system is described in terms of a single irreducible representation.

Examples of $\hat{U}(1)$ Kac–Moody primary operators are given by the **vertex operators**

$$V_\alpha(z) \sim\, : e^{i\alpha\varphi(z)} :\, . \tag{7.21}$$

It can be shown that they renormalize onto cut-off-independent operators, so that the formal treatment we carry out here is justified. It is enough here to consider only the holomorphic part, since the locality and monodromy problem can be treated as in the preceeding chapters. In quantum field theory, conserved quantities imply corresponding Ward identities. For the current algebra, one has

$$\langle J(z)V_{\alpha_1}(z_1)\cdots V_{\alpha_n}(z_n)\rangle = \left(\sum_{i=1}^n \frac{\alpha_i}{z - z_i} \right) \langle V_{\alpha_1}(z_1)\cdots V_{\alpha_n}(z_n)\rangle \tag{7.22}$$

which might also be formally derived from the correlators given above. The U(1) **charge** of V_α is α. A general theorem in field theory states that conserved currents always retain their canonical dimension under renormalization. For large distances, we must have for the conserved $\hat{U}(1)$ current $J(z) \sim 1/z^2$, when inserted into correlation functions. Combining with (7.22), it follows that the total charge $\alpha_1 + \cdots + \alpha_n$ must vanish for physical correlation functions

$$\langle V_{\alpha_1}(z_1)\cdots V_{\alpha_n}(z_n)\rangle = 0 \quad \text{if} \quad \alpha_1 + \cdots + \alpha_n \neq 0. \tag{7.23}$$

This **neutrality condition** is more traditionally obtained with a careful study of the ultra-violet R dependence of exponentials of the free field φ. Using the definition (7.21) and the Wick theorem, the neutral two-point function is given by

$$\langle V_\alpha(z)\, V_{-\alpha}(0)\rangle = \frac{1}{z^{2\alpha^2}} \tag{7.24}$$

The conformal weight is $\Delta_\alpha^{\text{free}} = \alpha^2$. Considering the operator product $T(z')V_\alpha(z)$, it is easy to check that V_α is indeed primary and to rederive the conformal weight. A general n-point function of Vertex operators is given by the infinite product

$$\langle V_{\alpha_1}(z_1)\cdots V_{\alpha_n}(z_n)\rangle = \prod_{i<j}(z_i - z_j)^{2\alpha_i\alpha_j}\,\delta_{\alpha_1+\cdots+\alpha_n,0}. \tag{7.25}$$

We now have together all the building blocks we shall need for discussing minimal models in terms of the Coulomb gas. We describe this in the sequel, following the presentation of [169]. Field theories with $c = 1$ correspond to statistical mechanics systems with a conserved U(1) charge. Because of the presence of the marginal operator $J\bar{J}$, there is a line of fixed points, related to a line of critical models with continuously varying exponents. This will be taken up again later when discussing applications in the XY and Ashkin-Teller models or in oriented polymers, see Chaps. 11, 12 and 14.

7.2 Screened Coulomb Gas

The free Coulomb gas is defined at the value $c = 1$ of the central charge. The method used to reduce the degrees of freedom consists in a modification of the boundary conditions on the free field φ. Introduce a vertex operator with charge $-2\alpha_0$ and conformal weight $\Delta_{-2\alpha_0}^{\text{free}} = 4\alpha_0^2$ placed at infinity in each correlator and consider the limit

$$\ll V_{\alpha_1}(z_1)\cdots V_{\alpha_n}(z_n) \gg := \lim_{R\to\infty} R^{8\alpha_0^2}\langle V_{\alpha_1}(z_1)\cdots V_{\alpha_n}(z_n)\, V_{-2\alpha_0}(R)\rangle. \tag{7.26}$$

The limit is well-defined because the correlation function on the r.h.s. scales as $R^{-2\Delta_{-2\alpha_0}^{\text{free}}}$ for large R. These new correlation functions have the neutrality condition replaced by

$$\sum_{i=1}^{n}\alpha_i = 2\alpha_0. \tag{7.27}$$

In particular, the two-point functions of the modified theory have to satisfy this new charge condition

$$\ll V_\alpha(z)\, V_{2\alpha_0-\alpha}(0) \gg = \frac{1}{z^{2\alpha(\alpha-2\alpha_0)}}. \tag{7.28}$$

The conformal weight of the vertex operator V_α or $V_{2\alpha_0-\alpha}$ becomes

$$\Delta_\alpha = \Delta_{2\alpha_0-\alpha} = \alpha(\alpha-2\alpha_0). \tag{7.29}$$

The difficult part is now to show that a consistent theory comes out of this constraint. Both operators V_α and $V_{2\alpha_0-\alpha}$ must be shown to be equivalent and to have isomorphic Verma modules. A full discussion would lead to the BRST structure of this model but this introductory chapter is only supposed to give the recipe necessary to compute correlation functions.

The charge at infinity modifies some geometric properties of the vertices but preserves the conformal invariance of the model. The energy–momentum tensor must be modified to take the modifications into account. We also want that V_α remains a Virasoro primary operator. The operators of the $c = 1$ theory are sufficient to completely take the effects of the charge at infinity into account and the modified holomorphic part of the energy–momentum tensor is

$$T(z) = \; :JJ: (z) + 2\alpha_0 \partial_z J(z) = -\frac{1}{4} \; :\partial_z \varphi \partial_z \varphi: (z) + i\alpha_0 \partial_z^2 \varphi(z). \qquad (7.30)$$

It is straightforward to verify that both V_α and $V_{2\alpha_0-\alpha}$ are primary operators. The new term in the definition of T should not surprise because it is the only other possible conserved quantity with scaling dimension 2 of the theory. The factor $2\alpha_0$ is adjusted to keep the vertex operators primary. The modified T gives rise to a Virasoro algebra with central charge (see also Chap. 10)

$$c = 1 - 24\alpha_0^2. \qquad (7.31)$$

This is the first spectacular result: the charge at infinity results in a lowering of the central charge. The second spectacular result is the existence of non-trivial **screening current** with conformal weight unity (check it)

$$S_\pm(z) = V_{\alpha_\pm}(z) \quad \text{with} \quad \alpha_\pm = \alpha_0 \pm \sqrt{\alpha_0^2 + 1} \quad \text{i.e.} \quad \Delta_{\alpha_\pm} = 1. \qquad (7.32)$$

These currents are used to create charges which can screen the charge at infinity. They are defined by

$$Q_\pm = \frac{1}{2\pi i} \oint_C dz \, S_\pm(z). \qquad (7.33)$$

The choices for the contour of integration will be discussed later. The charges have scaling dimension zero and therefore they will not affect the conformal properties of the correlation functions. However, they cause a shift in the charge by a amount of α_+ or α_-. Suppose now that we restrict the subset of permissible vertex operators to the ones whose charge is quantized and belongs to the subset

$$\alpha_{r,s} = \frac{1}{2}(1 - r)\alpha_+ + \frac{1}{2}(1 - s)\alpha_- \quad \text{with} \quad r, s \in \mathbb{N}. \qquad (7.34)$$

We introduce as in Chap. 4 the parametrization $c = 1 - 6/m(m+1)$ and we compute the conformal weight of $V_{r,s} := V_{\alpha_{r,s}}$ in terms of r, s, and m. We recognize the celebrated Kac formula

$$\Delta_{\alpha_{r,s}} = \Delta_{r,s} = \frac{(r(m+1) - sm)^2 - 1}{4m(m+1)}. \qquad (7.35)$$

This is the beginning of the connection to the minimal models. For the sake of simplicity, we restrict the discussion to the integer values of m, related to the unitary minimal models. It is straightforward to extend the discussion to rational values $m = (p-q)/q$ where p, q are coprime integers. This latter case represents the space of minimal models. With $m \in \mathbb{N}$, the values of r and s are restricted to $1 \le r \le (m-1)$ and $1 \le s \le m$.[1] We denote $V_{\alpha_{r,s}} = V_{r,s}$ and $V_{2\alpha_0 - \alpha_{r,s}} = V_{-r,-s}$. With this notation, $S_+ = V_{-1,1}$ and $S_- = V_{1,-1}$.

We want now to associate the holomorphic part $\phi_{(r,s)}$ of a primary operator with the vertex operator $V_{r,s}$. By convention we often denote the scalar operator $\phi(z, \bar{z}) \sim \phi_{(r,s)}(z) \otimes \phi_{(r,s)}(\bar{z})$ simply by $\phi_{(r,s)}$. If, however, the spin is not zero and the primary operator is made out of different representations, $\phi(z, \bar{z}) \sim \phi_{(r,s)}(z) \otimes \phi_{(r',s')}(\bar{z})$, we will write $\phi_{(r,s)(r',s')}$. In the Ising model, the fermionic operators should be written $\psi = \phi_{(2,1)(1,1)}$ and $\bar{\psi} = \phi_{(1,1)(2,1)}$, while the energy-density operator can be simply identified by $\varepsilon = \phi_{(2,1)}$. We recall now that the symmetry of the Kac table imposes $\phi_{(r,s)} = \phi_{(m-r,m+1-s)}$. To associate rigorously vertex operators to unitary primary operators, we should understand the sequence of relations between $V_{r,s}$ and $V_{-r,-s}$ given by (since $m\alpha_+ + (m+1)\alpha_- = 0$)

$$V_{r,s} \approx V_{m-r,m+1-s} \approx Q_+^m Q_-^{m+1} V_{-r,-s} \approx V_{-r,-s}. \qquad (7.36)$$

The exact meaning of the symbol $\approx$ requires a long discussion of the BRST structure lying behind the **screened Coulomb gas** [224, 225, 97], see also the review [249]. Instead, we will show how to build consistently the correlation functions and argue about the consistency of this construction.

7.3 Minimal Correlation Functions

We begin this section by a the calculation of a particular four-point correlation function. We consider the scalar operator $\phi_{(1,2)}$ of any minimal theory, which is known to have a null-vector at level 2. The four-point function

$$\langle \phi_{(1,2)}(z_4, \bar{z}_4) \phi_{(1,2)}(z_3, \bar{z}_3) \phi_{(1,2)}(z_1, \bar{z}_1) \phi_{(1,2)}(z_2, \bar{z}_2) \rangle \qquad (7.37)$$

has both the holomorphic and anti-holomorphic parts constrained by second-order differential operators. Consider first the holomorphic part and with the equivalence $V_{r,s} \equiv V_{-r,-s}$ we investigate

$$V_{1,2}(z_4) V_{1,2}(z_3) V_{-1,-2}(z_1) V_{1,2}(z_2). \qquad (7.38)$$

The total charge of this combination is $2\alpha_0 - \alpha_-$. The extra charge $-\alpha_-$ must be screened by a charge Q_-, which will not affect the conformal properties

---

[1] In the general case: $1 \le r \le (q-1)$ and $1 \le s \le (p-1)$.

but only balances the charge deficit. Notice that we chose the indices of the vertices in (7.38) to keep the charge deficit as small as possible. Our aim is to establish the relation

$$\langle \phi_{(1,2)}(z_4)\phi_{(1,2)}(z_3)\phi_{(1,2)}(z_1)\phi_{(1,2)}(z_2)\rangle \sim \tag{7.39}$$
$$\ll Q_- V_{1,2}(z_4)V_{1,2}(z_3)V_{-1,-2}(z_1)V_{1,2}(z_2) \gg .$$

It is now important to specify the contour of integration due to the screening charge Q_-. We write explicitly the r.h.s. of (7.39) using the free Coulomb gas $n-$point function (7.25) and the definition (7.26).

$$\ll Q_- V_{1,2}(z_4)V_{1,2}(z_3)V_{-1,-2}(z_1)V_{1,2}(z_2) \gg$$
$$= \frac{1}{2\pi i} \oint_C dz \ll V_{1,-1}(z)V_{1,2}(z_4)V_{1,2}(z_3)V_{-1,-2}(z_1)V_{1,2}(z_2) \gg$$
$$= (z_{23}z_{24}z_{34})^{2\alpha_{1,2}^2} (z_{12}z_{13}z_{14}))^{2\alpha_{1,2}(\alpha_{1,2}-2\alpha_0)} \left(\frac{1}{2\pi i}\right)$$
$$\times \oint_C dz (z-z_4)^{2\alpha-\alpha_{1,2}}(z-z_3)^{2\alpha-\alpha_{1,2}}(z-z_2)^{2\alpha-\alpha_{1,2}}(z-z_1)^{2\alpha-(2\alpha_0-\alpha_{1,2})}. \tag{7.40}$$

As in the previous chapters, we consider instead the function of the projective invariant parameter $\eta = (z_1 - z_2)(z_3 - z_4)/(z_3 - z_2)(z_1 - z_4)$ defined in the usual limit given in (5.11). We obtain a term proportional to

$$[\eta(1-\eta)]^{-2\Delta_{1,2}} \int_C dt\, (t-1)^{2\alpha-\alpha_{1,2}} t^{2\alpha-\alpha_{1,2}} (t-\eta)^{2\alpha-(2\alpha_0-\alpha_{1,2})}. \tag{7.41}$$

We know from the last chapters how the differential equation for such two-dimensional family of functions reduces to an hypergeometric differential equation. The integral of (7.41) is the general *Euler integral representation* of the hypergeometric functions [574]. In particular, the hypergeometric function $_2F_1(a,b,c;\eta)$ can be written as the integral (see exercises)

$$_2F_1(a,b,c;\eta) = \frac{\Gamma(c)}{\Gamma(b)\Gamma(c-b)} \int_1^\infty dt\, (t-1)^{c-b-1} t^{a-c} (t-\eta)^{-a}. \tag{7.42}$$

The choice of the contour determines which solution we consider. If we choose $C = [1,\infty[$, the integral is regular in η in the vicinity of zero. However, if we take $C = [0,\eta]$, the integral is singular for $\eta \to 0$ and behaves like $\eta^{1+4\alpha-\alpha_0} = \eta^{\Delta_{1,3}}$. This can be seen after the appropriate change of variable $t = t'\eta$. We obtain

$$\eta^{\Delta_{1,3}} \int_0^1 dt'\, (\eta t' - 1)^{2\alpha-\alpha_{1,2}} t'^{2\alpha-\alpha_{1,2}} (t'-1)^{2\alpha-(2\alpha_0-\alpha_{1,2})} \tag{7.43}$$

and the integral itself is now regular at $\eta = 0$. Thus, the two choices for the contour, $C = [1,\infty[$ and $C = [0,\eta]$, define a basis of solutions which diagonalize the monodromy matrix $\Omega^{(0)}$. Thus the OPE is given by

$$\phi_{(1,2)} \cdot \phi_{(1,2)} = \phi_{(1,1)} + \phi_{(1,3)} = \mathbf{1} + \phi_{(1,3)}. \tag{7.44}$$

This OPE and its corresponding singular behaviours could be calculated from the differential equation given by the level 2 null-vector and along Riemann's method. We made several calculations of this type in the last chapter, so we will not repeat them here. The uniqueness of the possible second-order differential equation with a given singular behaviour shows that the Coulomb gas realization discussed here makes sense.

The solutions giving the conformal blocks in the t-channel are obtained with the linearly related choice of contours $C =]-\infty, 0]$ and $C = [\eta, 1]$. The two last possible choices $C = [0, 1]$ and $C = [\eta, \infty[$ diagonalize $\Omega^{(\infty)}$ and give the u-channel.

A discussion of the singular behaviours related to specific choices of contours can be found in any good table of integrals. Therefore we do not discuss any further this example, but we rather make a general discussion on the construction of the general four-point function of scalar operators

$$\langle \phi_{(r_4,s_4)}(z_4, \bar{z}_4) \phi_{(r_3,s_3)}(z_3, \bar{z}_3) \phi_{(r_1,s_1)}(z_1, \bar{z}_1) \phi_{(r_2,s_2)}(z_2, \bar{z}_2) \rangle. \tag{7.45}$$

A vertex realization requires to replace each operator $\phi_{(r,s)}$ either by $V_{r,s}$ or $V_{-r,-s}$. The charge has to be balanced by a suitable number of screening charges, say r charges Q_+ and s charges Q_-, and for example we can calculate

$$\ll Q_+^r Q_-^s V_{r_4,s_4}(z_4) V_{r_3,s_3}(z_3) V_{-r_1,-s_1}(z_1) V_{r_2,s_2}(z_2) \gg. \tag{7.46}$$

The contours must be consistently ordered to give a well-defined basis of solutions. The number of screening charges depends on the choice of vertices, $V_{r,s}$ or $V_{-r,-s}$. This means that we have several possible contour integral realizations for the same correlation function. This reflects the fact that a primary operators has many null-vectors, which define many compatible differential equations for the same conformal blocks. The discussion of the compatibility of the different contour integral realizations has been discussed by many authors. We do not enter into more detail. Notice that there are infinitely many contour integral realizations due to the fact that

$$m\alpha_+ + (m+1)\alpha_- = 0 \tag{7.47}$$

and therefore the operator $Q_+^m Q_-^{m+1}$ can be added as many times as wanted into any correlation function. It is neutral with respect to both the conformal algebra and the charge balance.

The main interest of the contour integral representation is that it provides a tool to compute the monodromy matrices, and therefore the OPE coefficients. In any expression, which involves one or several contours, we have for each contour two independent choices for the path of integrations. The first two ones discussed in this section gave the singular behaviour at the origin. The other ones can be obtained by deformation of the former ones. This technical question is simple to implement once we know how to deform contours for a single integral.

As an application, we quote the result for the simplest monodromy invariant four-point correlation function of four scalar primary operators in any minimal model, one of which must be a $\phi_{(1,2)}$ (Similar explicit expressions have also been derived for correlators containing a $\phi_{(1,3)}$). Up to an overall normalization [205, 206]

$$\langle \phi_{(r_1,s_1)}(0)\phi_{(1,2)}(z)\phi_{(r_3,s_3)}(1)\phi_{(r_4,s_4)}(\infty)\rangle$$

$$\sim |z|^{4\alpha_{2,1}\alpha_{r_1,s_1}}\,|1-z|^{4\alpha_{2,1}\alpha_{r_3,s_3}}$$

$$\times \left[\frac{\sin(\pi b)\sin(\pi(a+b+c))}{\sin(\pi(a+c))}|I_1(z)|^2 + \frac{\sin(\pi a)\sin(\pi c)}{\sin(\pi(a+c))}|I_2(z)|^2\right], \quad (7.48)$$

where

$$I_1(z) = \int_1^\infty dw\, w^a(w-1)^b(w-z)^c \;,\; I_2(z) = \int_0^z dw\, w^a(1-w)^b(z-w)^c$$

$$(7.49)$$

are hypergeometric functions according to (7.42) and, for unitary minimal models

$$a := 2\alpha_-\alpha_{r_1,s_1} \;,\; b := 2\alpha_-\alpha_{r_3,s_3} \;,\; c := 2\alpha_-\alpha_{2,1} \;,\; \alpha_\pm = \pm\left(\frac{m+1}{m}\right)^{\pm 1/2},$$

$$(7.50)$$

where the $\alpha_{r,s}$ are given in (7.34) and m characterizes as usual the central charge. The neutrality condition is satisfied iff $r_4 = r_1 + r_2 + r_3 - 2$ and $s_4 = s_1 + s_2 + s_3 - 4$.

7.4 Minimal Algebras and OPE Coefficients

In this last section, we would like to survey some general results. With the Coulomb gas approach, we can study the solution to the monodromy problem and extract fundamental closed operator algebras. The simplest case is when all operators are spinless (scalars). The OPE of the operators $\{\phi_{(r,s)}(z,\bar{z})\}$ is

$$\phi_{(r_1,s_1)}\cdot\phi_{(r_2,s_2)} = \sum_{\substack{r_3=|r_1-r_2|+1, r_3=(r_1+r_2-1)\bmod(2) \\ s_3=|s_1-s_2|+1, s_3=(s_1+s_2-1)\bmod(2)}}^{\substack{\min(r_1+r_2-1,2m-r_1-r_2-1) \\ \min(s_1+s_2-1,2m-s_1-s_2+1)}} \phi_{(r_3,s_3)}. \quad (7.51)$$

The indices r_3 and s_3 run from their lower bounds to their upper ones with the condition that they have the same parity as $r_1 + r_2 - 1$ and $s_1 + s_2 - 1$ respectively. The largest complete operator algebra is given by the set of all scalar operators $\{\phi_{(r,s)}(z,\bar{z})\}$. This algebra will be called in Chap. 11 the A-algebra. However, this algebra contains several closed subalgebras like the one generated by the two operators $\{\phi_{(1,1)}, \phi_{(m-1,1)}\}$. Another interesting closed subalgebra is generated by the odd-indexed operators $\{\phi_{(r,s)} | r, s \text{ both odd}\}$. In Chap. 8, (7.51) will be used to relate the (scalar) primary operators $\phi_{(r,s)}$ with the composite fields ϕ^n in some continuum field theories.

The operator product coefficients themselves are rather complicated. We quote the full formula for reference. The *non-zero* operator product coefficients of the OPE (7.51) are [207]

$$[C_{(r_1,s_1),(r_2,s_2)}^{(r_3,s_3)}]^2 = \frac{D_{(r_1,s_1),(r_2,s_2)}^{(r_3,s_3)} D_{(r_1,s_1),(r_3,s_3)}^{(r_2,s_2)}}{D_{(r_1,s_1),(r_1,s_1)}^{(1,1)}}, \tag{7.52}$$

where the coefficients D are defined by

$$D_{(r_1,s_1),(r_2,s_2)}^{(r_3,s_3)} = \rho^{4(s'-1)(r'-1)} \prod_{j=1}^{r'-1} \prod_{i=1}^{s'-1} \frac{1}{(i-\rho j)^2} \prod_{j=1}^{r'-1} \frac{\Gamma(j\rho)}{\Gamma(1-j\rho)}$$

$$\times \prod_{j=1}^{s'-1} \frac{\Gamma(j\rho')}{\Gamma(1-j\rho')} \cdot \prod_{j=0}^{r'-2} \prod_{i=0}^{s'-2} \left(\frac{1}{[(s_1-1-i)-\rho(r_1-1-j)]} \right.$$

$$\left. \cdot \frac{1}{[(s_2-1-i)-\rho(r_2-1-j)][(s_3+1+i)-\rho(r_3+1+j)]} \right)^2$$

$$\times \prod_{i=0}^{s'-2} \left(\frac{\Gamma(-\rho'(s_1-1-i)+r_1)\Gamma(-\rho'(s_2-1-i)+r_2)}{\Gamma(1+\rho'(s_1-1-i)-r_1)\Gamma(1+\rho'(s_2-1-i)-r_2)} \right.$$

$$\left. \cdot \frac{\Gamma(\rho'(s_3+1+i)-r_3)}{\Gamma(1-\rho'(s_3+1+i)+r_3)} \right) \cdot \prod_{j=0}^{r'-2} \left(\frac{\Gamma(-\rho(r_1-1-j)+s_1)}{\Gamma(1+\rho(r_1-1-j)-s_1)} \right.$$

$$\left. \cdot \frac{\Gamma(-\rho(r_2-1-j)+s_2)\Gamma(\rho(r_3+1+j)-s_3)}{\Gamma(1+\rho(r_2-1-j)-s_2)\Gamma(1-\rho(r_3+1+j)+s_3)} \right) \tag{7.53}$$

with the parameters

$$r' = \frac{1}{2}(r_1+r_2-r_3+1), \qquad s' = \frac{1}{2}(s_1+s_2-s_3+1),$$

$$\rho = \alpha_+^2 = \frac{m+1}{m}, \qquad \rho' = \alpha_-^2 = \frac{m}{m+1} \tag{7.54}$$

and where $\Gamma(n)$ is the Euler Γ-function. The convention on the product is that if $r' < 2$ or $s' < 2$, then $\prod_{i=0}^{-1} h(i) = 1$, whatever $h(i)$ is. Care is required in deciding whether the indices of $\phi_{(r,s)}$ or those of $\phi_{(m-r,m+1-s)}$ should be taken, since r' and s' must be integers. In practice, this does not impede the calculation of the operator product coefficients. The OPE coefficients satisfy several symmetry properties. Although not obvious from the previous formulas, they are fully symmetric in the indices

$$C_{(r_1,s_1),(r_2,s_2)}^{(r_3,s_3)} = C_{(r_1,s_1),(r_3,s_3)}^{(r_2,s_2)} = C_{(r_2,s_2),(r_1,s_1)}^{(r_3,s_3)} \tag{7.55}$$

and for rational values of m, they have the property

$$\mathbf{C}^{(m-r_3,m+1-s_3)}_{(m-r_1,m+1-s_1),(r_2,s_2)} = \mathbf{C}^{(r_3,s_3)}_{(r_1,s_1),(r_2,s_2)}. \tag{7.56}$$

For packages for the calculation of the OPE coefficients see [585, 586].

Let us finish with some comments on non-scalar operators. Algebras with integer spin operators are a consequence of the fact that the following OPE are identical

$$\phi_{(r_1,s_1)(r_1,s_1)} \cdot \phi_{(r_2,s_2)(r_2,s_2)} = \phi_{(r_1,m+1-s_1)(r_1,m+1-s_1)} \cdot \phi_{(r_2,m+1-s_2)(r_2,m+1-s_2)}. \tag{7.57}$$

Natural non-scalar operators are given by $\phi_{(r,m+1-s)(r,s)}$ because their spins are always integer or half-integer.

$$\Delta_{r,m+1-s} - \Delta_{r,s} = \frac{(m-2r)(m+1-2s)}{4}. \tag{7.58}$$

If for example $m = 1 \bmod(4)$, then there is a closed local algebra made out of the operators $\{\phi_{(r,s)}|r, s \text{ odd}\} \cup \{\phi_{(r,m+1-s)(r,s)}|r, s \text{ odd}\}$. This algebra will appear in Chap. 11 as D-algebra. The OPEs between non-scalar operators are directly comparable with those for their scalar relatives.

$$\phi_{(r_1,m+1-s_1)(r_1,s_1)} \cdot \phi_{(r_2,m+1-s_2)(r_2,s_2)} = \phi_{(r_1,s_1)(r_1,s_1)} \cdot \phi_{(r_2,s_2)(r_2,s_2)}. \tag{7.59}$$

The non-scalar operators imply a $\mathbb{Z}_2$ grading of the operator algebra. Such algebras often implies the doubling of some scalar operators. In the particular example of the three-state Potts model ($m = 5$), the Kac Table 4.2 shows that the D-algebra contains two operators with highest weights $(2/3, 2/3)$ and two operators with highest weights $(1/15, 1/15)$. In this example, we did not consider the half-integer spin operators given by $\{\phi_{(m-r,s)(r,s)}|r, s \text{ even}\}$. Closed operator algebras can be obtained with such operators if one imposes the invariance of the correlation functions under a subgroup of the monodromy group. This was illustrated in the case of the Ising model in Chap. 6.

It is finally important to notice that the equivalence of the OPEs of two non-scalar operators and of their two scalar relatives does not imply the equality of their expansion coefficients. We have seen that in the Ising model where we relaxed the condition of strict locality. In that example, we encountered fermion representations. The energy-density operator $\varepsilon = \phi_{(2,1)(2,1)}$ was in the local sector and the two fermions $\psi = \phi_{(2,1)(1,1)}$ and $\bar{\psi} = \phi_{(1,1)(2,1)}$ in the semi-local sector. The OPE of the fermions were a direct consequence of the OPE of the energy-density operator. However, the numerical values of the OPE coefficients were different.

Some of these coefficients were measured directly in a Monte Carlo simulation of the Ising and three-states Potts models. In Table 7.1, some OPE coefficients for the order parameter and energy densities as found from (7.52) are compared with numerical estimates [56]. The agreement is quite good

Table 7.1. Numerical check of some OPE coefficients in the Ising and Potts models. Simulation results for $\mathbf{C}^{\varepsilon}_{\varepsilon,\varepsilon}$ are consistent with a zero value in both models (after [56])

model	OPE coefficient	predicted	simulation
Ising	$\mathbf{C}^{\sigma}_{\sigma,\varepsilon}$	1/2	0.54 ± 0.05
	$\mathbf{C}^{\varepsilon}_{\varepsilon,\varepsilon}$	0	$\lesssim 0.18$
Potts-3	$\mathbf{C}^{\overline{\sigma}}_{\sigma,\varepsilon}$	0.546	0.61 ± 0.06
	$\mathbf{C}^{\varepsilon}_{\varepsilon,\varepsilon}$	0	$\lesssim 0.02$
	$\mathbf{C}^{\sigma}_{\sigma,\sigma}$	1.092	1.16 ± 0.14

and does provide evidence that the operator product hypothesis is apparently satisfied in simple statistical systems. More extensive and much more precise tests can be performed through calculating finite-size corrections and finite-size scaling functions, see Chap. 13.

Exercises

1. For the free boson model, consider the normal ordering procedure

$$T(z) := \lim_{\epsilon \to 0} \left(J(z + a\epsilon)J(z - \epsilon) - \frac{1}{2(a+1)^2\epsilon^2} \right),$$

where $a \neq -1$ is an arbitrary constant. Show that the OPEs (7.13, 7.19) are independent of a.

2. Consider the hypergeometric differential equation $Dy(x) = 0$, with

$$D := x(1-x)\frac{\partial^2}{\partial x^2} + \left(c - (a+b+1)x \right)\frac{\partial}{\partial x} - ab.$$

Since Fuchs' theorem applies, the two solutions can be found in the form of a Frobénius series. Show in particular that

$$y(x) = {}_2F_1(a, b, c; x) := \sum_{k=0}^{\infty} \frac{(a)_k(b)_k}{k!(c)_k}x^k \quad , \quad (a)_k := \frac{\Gamma(a+k)}{\Gamma(a)}$$

is the only solution with $y(0) = 1$. Verify that the action of D on $f(t) := t^{b-1}(1-t)^{c-b-1}(1-tx)^{-a}$ is a total derivative with respect to t. Thus,

$$y(z) = y_0 \int_C dt\, t^{b-1}(1-t)^{c-b-1}(1-tz)^{-a}$$

is a solution of $Dy(x) = 0$, if the contour C is either closed or else has two of the zeroes of $f(t)$ as endpoints. Prove (7.42), and use this to show that ${}_2F_1(a, b, c; x) = (1-x)^{c-a-b}{}_2F_1(c-a, c-b, c; x)$.

3. Consider the unitary minimal model with $m = 4$. From the Kac formula, write down the complete set of primary operators. Find all subalgebras of the operator product algebra, using (7.51).

4. Write down the simplest four-point correlation function $\langle \phi\phi\phi\phi \rangle$ of four scalar primary operators, if one of them is a $\phi_{2,1}$ of a minimal model.

8. The Hamiltonian Limit and Universality

We have already discussed in Chap. 1 the close analogy between statistical mechanics and field theory. Reinterpreting the transfer matrix as an Euclidean time evolution operator, we were led to define a corresponding quantum Hamiltonian. While in general, this will have a rather complicated form, there is a limit procedure, called the **Hamiltonian limit**, to simplify the structure of the transfer matrix considerably without affecting the universal properties of the critical behaviour under study [582, 242]. As compared to the isotropic transfer matrix, the resulting quantum Hamiltonian is much sparser, which is useful for numerical investigations. The quantum Hamiltonian contains the information about the universal critical behaviour in condensed form. To illustrate this, we show that for the Ising spin system and for the scalar ϕ^4 field theory, which are in the same universality class, the *same* quantum Hamiltonian is obtained. For reviews, see [411, 312].

To make the relationship to the conformal techniques of the previous chapters explicit, we show that H can be written in terms of the generators of the Virasoro algebra, which therefore acts as a dynamical symmetry.

8.1 Hamiltonian Limit in the Ising Model

We now show how to derive explicitly the *quantum* Hamiltonian for a statistical system (see Chap.1), to be obtained via the Hamiltonian limit. Consider the 2D Ising model with the classical Hamiltonian

$$\mathcal{H} = -\sum_{r,t} \left\{ J_s \left[\cos(\Theta_{r,t} - \Theta_{r+1,t}) - 1 \right] + J_\tau \left[\cos(\Theta_{r,t} - \Theta_{r,t+1}) - 1 \right] \right\}, \quad (8.1)$$

where the lattice variable Θ can take the values 0 or π. This can be easily rewritten with the ordinary Ising spins $\sigma = \pm 1$, but the formulation chosen here is more readily extended to more general systems like a Potts model. We explicitly allow for distinct couplings J_s, J_τ between the Ising spins in the "time" and "space" directions. Take a square lattice with $N \times M$ sites, where N is the number of "spatial" and M the number of "temporal" sites. Then, with periodic boundary conditions in "time", the partition function is

$$\mathcal{Z} = \operatorname{tr} T^M, \quad (8.2)$$

where $T = \exp(-\tau H)$ is the transfer matrix introduced in Chap. 1, see (1.84). We now want to take the limit $\tau \to 0$ in a peculiar way. We have

$$T = e^{-\tau H} = 1 - \tau H + \mathcal{O}(\tau^2). \tag{8.3}$$

T is a $2^N \times 2^N$ matrix and acts between pairs of configurations $\{\Theta_i\}$, $\{\tilde{\Theta}_i\}$, with $i = 1, \ldots, N$.

For the *diagonal elements* of T, let $\Theta_i = \tilde{\Theta}_i$ for all $i = 1, \ldots, N$. Then

$$\begin{aligned}
T_{\text{diag}} &= \exp\left(\beta J_s \sum_{i=1}^{N} [\cos(\Theta_i - \Theta_{i+1}) - 1]\right) \\
&= 1 + \tau \lambda^{-1} \sum_{i=1}^{N} [\cos(\Theta_i - \Theta_{i+1}) - 1] + \mathcal{O}(\tau^2),
\end{aligned} \tag{8.4}$$

where

$$\beta J_s = \tau \lambda^{-1}. \tag{8.5}$$

Consider now the *first off-diagonal element*. Pick k such that $\Theta_k \neq \tilde{\Theta}_k$ and the rest of the two configurations is equal: $\Theta_i = \tilde{\Theta}_i$ for all $i \neq k$. We must have $\Theta_k - \tilde{\Theta}_k = \pm\pi$. Then the first class of off-diagonal elements of T is

$$\begin{aligned}
T_{\text{off},1} &= e^{-2\beta J_\tau} \cdot [\text{ diagonal terms }] \\
&= e^{-2\beta J_\tau}(1 + \mathcal{O}(\tau)) \\
&= \tau + \mathcal{O}(\tau^2),
\end{aligned} \tag{8.6}$$

where we have put

$$e^{-2\beta J_\tau} = \tau. \tag{8.7}$$

Consider the *second off-diagonal element*. Pick two distinct k_1, k_2 such that the corresponding lattice variables are distinct, while all the others are equal:

$$\Theta_{k_a} \neq \tilde{\Theta}_{k_a} \text{ if } a = 1, 2 \quad \text{and } \Theta_i = \tilde{\Theta}_i \text{ for all } i \neq k_a. \tag{8.8}$$

Then the corresponding off-diagonal elements of T are

$$T_{\text{off},2} = e^{-2(2\beta J_\tau)}(1 + \mathcal{O}(\tau)) = \mathcal{O}(\tau^2). \tag{8.9}$$

Similarly, one sees that the nth *off-diagonal element* is

$$T_{\text{off},n} = e^{-n(2\beta J_\tau)}(1 + \mathcal{O}(\tau)) = \mathcal{O}(\tau^n) \tag{8.10}$$

provided that the direction-dependent couplings J_s and J_τ are chosen according to (8.5) and (8.7). This gives the relation

$$e^{-2\beta J_\tau} = \lambda \beta J_s, \tag{8.11}$$

where λ is a free parameter.

We now let $\tau \to 0$. If (8.5) and (8.7) are satisfied, the resulting **quantum Hamiltonian** only contains, besides diagonal terms, contributions which involve a *single* spin flip. This is a great simplification as compared to the usual transfer matrix T, which in general contains terms involving up to N simultaneous spin flips. So H is a much sparser matrix and should be expected to be easier to treat, in particular in the context of numerical diagonalization. To achieve this, one is forced into an extremely *anisotropic* limit

$$\beta J_{\mathrm{s}} \to 0 \ , \ \beta J_\tau \to \infty \qquad (8.12)$$

but such that the parameter λ of (8.11) remains fixed. This limit is called the **Hamiltonian limit**. Recalling the results of Chap. 1, the $\tau \to 0$ limit can also be understood as a continuum limit in the time direction, while the discrete lattice in the space directions is retained. If the system under study has an isotropic RG fixed point with sufficiently local interactions, the full information about the universal critical behaviour of the system is preserved when taking the Hamiltonian limit. On the other hand, due to the freedom in defining λ in (8.11), the normalization of H is arbitrary. For conformal invariance applications, we must fix this normalization in a sensible way.

We proceed to write down H explicitly. Introduce an operator $\hat{\Theta}_n$ with eigenvalues $\Theta_n = 0, \pi$ and eigenstates $|\Theta_n\rangle$ and a conjugate operator $\hat{p}_n$

$$\hat{\Theta}_n |\Theta_n\rangle = \Theta_n |\Theta_n\rangle$$
$$\exp\left(\mathrm{i} m\, \hat{p}_n\right)|\Theta_n\rangle = |\Theta_n + m\pi\rangle, \qquad (8.13)$$

where m is an integer. Then, dropping an additive constant

$$H = -\sum_{n=1}^{N} \left[\lambda^{-1} \cos\left(\hat{\Theta}_n - \hat{\Theta}_{n+1}\right) + 2\cos\hat{p}_n \right]. \qquad (8.14)$$

This can be alternatively expressed in terms of Pauli matrices

$$\sigma^z(n) = \exp \mathrm{i}\hat{\Theta}_n = \begin{pmatrix} 1 & 0 \\ 0 & -1 \end{pmatrix}$$
$$\sigma^+(n) = \exp \mathrm{i}\hat{p}_n = \begin{pmatrix} 0 & 1 \\ 0 & 0 \end{pmatrix}. \qquad (8.15)$$

Writing $\sigma^\pm = (\sigma^x \pm \mathrm{i}\sigma^y)/2$, we have $2\cos\hat{p}_n = \sigma^x(n)$ and the quantum Hamiltonian of the 2D Ising model becomes

$$H = -\sum_{n=1}^{N} \left[t\sigma^x(n) + \sigma^z(n)\sigma^z(n+1) \right]. \qquad (8.16)$$

This Hamiltonian will be one of the main objects of study. One often refers to H as the **Ising quantum chain** or the **(1+1)D Ising model**. It will serve as a starting point for the application of conformal invariance in the 2D Ising model. The **transverse field** $t = \lambda^{-1}$ can be interpreted in two different ways:

1. when ground state properties of H are studied, t corresponds to the temperature of the 2D classical Ising spin model.
2. when H is used in turn to construct Boltzmann weights for a 1D quantum chain, t becomes a transverse magnetic field, which does not break the global symmetry of the model.

The correspondence between the eigenvalues and eigenvectors of H with the quantities of interest in statistical mechanics is clear from Tables 1.8 and 1.9. In particular, the E_i are just the eigenvalues of H and the inverse correlation lengths are simply given by the energy gaps $\xi_i^{-1} = E_i - E_0$. This is very convenient, because of (3.49) the correlation lengths are those quantities which are the easiest to describe from the point of view of conformal invariance. The generalization to more than one space dimension or to more general models is straightforward. Before analysing the spectrum of H in detail, we shall give a second method to derive H.

8.2 Hubbard–Stratonovich Transformation

There are many possible realizations of a given universality class. As a further illustration of the correspondence between statistical mechanics and field theory we give a reformulation of the Ising model in terms of a scalar ϕ^4 field theory, following [G1].

Consider a d-dimensional hypercubic lattice with $\mathcal{N}$ sites. The Ising model partition function can be written as

$$Z = \frac{1}{2^{\mathcal{N}}} \sum_{\{\sigma\}} \exp\left(\sum_{(a,b)} K_{ab}\, \sigma_a \sigma_b \right) \quad, \quad \sigma_a^2 = 1, \tag{8.17}$$

where $\hat{K}$ is a matrix of coupling constants. Define $\hat{P} := P_0 \cdot \mathbf{1} + \hat{K}$, where P_0 is some constant. If one chooses P_0 such that $P_0 \geq \max_a \sum_b K_{ab}$, then $\hat{P}$ is positive definite. Define a vector of spin variables $\boldsymbol{\sigma} = (\sigma_1, \dots, \sigma_{\mathcal{N}})$. Then

$$Z = e^{-\mathcal{N}P_0/2}\, 2^{-\mathcal{N}} \sum_{\{\sigma\}} \exp\left(\frac{1}{2}\boldsymbol{\sigma}\hat{P}\boldsymbol{\sigma} \right). \tag{8.18}$$

Now, consider a quadratic form of continuous variables y_a

$$Q(y) := \sum_{a,b=1}^{\mathcal{N}} Q_{ab}\, y_a y_b \tag{8.19}$$

and recall the identity (where $\boldsymbol{y} = (y_1, \dots, y_{\mathcal{N}})$)

$$\int_{-\infty}^{\infty} d^{\mathcal{N}}\boldsymbol{y}\, \exp\left(-\frac{1}{2}\boldsymbol{y}\hat{Q}\boldsymbol{y} \right) = (2\pi)^{\mathcal{N}/2} (\det \hat{Q})^{-1/2}. \tag{8.20}$$

Under a change of variables $y = s + \hat{Q}^{-1}\sigma$, one has

$$Q(y) = s\hat{Q}s + 2s \cdot \sigma + \sigma\hat{Q}^{-1}\sigma. \tag{8.21}$$

Now, if one chooses $\hat{Q} = \hat{P}^{-1} = (P_0 \cdot 1 + \hat{K})^{-1}$, from (8.20) it follows that

$$(2\pi)^{\mathcal{N}/2}(\det \hat{P})^{1/2} = \exp\left(-\frac{1}{2}\sigma\hat{P}\sigma\right)\int_{-\infty}^{\infty} d^{\mathcal{N}}s \, \exp\left(-\frac{1}{2}s\hat{Q}s - \sum_{a=1}^{\mathcal{N}} s_a\sigma_a\right) \tag{8.22}$$

and consequently

$$\mathcal{Z} = e^{-f_0}2^{-\mathcal{N}}\sum_{\{\sigma\}}\int_{-\infty}^{\infty} d^{\mathcal{N}}s \, \exp\left[-\sum_{(a,b)}{}' Q_{ab}s_a s_b - \frac{1}{2}\sum_a Q_{aa}s_a^2 - \sum_a s_a\sigma_a\right], \tag{8.23}$$

where $a \neq b$ in the sum and $e^{-f_0} = e^{-\mathcal{N}P_0/2}(2\pi)^{-\mathcal{N}/2}(\det \hat{P})^{-1/2}$ is some smooth background term. The sum $\sum_{\{\sigma\}}$ can now be carried out and

$$\mathcal{Z} = e^{-f_0}\int_{-\infty}^{\infty} d^{\mathcal{N}}s \, \exp\left[-\sum_{(a,b)}{}' Q_{ab}s_a s_b - \sum_a\left(\frac{1}{2}Q_{aa}s_a^2 - \ln\cosh s_a\right)\right]. \tag{8.24}$$

The exponential in (8.24) contains two terms. The first one describes the interactions between the continuous spin variables s. These are related to the original interaction matrix $\hat{K}$

$$-P_0^2\hat{Q} = -P_0 \cdot 1 + K - P_0^{-1}\hat{K}^2 + P_0^{-2}\hat{K}^3 + \cdots. \tag{8.25}$$

The second term in the exponential is a weight factor $w(s_a)$ for the continuous spins

$$w(s_a) = \frac{1}{2}Q_{aa}s_a^2 - \ln\cosh s_a = \frac{1}{2}(Q_{aa} - 1)s_a^2 + \frac{1}{12}s_a^4 + \cdots \tag{8.26}$$

We now make the connection with a scalar continuum ϕ^4 theory. Let $\hat{K}$ describe nearest neighbor interactions. Then, from (8.25) the higher powers of $\hat{K}$, which do contribute to $\hat{Q}$ and describe further interactions terms of longer range, become increasingly suppressed with powers of P_0. Therefore the presence of these terms will not change the predominantly ferromagnetic character of the interactions and they will be irrelevant. Going to a continuum scalar field $\phi(x)$, the interaction terms due to $\hat{K}$ will create a term $\sim (\partial_\mu\phi)^2$, while the other terms, due to the higher powers of $\hat{K}$, will create higher derivatives. At least close to the upper critical dimension $d^* = 4$, standard power counting arguments [G6, G9] can be used to show that these further contributions are irrelevant. The same is true for the higher powers of $s_a \sim \phi(x)$ neglected in the weight factor. One thus expects that the critical

behaviour of the model should be the same as the one of a scalar ϕ^4 theory
with a generating function

$$\mathcal{Z} \sim \int \mathcal{D}\phi \, e^{-\int dr \, \mathcal{L}}, \quad \mathcal{L} = \frac{1}{2}(\partial_\mu\phi)^2 + \frac{m^2}{2}\phi^2 + \frac{g}{4!}\phi^4, \quad (8.27)$$

where $\mathcal{L}$ is the Lagrangian. Of course, the last step in the derivation going
from (8.24) to (8.27) is at best heuristic. However, in the next section we
shall see that even for $d = 2$, we recover from (8.27) the same quantum
Hamiltonian as obtained for the Ising spin system.

8.3 Hamiltonian Limit of the Scalar ϕ^4 Theory

We now pass again from the Lagrangian formulation of the ϕ^4 theory as
given in (8.27) to the Hamiltonian formulation, following the work of Brower,
Furman and Subbarao [108]. We discretize the space direction(s), but the
time direction is continuous. For notational simplicity, take $d = 2$. From a
Legendre transformation one obtains in the canonical way the Hamiltonian

$$H =: \sum_n \left[\frac{1}{2}\pi^2(n) + V(\phi(n)) + \lambda \left(\phi(n+1) - \phi(n)\right)^2 \right] : \quad (8.28)$$

where $\pi(n)$ is the canonically conjugate momentum to $\phi(n)$

$$\pi^2(n) = -\frac{\partial^2}{\partial\phi^2(n)} \quad (8.29)$$

and the potential $V(\phi) = \frac{1}{2}m^2\phi^2 + \frac{1}{4!}g\phi^4$. For $m^2 < 0$ and $g > 0$, V has
the well-known double-well structure (see Fig. 8.1), where the two minima
correspond to the Ising states $s = \pm 1$.[1]

First, consider the case $\lambda = 0$, restricting attention to a single-site Hamiltonian. Then one has a 1D Schrödinger equation for an anharmonic oscillator.
In the limit $m^2 \to -\infty$, the two lowest levels become (almost) degenerate,
with an exponentially small splitting $\sim \exp m^2$. There is a large gap with all
other levels. The main contribution to $\mathcal{Z}$ comes only from two eigenstates,
which are concentrated around the Ising values $s = \pm 1$.

Next, reintroduce a small intersite coupling $\lambda \geq 0$. Then the two lowest
eigenstates will become degenerate if $-m^2$ is large enough. Since we are
only interested in the critical behaviour, which is dominated by the lowest
eigenvalues of H, it is sufficient to *truncate the Hilbert space just to these two
lowest lying states!* Let ε_0 and ε_1 be the energies of these two states and let

$$|\varepsilon_0\rangle = \begin{pmatrix} 0 \\ 1 \end{pmatrix} \quad \text{and} \quad |\varepsilon_1\rangle = \begin{pmatrix} 1 \\ 0 \end{pmatrix} \quad (8.30)$$

---

[1] We stress that the weight factor $w(s_a)$ in (8.26) has the same form, with $m^2 = Q_{aa} - 1$.

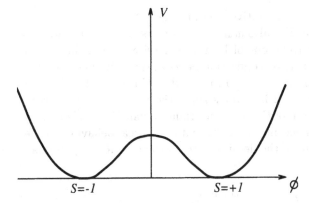

Fig. 8.1. Effective ϕ^4 potential [169]

be the corresponding eigenstates. Due to the truncation, H will contain only diagonal terms and those corresponding to a single "flip", i.e. a tunneling between $|\varepsilon_0\rangle$ and $|\varepsilon_1\rangle$,

$$H = H_{\text{diag}} + H_{\text{flip}}. \tag{8.31}$$

Writing $\varepsilon = \varepsilon_1 - \varepsilon_0$, the diagonal term is

$$H_{\text{diag}} = \sum_n \left[\frac{\varepsilon}{2} (1 + \sigma^z(n)) + \varepsilon_0 \right], \tag{8.32}$$

where σ^z is a Pauli matrix. For the flip term, observe that the tunneling between the two states is given by an operator with elements $\langle \varepsilon_i | \phi(n) | \varepsilon_j \rangle$, which vanishes if $i = j$. Therefore the tunneling operator must be, in the basis given by (8.30), proportional to σ^x. Using the identity

$$(\phi(n+1) - \phi(n))^2 = \phi^2(n+1) + \phi^2(n) - 2\phi(n+1)\phi(n) \tag{8.33}$$

and from (8.28) we have

$$H = \sum_n \left[\frac{\varepsilon}{2}(1 + \sigma^z(n)) + \Lambda(1 - \sigma^x(n+1)\sigma^x(n)) \right] \tag{8.34}$$

with $\Lambda = 2\lambda |\langle \varepsilon_0 | \phi(n) | \varepsilon_1 \rangle|^2$. After a change of basis $\sigma^x \to \sigma^z, \sigma^z \to -\sigma^x$ and setting $t = \varepsilon/(2\Lambda)$, the Hamiltonian is indeed, up to normalization and an additive constant, identical to the (1+1)D Ising model Hamiltonian (8.16).

We remark that the derivation of the Hamiltonian limit uses features analogous to the Ising spin model derivation.

1. One has to take the simultaneous limit $\lambda \to 0, m^2 \to -\infty$, such that t is kept fixed (compare with (8.11)).
2. Make the truncation to single-flip terms.
3. This is justified because (in the limit considered here) the minima of $V(\phi)$ are very deep and the critical behaviour of the model only depends on the lowest eigenvalues.

We summarize this in the commutative diagram Fig. 8.2.[2]

Therefore, the quantum Hamiltonian formulation is a convenient way to describe the long-range correlations of both the classical Ising spin system and the ϕ^4 field theory. The close connection between these two systems is in fact much more general and gives rise to the Landau–Ginzburg classification briefly described below. Fig. 8.2 illustrates again the universality properties of the critical point spectrum of the quantum Hamiltonian H, see also (8.35) below. In particular, it has become apparent that the critical behaviour of the system is quite independent of the detailed form of the interaction potential.

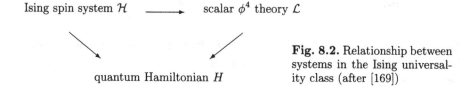

Ising spin system $\mathcal{H}$ ⟶ scalar ϕ^4 theory $\mathcal{L}$

quantum Hamiltonian H

Fig. 8.2. Relationship between systems in the Ising universality class (after [169])

8.4 Hamiltonian Spectrum and Conformal Invariance

We have seen that the conformal invariance results (3.49) and (3.54) were derived in the strip geometry with periodic boundary conditions. Taking the Hamiltonian limit, however, automatically involves going to the strip geometry. Therefore, with the normalization of H properly chosen, we can directly relate the entire spectrum of H, calculated precisely at the critical point $T = T_c$, to universal exponents x_i and the central charge c. On the other hand, we have already seen that the exponents x_i were obtained as eigenvalues of the states of the Verma module under the Virasoro generators $L_0, \bar{L}_0$. Consequently, the quantum Hamiltonian (with periodic boundary conditions!) can be written as being *a part of the Virasoro algebra* (2.65)

$$\boxed{H = \tfrac{2\pi}{N}\left(L_0 + \bar{L}_0\right) - \tfrac{\pi c}{6}\tfrac{1}{N} + \quad \text{const.}} \quad , \qquad (8.35)$$

where the transverse width of the strip is $L = Na$ and a is the spatial lattice constant. We do not specify explicitly here the non-universal bulk free energy.

Equation (8.35) can be used in two ways. First, one can use information on the critical spectrum of H to obtain representations of the Virasoro algebra. Conversely, knowing the Virasoro algebra representations, the *entire* critical point spectrum of H can be computed. Conformal invariance acts therefore as a **dynamical symmetry** of a critical statistical system. In particular,

---

[2] The equivalence between the temperature T in the classical Ising model and the transverse field t in the quantum Ising system has recently been demonstrated experimentally in the dipolar-coupled (leading to long-range interactions) Ising ferromagnet LiHoF$_4$ [E12].

(8.35) gives a necessary and sufficient criterion for conformal invariance of the critical point of a given two-dimensional lattice Hamiltonian.

Rather than obtaining (8.35) by combination of the results of the representation theory of the Virasoro algebra and the relation between energy gaps and the critical exponent, we shall now derive it directly. In order to make contact between the Virasoro generators $L_0, \bar{L}_0$ defined in the infinite plane and the quantum Hamiltonian H defined in the setting of an infinitely long strip of width L, we use as in Chap. 3 the conformal transformation $w = u + iv = (L/2\pi)\ln z$. In the Hamiltonian limit, $L = Na$, where N is the number of sites and a is the lattice constant. Consider the energy–momentum tensor $T(w)$ on the strip. The quantum Hamiltonian is

$$H = \frac{1}{2\pi} \int_0^L dv \left(T(w) + \bar{T}(\bar{w}) \right) \tag{8.36}$$

which is simply the total energy expressed through the energy–momentum tensor. On the other hand, we can formally expand $T(z)$ in the plane

$$T(z) = \sum_{n=-\infty}^{\infty} L_n z^{-n-2}. \tag{8.37}$$

Neglecting the contribution of the conformal anomaly for the moment, we have from (2.66) for the logarithmic map $w = u + iv = (L/2\pi)\ln z$

$$T(w)dw = \left(\frac{dz}{dw} \right)^2 T(z) \frac{dw}{dz} dz = \frac{2\pi}{L} z\, T(z)dz \tag{8.38}$$

and it follows that, up to anomaly terms

$$\frac{1}{2\pi} \int_0^L dv\, T(w) = \frac{2\pi}{L^2} \sum_{n=-\infty}^{\infty} L_n \int_0^L dv \exp\left(-n\frac{2\pi}{L}(u + iv) \right) = \frac{2\pi}{L}L_0. \tag{8.39}$$

The same argument applies to $\bar{T}(\bar{w})$. Combining the two terms and taking care of the conformal anomaly by adding the conformal result (3.54) for the ground state energy, we recover indeed (8.35). Similarly, the eigenvalues of the momentum operator P can be found from the relation

$$\boxed{P = \tfrac{2\pi}{N} \left(L_0 - \bar{L}_0 \right)} . \tag{8.40}$$

It is instructive to derive (8.35,8.40) in yet another way. The energy–momentum tensor can be formally expanded in terms of the Virasoro generators both in the plane (PL) and on the strip (ST), as

$$T(z) = \sum_{n=-\infty}^{\infty} L_n^{PL} (z - z_0)^{-n-2} , \quad T(w) = \sum_{m=-\infty}^{\infty} L_m^{ST} (w - w_0)^{-m-2}. \tag{8.41}$$

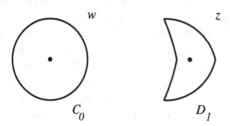

Fig. 8.3. Transformation of contours under $w = \ln z$

Now consider a small circle C_0 around $w_0 = 0$. In the plane, this is transformed into the closed curve D_1 around $z_0 = 1$, as sketched in Fig. 8.3. Integrating (8.38), we have

$$\oint_{C_0} dw\, T(w) = \oint_{D_1} dz\, z T(z) \frac{2\pi}{L} \tag{8.42}$$

and both sides can be calculated with the help of the above expansions

$$\oint_{C_0} dw \sum_m L_m^{\mathrm{ST}} w^{-m-2} = 2\pi i\, L_{-1}^{\mathrm{ST}}$$

$$= \frac{2\pi}{L} \oint_{D_1} dz \sum_n L_n^{\mathrm{PL}} (z-1)^{-n-2} z$$

$$= \frac{2\pi}{L} \oint_{D_1} dz \sum_n L_n^{\mathrm{PL}} (z-1)^{-n-1} + \underbrace{\frac{2\pi}{L} \oint_{D_1} dz\, T(z)}_{=0}$$

$$= 2\pi i \frac{2\pi}{L} L_0^{\mathrm{PL}} \tag{8.43}$$

since $T(z)$ is analytic. Add the contribution of the conformal anomaly to get

$$L_{-1}^{\mathrm{ST}} = \frac{2\pi}{L} \left(L_0^{\mathrm{PL}} - \frac{c}{24} \right) \ , \quad \bar{L}_{-1}^{\mathrm{ST}} = \frac{2\pi}{L} \left(\bar{L}_0^{\mathrm{PL}} - \frac{c}{24} \right). \tag{8.44}$$

Combing these, (8.35) and (8.40) are recovered. In other words, the generator of the dilatations in the infinite plane, $L_0 + \bar{L}_0$, is transformed into the generator of time translations, $L_{-1} + \bar{L}_{-1}$, in the strip geometry. Since in quantum mechanics, the quantum Hamiltonian H is shown to be the infinitesimal generator of time translations, this result is physically sensible.

8.5 Temperley–Lieb Algebra

In this section, we reconsider the self-duality of the q-states Potts model, already derived in Chap. 1 and use the opportunity to briefly introduce the **Temperley–Lieb algebra** [584, 493] which plays an important role in integrable systems. The book [450] provides a very detailed and exhaustive introduction and review to *word algebras*, of which the Temperley–Lieb algebras are but the simplest example. With the modest goal stated above, we follow here the presentation of [169].

Reconsider the q-states Potts model as defined in Chap. 1. One may generalize it by allowing for different coupling constants J_s along the (horizontal) "space" direction and J_τ along the (vertical) "time" direction. Let $K_s = J_s/T$ and $K_\tau = J_\tau/T$. The partition function is

$$\mathcal{Z} = \sum_{\{\Theta\}} \exp \left(\sum_{\langle i,j\rangle_\tau} K_\tau \delta(\Theta_i - \Theta_j) + \sum_{\langle i,j\rangle_s} K_s \delta(\Theta_i - \Theta_j) \right). \qquad (8.45)$$

where the two sums $<\ >_{\tau,s}$ run over pairs of nearest neighbours in the time and space directions. The spin variable Θ_i may take the values $0, \ldots, q-1$ and the lattice is assumed to have M temporal rows and N spatial sites.

We now write the partition function in terms of the transfer matrix. Let $\Theta = \{\Theta_1, \ldots, \Theta_N\}$ denote the spin configuration of a entire row at time t and let similarly be Θ' stand for the configuration for the neighbouring row at time $t + 1$. Define two $q^N \times q^N$ matrices $\hat{V}$ and $\hat{W}$ via their elements

$$V_{\Theta,\Theta'} := \exp \left(K_s \sum_{i=1}^{N} \delta(\Theta_i - \Theta_{i+1}) \right) \prod_{j=1}^{N} \delta(\Theta_j - \Theta'_j)$$

$$W_{\Theta,\Theta'} := \exp \left(K_\tau \sum_{i=1}^{N} \delta(\Theta_i - \Theta'_i) \right). \qquad (8.46)$$

For periodic boundary conditions in the "time" direction, one has

$$\mathcal{Z} = \mathrm{tr} \left(\hat{V}\hat{W} \right)^M. \qquad (8.47)$$

We want to rewrite $\mathcal{Z}$ such that the *duality transformation* between the high- and low-temperature regions becomes explicit in an algebraic way. This is an alternative to the graph-theoretical arguments used in Chap. 1.

In order to simplify the explicit matrices to be written down, we shall restrict to the Ising model special case $q = 2$ throughout. The extension to general q is left as a straightforward exercise, see also [450]. Write

$$W_{\Theta,\Theta'} = \prod_{i=1}^{N} \exp \left(K_\tau \delta(\Theta_i - \Theta'_i) \right) =: \prod_{i=1}^{N} \hat{w}(i), \qquad (8.48)$$

where $\hat{w}(i)$ is a $2^N \times 2^N$ matrix which connects the states of each row

$$w(i) = \mathbf{1} \otimes \cdots \otimes \mathbf{1} \otimes \overset{i}{\overbrace{\mathrm{w}}} \otimes \cdots \mathbf{1} \qquad (8.49)$$

and where $\mathbf{1}$ is the identity matrix in 2 dimensions. The matrix w acts between the sites i and i'. Since it is site-independent, however, it is enough to write it as a 2×2 matrix

$$\mathbf{w} = \begin{pmatrix} e^{K_\tau} & 1 \\ 1 & e^{K_\tau} \end{pmatrix} \tag{8.50}$$

(for the q-states Potts model, w would become a $q \times q$ matrix) and define

$$\Upsilon_\tau := \frac{1}{\sqrt{2}} (e^{K_\tau} - 1) \tag{8.51}$$

(again, for the q-states Potts model, replace $\sqrt{2}$ by $\sqrt{q}$). Consequently

$$\hat{w}(i) = \sqrt{2} \, (\Upsilon_\tau \mathbf{1}_N + e_{2i-1}), \tag{8.52}$$

where $\mathbf{1}_N = \mathbf{1} \otimes \cdots \otimes \mathbf{1}$ is the identity acting on all N sites in a row. Here

$$e_{2i-1} := \frac{1}{\sqrt{2}} \mathbf{1} \otimes \cdots \otimes \overbrace{\begin{pmatrix} 1 & 1 \\ 1 & 1 \end{pmatrix}}^{i} \otimes \cdots \mathbf{1} \tag{8.53}$$

will play an important role later. The matrix $\hat{V}$ is treated analogously

$$V_{\Theta,\Theta'} = \prod_{i=1}^{N} \delta(\Theta_i - \Theta'_i) \, \delta(\Theta_{i+1} - \Theta'_{i+1}) \exp\left(K_s \delta(\Theta_i - \Theta_{i+1})\right) =: \prod_{i=1}^{N} \hat{v}(i) \tag{8.54}$$

and from its definition, one can see that $\hat{v}(i)$ is different from the unit matrix only on the subspace connecting the sites i and $i+1$ to the sites i' and $i'+1$. The two delta functions $\delta(\Theta_i - \Theta'_i)\delta(\Theta_{i+1} - \Theta'_{i+1})$ in (8.54) show that

$$\hat{v}(i) = \mathbf{1} \otimes \cdots \otimes \overbrace{\mathbf{v}}^{i\otimes(i+1)} \otimes \cdots \otimes \mathbf{1} \;, \quad \mathbf{v} := \begin{pmatrix} e^{K_s} & 0 & 0 & 0 \\ 0 & 1 & 0 & 0 \\ 0 & 0 & 1 & 0 \\ 0 & 0 & 0 & e^{K_s} \end{pmatrix} \tag{8.55}$$

is diagonal. As before, a new weight factor

$$\Upsilon_s := \frac{1}{\sqrt{2}} (e^{K_s} - 1) \tag{8.56}$$

is defined and we have

$$\hat{v}(i) = (\mathbf{1}_N + \Upsilon_s e_{2i}), \tag{8.57}$$

where the generators e_{2i} read

$$e_{2i} := \sqrt{2} \, \mathbf{1} \otimes \cdots \otimes \overbrace{\begin{pmatrix} 1 & 0 & 0 & 0 \\ 0 & 0 & 0 & 0 \\ 0 & 0 & 0 & 0 \\ 0 & 0 & 0 & 1 \end{pmatrix}}^{i\otimes(i+1)} \otimes \cdots \otimes \mathbf{1}. \tag{8.58}$$

We now come to an essential stage of this exercise. It turns out that the generators e_i defined in (8.53,8.58) satisfy a common set of algebraic rules which is called a **Temperley–Lieb algebra** [584]

$$
\begin{aligned}
e_i^2 &= \bar{\beta} e_i , & \bar{\beta} &= \sqrt{q} \\
e_i e_j &= e_j e_i , & |i - j| &\geq 2 . \\
e_i e_{i\pm 1} e_i &= e_i
\end{aligned}
\tag{8.59}
$$

Although from the explicit expressions given above only the special case $q = 2$ can be checked directly, the generalization goes through for any value of q. The Temperley–Lieb algebra is not a Lie algebra, since there are no commutators. Rather, it represents a different class of algebras which are called *word algebras*. There is a rich knowledge of their representations and other properties which can be found in [450]. Since our purpose here is much more modest, we continue with the q-state Potts model and rewrite the partition function in terms of the Temperley–Lieb generators e_i (again, we leave the extension of the above calculation to general q to the reader)

$$
\mathcal{Z} = \left(\prod_{i=1}^{N} \sqrt{q} \left(\varUpsilon_\tau \mathbf{1}_N + e_{2i-1} \right) \left(\mathbf{1}_N + \varUpsilon_s e_{2i} \right) \right)^{M} .
\tag{8.60}
$$

The Temperley–Lieb algebra (8.59) is invariant under the mapping $e_i \to e_{i+1}$, which exchanges the roles played by the even-indexed and the odd-indexed generators. While that does not affect the algebraic relations between the generators, it relates the partition function $\mathcal{Z} = \mathcal{Z}(\varUpsilon_\tau, \varUpsilon_s)$ to another partition function with modified Boltzmann weights. In the limit of an infinitely large lattice, the mapping $e_i \to e_{i+1}$ does not create problems with any boundary terms which might change the value of the partition function. Extracting a factor $(\varUpsilon_\tau \varUpsilon_s)^{NM}$, one finds

$$
\mathcal{Z}(\varUpsilon_\tau, \varUpsilon_s) = (\varUpsilon_\tau \varUpsilon_s)^{NM} \, \mathcal{Z}(1/\varUpsilon_s, 1/\varUpsilon_\tau)
\tag{8.61}
$$

and consequently, one recovers the self-duality relation

$$
f(\varUpsilon_\tau, \varUpsilon_s) = - \ln(\varUpsilon_\tau \varUpsilon_s) + f(1/\varUpsilon_s, 1/\varUpsilon_\tau)
\tag{8.62}
$$

for the free energy density defined as

$$
f = - \lim_{N,M \to \infty} \frac{1}{NM} \ln \mathcal{Z}.
\tag{8.63}
$$

In particular, it is easy to check that for $K_\tau = K_s = K$, one recovers the isotropic self-duality relation (1.51) and in particular, the result (1.53), viz. $K_c = \ln(\sqrt{q} + 1)$ for the critical point. More generally, the self-dual line $\varUpsilon_\tau \varUpsilon_s = (e^{K_\tau} - 1)(e^{K_s} - 1)/q = 1$ separates a disordered high-temperature phase, with $\varUpsilon_\tau \varUpsilon_s < 1$ from an ordered low-temperature phase with $\varUpsilon_\tau \varUpsilon_s > 1$. Along this entire line, the anisotropic q-states Potts model is critical.

The relationship of a given statistical model with a Temperley–Lieb algebra through its partition function gives a clear indication that the model is integrable. Integrability is independent of whether the model may or may not be critical. For a more precise formulation of this statement, making use of the even more general Hecke algebras, we refer to an important paper by Jones [374]. If a given model can be shown to contain a Temperley–Lieb algebra, the calculation of its partition function may be reduced to a problem of the representation theory of that algebra and existing general techniques and results may then be applied [450]. We mention a simple application to the example of the q-state Potts model. One can show in general that hermitian representations of the Temperley–Lieb algebras with $\bar{\beta} < 2$ only exist if

$$\bar{\beta} = 2 \cos \left(\frac{\pi}{m+1} \right), \tag{8.64}$$

where m is a positive integer. Since $\bar{\beta}^2 = q$, letting $m = 3, 5, \infty$, one recovers the $q = 2, 3, 4$-states Potts models. This suggests an important difference between these three models and the q-states Potts models with $q > 4$, which in fact are known to undergo a first-order transition with a finite latent heat [G19]. Furthermore, it also suggests the presence of many more integrable spin models corresponding to Potts models with certain non-integral values of q. In fact, the models characterized by (8.64) can be shown to be conformally invariant at criticality, having central charges $c = 1 - 6/(m(m+1))$. Their partition functions can be described in terms of RSOS models.

Temperley–Lieb algebras can be used to relate the spectra of the quantum Hamiltonians of quite distinct systems. In Chap. 15, we shall map in this way the q-states Potts model onto the XXZ quantum chain. This allows to make contact with Bethe-ansatz techniques to solve the model explicitly.

8.6 Laudau–Ginzburg Classification

We have seen earlier that the Ising model quantum Hamiltonian could be obtained from either an Ising spin system or a scalar ϕ^4 field theory. This universality suggests that there might be a series of field theory realizations of conformally invariant systems and we shall now briefly mention a correspondence between the A_m-RSOS models and a class of scalar field theories. Such a realization might for example be useful to make explicit contact with ε-expansion related studies.

Consider the Landau–Ginzburg–Wilson (LGW) action of the form

$$S = \int d^2 r \left(\frac{1}{2} (\partial \phi)^2 + \sum_{n=1}^{2(m-1)} \lambda_n \phi^n \right). \tag{8.65}$$

For $\lambda_1 = \lambda_2 = \ldots = \lambda_{2m-3} = 0$, one has a multicritical point. These 2D critical systems have a central charge $c(m) = 1 - 6/(m(m+1))$ and the scaling dimensions of the fields are given by the Kac formula. The correspondence, due to A. Zamolodchikov [639], between the scaling fields ϕ^n and the conformal weights $\Delta_{r,s}$ is illustrated in Table 8.1 for the cases $m = 3, 4, 5$ which already shows the general structure. The index s runs horizontally from 1 at the left to m at the right and the index r run vertically from 1 below to $m - 1$ at the highest line. The conformal field $\phi_{1,1}$ always corresponds to the identity operator $\mathbf{1}$ and $\Delta_{1,1} = 0$. The LGW field ϕ^{2m-3} is redundant and can always be eliminated form the action through the equations of mation. To see the correspondence between the LGW fields ϕ^n and the conformal primary fields $\phi_{r,s}$, recall the symmetry relation $\Delta_{r,s} = \Delta_{m-r,m+1-s}$. The LGW fields, reading the table from the lower left corner, correspond to the conformal weights descending from the upper right corner.

Table 8.1. Correspondence between conformal primary operators and the LGW fields ϕ^n, for $m = 3, 4, 5$

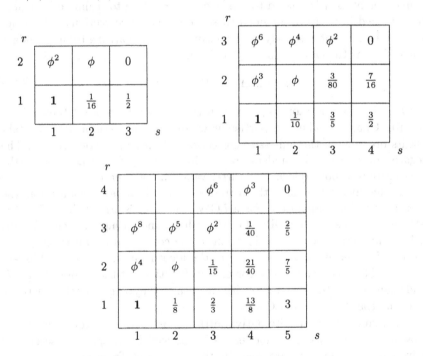

r			
2	ϕ^2	ϕ	0
1	1	$\frac{1}{16}$	$\frac{1}{2}$
	1	2	3 s

r				
3	ϕ^6	ϕ^4	ϕ^2	0
2	ϕ^3	ϕ	$\frac{3}{80}$	$\frac{7}{16}$
1	1	$\frac{1}{10}$	$\frac{3}{5}$	$\frac{3}{2}$
	1	2	3	4 s

r					
4			ϕ^6	ϕ^3	0
3	ϕ^8	ϕ^5	ϕ^2	$\frac{1}{40}$	$\frac{2}{5}$
2	ϕ^4	ϕ	$\frac{1}{15}$	$\frac{21}{40}$	$\frac{7}{5}$
1	1	$\frac{1}{8}$	$\frac{2}{3}$	$\frac{13}{8}$	3
	1	2	3	4	5 s

Let us make this explicit through a few examples. Begin with the identity operator $\phi_{1,1} = \mathbf{1}$ in the lower left corner. Because of the relation

$$\phi_{r,s} = \phi_{m-r,m+1-s} \tag{8.66}$$

this should be identified with the operator $\phi_{m-1,m}$ in the upper right corner where the conformal weight $\Delta = \Delta_{1,1} = \Delta_{m-1,m} = 0$ is shown. Next, we move from $\phi_{1,1}$ in the lower left corner one field to the right and find the operator $\phi_{1,2}$. Correspondingly, because of (8.66), we must also move one field to the left from the upper right corner and find the operator $\phi_{m-1,m-1}$. We can then read off that in the LGW formulation, this yields the identification $\phi_{1,2} = \phi_{m-1,m-1} = \phi^{m-2}$ with conformal weight $\Delta = \Delta_{1,2} = \Delta_{m-1,m-1} = 1/16, 1/10, 1/8$ for $m = 3, 4, 5$, respectively. Next, we move from this operator one field up. We find the operator $\phi_{2,2}$. One the other side of the table, we have to move one field down. Then we have the identification $\phi_{2,2} = \phi_{m-2,m-1} = \phi$. The conformal weight is then $\Delta = \Delta_{2,2} = \Delta_{m-2,m-1} = 1/16, 3/80, 1/40$ for $m = 3, 4, 5$, respectively.

Since the LGW action is invariant under the $\mathbb{Z}_2$-symmetry $\phi \to -\phi$, one expects these models to coincide with the A series in Table 11.1. This is the case. In particular, the least relevant of the relevant operators corresponds to the conformal field $\phi_{1,3} \sim \phi^{2m-4}$, while the field $\phi_{3,1} \sim \phi^{2m-2}$ corresponds to the most relevant of the irrelevant primary fields. Thus, the associated conformal weight $\Delta_{3,1}$ is related to the leading correction to scaling exponent ω [610] defined in Chap. 3, provided these *non-analytic* corrections are asymptotically larger that the analytic corrections to scaling arising from secondary fields. That is the case for for $m \geq 4$. Then

$$\omega = 2(\Delta_{3,1} - 1) \ , \quad m \geq 4. \tag{8.67}$$

For $m = 3$, the operator $\phi_{3,1}$ is not in the Kac table. More details will be given in Chap. 13, where in particular we discuss an attempt [94] to establish its presence in the finite-size corrections of an Ising lattice system. The negative outcome of this underscores the importance of the Kac table for the classification of both relevant and irrelevant scaling operators.

The justification [639] of the correspondence in Table 8.1 comes from the operator product expansion. For a LGW field ϕ, it is trivial that $\phi^a \cdot \phi^b = \phi^{a+b}$ in the sense of the OPE. On the other hand, using the general OPE (7.51) and the proposed identification of the conformal operators $\phi_{r,s}$ with the LGW fields ϕ^a, the consistency of the identification can be established. As a corollary, the multipoint correlation functions of the composite LGW fields can be written down explicitly in terms of hypergeometric functions, following the techniques of Chap. 7.

We remark that an alternative lattice construction of these models is provided by the A_m-RSOS models as introduced in Chap. 1 whose critical point partition functions are given by the A series of Table 11.1.

Exercise. For the three-states Potts model, derive the quantum Hamiltonian H explicitly and find the associated continuum field theory via the Hubbard–Stratonovich transformation. Include a symmetry-breaking magnetic field.

9. Numerical Techniques

We now discuss the techniques necessary for a numerical diagonalization of a given quantum Hamiltonian as obtained in Chap. 8. First, the problem of diagonalizing H can be broken into the separate consideration of block-diagonal parts by using a few simple symmetry properties. In connection with conformal invariance, the systematic use of translation invariance allows to fix the normalization of H in a unique way. Numerical diagonalization and finite-lattice extrapolation algorithms will be described. In doing so, we shall merely mention a few selected techniques which have proved to work reliably in the context of critical quantum chains and do not attempt to give a systematic overview on the numerous numerical algorithms which exist in the literature. These methods can be applied independently of the conformal invariance of the model, but throughout this chapter, we shall take H to be defined on a quantum *chain* of N sites.

9.1 Simple Properties of Quantum Hamiltonians

For the purposes of this section, we shall use the example of the 2D p-states vector Potts model (see Chap. 1), but the presentation is readily extended to any system with p states per site. The classical Hamiltonian is

$$\mathcal{H} = -\beta \sum_{r,t} \left\{ J_s \left[\cos(\Theta_{r,t} - \Theta_{r+1,t}) - 1 \right] + J_\tau \left[\cos(\Theta_{r,t} - \Theta_{r,t+1}) - 1 \right] \right\}, \tag{9.1}$$

where the spin variables $\Theta = (2\pi/p)\tilde{r}$, with $\tilde{r} = 0, \ldots, p-1$, are attached to each site $r = (r,t)$. The Hamiltonian limit of (9.1) can be done in complete analogy with the calculations of Chap. 8. For a quantum chain with N sites, the quantum Hamiltonian is in the limit $\beta J_s \to 0, \beta J_\tau \to \infty$ with λ fixed [572], up to an additive constant

$$H = -2 \sum_{n=1}^{N} \left\{ \lambda \cos(\hat{\Theta}_n - \hat{\Theta}_{n+1}) + \cos \hat{p}_n \right\}. \tag{9.2}$$

The limit taken implies that we are considering the geometry of an infinitely long strip of finite width $L = Na$, where a is the lattice constant. Expressed with variables

$$\sigma_n := \exp i\hat{\Theta}_n = \begin{pmatrix} 1 & & & & \\ & \omega & & & \\ & & \omega^2 & & \\ & & & \ddots & \\ & & & & \omega^{p-1} \end{pmatrix}, \tag{9.3}$$

where $\omega = \exp(2\pi i/p)$ and

$$\Gamma_n := \exp i\hat{p}_n = \begin{pmatrix} 0 & & & & 1 \\ 1 & 0 & & & \\ & 1 & 0 & & \\ & & \ddots & \ddots & \\ & & & 1 & 0 \end{pmatrix}, \tag{9.4}$$

the quantum Hamiltonian becomes

$$H = -\sum_{n=1}^{N} \left\{ \lambda \left(\sigma_n \sigma_{n+1}^+ + \sigma_n^+ \sigma_{n+1} \right) + \Gamma_n + \Gamma_n^+ \right\} \tag{9.5}$$

which has the dihedral group $\mathbb{D}_p$ as global symmetry.[1] More general choices of the couplings are possible and lead to quantum Hamiltonians describing other universality classes. We now give a few properties of the $p^N \times p^N$ matrix H (9.5).

Charge symmetry. H commutes with the $\mathbb{Z}_p$ *charge operator*

$$\hat{Q} := \prod_{n=1}^{N} \Gamma_n. \tag{9.6}$$

$\hat{Q}$ has the eigenvalues ω^Q. The corresponding eigenspaces are called **charge sectors**, $Q = 0, 1, \ldots, p-1$.

Parity. The *parity operator* $\hat{\Pi}$ is defined by

$$\hat{\Pi}\sigma_n\hat{\Pi}^+ = \sigma_{N-n+1} \;,\; \hat{\Pi}\Gamma_n\hat{\Pi}^+ = \Gamma_{N-n+1} \tag{9.7}$$

and has eigenvalues $\Pi = \pm 1$. The eigenspaces of $\hat{\Pi}$ are called **parity sectors**.

Charge conjugation. The *charge conjugation* operator

$$\hat{C} := \prod_{n=1}^{N} \hat{c}_n \;,\; \hat{c}_n = \begin{pmatrix} 1 & & & & & 0 \\ & 0 & & & 0 & 1 \\ & & \ddots & 1 & & \\ & & 1 & & & \\ & 0 & 1 & & & \ddots \\ 0 & 1 & & & & 0 \end{pmatrix} \tag{9.8}$$

---

[1] Recall the definition of $\mathbb{D}_p$ [293]: it is a finite group with a p-fold axis and a system of orthogonal two-fold axis and contains $2p$ elements. It can be generated from two elements a, b such that $a^p = \mathbf{1}$, $b^2 = \mathbf{1}$ and $baba = \mathbf{1}$. (The dihedral group $\mathbb{D}_p$ should not be confused with the semi-simple Lie algebra D_p of rank p.)

has the eigenvalues $C = \pm 1$. $\hat{C}$ transforms the charge sector $Q = q$ into $Q = p - q(\mathrm{mod}\,p)$ and vice versa. One has

$$\hat{C}\sigma_n\hat{C}^+ = \sigma_n^+ \ , \quad \hat{C}\Gamma_n\hat{C}^+ = \Gamma_n^+ . \tag{9.9}$$

Invariance under charge conjugation is a property of the Potts Hamiltonian (9.5), but not necessarily of more general Hamiltonians with complex couplings. Consequently, H of (9.5) has the same spectrum in the two charge sectors linked by $\hat{C}$. In the sector $Q = 0$ (and also $Q = p/2$ if p is even), the states can be labeled by the values of the C-parity and are thus cast into **charge conjugation sectors**.

Algebra of σ_n and Γ_m. The matrices σ_n and Γ_m satisfy the algebra

$$\begin{aligned}
\sigma_n\Gamma_m &= \delta_{nm}\,\omega\Gamma_m\sigma_n + (1 - \delta_{nm})\,\Gamma_m\sigma_n \\
\sigma_n^p &= \Gamma_n^p = 1.
\end{aligned} \tag{9.10}$$

This algebra is left invariant under the global transformation

$$\begin{aligned}
\Gamma &\to U\Gamma U^+ = \sigma \\
\sigma &\to U\sigma U^+ = \Gamma^+.
\end{aligned} \tag{9.11}$$

Also, the matrix form (9.8) of $\hat{C}$ is kept invariant under this transformation. Sometimes, the application of (9.11) to the Hamiltonian (9.5) is referred to as the change into the *high-temperature basis*.

Duality. Another transformation, which leaves the algebra (9.10) invariant, is the *duality transformation*

$$\begin{aligned}
\sigma_n &\to \tilde{\sigma}_n := \Gamma_{n+1}\Gamma_n^+ \quad \text{if } n < N \\
\sigma_N &\to \tilde{\sigma}_N := \Gamma_N^+ \\
\Gamma_n &\to \tilde{\Gamma}_n := \prod_{l=1}^{n}\sigma_l.
\end{aligned} \tag{9.12}$$

Note the peculiar treatment of the boundary spin. Up to boundary terms (which means, in the $N \to \infty$ limit), the p-state Potts model satisfies the **self-duality** property

$$H(\lambda) = \lambda H(1/\lambda). \tag{9.13}$$

This is the analogue of self-duality as derived in Chaps. 1 and 8. To see this, one has to apply the Hamiltonian limit to the self-duality condition (8.62). In order to have self-duality for N finite, special boundary conditions are needed. If the model has only one critical point, then by self-duality it has to be at $\lambda_c = 1$.

To summarize the contents of this section, consider the most general $\mathbb{Z}_p$ symmetric quantum Hamiltonian H of a model with p states per site and N sites

$$H = -\sum_{n=1}^{N}\sum_{k=1}^{p}\left[a_k(n)\sigma_n^k\sigma_{n+1}^{p-k} + b_k(n)\Gamma_n^k\right],\qquad(9.14)$$

where $a_k(n)$ and $b_k(n)$ are (possibly site-dependent) couplings. Particular choices for the couplings reflect themselves in symmetry properties of H as shown in Table 9.1.

Table 9.1. Symmetry properties of a p-state $\mathbb{Z}_p$ symmetric quantum Hamiltonian

Symmetry	Condition	
$H = H^+$	$a_k(n) = a_{p-k}^*(n)$	$b_k(n) = b_{p-k}^*(n)$
$[H, \hat{\Pi}] = 0$	$a_k(n) = a_k(N-n+1)$	$b_k(n) = b_k(N-n+1)$
$[H, \hat{C}] = 0$	$a_k(n) = a_{p-k}(n)$	$b_k(n) = b_{p-k}(n)$
H self-dual	$a_k(n) = \lambda b_k(n)$	

9.2 Some Further Physical Quantities and their Critical Exponents

We now define a few additional physical quantities. Since λ can be seen to correspond to a temperature-like variable, the analogon of the *specific heat* can be defined as

$$C = -\frac{1}{N}\frac{\mathrm{d}^2 E_0}{\mathrm{d}\lambda^2} \sim (\lambda - \lambda_c)^{-\alpha}.\qquad(9.15)$$

To define an analogon of a *magnetic susceptibility*, we have to add a symmetry-breaking term, e.g. $hH_1 = -h\sum_n(\sigma_n + \sigma_n^+)$, to H and to calculate the ground state energy $E_0(h)$. Then the susceptibility per spin is

$$\chi = -\frac{1}{N}\frac{\mathrm{d}^2 E_0}{\mathrm{d}h^2}\bigg|_{h=0} \sim (\lambda - \lambda_c)^{-\gamma}.\qquad(9.16)$$

Turning to the *order parameter*, one faces the problem that a double limit: (i) $N \to \infty$ with $T \le T_c$ and only afterwards (ii) $T \to T_c$; is needed for its definition. Carrying out the double limit is computationally difficult. A trick first used by Yang [629] to calculate the spontaneous magnetization in the Ising model can be used to circumvent this difficulty. Write the total Hamiltonian $H_{\text{tot}} = H + hH_1$. Close to criticality ($\lambda \to \lambda_c, h \to 0$), one has a very small energy gap $\xi^{-1} = E_1 - E_0 \to 0$. Now, suppose that even for finite N, ξ^{-1} is exponentially small and therefore treat E_1 and E_0 as degenerate. In first order degenerate perturbation theory in h, the magnetization per site M will be obtained from the smallest eigenvalue of the matrix [597, 295]

$$M = -\frac{\partial f^{\rm sin}}{\partial h}\bigg|_{h=0} = \frac{1}{N} \begin{pmatrix} \langle 0|H_1|0\rangle & \langle 0|H_1|1\rangle & \cdots \\ \langle 1|H_1|0\rangle & \langle 1|H_1|1\rangle & \cdots \\ \vdots & \vdots & \ddots \end{pmatrix}, \tag{9.17}$$

where $|0\rangle, |1\rangle, \ldots$ are the eigenstates of the smallest eigenvalues in the charge sectors $Q = 0, 1, \ldots$. Since H_1 is connecting different charge sectors and breaks the global symmetry, its diagonal elements vanish. The 'order parameter' defined this way can be used for either first-order or second-order transitions. In the latter case, it shows the correct finite-size scaling behaviour $M \sim N^{-\beta/\nu}$.

A similar problem arises in the definition of the *latent heat*. Here, we write

$$H_{\rm tot} = H + \vartheta V \quad \text{with } \vartheta = \lambda - \lambda_{\rm c}. \tag{9.18}$$

V is proportional to the energy density and does not link states in different charge sectors. Denote by $|0\rangle, |\varepsilon\rangle$ the eigenstates with the lowest eigenvalues in the charge sector $Q = 0$. Assume these two states to be degenerate even for N finite. The ground state energy is then, to lowest order in ϑ, obtained from the eigenvalues of $\mathcal{E} = E_0 \cdot 1 + \vartheta \mathcal{E}_1$, where

$$\mathcal{E}_1 = \begin{pmatrix} \langle 0|V|0\rangle & \langle 0|V|\varepsilon\rangle \\ \langle \varepsilon|V|0\rangle & \langle \varepsilon|V|\varepsilon\rangle \end{pmatrix}. \tag{9.19}$$

Let $e_\pm$ be the higher and lower eigenvalues of $\mathcal{E}_1$. Then the latent heat per site $\ell_{\rm h}$ is [296]

$$\ell_{\rm h} = \frac{e_+ - e_-}{N} = \frac{1}{N}\left[(\langle 0|V|0\rangle - \langle \varepsilon|V|\varepsilon\rangle)^2 + 4\langle 0|V|\varepsilon\rangle^2\right]^{1/2} \tag{9.20}$$

which can be used for both first-order and second-order transitions. While ℓ_h remains finite for a first order transition (as M does), for a second-order transition one has $\ell_h \sim N^{-(1-\alpha)/\nu}$. For the p-state Potts model, it was checked [296] that ℓ_h reproduces the exactly known result (e.g. [G19]) for the latent heat.

These expressions, although they require the calculation of eigen*vectors*, provide direct access to the scaling dimensions of the order parameter (x_σ) and the energy density (x_ε)

$$\boxed{x_\sigma = \beta/\nu \;, \quad x_\varepsilon = (1-\alpha)/\nu} \;. \tag{9.21}$$

While this is the only known way to calculate β directly from finite-size data, α could also be obtained from the specific heat. If α is small, however, background terms may inhibit a precise determination and using ℓ_h becomes a useful alternative[2].

---

[2] This situation occurs for example in the (2+1)D Ising model, where $\alpha \simeq 0.11$ [303].

Given some quantity X with a finite-size scaling behaviour $X_N \sim N^{-x/\nu}$, finite-size estimates for the exponent x/ν can be obtained by two ways, namely either from

$$\left(\frac{x}{\nu}\right)_N := -N\left(\frac{X_N}{X_{N-1}} - 1\right) \tag{9.22}$$

or else from

$$\left(\frac{x}{\nu}\right)_N := -\frac{\ln X_N - \ln X_{N-1}}{\ln N - \ln(N-1)}. \tag{9.23}$$

Practical experience suggests that for most systems with power-law singularities, (9.23) produces smaller correction-to-scaling terms than (9.22), while the opposite seems to be true for Kosterlitz–Thouless-like systems.

9.3 Translation Invariance

We go back to the quantum Hamiltonian (9.5). Translation invariance, if permitted by the interactions contained in H and by the boundary conditions, is the most effective means to break H into smaller blocks which can be diagonalized separately. We shall here, in the context of p-state systems like the Potts model, consider **toroidal boundary conditions**

$$\sigma_{N+1} = \exp\left(\frac{2\pi i \tilde{Q}}{p}\right) \sigma_1 \tag{9.24}$$

with $\tilde{Q} = 0, 1, \ldots, p-1$. The special case $\tilde{Q} = 0$ gives periodic boundary conditions. (Note that these boundary conditions are not compatible with self-duality for N finite.) With these boundary conditions, H is broken into the blocks $H_Q^{(\tilde{Q})}$ which can be diagonalized separately (if terms breaking the global symmetry are absent). At the self-dual point $\lambda = 1$, one has

$$H_Q^{(\tilde{Q})} = H_{\tilde{Q}}^{(Q)}. \tag{9.25}$$

Similarly, for the case of free boundary conditions, one has the sectors $H_Q^{(F)}$.

Translation invariance is expressed by the existence of some translation operator $T_{\tilde{Q}}$, commuting with H. Its eigenvalues are related in a well-known way to the discretized lattice momentum $\mathcal{P}$. It is therefore useful to write the quantum Hamiltonian not in the basis of individual spins spanned by the states $|\alpha_1, \ldots, \alpha_N\rangle$, where the α_k give the values of an individual spin variable at site k (with p possible values per site), but in a basis of states $|\mathcal{P}, Q, \ldots\rangle$ in which the translation operator $T_{\tilde{Q}}$ is diagonal. For example, if the global symmetry group of the quantum Hamiltonian is the abelian group $\mathbb{Z}_p$, the eigenstates of H with a **lattice momentum** $\mathcal{P}$ on a chain of N sites are [260]

$$\aleph(N)^{1/2}|\mathcal{P},Q,\ldots\rangle = |\alpha_1,\alpha_2,\ldots,\alpha_N\rangle$$

$$+ \exp\left(\frac{2\pi i \mathcal{C}_1}{pN}\right)|\alpha_2,\alpha_3,\ldots\alpha_N,\alpha_1\rangle$$

$$+ \ldots$$

$$+ \exp\left(\frac{2\pi i \mathcal{C}_{N-1}}{pN}\right)|\alpha_N,\alpha_1,\alpha_2,\ldots,\alpha_{N-1}\rangle, \quad (9.26)$$

$$\mathcal{C}_k = \tilde{Q}N\left(\alpha_1+\ldots+\alpha_k\right)-(Q\tilde{Q}+\mathcal{P}p)k \quad (9.27)$$

and $\aleph(N)$ is the number of states $|\alpha_1,\ldots,\alpha_N >$ appearing in the translation-invariant state $|\mathcal{P},Q,\ldots\rangle$. Going to this basis amounts to a prediagonalization which reduces the matrices to be considered by a factor of order $\mathcal{O}(N)$.

Since N is finite, the eigenvalues of $T_{\tilde{Q}}$ will be quantized such that $\mathcal{P}$ is an integer. On the other hand, the Virasoro generator $L_0 - \bar{L}_0$ measures the spin s of a certain scaling operator ϕ. The relation between s and the lattice momentum $\mathcal{P}$ was already deduced in (3.46) and (8.40). If Δ and $\bar{\Delta}$ are the conformal weights of ϕ, the scaling dimension x_ϕ and the spin s_ϕ are given by

$$x_\phi = \Delta + \bar{\Delta} \;,\quad s_\phi = \Delta - \bar{\Delta} \quad (9.28)$$

and the lattice momentum $\mathcal{P}$ can be read off as

$$\mathcal{P} = \Delta - \bar{\Delta} + \frac{1}{p}\tilde{Q} = L_0 - \bar{L}_0 + \frac{1}{p}\tilde{Q} \quad (9.29)$$

for either a primary or a secondary field.

In connection with (8.35), this can be used to fix the **normalization** of the lattice Hamiltonian H. The momentum eigenvalues are $(2\pi/N)\mathcal{P}$. Thus the energy levels of a set of states belonging to the same **conformal tower**, should be integer spaced if $\mathcal{P} \rightarrow \mathcal{P} \pm 1$. If two states belong to different conformal towers, their energy spacing is not simply related. In any case, for a given value of $\mathcal{P}$, there will be minimal energy, in conformal units: $E_{min}(\mathcal{P}) - E_0 = |\mathcal{P}|$. This means that all these states must be secondary to the unit operator. In particular, due to the presence of the energy–momentum tensor $T(z)$, there will be states with $\mathcal{P} = \pm 2$ and $E(\mathcal{P}) - E_0 = 2$. This fixes the normalization of H. It follows that *a linear dispersion relation is a necessary condition for conformal invariance.*

9.4 Diagonalization

In order to diagonalize a lattice Hamiltonian exactly, one of the best methods is provided by the **Lanczos algorithm**, see [371, 397, 544, 189]. Let H be a real symmetric $n \times n$ matrix, but the algorithm is readily adapted to complex hermitian matrices as well. One wants to find a basis $v_1,\ldots,v_n$ such that H

is tridiagonal. A convenient representation of the main loop which is easy to implement is

$$
\begin{aligned}
w' &= w + H v_i & & [v_i, w] \\
& & & \downarrow \\
\alpha_i &= v_i \cdot w' & & [v_i, w'] \\
& & & \downarrow \\
w'' &= w' - \alpha_i v_i & & [v_i, w''] \\
\beta_i &= (w'' \cdot w'')^{1/2} & & \downarrow \\
v_{i+1} &= w'' / \beta_i & & [v_i, v_{i+1}] \\
& & & \downarrow \\
& \mathrm{swap}(v_i, v_{i+1}) & & [v_{i+1}, v_i] \\
& & & \downarrow \\
w &= -\beta_i v_i & & [v_{i+1}, w]
\end{aligned}
\tag{9.30}
$$

where v_i, w, w', w'' are n-component vectors. Having finished these steps, one jumps back to the beginning. The **initial conditions** are $w = 0$ and v_1 has to be chosen as a convenient, normalized **starting vector**. It should have a sizable overlap with the true ground state eigenvector, but otherwise, the actual choice of v_1 is normally not very important. The square brackets indicate which vectors take the memory locations $[A, B]$ in storage. This algorithm is (possibly apart from the matrix multiplication $H v_i$) vectorizable and parallelizable [189] and one makes full use of the sparse structure of H.

In the basis of the v_i $(i = 1, \dots, n)$, H takes the tridiagonal form

$$
H = \begin{pmatrix}
\alpha_1 & \beta_1 & & & & \\
\beta_1 & \alpha_2 & \beta_2 & & & \\
& \ddots & \ddots & \ddots & & \\
& & \ddots & \ddots & \ddots & \\
& & & \ddots & \alpha_{n-1} & \beta_{n-1} \\
& & & & \beta_{n-1} & \alpha_n
\end{pmatrix}, \qquad \beta_n = 0.
\tag{9.31}
$$

However, if one only computes a *truncated* tridiagonal matrix using only m steps of the Lanczos algorithm, the eigenvalues of the $m \times m$ principal submatrix of (9.31) converge extremely rapidly to those of H. It is this observation which makes the Lanczos algorithm useful for practical calculations. Typically, one has values of $n \sim 10^4 - 10^5$. Using $m \simeq 20 - 100$ Lanczos steps, one typically obtains the first few lowest eigenvalues of H to a precision of 8 digits or more (precise statements are of course model-dependent). These rates of convergence generically only depend very weakly on the model parameters and anything comparable to the critical slowing-down seen in Monte Carlo simulations is completely absent.

For further illustration, we present a sample Fortran code implementing the Lanczos procedure (here and below, continuation lines are not explicitly indicated). The statements commented out are only needed when re-orthogonalization of the v_i, which is absent in (9.30), is required (the reference to the variables vs and id in the subroutine statement should normally be omitted). We come back to this below. Without re-orthogonalization, this scheme has the advantage that storage is minimized in the sense that besides the elements of H, only two more long vectors are needed. All calculations must be done in double precision.

```
      subroutine lancz(v1,v2,vs,id,lang,n,elem,diag,index,
                       nit,a,b,ndone)
      implicit real*8 (a-h,o-z)
      dimension v1(lang),v2(lang),a(250),b(250)
      dimension elem(0:23,lang),index(0:23,lang),diag(lang)
c     dimension vs(id,250),cc(250)
c
c     do 10 i1=1,lang
c10   vs(i1,1)=v1(i1)
      do 100 j=1,min(nit,40)
        call hamil(v1,v2,lang,n,elem,diag,index)
        a(j)=ddot(lang,v1,1,v2,1)
        aj=-a(j)
        call daxpy(lang,aj,v1,1,v2,1)
c       call dgemv('T',lang,j,1.d0,vs,id,v2,1,0.d0,cc,1)
c       call dgemv('N',lang,j,-1.d0,vs,id,cc,1,1.d0,v2,1)
        b(j)=dnrm2(lang,v2,1)
        if (b(j).ne.0.d0) then
          bj=1.d0/b(j)
          call dscal(lang,bj,v2,1)
c         do 110 k=1,lang
c110      vs(k,j+1)=v2(k)
          call dswap(lang,v1,1,v2,1)
          bj=-b(j)
          call dscal(lang,bj,v2,1)
        else
          ndone=j
          return
        endif
100   continue
      ndone=nit
      return
      end
```

Here, v1,v2 are the state vectors, n is the number of sites, lang the dimension of the Hilbert space, nit the number of requested and ndone the number of actually performed Lanczos iterations, index contains the pointers to the non-diagonal matrix elements of H, which are themselves specified in elem and diag contains the diagonal matrix elements. The routine also contains a safeguard against the vanishing of one of the β_i. Initially, the vector **v1** contains the normalized starting vector and **v2** is set to zero. On exit, these vectors are overwritten and the elements of the tridiagonal matrix are returned in the vectors **a,b**, respectively. The diagonalization of the resulting symmetric tridiagonal matrix of size **ndone** is trivial and must be done by the calling program. The choice of the dimensions is set up for chains with at most 24 sites and for up to 250 Lanczos iterations. The basic linear algebra operations are performed through routines of the BLAS package.

The code for the subroutine **hamil** where H is applied to the vector **v1** is given below. On exit, the result is returned in **v2**.

```
      subroutine hamil(v1,v2,lang,n,elem,diag,index)
      implicit real*8 (a-h,o-z)
      dimension v1(lang),v2(lang)
      dimension elem(0:23,lang),index(0:23,lang),diag(lang)
      do 10 j=1,lang
        do 20 k=n-1,0,-1
          v2(j)=v2(j)+elem(k,j)*v1(index(k,j))
20      continue
        v2(j)=v2(j)+diag(j)*v1(j)
10    continue
      return
      end
```

As it stands, the code is written for an Ising quantum chain with a quantum Hamiltonian as in (9.32), but extensions to other models are straightforward. Also, it is used that in each column of H, there are exactly N off-diagonal matrix elements. If necessary, it is easy to rewrite the routine such that only those portions of **elem** and **index** which are currently used are kept in the main storage.

In order to calculate the matrix elements, recall that to each spin configuration, one may associate a number j through its binary representation, where a bit 0 corresponds to a spin up, say, and a bit 1 to a spin down. When translation invariance holds, one must generate through (9.26) an ordered list of *classes*, labeled by j (usually the smallest of the numbers characterizing the configurations of a class), containing the configurations j, j', j", ... which can be mapped onto j by a translation. H then acts on the *classes* labeled by j. For each allowed value of j, **index(k,j)** contains the number of the class which is obtained from j by reversing one of its spins and **elem(k,j)** then contains the corresponding matrix element, found from (9.26). Furthermore, in **hamil** the symmetry of H is exploited to spare an extra index calculation. Note that **v2** is *not* reset by **hamil** and that the inner index in both **elem,index** starts from 0 rather than 1, in order to make use of the more efficient decrements and tests against zero.

We now discuss the Lanczos algorithm *without* re-orthogonalization, which may be sufficient when only eigenvalues, but no eigenvectors, are needed. If one is only interested in the leading critical exponents, it is sufficient to obtain the lowest two eigenvalues of H, corresponding to states in different charge sectors. Then one never encounters a numerical instability of the Lanczos algorithm due to rounding errors, to be discussed.

Choosing a "good" starting vector v_1, one gets a sequence of estimates of the eigenvalues and the corresponding eigenvectors. While the second eigenvector is in principle orthogonal to the first, rounding errors will make it to get a significant overlap with the first eigenvector after sufficiently many iterations. When this occurs, the estimates for the second eigenvalue will cross over to the first, producing an artificial degeneracy. One can prevent this by re-orthogonalizing the set of the v_i sufficiently often, but this may require much memory to keep all the vectors needed.

Practical experience suggests the following. Suppose that one is interested in the higher eigenvalues but not in the eigenvectors. Then one observes that the apparent convergence of the higher eigenvalue estimates increases after each cross-over of an eigenvector into the ground state. This cross-over usually occurs in a fairly regular fashion and stable estimates are reached again after just a few (up to about 5) iterations, then remaining stable for another $\sim 30-60$ iterations until the next cross-over. We illustrate this with an example. In Table 9.2 we give the first few eigenvalues of the Ising model quantum Hamiltonian

$$- H = -10 + \sum_{n=1}^{N} [t\sigma^x(n) + h\sigma^z(n) + \sigma^z(n)\sigma^z(n+1)] \tag{9.32}$$

with $t = 1$, $h = 0.3$, $N = 13$ sites and periodic boundary conditions as found for various sizes m of the Lanczos submatrix. We observe a systematic evolution of the eigenvalue estimates E_i with the number of iteration steps m. The first eigenvalue E_0 converges very rapidly. However, we see that already for E_1 there is some apparent convergence to some limit for moderate values of m, before the cross-over to another value (in this case E_0) mentioned above sets in, which is due to the numerical rounding errors. We also observe that for the very value of m where the cross-over takes place in E_1, also for E_2 a similar cross-over is seen. The estimates E_2 now converge to the true second eigenvalue, and similarly for the higher ones. In this particular example, after 60 iterations E_0, E_2, E_4 and E_5 will be accepted as genuine eigenvalues of $-H$, while E_1 and E_3 will be rejected. This trick allows a faster determination of the higher eigenvalues which are needed for conformal invariance applications. Some caution is in order, however. The procedure described yields artificial degeneracies not only for the ground state, but also for the higher states. The question whether an eigenvalue is degenerate or not must be answered by looking at the evolution of the eigenvalue estimates

Table 9.2. Convergence of the Lanczos algorithm without re-orthogonalization in the Ising model (9.32), with $t = 1$, $h = 0.3$ and $N = 13$ sites, as a function of the number m of iterations.

m	E_0	E_1	E_2	E_3	E_4	E_5
30	9.795206293	6.961613590	5.253109231	4.289888075	3.579039982	2.929548392
33	9.795206293	6.961613590	5.253109251	4.289952969	3.581189795	3.232102549
36	9.795206293	6.961613590	5.253109251	4.289960211	3.582247409	3.333511487
39	9.795206293	6.961613590	5.253109251	4.289960793	3.582515210	3.357819874
42	9.795206293	6.961613590	5.253109251	4.289960816	3.582540980	3.360748774
45	9.795206293	9.721013310	6.961613590	5.253109251	4.289960815	3.582539177
48	9.795206293	9.795145787	6.961613590	5.253109251	4.289960815	3.582539476
51	9.795206293	9.795206176	6.961613590	5.253109251	4.289960815	3.582539482
54	9.795206293	9.795206293	6.961613590	5.253109251	4.289960815	3.582539483
57	9.795206293	9.795206293	6.961613590	5.253109251	4.289960815	3.961362436
60	9.795206293	9.795206293	6.961613590	6.949143430	5.253109251	4.289960815

with the number of Lanczos steps m.[3] The number of iterations necessary for a certain precision of some eigenvalue is model dependent, but usually does not exceed $m \sim \mathcal{O}(200)$ for the first few eigenvalues.

Eigenvectors might also be obtained from this scheme, without using re-orthogonalization. However, the straightforward application has the disadvantage that the desired eigenvectors are linear combinations $\sum_i c_i v_i$ of the intermediate vectors v_i, and where the c_i are only known *after* diagonalizing the tridiagonal matrix. Keeping these vectors in memory might be prohibitive. A simple workaround is to run the Lanczos subroutine *twice*, once to obtain the coefficients c_i and a second time to accumulate the eigenvectors, *without* storing the intermediates v_i [176]. This will work as long as rounding errors are small enough for the v_i to stay effectively orthogonal. Typically, ground state eigenvectors can be found this way, but this becomes increasingly problematic for the higher eigenvectors.

Obtaining higher eigenvectors with the Lanczos algorithm does require to re-orthogonalize the v_i, which then have to be kept, leading to increased memory requirements. To build this feature into the Fortran code of `lancz`, we simply include the statements commented out so far. An extra array `vs(id,ilan)`, where `id` is the maximal dimension of the Hilbert space and `ilan` the maximal number of Lanczos iterations, and a vector `cc(ilan)` are needed. It can be seen that for `ilan` (which is 250 in our example) even moderately large, the storage needed for the intermediate vectors rapidly becomes larger than the storage required for the Hamiltonian matrix itself. For that reason, sometimes only a partial re-orthogonalization with respect to the lowest few vectors v_i is performed.

---

[3] All in all, it might be safer not to rely on the apparent degeneracies produced by the Lanczos algorithm without re-orthogonalization.

To complete the discussion, we give an example for a driving routine for the Lanczos algorithm with full re-orthogonalization, which yields a selected eigenvalue and -vector.

```
subroutine eigen(lang,n,elem,diag,index,nit,ndone,ivec,
                 ev,v1)
implicit real*8 (a-h,o-z)
dimension v1(lang),v2(lang),a(250),b(250)
dimension elem(0:23,lang),index(0:23,lang),diag(lang)
parameter(id=1111)
dimension vs(id,250),work(2*250-2),z(250,250)
do j=1,lang
  v1(j)=0.d0
  v2(j)=0.d0
enddo
v1(1)=1.d0
call lancz(v1,v2,vs,id,lang,n,elem,diag,index,nit,
           a,b,ndone)
call dstev('V',ndone,a,b,z,250,work,info)
ev=a(ivec)
do k=1,ndone
  b(k)=z(k,ivec)
enddo
call dgemv('N',lang,ndone,1.d0,vs,id,b,1,0.d0,v1,1)
return
end
```

This routine takes as arguments the previously calculated matrix elements, the Hilbert space dimension **lang**, the system size **n** and the number of desired iterations **nit**. Furthermore, **ivec** = 1, 2, . . . specifies whether the first, second, . . . eigenvalue and eigenvector is desired. On exit, that eigenvalue is returned in **ev** and the corresponding normalized eigenvector in **v1**. The intermediate vectors v_i are kept in the array **vs** and the parameter **id** is used to declare its set-up (we must have **lang** ≤ **id**). In this example, up to 250 Lanczos iterations are possible. The diagonalization of the tridiagonal matrix is performed by the LAPACK routine **dstev**, with eigenvalues in ascending order.[4] For brevity, we have used here the simplest possible starting vector. In practice, it is advisable to experiment what type of starting vector gives quickly converging results. Of course, it is straightforward to extract more than one eigenvalue and/or eigenvector.

An alternative method is provided by the **BMKD algorithm** [85] which converges somewhat slower, but does have the advantage of being numerically extremely stable.[5] One writes, beginning from a normalized starting vector

---

[4] See http://www.netlib.org/lapack/lug/node0.html for documentation.

[5] The so-called power method or vector iteration [371, 397, 189] does yield the eigenvalue of the largest absolute value and the corresponding eigenvector, but

v_1 (we assume H to be real symmetric for notational simplicity)

$$Hv_1 = e_1 v_1 + d_1 v_2 \quad \text{with } v_2 \cdot v_1 = 0 \quad \text{and } v_2 \cdot v_2 = 1 \qquad (9.33)$$

from which the vector v_2 and e_1 and d_1 are calculated. Let $e_2 = v_2 H v_2$ and form the matrix of H in the subspace spanned by v_1, v_2

$$\check{h} := \begin{pmatrix} e_1 & d_1 \\ d_1 & e_2 \end{pmatrix}. \qquad (9.34)$$

If $d_1 \neq 0$, then one of the eigenvalues of $\check{h}$ is below and the other above e_1. To find the lowest eigenvalue of H, take the eigenvector of $\check{h}$ corresponding to its lower eigenvalue and let it be the v_1 in the next iteration step. This gives a reliable estimate for the lowest eigenvector. Combinations and hybridizations with the Lanczos algorithm are possible.

Recent numerical studies increasingly also consider real *non-symmetric* quantum Hamiltonians. These can be brought to an upper Hessenberg form through the **Arnoldi algorithm**, see [33, 397, 189, 633]. The main loop may be written as follows

$$
\begin{aligned}
w &= H v_j \\
h_{ij} &= v_i \cdot w \qquad ; \quad i = 1, \ldots, j \\
w' &= w - \sum_{i=1}^{j} h_{ij} v_i \\
h_{j+1,j} &= (w' \cdot w')^{1/2} \\
v_{j+1} &= w'/h_{j+1,j}
\end{aligned}
\qquad (9.35)
$$

where v_i, w, w' are n-component vectors. A normalized starting vector v_1, which should not coincide with one of the eigenvectors of H, must be given. The numbers $h_{i,j}$ form an upper Hessenberg matrix. Rapidly converging estimates for the lowest eigenvalues can be found by diagonalizing the Hessenberg submatrix obtained after a moderate number m of iterations.

Practically, it turns out that the v_i should *always* be re-orthogonalized, since the Hessenberg matrix is more sensitive to rounding errors than the tridiagonal matrix encountered with the Lanczos algorithm. Finally, it is sometimes even recommended to work with quadruple precision [633].

9.5 Extrapolation

We now turn to the important and difficult question of how to extrapolate a given sequence $\{f_N\}$ of finite-size estimates towards its limit f as $N \to \infty$ for critical temperatures, exponents or amplitudes. An example for a nicely converging sequence for the critical temperature in a layered Ising model was already presented in Fig. 3.2 and an experimental example was shown in

---

its convergence is much slower than for the techniques discussed here and cannot be recommended, see [85] for a nice illustration.

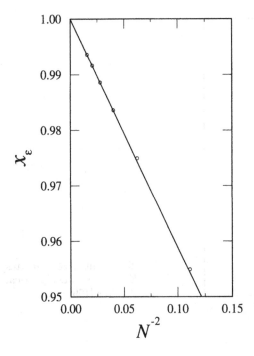

Fig. 9.1. Finite-size estimates of the exponent x_ε in the (1+1)D Ising model (after [169])

Fig. 3.5. In Fig. 9.1 we show an example of what one would like to have for the convergence of critical exponents. The data are for the scaling dimension x_ε of the energy density of the (1+1)D Ising model. They were obtained from the universal finite-size scaling amplitude $S_\varepsilon(0,0) = 2\pi x_\varepsilon$, derived as (3.49) in Chap. 3. In order to get an idea of the convergence of the finite-size data, one may compare with the asymptotic behaviour

$$x_\varepsilon \simeq 1 - \frac{\pi^2}{24}\frac{1}{N^2} + \dots \qquad (9.36)$$

valid for $N \to \infty$, which is given as the full line in Fig. 9.1. Clearly, the data must converge towards the exact result $x_\varepsilon = 1$ for large lattices, but it is another matter whether the *asymptotically valid* expressions can be used for N finite and small (usually, $N \leq 20$ for quantum chains). Any extrapolation procedure has therefore *by necessity to assume* that the asymptotic behaviour somehow applies to the values of N within reach. This is apparently satisfied for the Ising model example of Fig. 9.1, even for the small lattices taken into account. However, this does by no means imply that a similarly rapid convergence is the rule.

A very clear and instructive warning comes from the 3D spherical model [440]. In Fig. 9.2, we give finite-size estimates of the inverse critical temperature $K_c = J/T_c$ as calculated from phenomenological renormalization.

The asymptotic behaviour of $K_c(N)$ is, with $K_c = 0.25274\dots$ [440]

$$K_c(N) - K_c(\infty) = 0.027807 \cdot N^{-2} + \dots \qquad (9.37)$$

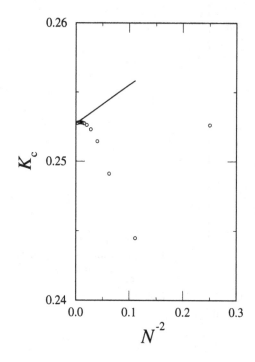

Fig. 9.2. Finite-size estimates for K_c in the 3D spherical model (after [169, 440])

and is given by the full line. It is clearly seen that up to lattice sizes $N \simeq 11$ (a cubic lattice of the shape $N \times N \times \infty$ with periodic boundary conditions was used) the behaviour of the finite-size data *has nothing to do* with the asymptotic expectation. Even the sign of the leading correction term is different ! We stress that this kind of phenomenon is not at all restricted to the spherical model. In order to illustrate the implications of this result, we remark that the largest square lattice of any model ever calculated exactly until today is the 6×6 lattice for the spin-1/2 quantum Heisenberg antiferromagnet [559]. Using the full space group and periodic boundary conditions, the size of the Hilbert space to be treated is 15804956 and there are $\sim 1.17 \cdot 10^9$ non-vanishing matrix elements. For the (2+1)D Ising model, the 6×6 lattice was calculated using Monte Carlo methods [513].

Generally, turning to algorithms devised for the extrapolation versus $N \rightarrow \infty$ of a finite-lattice sequence, "... *the method should be able to represent the underlying singular structure of the function under investigation*" (Guttmann [289]). This means that *there is nothing like a straightforward and foolproof extrapolation method*. The use of an extrapolation algorithm does require care and a considerable amount of judgement. By necessity, this implies that the final estimates for $N \rightarrow \infty$ are somewhat subjective. In particular, this holds for the estimated *accuracy* of the final results since the estimates eventually obtained will almost always contain systematic errors. If at all possible, one should therefore try to obtain a second estimate of the same quantity by a different method. The variation of the results thus obtained should give a fair idea on the lower bound of the extrapolation error.

After these remarks, we now turn to describe how finite-lattice extrapolations can be done in practice. We merely mention some of the most basic ideas and refer to the reviews of Guttmann [289] and of Weniger [612][6] for more detailed information. We need some terminology on the asymptotic convergence behaviour of finite-lattice sequences. Let f_N be a convergent sequence with $f = \lim_{N \to \infty} f_N$. Then consider

$$\rho := \lim_{N \to \infty} \frac{f_{N+1} - f}{f_N - f}. \tag{9.38}$$

There are two principal situations to be distinguished:

1. **Linear convergence.** By definition, $0 < |\rho| < 1$ in this situation. As an example, take for $N \to \infty$

$$f_N = f + a \exp(-bN) + \dots, \tag{9.39}$$

where a and $b > 0$ are constants. For statistical systems, this corresponds to systems far away from criticality.

2. **Logarithmic convergence.** This case is defined by $\rho = 1$. Examples of this are given by, for $N \to \infty$

$$f_N = f + a_1 N^{-\omega_1} + a_2 N^{-\omega_2} + \dots \tag{9.40}$$

$$f_N = f + a \exp(-bN^\omega) + \dots \ ; \quad \omega < 1, \tag{9.41}$$

where a_1, a_2, a and $b, \omega_1, \omega_2, \omega > 0$ are constants. The first of these sequences is typically realized for critical systems and most of the determinations of critical temperature, exponents and amplitudes fall into this class. The cross-over between the two regimes of linear and logarithmic convergence may pose yet additional problems.

The Germain-Bonne theorem [267, 612] allows to construct extrapolation schemes which will accelerate the convergence of *any* linearly converging sequence, at least asymptotically for $N \to \infty$. On the other hand, one can show that there is nothing of the kind for logarithmically convergent sequences. Every algorithm which works apparently well on some types of logarithmically convergent sequences will fail on some other. For example, many extrapolation algorithms which improve the convergence of logarithmically converging sequences when $\omega_1, \omega_2, \dots$ are positive *integers*, will cease to do so if non-integer values for the ω_i arise. We now present two widely used extrapolation algorithms designed for the extrapolation of logarithmically converging sequences, in particular for sequences of the type of (9.40), which have been useful in the context of critical phenomena.

---

[6] These reviews deal mainly with the extrapolation of sequences obtained from series expansions. These appear to have a structure slightly different from finite-size sequences ($\omega_{1,2}$ in (9.40) are integers for series data) which should play a role in the selection of a convenient algorithm.

9.5.1 VBS Algorithm

This algorithm was first proposed by van den Broeck and Schwartz [601] and introduced into the physics literature by Barber and Hamer [47, 294]. Consider a **consecutive sequence** of lattice sizes with constant increment between sizes such as $N = 1, 2, 3, 4, \ldots$ or $2, 4, 6, 8 \ldots$, but avoid sequences like $2, 6, 3, \ldots$ or $1, 2, 3, 7, 8, 11, \ldots$. Form approximants $[N, M]$, arranged as a triangular table

$$
\begin{array}{lllll}
[0,0] & & & \\
[1,0] & [1,1] & & \\
[2,0] & [2,1] & [2,2] & \\
[3,0] & [3,1] & & \\
[4,0] & & \ddots &
\end{array}
\tag{9.42}
$$

where the $[N, M]$ are defined recursively

$$
\begin{aligned}
&[N, -1] = \infty, \\
&[N, 0] = f_N, \\
&([N, M+1] - [N, M])^{-1} + \alpha\, ([N, M-1] - [N, M])^{-1} \\
&\quad = ([N+1, M] - [N, M])^{-1} + ([N-1, M] - [N, M])^{-1}.
\end{aligned}
\tag{9.43}
$$

The original sequence $f_N = [N, 0]$ is transformed into a new one, e.g. $[N, 1]$, which is expected to converge faster towards the desired limit f. For $\alpha = -1$, one can show that [47]

$$
[N, 2] = f + \mathcal{O}(N^{-\omega'}) \quad \text{where } \omega' = \min(\omega_1 + 2, \omega_2)
\tag{9.44}
$$

for a logarithmically convergent sequence of the type (9.40). The algorithm arises from a generalization of the Padé table, to which it reduces for $\alpha = 1$. The final estimate f should in principle be independent of the choice of α, but practically, performance may change considerably. A possible choice is $\alpha_M = -(1-(-1)^M)/2$. The consideration of a set of test sequences of the type (9.40), some of with non-integer ω_1, ω_2, suggests that this algorithm converges faster than many other conventional extrapolation algorithms (which only allow for ω_1, ω_2 to be integer or half-integer at most) [47].

The algorithm may converge to a spurious limit, if either the sequence to be extrapolated is non-monotonous or else the correction coefficients a_i have different signs. To rectify this, one might replace in the finite-size estimators (9.22) and (9.23) N by $N+\varepsilon$ where ε is a free parameter (*N*-**shift technique** [294]). The final result should be independent of ε, so one will look for the regions of stability with respect to it.

9.5.2 BST Algorithm

A special case of this algorithm was proposed by Bulirsch and Stoer [113] as a variant of Romberg integration, but applications to critical phenomena are relatively recent [307]. Consider a sequence h_N ($N = 0, 1, 2, \ldots$) converging to zero as $N \to \infty$. In many applications, $h_N = 1/N$, but there is no need to have a consecutive sequence here. Form a table of extrapolants

$$
\begin{array}{cccc}
T_0^{(0)} & & & \\
& T_1^{(0)} & & \\
T_0^{(1)} & & T_2^{(0)} & \\
& T_1^{(1)} & & \\
T_0^{(2)} & & \ddots &
\end{array}
\tag{9.45}
$$

where

$$
T_{-1}^{(N)} = 0
$$

$$
T_0^{(N)} = T(h_N) = f_N \tag{9.46}
$$

$$
T_m^{(N)} = T_{m-1}^{(N+1)} + \left(T_{m-1}^{(N+1)} - T_{m-1}^{(N)} \right)
$$

$$
\times \left[\left(\frac{h_N}{h_{N+m}} \right)^\omega \left(1 - \frac{T_{m-1}^{(N+1)} - T_{m-1}^{(N)}}{T_{m-1}^{(N+1)} - T_{m-2}^{(N+1)}} \right) - 1 \right]^{-1}
$$

with ω as a free parameter. This algorithm arises from approximating the function $T(h)$ by a sequence of rational functions in the variable h^ω. One expects that $T_1^{(N)}$ converges faster than $T_0^{(N)}$ to the desired limit $f = T(0)$ and so on.

For a logarithmically converging sequence of the type (9.40) and if one chooses $\omega = \omega_1$, one can show that [307]

$$
T_1^{(N+1)} = f + \mathcal{O}(N^{-\omega'}) \quad \text{where } \omega' = \min(\omega_1, 2\omega_2). \tag{9.47}
$$

Practically, one tries to choose ω such as to minimize simultaneously the differences $T_m^{(N+1)} - T_m^{(N)}$. It is also worth mentioning that the pseudoconverge of the VBS algorithm does not appear to show up for the BST algorithm and the N-shift technique is not really needed here. Furthermore, the BST algorithm appears to be much less sensible to rounding errors in the data than the VBS algorithm.

Comparing the two algorithms, we note that both tend to "absorb" the leading correction into the sequence. The VBS algorithm does so in two iteration steps, while the BST algorithm can do so (with a nearly optimal ω) already in one step. The algorithms will show their full strength only if they can be iterated several times on the sequence $\{f_N\}$. The application of both is best acquired by practical experience. Here, we just present one example how a relatively short sequence may be extrapolated. The data are estimates [313] for the critical point t_c of the following quantum Hamiltonian

$$H = -\frac{1}{2}\sum_{j=1}^{N}\left[t\sigma_j^z + \sigma_j^x\sigma_{j+1}^x + \sigma_j^y\sigma_{j+1}^y - \sigma_j^z\sigma_{j+1}^z\right.$$
$$\left. + \mathrm{i}\left((1 - \sigma_j^z)\sigma_{j+1}^y + \sigma_j^y(1 - \sigma_{j+1}^z)\right)\right] \tag{9.48}$$

which can be shown to be in the same universality class as directed percolation [129, 108, 399].

Table 9.3. Finite-lattice extrapolation with the VBS algorithm, $\alpha = -1$

N				
8	2.6138388			
9	2.6125490	2.6081970		
10	2.6115541	2.6079750	2.6003815	
11	2.6107756	2.6077709	2.6059171	2.6059186
12	2.6101573	2.6075978	2.6059186	2.6059171
13	2.6096593	2.6074498	2.6068248	
14	2.6092529	2.6073364		
15	2.6089176			

Table 9.4. Finite-lattice extrapolation with the BST algorithm, $\omega = 1.94$

2.6138388								
	2.6075368							
2.6125490		2.6063534						
	2.6071762		2.6064220					
2.6115541		2.6063865		2.6064067				
	2.6069493		2.6064116		2.6064131			
2.6107756		2.6063975		2.6064099		2.6064068		
	2.6068001		2.6064107		2.6064121		2.6064107	
2.6101573		2.6064029		2.6064029		2.6064103		
	2.6066987		2.6064029		2.6064109			
2.6096593		2.6064029		2.6064029				
	2.6066275		2.6064321					
2.6092529		2.6064141						
	2.6065788							
2.6089176								

In Table 9.3, we display the iterated sequences as obtained with the VBS algorithm, where we have taken $\alpha = -1$. It is necessary to vary α and look for optimal convergence. In Table 9.4, the corresponding results for the BST algorithm are shown, where $\omega = 1.94$ was used. If the chosen values of α and ω were indeed the final ones, we would read off $t_c^{(\mathrm{VBS})} \simeq 2.606$ and $t_c^{(\mathrm{BST})} \simeq 2.60641$. From these tables, the reader should get some idea on how the iterated finite-size sequences behave. We recommend to investigate the structure of these extrapolation tables, while varying the parameters α and ω. For a detailed description of the many other extrapolation algorithms and further numerical examples, we refer to the literature, e.g. [289, 612, 48, 312] and references therein.

9.6 The DMRG Algorithm

Obviously, the techniques presented so far are severely constrained by the *exponential* growth of both memory and CPU time requirements with the system size N. This limits the applicability of these techniques to systems of at most $\mathcal{O}(10 - 20)$ sites. Recently, enormous progress has been achieved by White [615, 616, 617] through the construction of the so-called **density matrix renormalization group algorithm**, often abbreviated to *DMRG algorithm*. This new scheme permits to obtain low-lying eigenvalues of quantum Hamiltonians on chains of a few hundred sites at the least, which precisions which in favorable cases can be brought down to machine accuracy. Since it is likely to become the standard method in numerical diagonalization studies for some time to come, we shall give a brief introduction to it, following the original work of White [615, 616]. For a recent review, see [265].

To fix ideas, consider an Ising quantum chain with free boundary conditions with

$$H = -t \sum_{n=1}^{N} \sigma^x(n) - \sum_{n=1}^{N-1} \sigma^z(n)\sigma^z(n+1). \tag{9.49}$$

The associated Hilbert space has dimension $\mathcal{N} = 2^N$. One is therefore interested in working with a reduced Hilbert space of dimension $m \ll \mathcal{N}$. It might appear natural to use as basis states the m lowest eigenvectors of H and then project H onto this subspace through the operation

$$H' := \hat{O}H\hat{O}^\dagger, \tag{9.50}$$

where $\hat{O}$ is an $m \times \mathcal{N}$ matrix such that the rows of $\hat{O}$ are the m lowest eigenstates of H. In the same way, other operators may be projected, e.g. $\sigma_n^{z\prime} = \hat{O}\sigma_n^z\hat{O}^\dagger$.

The following iteration procedure then appears suggestive. Break the chain into finite identical blocks B with block Hamiltonians H_B. Join two blocks to form a superblock BB with Hamiltonian $H_{BB} = H_{B_1} + H_S + H_{B_2}$, where the H_{B_i} are the block Hamiltonians for the two blocks and H_S is the surface term describing the interactions between the two blocks. Diagonalize the matrix H_{BB} to obtain the m lowest eigenvectors which make up the matrix $\hat{O}$. Project $BB \to B'$ via (9.50) and replace B by B' and repeat until convergence is achieved. However, practical experience has shown that the accuracies obtainable by this approach and variants thereof are slight.

The innovation of the DMRG method involves the details how the formation of the superblock BB is done as well as the choice of the m vectors which generate the projection onto the new block B'. It is carried out in two steps, referred to as 'infinite-system' and 'finite-system' algorithm [616]. One obtains an estimate for a single selected eigenvalue and its eigenvector.

The main loop of the **infinite-system algorithm** is summarized in Table 9.5. One starts with four initial blocks, each consisting of a single site.

Table 9.5. Infinite-system DMRG algorithm (after [616])

1.	determine the (sparse) superblock Hamiltonian H_{BB}.
2.	diagonalize H_{BB} and find the target state $\psi = \psi(i_1, i_2, i_3, i_4)$.
3.	calculate the density matrix $\rho(i_1, i_2; i_1', i_2') = \sum_{i_3, i_4} \psi(i_1, i_2, i_3, i_4)\psi(i_1', i_2', i_3, i_4)$.
4.	find the m largest eigenvalues w_α of ρ and their eigenvectors u_α.
5.	determine the matrix representations of H (and other operators) for the two-block system 1-2.
6.	form the matrix $\hat{O}$ from the m largest eigenvectors u_α and project the system 1-2 into the new block 1 using (9.50).
7.	replace the old block 1 by the new block 1.
8.	replace the old block 4 by the reflection of the new block 1.

Symbolically, the superblock for this four-site system is $BB = B_1 \bullet \bullet \bar{B}_1$, where $\bar{B}$ is the reflection of the block B and $\bullet$ indicates a single site. The block Hamiltonian H_B and the matrices for other operators, if needed, are calculated and stored. The superblock Hamiltonian is, as before, $H_{BB} = H_{B_1} + H_S + H_{\bar{B}_1}$. H_S contains here the Hamiltonian of the central two sites making up the blocks $B_{2,3}$ plus the interaction of these with the two blocks B_1 and $B_4 = \bar{B}_1$. It is only necessary to keep H_{B_1} explicitly in storage. The diagonalization of the still relatively sparse matrix H_{BB} can be achieved via the Lanczos (or else Davidson [178]) algorithm with reorthogonalization. If the kth eigenvalue of H is desired, its corresponding eigenvector ψ (also referred[7] to as *target state*), is obtained, with components $\psi(i_1, i_2, i_3, i_4)$ referring to the four blocks. From this, the density matrix

$$\rho(i_1, i_2; i_1', i_2') = \sum_{i_3, i_4} \psi(i_1, i_2, i_3, i_4)\psi(i_1', i_2', i_3, i_4) \tag{9.51}$$

is obtained by summing of the degrees of freedom of the third and forth block. Indeed, ρ is the reduced density matrix of the subsystem made out of the blocks 1 and 2, while the blocks 3 and 4 are considered as "environment". Since for the (ordered) eigenvalues w_α it is known that $\sum_\alpha w_\alpha = 1$, the deviation

$$Q_m = 1 - \sum_{\alpha=1}^{m} w_\alpha \tag{9.52}$$

---

[7] It turns out that using more than one target state merely leads to a loss of accuracy [616].

measures the accuracy of the truncation of the vector space when going from the double block B_{12} to the new block B_1' via the projection (9.50). Having thus completed one passage through the main loop, the next iteration is prepared by considering the new superblock $BB = B_1 \bullet \bullet \bar{B}_1$, that is by adding two new sites between the blocks B_1 and $B_4 = \bar{B}_1$ found from the previous iteration. One now goes back to the beginning of Table 9.5. Each iteration step thus adds two more sites to the system and one obtains a sequence of estimates for the targeted eigenvalue for $N = 4, 6, 8, \ldots$ sites, respectively. If an expectation value of an observable $\mathcal{O}$ is requested, it is necessary to update its elements each time the projection to the new block B_1' is carried out.

In order to improve on these estimates further,[8] the **finite-system algorithm** can be used. It consists of one or several hyperiterations $I = 1, 2, 3 \ldots$ and is summarized in Table 9.6. It yields for a N-site system a selected eigenvalue and the eigenvector of H. In contrast to the infinite-system algorithm,

Table 9.6. Finite-system DMRG algorithm (after [616])

1.	(first half of $I = 1$.) perform $N/2 - 1$ steps of the infinite-size algorithm to build up the N-sites lattice. After each iteration, store the Hamiltonian for block 1 as size-dependent matrix $B_\ell, \ell = 1, \ldots, N/2$.
2.	(second half of $I = 1$.) Set $\ell = N/2$ and use B_ℓ as block 1 and $\bar{B}_{N-\ell-2}$ as block 4.
3.	perform the steps 1-6 of Table 9.5.
4.	replace $B_{\ell+1}$ by the new block 1.
5.	replace block 4 by $\bar{B}_{N-\ell-2}$, obtained from the first half of $I = 1$.
6.	if $\ell < N - 3$, increment ℓ and go to step 3.
7.	(hyperiteration I, with $I \geq 2$.) initialize blocks $1, 2, 3$ as a single site and block 4 as $\bar{B}_{N-3}$, as found from hyperiteration $I - 1$ and set $\ell = 1$.
8.	perform the steps 1-6 of Table 9.5.
9.	replace $B_{\ell+1}$ by the new block 1.
10.	replace block 4 by $\bar{B}_{N-\ell-2}$.
11.	if $\ell < N - 3$, increment ℓ and go to step 8. Otherwise, increment I and start a new hyperiteration at step 7.

the blocks B_1 to B_{N-3} must be stored and the infinite-system algorithm is used to initialize $B_1, \ldots, B_{N/2}$. For $I = 1$, the system $B_{N/2-1} \bullet \bullet \bar{B}_{N/2-1}$ is used to find $B_{N/2}$ which in turn serves to find $B_{N/2+1}$ by considering the system $B_{N/2} \bullet \bullet \bar{B}_{N/2-2}$ such that the total number of sites remains constant. For $I \geq 2$, one starts with the superblock $B_1 \bullet \bullet \bar{B}_{N-3}$ (where B_1 corresponds to a single site and it thus known) and proceeds to replace the old B_ℓs with

---

[8] The relative error $(E_{\mathrm{DMRG}} - E_{\mathrm{exact}})/E_{\mathrm{exact}}$ obtained from the infinite-system algorithm increases with N, as may be easily checked in the Ising model. As a rule, the infinite-system algorithm is not precise enough if excited states at criticality are required.

$\ell = 2, \ldots, N - 3$, step by step. In practice, it turns out that going beyond $I = 3$ does not lead to further improvements in accuracy. In the last hyperiteration, one might stop after having diagonalized $B_{N/2-1} \bullet \bullet \bar{B}_{N/2-1}$ and use the resulting eigenvector for the calculation of the averages of interest.

What has one gained by applying this quite sophisticated procedure? In the Ising model case, for the diagonalization of a quantum chain of N sites, the time required goes as $m_{\text{Lan}} 2^{2N}$, where m_{Lan} is the number of Lanczos iterations, and the memory needed goes as $N 2^{2N}$ (without reorthogonalization). On the other hand, for the infinite-system DMRG method, a passage through Table 9.5 requires a time of order $\sim m_{\text{Lan}} m^4 + m^6$ and memory of order $\sim m_{\text{Lan}} m^2$. However, since for $2N$ sites N iterations are needed, the time needed takes an extra factor of N. Thus, an exponential growth of both time and memory requirements has been converted into merely a *linear* dependence on the system size N (for the finite-system algorithm, both of these numbers are to be multiplied with N). A few requirements have to be met to make the DMRG applicable and effective.

1. H must not contain long-range interactions. Already the consideration of nearest-neighbour interactions requires that the operator describing the interactions between blocks and 1 and 2 is kept explicitly in memory, besides the Hamiltonian to be diagonalized. For each further interaction term an extra matrix must be kept.

2. As it stands, the method applies to 1D systems. It is relatively straightforward to extend it to quantum ladders, but fully two-dimensional systems are still difficult to treat, in spite of recent progress in the critical *quantum* (2+1)D Ising model [175].

3. The method works best for obtaining ground states and when these are separated from the excited states by a finite (and large) energy gap. Applications to critical phenomena, while feasible, require larger values of m to give precise results than non-critical situations [433]. For a recent study on critical exponents, amplitudes and density profiles in the 2D classical Potts model, see [154].

4. The convergence depends crucially on the boundary conditions. While for example surface fields are unproblematic, periodic boundary conditions are much more difficult to treat [616].

5. As formulated here, only an even number of sites N can be treated.

6. The DMRG algorithm is a variational method. If it converges, the desired eigenstates can be written as matrix product states, as explained in detail in [489].

It is straightforward, if somewhat tedious, to convert the steps of Tables 9.5,9.6 into a program. As an illustration, we list in Table 9.7, for the lowest five levels $E_i, i = 0, \ldots 4$ of the Ising quantum chain (9.49), results for the accuracies $\Delta E_i = E_{i,\text{DMRG}} - E_{i,\text{exact}} > 0$ and the truncation error Q_m. These data were obtained by using the corresponding eigenstate of H_{BB} as target state. First of all, it is clear that the precision of the eigenvalue esti-

Table 9.7. Convergence of the DMRG algorithm for the quantum Ising chain (9.49), at $t = 1$ and for $N = 48$ sites, with $I = 3$ hyperiterations

level	0		1		2		3		4	
m	$\lg \Delta E_0$	$\lg Q_m$	$\lg \Delta E_1$	$\lg Q_m$	$\lg \Delta E_2$	$\lg Q_m$	$\lg \Delta E_3$	$\lg Q_m$	$\lg \Delta E_4$	$\lg Q_m$
4	-4.33	-4.92	-3.53	-4.38	-2.56	-3.58	-2.23	-3.57	-1.76	-3.62
6	-5.72	-6.25	-4.61	-5.44	-3.04	-4.15	-2.75	-4.21	-2.15	-2.76
8	-6.22	-6.81	-5.18	-6.34	-4.02	-5.68	-4.16	-4.63	-2.90	-5.16
10	-7.46	-7.88	-6.21	-7.54	-4.44	-6.30	-4.63	-6.20	-3.33	-5.78
12	-7.88	-8.84	-6.63	-8.45	-5.19	-7.73	-5.68	-7.37	-3.94	-6.82
14	-8.95	-9.93	-7.52	-9.21	-5.53	-7.97	-5.95	-7.84	-3.99	-7.28
16	-9.08	-10.48	-7.59	-9.75	-5.59	-8.45	-6.01	-8.15	-4.38	-7.62

mates improves quickly (indeed, roughly exponentially) with increasing m. As a function of m, the truncation error Q_m behaves similarly and furthermore, varies relatively little with the eigenvalue to be estimated. Furthermore, for a given level, Q_m and ΔE are roughly proportional to each other [433]. However, the convergence of the eigenvalue estimates themselves depends quite strongly on the level considered. As a function of m, the convergence is quite non-uniform. This is so because when increasing m, whole clusters of excited states start to be included into the calculation but a significant gain in accuracy can only be expected if m is made large enough that the *whole* cluster is taken into account. For $t > 1$, convergence becomes much faster, but in the disordered phase $t < 1$ the situation is more intricate [433]. Critical point estimates for correlators are complicated functions of the model and the iteration parameters, see [433]. Finally, we remark that the Ising model estimates apparently converge better than more general spin models, see [154], and refer to [330, 395] for extensions to non-hermitian matrices.

Exercises

1. Consider the quantum chain with Hamiltonian

$$H = -\sum_n \left[\lambda \prod_{\ell=0}^{m-1} \sigma_{n+\ell}^z + \sigma_n^x \right].$$

Neglecting boundary effects, show that H is self-dual [590, 498]. See [620] for the operator content for $m = 3$.

2. Consider the Ising quantum chain

$$H = -\sum_{n=1}^N \left[t\sigma_n^z + \frac{1+\eta}{2} \sigma_n^x \sigma_{n+1}^x + \frac{1-\eta}{2} \sigma_n^y \sigma_{n+1}^y \right]$$

with *anti*periodic boundary conditions $\sigma_{N+1}^{x,y} = -\sigma_1^{x,y}$. Let $\tilde{T} := \sigma_1^z \hat{T}$, where $\hat{T}$ is the usual translation operator. Show that $[H, \tilde{T}] = 0$, see [303].

3. If σ_n, Γ_n are defined as in (9.3,9.4), show that the Hamiltonians

$$H_1 := -\sum_n \sum_{k=1}^{p-1} b_k [\Gamma_n^k \Gamma_{n+1}^{p-k} + \lambda \sigma_n^k \sigma_{n+1}^{p-k}], \quad H_2 := -\sum_n \sum_{k=1}^{p-1} b_k [\lambda \Gamma_n^k \Gamma_{n+2}^{p-k} + \sigma_n^k]$$

are dual to each other, where b_k and λ are constants [656]. Take $p = 2$.

4. Consider the quantum Ising chain. Write a program which diagonalizes H for suitably large chains. Use phenomenological renormalization (see Chap. 3) to obtain finite-size estimates for the critical point. Extrapolate your results to $N \to \infty$ and compare with the exact result $t_c = 1$, which can for example be obtained from self-duality. Extract the shift exponent λ and some other exponent from the finite-size scaling of some other observable. Compare the convergence of your finite-size data for different boundary conditions (e.g. periodic vs. free vs. antiperiodic).

5. Study the convergence of the DMRG algorithm for the quantum Ising chain with a surface magnetic field h_1 (see Chap. 15 for the exact diagonalization of H).

6. Write a program to calculate the critical two-point function $\langle \phi(r)\phi(0) \rangle$ on the cylinder for some spin model. Fig. 9.3 illustrates this for the four-states Potts model.

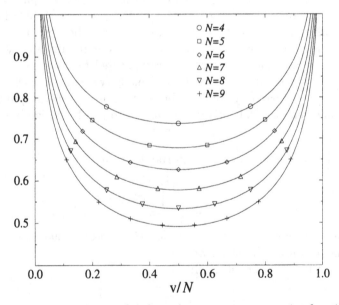

Fig. 9.3. Finite-size estimates for the order parameter two-point function in the 4-states Potts model on a cylinder with N sites and comparison with the conformal prediction (3.45) $\langle \phi(v)\phi(0) \rangle \sim ((N/2\pi) \sin(2\pi v/N))^{-2x_\phi}$, where v is the distance along the finite direction [160]

10. Conformal Invariance in the Ising Quantum Chain

At last, we shall give the first explicit example of the application of conformal invariance methods in the context of a specific lattice model. We choose the 2D Ising model as an example. Of course, the model was solved by Onsager back in 1944 and we shall not learn anything about this model which has not yet been deduced long ago by different means [G22, G19]. However, since all quantities of interest can be worked out explicitly, we get in this example absolute control of the critical behaviour of the model. This is useful to see how the techniques of conformal invariance work and one is not disturbed by convergence problems to reach the $N \to \infty$ limit. In particular, recalling the treatment of the same model in the language of continuum field theory in Chap. 6, this example further illustrates the general machinery used for applying conformal invariance and useful insight will be obtained by comparing these two different approaches to the same problem. For the convenience of the reader, the main steps to be followed applying conformal invariance to any given lattice model will be summarized at the end of this chapter.

We shall work with the following quantum Hamiltonian

$$H = -\frac{1}{\zeta} \sum_{n=1}^{N} \left[t\sigma^z(n) + \frac{1+\eta}{2}\sigma^x(n)\sigma^x(n+1) + \frac{1-\eta}{2}\sigma^y(n)\sigma^y(n+1) \right],$$
(10.1)

where N is the number of sites, $\sigma^x, \sigma^y, \sigma^z$ are Pauli matrices, t is the transverse field, η an irrelevant coupling introduced to check explicitly for universality and ζ is the normalization of H to be determined such that conformal invariance results can be applied. While this quantum Hamiltonian has been studied in detail for a long time (see e.g. [393, 473, 503, 158]), the systematic application of conformal invariance techniques was done first in [256].

10.1 Exact Diagonalization

10.1.1 General Remarks

We begin by specifying the problem completely. The boundary conditions are

$$\sigma(N+1) = (-1)^{\tilde{Q}}\sigma(1),$$
(10.2)

where σ stands for σ^x, σ^y and $\tilde{Q} = 0, 1$. The charge operator, commuting with H, is

$$\hat{Q} = \frac{1}{2}\left(1 - \prod_{n=1}^{N}\sigma^z(n)\right) \tag{10.3}$$

and has eigenvalues $Q = 0, 1$. H is thus decomposed into the blocks $H_Q^{(\tilde{Q})}$. The global symmetry depends on the value of η. If $\eta = 0$, then H is invariant under a global U(1) transformation

$$\begin{pmatrix} \sigma^x \\ \sigma^y \\ \sigma^z \end{pmatrix} \rightarrow \begin{pmatrix} \bar{\sigma}^x \\ \bar{\sigma}^y \\ \bar{\sigma}^z \end{pmatrix} \begin{matrix} := \\ := \end{matrix} \begin{pmatrix} \cos\theta & -\sin\theta \\ \sin\theta & \cos\theta \\ & 1 \end{pmatrix} \begin{pmatrix} \sigma^x \\ \sigma^y \\ \sigma^z \end{pmatrix}, \tag{10.4}$$

where θ is an arbitrary real number. If $\eta \neq 0$, one only has a global $\mathbb{Z}_2$ invariance

$$\begin{pmatrix} \sigma^x \\ \sigma^y \\ \sigma^z \end{pmatrix} \rightarrow \begin{pmatrix} \bar{\sigma}^x \\ \bar{\sigma}^y \\ \bar{\sigma}^z \end{pmatrix} \begin{matrix} := \\ := \end{matrix} \begin{matrix} - \\ \end{matrix}\begin{pmatrix} \sigma^x \\ \sigma^y \\ \sigma^z \end{pmatrix}. \tag{10.5}$$

In the sequel, we shall only consider the case $\eta \neq 0$.

10.1.2 Jordan–Wigner Transformation

Introduce the raising and lowering operators

$$\sigma^{\pm}(n) := \frac{1}{2}\left(\sigma^x(n) \pm i\sigma^y(n)\right). \tag{10.6}$$

The σ-variables commute on different sites and anticommute at the same site. This hybrid character of the variables is avoided by passing to truly fermionic variables c_n^+, c_n by the celebrated Jordan-Wigner transformation

$$\sigma^+(n) = \exp\left(i\pi\sum_{\ell=1}^{n-1} c_\ell^+ c_\ell\right) c_n^+$$

$$\sigma^-(n) = c_n \exp\left(i\pi\sum_{\ell=1}^{n-1} c_\ell^+ c_\ell\right). \tag{10.7}$$

The c's satisfy anticommutation relations

$$\{c_n^+, c_m\} = \delta_{n,m} \ , \quad \{c_n, c_m\} = 0. \tag{10.8}$$

Let $\mathcal{N} := \sum_{\ell=1}^{N} c_\ell^+ c_\ell$ denote the **fermion number operator**. $\mathcal{N}$ is even in the sector $Q = 0$ and odd in the sector $Q = 1$. Then the Hamiltonian (10.1) takes the form (see exercises)

$$H = -\frac{2}{\zeta}\left\{\sum_{n=1}^{N}(-t)\left(c_n^+c_n - \frac{1}{2}\right)\right.$$

$$+\frac{1}{2}\sum_{n=1}^{N-1}\left[c_n^+c_{n+1} + c_nc_{n+1}^+ + \eta c_n^+c_{n+1}^+ + \eta c_nc_{n+1}\right]$$

$$\left.-\frac{1}{2}e^{i\pi\mathcal{N}}\left[c_N^+c_1 + c_Nc_1^+ + \eta c_N^+c_1^+ + \eta c_Nc_1\right]\right\} \tag{10.9}$$

$$= -\frac{2}{\zeta}\sum_{n,m=1}^{N}\left[c_n^+A_{nm}c_n + \frac{1}{2}\left(c_n^+B_{nm}c_m^+ + c_nB_{nm}c_m\right)\right] + \text{const.} \tag{10.10}$$

which, when restricted to the blocks $H_Q^{(\tilde{Q})}$, is a quadratic form in the c's and the integrability of the model is already clear. This result only depends on the nearest-neighbor nature of the interactions.[1] The matrices $\hat{A}$ and $\hat{B}$ are, with $\mathcal{Q} := 1 - 2Q$ and for $\tilde{Q} = 0$

$$\hat{A} = \begin{pmatrix} -t & \frac{1}{2} & & & & & -\frac{\mathcal{Q}}{2} \\ \frac{1}{2} & -t & \frac{1}{2} & & & & \\ & \frac{1}{2} & -t & \ddots & & & \\ & & \ddots & \ddots & \ddots & & \\ & & & \ddots & -t & \frac{1}{2} \\ -\frac{\mathcal{Q}}{2} & & & & \frac{1}{2} & -t \end{pmatrix}, \tag{10.11}$$

$$\hat{B} = \frac{\eta}{2}\begin{pmatrix} 0 & 1 & & & & \mathcal{Q} \\ -1 & 0 & 1 & & & \\ & -1 & 0 & \ddots & & \\ & & \ddots & \ddots & \ddots & \\ & & & \ddots & 0 & 1 \\ -\mathcal{Q} & & & & -1 & 0 \end{pmatrix}. \tag{10.12}$$

If $\tilde{Q} = 1$ (antiperiodic boundary conditions), replace $\mathcal{Q} \to -\mathcal{Q}$.

10.1.3 Diagonalization of a Quadratic Form

Equation (10.10) is most efficiently diagonalized by the well-known technique of Lieb, Schultz and Mattis [436]. In particular, *the technique is directly applicable for quite general nearest-neighbor interactions, including for example*

---

[1] Generalizations to site-dependent interaction parameters t_n and η_n will only affect the precise form of the matrices $\hat{A}$, $\hat{B}$ but not the diagonalization technique as such.

quasiperiodic and random ones. One wants to rewrite the Hamiltonian in a diagonal form, using new fermionic variables a_k, a_k^+

$$H = \sum_k \Lambda_k \left(a_k^+ a_k - \frac{1}{2} \right),$$ (10.13)

where the Λ_k are the **one-fermion energies**. One makes the ansatz

$$a_k = \sum_n \left(g_{kn} c_n + h_{kn} c_n^+ \right)$$

$$a_k^+ = \sum_n \left(g_{kn} c_n^+ + h_{kn} c_n \right),$$ (10.14)

where the g_{kn} and h_{kn} can be chosen to be real. In order that the a_k do satisfy the fermionic anticommutation relations, we have to require that

$$\sum_n \left(g_{kn} g_{k'n} + h_{kn} h_{k'n} \right) = \delta_{kk'}$$

$$\sum_n \left(g_{kn} h_{k'n} - g_{k'n} h_{kn} \right) = 0.$$ (10.15)

Using the relation $[a_k, H] - \Lambda_k a_k = 0$, one finds

$$\Lambda_k g_{kn} = \sum_m \left(g_{km} A_{mn} - h_{km} B_{mn} \right)$$

$$\Lambda_k h_{kn} = \sum_m \left(g_{km} B_{mn} - h_{km} A_{mn} \right).$$ (10.16)

From the matrices $\hat{g}$ and $\hat{h}$, build $2N$ vectors $\boldsymbol{\Phi}_k$ and $\boldsymbol{\Psi}_k$

$$(\boldsymbol{\Phi}_k)_i = g_{ki} + h_{ki} \ , \quad (\boldsymbol{\Psi}_k)_i = g_{ki} - h_{ki}$$ (10.17)

and we obtain

$$\boldsymbol{\Phi}_k \left(\hat{A} - \hat{B} \right) = \Lambda_k \boldsymbol{\Psi}_k \ , \quad \boldsymbol{\Psi}_k \left(\hat{A} + \hat{B} \right) = \Lambda_k \boldsymbol{\Phi}_k.$$ (10.18)

Now, the one-fermion energies Λ_k are obtained from the eigenvalues of an $N \times N$ matrix

$$\widehat{\mathcal{M}} \boldsymbol{\Phi}_k := \left(\hat{A} + \hat{B} \right) \left(\hat{A} - \hat{B} \right) \boldsymbol{\Phi}_k = \Lambda_k^2 \boldsymbol{\Phi}_k.$$ (10.19)

The achievement of the method is obvious: the problem of diagonalizing a $2^N \times 2^N$ matrix has been reduced to the eigenvalue problem of a $N \times N$ matrix. Even in cases where the eigenvalues of $\widehat{\mathcal{M}}$ cannot be written down analytically, numerical studies are considerably simplified. The price to pay is that one only obtains the one-particle energies directly. Multiparticle states have to be built up by linear superposition of the one-particle ones.

The matrix $\widehat{\mathcal{M}}$ takes the form $\widehat{\mathcal{M}} = (4/\zeta^2)\widehat{\mathcal{M}}_0$ where (for $\tilde{Q} = 0$)

$$
\widehat{\mathcal{M}}_0 =
\begin{pmatrix}
t^2 + \frac{1+\eta^2}{2} & -t & \frac{1-\eta^2}{4} & & -\frac{1-\eta^2}{4}Q & tQ \\
-t & t^2 + \frac{1+\eta^2}{2} & -t & \ddots & & -\frac{1-\eta^2}{4}Q \\
\frac{1-\eta^2}{4} & -t & \ddots & \ddots & \ddots & \\
 & \ddots & \ddots & \ddots & \ddots & \\
 & & \ddots & \ddots & \ddots & \ddots & \frac{1-\eta^2}{4} \\
-\frac{1-\eta^2}{4}Q & & \ddots & \ddots & \ddots & -t \\
tQ & -\frac{1-\eta^2}{4}Q & & \frac{1-\eta^2}{4} & -t & t^2 + \frac{1+\eta^2}{2}
\end{pmatrix}
$$

$$(10.20)$$

Since $\hat{A}$ is symmetric and $\hat{B}$ is antisymmetric, $\widehat{\mathcal{M}}$ is symmetric and positive semidefinite. Writing (10.19) in components of $\boldsymbol{\Phi}_k$, we recover the discretized Schrödinger equation of a spinless electron in a potential (in our case just a constant one) given by the interactions of the Hamiltonian (10.1).

10.1.4 Eigenvalue Spectrum and Normalization

The eigenvalue problem is now solved by the ansatz $(\boldsymbol{\Phi}_k)_r = e^{ikr}$ and gives, as first obtained by Katsura [393]

$$
\Lambda_k^2 = \left(\frac{2}{\zeta}\right)^2 \left[(t - \cos k)^2 + \eta^2 \sin^2 k\right]. \tag{10.21}
$$

The critical point is determined by the vanishing of the lowest energy gap $\xi^{-1} = E_1 - E_0 = \Lambda_{k \to 0}$ in the $N \to \infty$ limit

$$
\Lambda_{k \to 0} \simeq \frac{2}{\zeta} \left[(t-1)^2 + ((t-1) + \eta^2)k^2 + \mathcal{O}(k^4)\right]^{1/2} \tag{10.22}
$$

from which we read off the longest wavelength mode ($k \to 0$) behaviour $\xi^{-1} \sim |t - 1|$ which yields for the critical point t_c and the exponent ν

$$
t_c = 1 \ , \ \nu = 1 \tag{10.23}
$$

and we see that ν is independent of η, as expected from universality.

The full phase diagram is shown in Fig. 10.1 [53, 582] and we concentrate ourselves on the region $0 \leq \eta \leq 1$. For $t > 1$, the model is paramagnetic. If $t < 1$ and $t^2 + \eta^2 > 1$, we have an ordered ferromagnetic phase.[2] This phase is separated by a **disorder line** at $t_{\text{dis}}^2 + \eta_{\text{dis}}^2 = 1$ from an ordered oscillating phase in the region $t^2 + \eta^2 < 1$. These phases can be distinguished through the long-distance behaviour of the (connected) spin-spin correlation function $C(R) = \langle \sigma^x(R)\sigma^x(0)\rangle - \langle \sigma^x(R)\rangle\langle \sigma^x(0)\rangle$, namely [53]

---

[2] Only the two phases P, F have an analogon in the classical 2D Ising model [582].

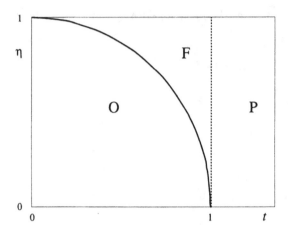

Fig. 10.1. Ground state phase diagram of the spin-$\frac{1}{2}$ quantum Ising chain (10.1). P is the paramagnetic phase, F the ordered ferromagnetic phase and O the ordered oscillatory phase. Along the dashed line, there is a transition in the 2D Ising universality class for $\eta \neq 0$. The disorder line is indicated by the full line

$$C(R) \sim \begin{cases} R^{-1/2} \exp(-R/\xi) & ; \text{phase P} \\ R^{-2} \exp(-R/\xi) & ; \text{phase F} \\ R^{-2} \exp(-2R/\xi) \operatorname{Re} B \exp(\mathrm{i}KR) & ; \text{phase O} \end{cases} \qquad (10.24)$$

where the correlation length ξ is a known function of t and η in each of the three phases, B is a constant and K is a wave vector which vanishes when the disorder line is approached. Finally, for $\eta = 0$, $t < 1$, one has an incommensurate phase. Note that for $\eta = 1$, t_c could have also been obtained from self-duality.[3] Here, we shall be interested exclusively in the critical line $t = t_c = 1$ which for $\eta \neq 0$ is in the Ising universality class.

Precisely *at* the critical point $t = t_c$, one has the **dispersion relation** $\Lambda_k \simeq 2\zeta^{-1}\eta|k|$. This is indeed a *linear* dispersion relation as required for conformal invariance. We now *choose* units of energy and momentum such that the speed of sound $v_s = 2\zeta^{-1}\eta$ becomes unity in these units. This fixes the correct conformal normalization of H [256]

$$\zeta = 2\eta. \qquad (10.25)$$

Finally, we discuss how the lattice momenta k should be discretized. The Jordan-Wigner transformation had introduced a non-local boundary term (see (10.9)) which causes a mixture of periodic and antiperiodic boundary conditions in the fermionic language. This leads to the discretization

---

[3] Self-duality can also be derived for the XY Hamiltonian $\tilde{H}(\eta, q) := -\frac{1}{2}\sum_{n=1}^{N-1}[\eta\sigma_n^x\sigma_{n+1}^x + \eta^{-1}\sigma_n^y\sigma_{n+1}^y + q\sigma_n^z + q^{-1}\sigma_{n+1}^z]$, where η, q are complex constants. It can be shown that $\tilde{H}(\eta, q) = \tilde{H}(q, \eta)$, up to a similarity transformation [335, 336].

$$\text{if } Q + \tilde{Q} = 0 \text{ mod } 2 \quad k = \frac{2m+1}{N}\pi \ ; \ m = 0, 1, \ldots, N-1$$

$$\text{if } Q + \tilde{Q} = 1 \text{ mod } 2 \quad k = \frac{2m}{N}\pi \qquad ; \ m = 0, 1, \ldots, N-1 \qquad (10.26)$$

and the one-fermion energies are, with the above choice for k

$$\Lambda_k = \frac{1}{\eta}\left[(t-1)^2 + 4(t-1+\eta^2)\sin^2\frac{k}{2} - 4(\eta^2-1)\sin^4\frac{k}{2}\right]^{1/2}. \qquad (10.27)$$

This gives the complete solution for the spectrum of H (10.1).

10.2 Character Functions

We now state explicitly the prediction of conformal invariance for the critical point spectrum, making use of the representation theory of the Virasoro algebra (2.65) developed earlier. The basic objects of conformal theory are the *primary operators*. For unitary minimal models, their conformal weights were seen to be given by the Kac formula, see (4.42) and (4.43)

$$c = 1 - \frac{6}{m(m+1)} \ ; \quad m = 3, 4, 5, \ldots$$

$$\Delta = \Delta_{r,s} = \Delta_{m-r,m+1-s} = \frac{[(m+1)r - ms]^2 - 1}{4m(m+1)} \ ; \qquad (10.28)$$

$$1 \le r \le m-1 \ ; \ 1 \le s \le m$$

and we denote the primary operator corresponding to $\Delta_{r,s}$ by $\phi_{r,s}$. Higher states, corresponding to secondary operators, are obtained by acting with L_{-i} on the primary highest weight states $|\Delta_{r,s}\rangle$ (see (4.22))

$$L_{-i_1}L_{-i_2}\ldots|\Delta_{r,s}\rangle \quad i_1 \ge i_2 \ge \ldots > 0. \qquad (10.29)$$

If $i_1 + i_2 + \ldots = I$, the number $d(\Delta_{r,s}, I)$ of excited states of *level I* is read from the *Virasoro character* [539], see (4.48)

$$\chi_{r,s}(q) = q^{-c/24}P(q)\sum_{k=-\infty}^{\infty}\left(q^{\Delta_{2mk+r,s}} - q^{\Delta_{2mk+r,-s}}\right)$$

$$= q^{\Delta_{r,s}-c/24}\sum_{I=0}^{\infty}d(\Delta_{r,s}, I)q^I, \qquad (10.30)$$

where

$$P(q) = \prod_{n=1}^{\infty}(1-q^n)^{-1} = \sum_{I=0}^{\infty}p(I)q^I \qquad (10.31)$$

is the generating function for the number of partitions $p(I)$ of the integer I.

Table 10.1. Calculation of the degeneracies $d(0, I)$ of the operator $\phi_{1,1}$ for $m = 3$

I	0	1	2	3	4	5	6	7	8	9	10	11	12	13	14	15
0^+	1	1	2	3	5	7	11	15	22	30	42	56	77	101	135	176
1^-		-1	-1	-2	-3	-5	-7	-11	-15	-22	-30	-42	-56	-77	-101	-135
6^-							-1	-1	-2	-3	-5	-7	-11	-15	-22	-30
11^+												1	1	2	3	5
13^+														1	1	2
$d(0, I)$	1	0	1	1	2	2	3	3	5	5	7	8	11	12	16	18

As an exercise, let us use the generating function $\chi_{r,s}(q)$ to calculate the first few $d(\Delta_{r,s}, I)$. Write $\chi_{r,s}(q) = \chi^+_{r,s}(q) - \chi^-_{r,s}(q)$ in an obvious notation. To each k in (10.30) there are two new null states in the Verma module, one coming from $\chi^+_{r,s}$ an done coming from $\chi^-_{r,s}$. The corresponding Δ-values must be equal to $\Delta_{r,s}$ up to an integer

$$\Delta_{2mk+r,\pm s} = \Delta_{r,s} + k^2 m(m+1) + k\left[(m+1)r \mp ms\right] + \frac{1}{2}rs(1 \mp 1). \quad (10.32)$$

We calculate the first few of these integers for the example $m = 3$. Then there are three primary operators, $\phi_{1,1}, \phi_{1,2}, \phi_{1,3}$. The integers involved in (10.32) are then up to the level $I = 20$

$$\phi_{1,1} : 0^+, 1^-, 6^-, 11^+, 13^+, 20^-, \ldots$$
$$\phi_{1,2} : 0^+, 2^-, 4^-, 10^+, 14^+, \ldots$$
$$\phi_{1,3} : 0^+, 2^-, 3^-, 7^+, 17^+, \ldots, \quad (10.33)$$

where the index labels whether the corresponding term arises in $\chi^+_{r,s}$ or $\chi^-_{r,s}$, respectively. The **degeneracies** $d(\Delta_{r,s}, I)$ are then calculated by summing all the contributions for all values of k, where those terms coming from $\chi^+_{r,s}$ will make a positive and those coming from $\chi^-_{r,s}$ a negative contribution. This is illustrated in Table 10.1 for the operator $\phi_{1,1}$. The first contribution, labeled by the first integer coming from (10.32) by 0^+, is nothing else than the contribution of the generic character (4.18), using the tabulated values [1] of the number of partitions $p(I)$ of I, obtained from $P(q)$. The second contribution, being negative since coming from $\chi^-_{1,1}$, repeats the same numbers $p(I)$, but shifted by one. This is nothing else than explicitly carrying out the Verma module construction described in Chap. 4. Summing all contributions arising up to level $I = 15$ as indicated in Table 10.1, the final degeneracies are obtained. Similarly, the corresponding degeneracies for the other primary operators are calculated. The results are collected in Table 10.2, and we recall that $\Delta_{1,1} = \Delta_{2,3} = 0, \Delta_{1,2} = \Delta_{2,2} = 1/16$ and $\Delta_{1,3} = \Delta_{2,1} = 1/2$.

From Table 10.2, recalling the relations between the spectra of the Hamiltonian ((8.35)) and of the momentum operator (see (9.29)) and the scaling

Table 10.2. Virasoro characters $d(\Delta, I)$ of the primary operators $\phi_{r,s}$ for $m = 3$.

Δ	\multicolumn{16}{c}{I}	r	s

Δ	0	1	2	3	4	5	6	7	8	9	10	11	12	13	14	15	r	s
0	1	0	1	1	2	2	3	3	5	5	7	8	11	12	16	18	1	1
1/16	1	1	1	2	2	3	4	5	6	8	10	12	15	18	22	27	1	2
1/2	1	1	1	1	2	2	3	4	5	6	8	9	12	14	17	20	1	3

dimensions and spin of the scaling fields, we can thus write down the conformal invariance prediction. We have two Virasoro algebras, the primary operators have *two real conformal weights* $\Delta_{r,s}, \overline{\Delta}_{\bar{r},\bar{s}}$ and the respective excitations are characterized by two integers $I, \bar{I}$. The states will occur at the scaled energy $\mathcal{E}$ and momentum $\mathcal{P}$ given by

$$
\begin{aligned}
\mathcal{E}_Q^{(\tilde{Q})}(\mathcal{P}, \alpha; N) &:= \frac{N}{2\pi}(E_i - E_0) = (\Delta_{r,s} + I) + \left(\overline{\Delta}_{\bar{r},\bar{s}} + \bar{I}\right) \\
\mathcal{P} &= (\Delta_{r,s} + I) - \left(\overline{\Delta}_{\bar{r},\bar{s}} + \bar{I}\right) + \mathcal{P}_{\text{shift}}
\end{aligned}
\tag{10.34}
$$

with degeneracy $d(\Delta_{r,s}, I) \cdot d(\overline{\Delta}_{\bar{r},\bar{s}}, \bar{I})$ and α labels the eigenvalues. As discussed in Chap. 9 in connection with translation invariance, the momentum shift is $\mathcal{P}_{\text{shift}} = \tilde{Q}/2$ for the Ising model, see (9.29). This completes the statement of the critical point prediction for the Hamiltonian spectrum.

In the sequel, we shall refer to the characters $\chi_{r,s}(q)\bar{\chi}_{r',s'}(q)$ also by $(\Delta_{r,s}, \overline{\Delta}_{r',s'})$ and will give the numerical values of the $\Delta_{r,s}$. For example, the $m = 3$ character $\chi_{1,2}(q)\bar{\chi}_{1,2}(q)$ will be denoted $(1/16, 1/16)$.

10.3 Finite-Size Scaling Analysis

We now turn to the explicit verification of conformal invariance in the Ising model at the critical point $t = t_c = 1$. At the same time, explicit expressions for the finite-size scaling functions and corrections to the leading finite-size scaling behaviour will be derived.

10.3.1 Ground State Energy

We first have to find the central charge c. To do so, we need the universal finite-size scaling amplitude $Y(0,0)$ of the ground state energy E_0. It is given by (we take periodic boundary conditions)

$$
\begin{aligned}
E_0^{(0)} &= -\sum_{k=0}^{N-1}\left[\sin^2\left(\frac{\pi(2k+1)}{2N}\right) - \frac{\eta^2 - 1}{\eta^2}\sin^4\left(\frac{\pi(2k+1)}{2N}\right)\right]^{1/2} \\
&= -\left[\sin^2\left(\frac{\pi}{2N}\right) - \frac{\eta^2 - 1}{\eta^2}\sin^4\left(\frac{\pi}{2N}\right)\right]^{1/2}
\end{aligned}
$$

$$-\sum_{k=1}^{N-1}\left[\sin^2\left(\frac{\pi(2k+1)}{2N}\right)-\frac{\eta^2-1}{\eta^2}\sin^4\left(\frac{\pi(2k+1)}{2N}\right)\right]^{1/2}$$

$$=: G_1 + G_2 \tag{10.35}$$

and

$$G_1 = -\frac{\pi}{2N} - \frac{\pi^3}{16N^3}\left(\frac{1}{\eta^2} - \frac{4}{3}\right) + \mathcal{O}(N^{-5}) \tag{10.36}$$

$$G_2 = -\sum_{\ell=0}^{\infty}(-1)^\ell\binom{1/2}{\ell}\left(1-\frac{1}{\eta^2}\right)^\ell\sum_{k=1}^{N-1}\left(\sin\frac{\pi(2k+1)}{2N}\right)^{2\ell+1}.$$

We now analyse G_2. This is not merely a technical exercise, but also allows some insight in the nature of the finite-size terms arising. This will become important in Chap. 13 when considering conformal perturbation theory. We use the identity [298]

$$\sum_{k=1}^{N-1}\left(\sin\frac{\pi(2k+1)}{2N}\right)^{2\ell+1} = \left(-\frac{1}{4}\right)^\ell\sum_{k=0}^{\ell}(-1)^k\binom{2\ell+1}{k}$$

$$\times\cos\left((l+\frac{1}{2}-k)\frac{\pi}{N}\right)\cot\left((l+\frac{1}{2}-k)\frac{\pi}{N}\right) \tag{10.37}$$

and expand in $1/N$. One then observes a clear separation of terms. The first term, of order $\mathcal{O}(N)$, receives contributions from all values of ℓ. This term is the *non-universal* bulk term and since each term in ℓ has a different η-dependence, it is not surprising to see that it will depend on all the finer details of the Hamiltonian. One might say that it is the presence of this term which complicates the exact solution of spin systems so much. The second term, of order $\mathcal{O}(1/N)$, only receives a contribution for $\ell = 0$, which is η-independent. We expect this term to be universal. The third term, of order $\mathcal{O}(1/N^3)$, receives contributions from $\ell = 0, 1$ and will be η-dependent. Since there is only *one* contribution depending on η, terms of this order will be the leading non-universal correction and their η-dependent coefficient can be identified with the non-universal metric factor of the leading correction operator.

These remarks are formally stated by the identities (see [298])

$$\sum_{k=0}^{\ell}(-1)^k\binom{2\ell+1}{k}\frac{1}{2\ell+1-2k} = (-1)^\ell 2^{4\ell}\frac{(\ell!)^2}{(2\ell+1)!}$$

$$\sum_{k=0}^{\ell}(-1)^k\binom{2\ell+1}{k}(2\ell+1-2k) = 0 \quad \text{if } \ell \geq 1$$

$$\sum_{k=0}^{\ell}(-1)^k\binom{2\ell+1}{k}(2\ell+1-2k)^3 = 0 \quad \text{if } \ell \geq 2. \tag{10.38}$$

Collecting terms, we find

$$G_2 = -\frac{N}{2} T_1(\eta) + \frac{5\pi}{12} \frac{1}{N} + \frac{67\pi^3}{960} \left(\frac{1}{\eta^2} - \frac{4}{3} \right) \frac{1}{N^3} + \mathcal{O}(N^{-5}), \qquad (10.39)$$

where

$$T_1(\eta) = \frac{2}{\pi} \left(1 + \frac{\arccos \eta}{\eta \sqrt{1 - \eta^2}} \right). \qquad (10.40)$$

The critical free energy density $f = E_0^{(0)}/N$ is thus, in the strip geometry and with *periodic* boundary conditions [302]

$$f = -\frac{1}{2} T_1(\eta) - \frac{\pi}{12} \frac{1}{N^2} + \frac{7\pi^3}{960} \left(\frac{1}{\eta^2} - \frac{4}{3} \right) \frac{1}{N^4} + \mathcal{O}(N^{-6}). \qquad (10.41)$$

We note the following:

1. The bulk free energy is η-dependent and thus non-universal. One can show by explicit calculation that it does not depend on the boundary condition $\tilde{Q} = 0, 1$, see (10.44) below and Chap. 15.
2. The leading *singular* term is $f^{\text{sin}} \sim N^{-2}$, in agreement with the finite-size scaling expectation. Further, we explicitly see that the finite-size scaling amplitude $Y(0,0) = -\pi/12$ is universal. Using (3.54), we read off the desired value of the central charge

$$c = \frac{1}{2}. \qquad (10.42)$$

   This value is in agreement with the determination of c from the low-temperature specific heat of the 1D ideal fermionic ultrarelativistic gas, see Chap. 3.
3. The leading correction term is η-dependent and thus non-universal.

The same techniques can also be applied for antiperiodic boundary conditions. Using the identity [298]

$$\sum_{k=1}^{N-1} \left(\sin \frac{\pi k}{N} \right)^{2\ell+1} = \left(-\frac{1}{4} \right)^\ell \sum_{k=0}^\ell (-1)^k \binom{2\ell+1}{k} \cot \left(\left(\ell + \frac{1}{2} - k \right) \frac{\pi}{N} \right) \qquad (10.43)$$

we obtain for the free energy density for *antiperiodic* boundary conditions in the strip geometry [302]

$$f_{\text{A}} = -\frac{1}{2} T_1(\eta) + \frac{\pi}{6} \frac{1}{N^2} - \frac{\pi^3}{120} \left(\frac{1}{\eta^2} - \frac{4}{3} \right) \frac{1}{N^4} + \mathcal{O}(N^{-6}). \qquad (10.44)$$

We remark that the leading bulk term, although manifestly non-universal, is seen to be independent of the boundary conditions considered. Furthermore, the leading singular term, of order $\mathcal{O}(N^{-2})$, is universal.

10.3.2 Operator Content

Since $c = 1/2$, we can apply the $m = 3$ predictions from Table 10.2 to obtain the conformal predictions for the energy levels and their degeneracies. We still have to form the *product* of two such representations, since there are two Virasoro algebras, with generators $L_n, \bar{L}_n$. For the example of the primary operator $(0,0)$ with $m = 3$ (or $c = 1/2$), this leads to the degeneracy pattern

$$
\begin{array}{ccccccccc}
\vdots & & \vdots & & \vdots \\
2 & & 0 & 1 & 0 & 2 \\
& 1 & & 0 & 0 & 1 \\
& & 1 & & 0 & 1 \\
& & & 0 & 0 \\
& & & & 1
\end{array}
\tag{10.45}
$$

which is beginning of the *conformal tower* of the identity operator $(0,0)$. In the same way, the beginning of the conformal tower of the energy density operator $(1/2, 1/2)$ can be found. On the other hand, from (10.27), taking $t = t_c = 1$, the leading finite-size scaling amplitude for each multiparticle state can be easily obtained by adding the appropriate combination of the one-particle energies Λ_k. The result for the block $H_0^{(0)}$ is shown in Fig. 10.2. Fig. 10.3 gives the result for the block $H_1^{(0)} = H_0^{(1)}$ and Fig. 10.4 displays the result for the block $H_1^{(1)}$. We first note in Fig. 10.2 that the boundary

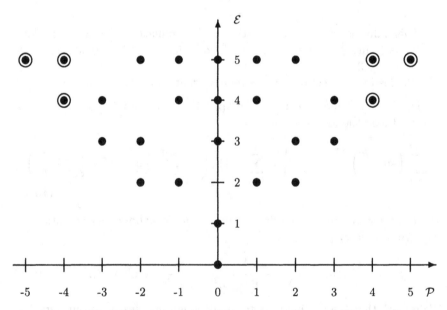

Fig. 10.2. Low-lying critical finite-size scaling spectrum for $H_0^{(0)}$ in the 2D Ising model. Double circles denote two-fold degenerate states ([169] after [G11, 302])

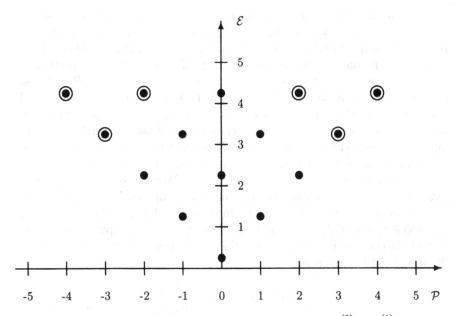

Fig. 10.3. Low-lying critical finite-size scaling spectrum for $H_1^{(0)} = H_0^{(1)}$ in the 2D Ising model

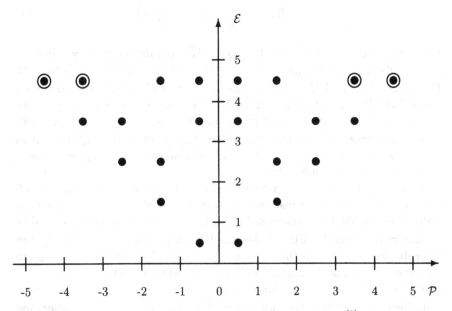

Fig. 10.4. Low-lying critical finite-size scaling spectrum for $H_1^{(1)}$ in the 2D Ising model

of the kinematically allowed region is indeed given by a linear dispersion relation $\mathcal{E} \sim |\mathcal{P}|$. Secondly, the energy spacings are indeed equidistant, as required from the structure of the Virasoro representations, and also integer-valued, which confirms the normalization of H from (10.25). In particular, one recognizes that the degeneracies of the states with $\mathcal{E} = |\mathcal{P}|$ are just those of a single Virasoro algebra and in fact agree with the values of $d(0, I)$ which can be read from Table 10.2. Turning to the quantitative comparison, we see that besides the $(0, 0)$-operator, there is at least one further primary operator present with exponent $x = 1$. Given the list of allowed values for the conformal weights from the Kac formula (10.28), this can be only $\Delta = \overline{\Delta} = 1/2$, which is the energy density $\varepsilon = (1/2, 1/2)$. Working out the degeneracy pattern of the higher states, the agreement is complete. For a quick check, consider again the degeneracies along the lines $\mathcal{E} = 1 + |\mathcal{P}|$. These agree with the degeneracies $d(1/2, I)$ of Table 10.2. Similar remarks appy to the other sectors. In conclusion, we have seen that the *entire operator content* of the 2D Ising model is given by

$$H_0^{(0)} = (0, 0) + \left(\frac{1}{2}, \frac{1}{2}\right)$$

$$H_1^{(0)} = H_0^{(1)} = \left(\frac{1}{16}, \frac{1}{16}\right)$$

$$H_1^{(1)} = \left(0, \frac{1}{2}\right) + \left(\frac{1}{2}, 0\right). \tag{10.46}$$

The interpretation is as follows. The block $H_0^{(0)}$ contains as primary operators the identity and the energy density (ε), both even under the $\mathbb{Z}_2$ symmetry. The sectors $H_1^{(0)}$ and $H_0^{(1)}$ contain the order (σ) and disorder (μ) operators, respectively. These cannot be written in a local form with respect to neither the energy density nor the pair of conjugate Majorana spinors $(\psi, \bar{\psi})$ contained the sector $H_1^{(1)}$. In particular, from the operator $(1/2, 1/2)$, we read off the exponent $x_\varepsilon = 1/2 + 1/2 = 1$ and similarly $x_\sigma = x_\mu = 1/8$ and $x_\psi = x_{\bar{\psi}} = 1/2$, as it should be.

It is of interest to compare these results with those obtained in Chap. 6 from the correlation functions using monodromy arguments. The various operator product algebras considered there appear when combining several of the sectors of the model. First, the sector $H_0^{(0)}$ by itself realizes the subalgebra generated by $\mathbf{1}$ and ε. In the fermionic language, this sector corresponds to the subspace of states generated by an even number of fermionic excitations from the vacuum. It is clear that a restriction to this subspace is consistent. Next, we combine the two sectors $H_0^{(0)}$ and $H_1^{(1)}$. The corresponding operator product algebra is now generated by $\mathbf{1}$, ε, ψ and $\bar{\psi}$. In the fermionic language, these two sectors do belong together, since they both have *antiperiodic* boundary conditions as can be seen from (10.26). For periodic boundary conditions of the spin variables, we combine the sectors $H_0^{(0)}$ and $H_1^{(0)}$, which give rise to the operator product algebra spanned by $\mathbf{1}, \varepsilon$ and σ. Finally, the full operator product algebra is realized by taking all sectors of the model together.

For an alternative discussion of the continuum limit of the Ising model from the point of view of conformal invariance, see [45].

10.3.3 Finite-Size Corrections

Corrections to the leading finite-size scaling behaviour can be worked out easily. We give here just the results for the lowest levels, but similar expressions can be obtained [302] for any desired state without much effort, and show the same general characteristics.

We have [302]

$$
\mathcal{E}_0^{(0)}(0, 1; N) = 1 + \frac{\pi^2}{8} \left(\frac{1}{\eta^2} - \frac{4}{3} \right) \frac{1}{N^2}
$$

$$
+ \frac{\pi^4}{16} \left[\frac{1}{8} \left(\frac{1}{\eta^2} - \frac{4}{3} \right)^2 + \frac{1}{3} \left(\frac{1}{\eta^2} - \frac{16}{15} \right) \right] \frac{1}{N^4} + \mathcal{O}(N^{-6})
$$

$$
\mathcal{E}_1^{(0)}(0, 1; N) = \frac{1}{8} - \frac{\pi^2}{128} \left(\frac{1}{\eta^2} - \frac{4}{3} \right) \frac{1}{N^2} + \mathcal{O}(N^{-4}) \tag{10.47}
$$

and $\mathcal{E}_1^{(1)}(-1/2, 1; N) = \mathcal{E}_1^{(1)}(1/2, 1; N) = \frac{1}{2}\mathcal{E}_0^{(0)}(0, 1; N)$. In the $N \to \infty$ limit, these scaled gaps converge towards the exponents x_ε, x_σ and x_ψ. We see again that the exponent estimates obtained for $N \to \infty$ are indeed universal. Note that only integer powers of $1/N$ appear. In addition, we observe that *all* leading finite-size correction terms (see (10.41)) have the *same* η-dependent prefactor, to leading order in $1/N$, while the next terms appear to contain a second-order effect arising from the leading correction plus a new contribution. These observations will be explained in Chap. 13.

10.3.4 Finite-Size Scaling Functions

The scaled energies do depend on the finite-size scaling variable, in this case $z = N(t - 1)$. The finite-size scaling functions for the inverse correlation lengths (see (3.30)) are, again just for the lowest levels but easy to write down for the entire spectrum (we have also included the leading finite-size corrections) [302]

$$
\frac{1}{2\pi} S_\varepsilon(z, 0) = \left[1 + \frac{z^2}{\pi^2 \eta^2} \right]^{1/2} + \frac{z}{2\eta^2} \left[1 + \frac{z^2}{\pi^2 \eta^2} \right]^{-1/2} \frac{1}{N} + \mathcal{O}(N^{-2})
$$

$$
\frac{1}{2\pi} S_\sigma(z, 0) = \frac{1}{8} + \frac{1}{4\pi} \frac{z}{\eta} + \frac{\ln 2}{4\pi^2} \frac{z^2}{\eta^2} + \frac{1}{2} R_{1\frac{1}{2}, 0} \left(\frac{z^2}{4\pi^2 \eta^2} \right) - \frac{1}{8} R_{1\frac{1}{2}, 0} \left(\frac{z^2}{\pi^2 \eta^2} \right)
$$

$$
+ \mathcal{O}(N^{-1}) \tag{10.48}
$$

and $S_\psi = S_\varepsilon/2$, where

$$R_{1\frac{1}{2},0}(x) := -4 \sum_{r=1}^{\infty} \left(\left(r^2 + x \right)^{1/2} - r - \frac{x}{2r} \right) \tag{10.49}$$

is a particular **remnant function** [228]. These are special functions frequently arising in the study of finite lattice sums. We merely list the behaviour for small (above) and large (below) values of x, respectively [228]

$$R_{1\frac{1}{2},0}(x) \simeq \begin{cases} \frac{\zeta(3)}{2}x^2 - \frac{\zeta(5)}{9}x^3 + \mathcal{O}(x^4) \\ x \ln x + 2 \left(C_E - \ln 2 - \frac{1}{2} \right) x + 2\sqrt{x} - \frac{1}{3} + \frac{2}{\pi}x^{-1/4}e^{-2\pi\sqrt{x}} \end{cases}$$
$$\tag{10.50}$$

and $C_E = 0.5772\ldots$ is Euler's contant and ζ is the Riemann zeta function.

Considering these results, we note that the leading term in $1/N$ only depends on the combination z/η, in agreement with the two-scale factor universality expectation [516] discussed in Chap. 3. We identify the non-universal metric factor $C_1 = 1/\eta$. The correction terms, however, do contain further η-dependent terms.

The finite-size scaling functions were calculated here using explicitly the technical simplifications of the Hamiltonian limit. It can be checked [117], however, that the *same* result is also found for the classical 2D Ising model (see exercises). This is in agreement with the expectation from universality that the results obtained for the finite-size scaling functions do not depend on taking the Hamitonian limit [117].

We can conclude that at least in this example, the expectations from finite-size scaling theory have been fully confirmed. This is remarkable, since the usual field-theoretic justification [107, G20] of finite-size scaling relies on the ε-expansion around the upper critical dimension $d^* = 4$, which is expected to break down in two dimensions. It will be shown in Chap. 13, using the n-point correlation functions obtained from conformal invariance (see Chaps. 5, 6), that the finite-size scaling functions and finite-size corrections can be reconstructed from knowing the operator content of a given lattice model through power series expansions in the finite-size scaling variables.

10.4 The Spin-1 Quantum Chain

Having seen how the operator content of a spin system may be extracted from finite-size data obtained from the critical point quantum Hamiltonian H (or equivalently the transfer matrix) through the example of the exactly solvable spin-1/2 Ising chain, we now briefly discuss the spin-1 quantum chain in order to give some impression to what extent this program can be carried out when only numerical data are available [338]. The quantum Hamiltonian is

$$H = -\frac{1}{\zeta}\sum_{n=1}^{N}\left[tS_n^z + \frac{1+\eta}{2}S_n^x S_{n+1}^x + \frac{1-\eta}{2}S_n^y S_{n+1}^y\right], \tag{10.51}$$

where the $S^{x,y,z}$ are spin-1 matrices

$$S^x = \frac{1}{\sqrt{2}}\begin{pmatrix} 0 & 1 & 0 \\ 1 & 0 & 1 \\ 0 & 1 & 0 \end{pmatrix}, \quad S^y = \frac{1}{\sqrt{2}}\begin{pmatrix} 0 & -i & 0 \\ i & 0 & -i \\ 0 & i & 0 \end{pmatrix}, \quad S^z = \begin{pmatrix} 1 & 0 & 0 \\ 0 & 0 & 0 \\ 0 & 0 & -1 \end{pmatrix} \tag{10.52}$$

and periodic boundary conditions are used. Translation, charge and parity symmetry can be used to obtain the low-lying eigenvalues for chains with up to $N = 14$ sites with the Lanczos algorithm. The topology of the ground state phase diagram is the same as in Fig. 10.1, in particular, the disorder line is still described through $t^2 + \eta^2 = 1$ [421]. In Table 10.3, we show extrapolated estimates for the location of the Ising critical line together with estimates for the correct normalization of H such that conformal invariance becomes applicable. Here, $t_c(\eta)$ was obtained from the phenomenological renormalization described in Chap. 3, while ζ was determined from the requirement that the lowest state with scaled momentum $\mathcal{P} = 2$ should be identified with the energy–momentum tensor and should have the scaled energy $\mathcal{E} = 2$ (see Chap. 9).

Table 10.3. Critical points $t_c(\eta)$ and conformal normalization $\zeta(\eta)$ for the spin-1 quantum Ising chain (10.51) along the Ising transition line. The numbers in brackets give the estimated uncertainty in the last digit [338]

η	0.05	0.1	0.15	0.3	0.5	0.7	1.0
t_c	1.002(1)	1.011(1)	1.0210(1)	1.0637(1)	1.1325(1)	1.2080(1)	1.32587(1)
ζ	–	0.170(1)	0.239(1)	0.4252(1)	0.6416(1)	0.8417(1)	1.12706(1)

Next, we have to determine the central charge. Since the (correctly normalized) critical point ground state energy $E_0(N) \simeq e_0 N - \frac{\pi c}{6}N^{-1} + \dots$, finite-size estimates $c(N)$ for the central charge can be obtained as follows. First, the value of e_0 is determined through numerical extrapolation. Then, $c(N)$ is calculated[4] from

$$c(N) = -\frac{6}{\pi}\left(\frac{1}{N}E_0(N) - e_0\right)N^2. \tag{10.53}$$

Numerical values of $c(N)$ for $\eta = 1$ are shown in Table 10.4. Extrapolation of the data (see Chap. 9) to $N \to \infty$ yields $c = 0.49999(1)$, very close to the expected $c = 1/2$.

---

[4] One might attempt to avoid the explicit determination of e_0 by determining $c(N)$ through the relation $c(N) = -\frac{6}{\pi}\left(\frac{1}{N}E_0(N) - \frac{1}{N-1}E_0(N-1)\right)/\left(N^{-2} - (N-1)^{-2}\right)$, but the convergence of this sequence is usually rather slow.

Table 10.4. Finite-size estimates for the central charge obtained for lattices with N sites for the spin-1 quantum Ising chain (10.51) with $\eta = 1$ and $\zeta = 1.12706$

N	$c(N)$
6	0.541993035
7	0.528042832
8	0.520437541
9	0.515653838
10	0.512406764
11	0.510089257
12	0.508372836
13	0.507064189
14	0.506042634

Having done this, we can compare the extrapolated estimates for the scaled energies $\mathcal{E}$ with the predicted operator content (10.46) and the resulting low-lying spectra (Figs. 10.2, 10.3). This is done in the Tables 10.5 and 10.6 for two values of η. Because H is parity-invariant, the spectrum does not depend on the sign of $\mathcal{P}$ and it is sufficient to consider $\mathcal{P} \geq 0$ only. For each scaled energy $\mathcal{E}$, the upper value corresponds to $\eta = 1$ and the lower value corresponds to $\eta = 0.3$. A * indicates that the level(s) was numerically seen to be present but the finite-size data did not converge sufficiently to allow a reliable estimate.

It can be seen that the estimates for the amplitudes $\mathcal{E}$ do not depend on η, in agreement with the expected universality. From the lowest gaps for $\mathcal{P} = 0$, it can be seen to what precision the numerical data are capable of reproducing the expected scaling dimensions $x_\epsilon = 1$ and $x_\sigma = 1/8$. Furthermore, the spectra obtained are within the numerical errors identical to the ones obtained above for the spin-1/2 case. In particular, for some of the higher levels it is possible to confirm the expected presence of double degenerate states. On

Table 10.5. Low lying scaled energies for the spin-1 Ising chain (10.51) in the $Q = 0$ charge sector. The first column gives the expected energies. For each value of $\mathcal{E}$, the upper value corresponds to $\eta = 1$ and the lower one to $\eta = 0.3$ (after [338])

5		*	5.03(3)		4.8(2)
		*	*		*
4	*	3.98	-	4.00(1)	3.95(3), 3.94(3)
	*	3.99(2)	-	3.9(1)	*
3	3.00(1)	-	3.000(1)	2.999(1)	-
	3.00(1)	-	3.001(1)	3.00(2)	-
2	-	2.00000(1)	2	-	-
	-	2.000(1)	2	-	-
1	1.00000(1)	-	-	-	-
	1.0002(2)	-	-	-	-
0	0	-	-	-	-
	0	-	-	-	-
$\mathcal{E}/\mathcal{P}$	0	1	2	3	4

Table 10.6. Low lying scaled energies for the spin-1 Ising chain in the $Q = 1$ charge sector. For each value of $\mathcal{E}$, the upper value corresponds to $\eta = 1$ and the lower one to $\eta = 0.3$ (after [338])

$\mathcal{E}$	0	1	2	3
$5\frac{1}{8}$		5.2(1), 5.18(2)		*
		*		*
$4\frac{1}{8}$	4.123(2)	-	4.128(2), 4.13(1)	-
	4.1(1)	-	4.13(1), *	-
$3\frac{1}{8}$	-	3.124(1)	-	3.125(1), 3.124(1)
	-	3.12(1)	-	3.1(1), 3.1(1)
$2\frac{1}{8}$	2.1249(1)	-	2.1251(2)	-
	2.126(2)	-	2.121(2)	-
$1\frac{1}{8}$	-	1.12501(1)	-	-
	-	1.1249(1)	-	-
$\frac{1}{8}$	0.12499(1)	-	-	-
	0.1249(1)	-	-	-
$\mathcal{E}/\mathcal{P}$	0	1	2	3

the other hand, it also transpires that the study of levels with $\mathcal{E} \gtrsim 5$ becomes numerically difficult.

10.5 The Virasoro Generators

Finally, after having found the complete operator content, we complete the discussion of the Ising model by giving the explicit construction of the generators of the Virasoro algebra in terms of the fermionic oscillators a_k. This also further illustrates the field-theoretic discussion presented in Chap. 6.

Since the discretization condition (10.26) selects the k to be positive, we can adopt the convention that

$$a_{-k} = a_k^+ \ , \quad k > 0 \tag{10.54}$$

and the anticommutation relations read simply

$$\{a_k, a_\ell\} = \delta_{k+\ell,0}. \tag{10.55}$$

Further, in the sequel we shall always multiply k by $N/2\pi$, so they become integer or half-integer. Then the Virasoro generators can be written as (for the sectors $H_0^{(0)}$ and $H_1^{(1)}$, the two other sectors are treated similarly)

$$L_n := \frac{1}{2} \sum_{k \in \mathbb{Z}+\frac{1}{2}} k : a_{n-k} a_k :, \tag{10.56}$$

where the dots :: denote the fermionic **normal ordering** and it is understood that the a_k with a negative index are written to the left

$$a_k a_\ell =: a_k a_\ell : + \delta_{k+\ell,0} \, \Theta(k - \ell) \tag{10.57}$$

and

$$\Theta(x) := \begin{cases} 1 & ; \ x > 0 \\ 0 & ; \ x \leq 0 \end{cases}. \tag{10.58}$$

A few comments are necessary here. We have derived earlier the lattice dispersion relation (10.21) and (10.26). In describing the critical behaviour, we are only interested in the lowest excitations of the systems. These lowest excitations come either from the smallest values of the (rescaled) k, or else from those values of k, which are close to the number of sites N. Since these two regimes are clearly distinct, we can treat them separately. In the construction of the L_n given above, we have just kept the linear term of the dispersion relation and only considered the small values of of k. Similarly, there is a second set of fermionic excitations, denoted by $\bar{a}_k$, describing those excited states for which the k-values are close to N. The set of Virasoro generators constructed from the $\bar{a}_k$ are the conjugate generators $\bar{L}_n$. Since the two sets of fermionic operators anticommute, it follows immediately that $[L_n, \bar{L}_m] = 0$. Using (10.54), it is an easy exercise to confirm that the L_n constructed indeed satisfy the hermiticity condition $L_n^+ = L_{-n}$. The commutator is

$$[L_n, L_m] = \frac{1}{4} \sum_{k,\ell} k\ell \, [a_{n-k} a_k, a_{m-\ell} a_\ell]. \tag{10.59}$$

This can be evaluated using the identity

$$[AB, CD] = A\{B, C\}D - \{A, C\}BD + CA\{B, D\} - C\{A, D\}B \tag{10.60}$$

to yield

$$\begin{aligned}
[L_n, L_m] = \frac{1}{4} \sum_{k,\ell} k\ell \, (&a_{n-k} a_\ell \delta_{k+m-\ell,0} - a_k a_\ell \delta_{n-k+m-\ell,0} \\
&+ a_{m-\ell} a_{n-k} \delta_{k+\ell,0} - a_{m-\ell} a_k \delta_{n-k+\ell,0}).
\end{aligned} \tag{10.61}$$

At this point one should be careful. Since this expression is singular, it is preferable not to perform an immediate translation on indices but first to reintroduce the normal ordering prescription via (10.57). Using the relation $\sum_k : a_{n-k} a_k := 0$ leads to

$$\begin{aligned}
[L_n, L_m] = \ &\frac{n-m}{2} \sum_k k : a_{n+m-k} a_k : \\
&+ \frac{1}{4} \sum_{k,\ell} \{ k\ell \, \delta_{k+\ell,0} \delta_{n+m-k-\ell,0} \, [\Theta(k-\ell) - \Theta(m-n+k-\ell)] \\
&\quad + k\ell \, \delta_{k-\ell+m,0} \delta_{k-\ell-n,0} \, [\Theta(m-\ell-k) - \Theta(n-\ell-k)] \} \\
= \ &(n-m) L_{n+m} + \frac{1}{24} (n^3 - n) \, \delta_{n+m,0}
\end{aligned} \tag{10.62}$$

and we obtain again the expected value of the central charge $c = 1/2$. Since the critical quantum Hamiltonian is $H = (N/2\pi)(L_0 + \bar{L}_0 - 1/24)$, we have an alternative proof that the critical two-dimensional Ising model is conformal invariant. We shall reuse this technique in Chap. 15 in connection with surface critical phenomena.

10.6 Recapitulation

It may be helpful to summarize what the logical steps to be followed are. This is an algorithm which can be followed in a model-independent way.[5]

1. One has to obtain finite-lattice data, thereby making the number of sites N large enough. The Ising quantum chain (10.1) was so simple that these could be obtained from the exact solution. For more complex models, one often will have to be satisfied with numerical data.
2. Use these data to locate the critical point, for example by the phenomenological renormalization procedure described in Chap. 3. If applicable, the critical point can also be found by analytic techniques, for example self-duality arguments.
3. Precisely at the critical point, the spectrum of the inverse correlation length finite-size scaling amplitudes should show a tower-like structure. The level spacings in the conformal tower(s) are integers. This can be used to fix the correct normalization factor ζ of the quantum Hamiltonian H. In this context, the conformal tower of the identity is particularly useful as its critical exponents are known beforehand.
4. Knowing the conformal normalization ζ, the central charge c can now be found from the ground state energy.
5. From the knowledge of the central charge c, at least for a minimal model, the finite list of possible primary operators and their scaling dimensions is now available. Using the conformal characters, the expected inverse correlation length spectrum can now be calculated in a straightforward way.
6. Starting from the lowest levels, identify the primary operators realized in the model. Compare with general results such as modular invariance to be discussed in the next chapter.

These calculations will rely on the capability to perform a precise numerical estimation of the $N \to \infty$ limit. This can be conveniently done with the diagonalization and extrapolation techniques described in Chap. 9.

Exercises

1. Consider an Ising quantum chain with N sites and periodic boundary conditions. Using the Jordan-Wigner transformation, show that

---

[5] While here we wrote everything in terms of the eigenvalues E_i of the quantum Hamiltonian H, from (1.84) it is clear that the techniques apply in the same way to the eigenvalues Λ_i of the transfer matrix T through $E_i \sim \ln \Lambda_i$.

$$c_n^+ c_n = \sigma^+(n)\sigma^-(n) \ , \ \ c_{n+1}^+ c_n = \sigma^+(n+1)\sigma^-(n) \ ,$$
$$c_{n+1}^+ c_n^+ = \sigma^+(n+1)\sigma^+(n) \ ,$$

if $n \neq N$ and

$$\sigma^+(N)\sigma^-(1) = -e^{i\pi\mathcal{N}}c_N^+ c_1 \ , \ \ \sigma^+(N)\sigma^+(1) = -e^{i\pi\mathcal{N}}c_N^+ c_1^+.$$

Use these to rewrite H (10.1) in terms of the fermionic operators c_n, c_n^+. Generalize to antiperiodic boundary conditions.

2. Verify the entries in Table 10.2.

3. Calculate explicitly the tensor product $(\Delta_1, \Delta_2) = (\Delta_1) \otimes (\Delta_2)$ of two unitary representations of the $c = 1/2$ Virasoro algebra. Find explicitly the critical point excitation spectrum of the sectors $H_Q^{(\tilde{Q})}$ and compare.

4. Analyse the quantum Hamiltonian

$$H = -\frac{1}{\zeta} \sum_{n=1}^{N} [\sigma^x(n)\sigma^x(n+1) + \sigma^y(n)\sigma^y(n+1)]$$

in the case when exactly $N/2$ particles are present (half-filled electron band) from the point of view of conformal invariance.

5. In the 2D Ising model with isotropic couplings and on an infinitely long strip of width L, the spin-spin correlation length is $\xi_\sigma^{-1} = K^* - K + (S_{2L}(\tau^2) - 2S_L(\tau^2))/2$, where K^* is the dual coupling,

$$S_N(\tau^2) = \sum_{k=1}^{N-1} \left| \operatorname{arcosh}\left(1 + \tau^2 + 2\sin^2(\pi k/N)\right) \right|$$

and $\tau = (1 - \sinh 2K)/\sqrt{\sinh 2K}$. In the finite-size scaling limit with τL fixed, re-derive (10.48) and identify the metric factor [117].

6. Consider the spherical model on a hypercubic lattice in $d = 2 + \epsilon$ dimensions. Work out the critical point spectrum of the correlation lengths. To what extent is the critical behaviour of the model at least formally described by 2D conformal invariance, in the $\epsilon \to 0$ limit?

7. Consider a set of bosonic oscillators b_n, which satisfy $b_n^+ = b_{-n}$, and the commutator $[b_n, b_m] = n\delta_{n+m,0}$. Define the normal ordering $::$ through

$$b_n b_m =: b_n b_m : + n\, \delta_{n+m,0}\, \Theta(n-m).$$

Let, with α a free (real) parameter

$$L_n := \frac{1}{2}\sum_{p \in \mathbb{Z}} : b_{n-p}b_p : +i\alpha n\, b_n. \qquad (10.63)$$

Verify the hermiticity condition $L_n^+ = L_{-n}$. Show that the L_n satisfy a Virasoro algebra of central charge $c = 1 + 12\alpha^2$. A useful identity is

$$[AB, CD] = A[B, C]D + [A, C]BD + CA[B, D] + C[A, D]B.$$

8. Derive (10.63) from the energy–momentum tensor (7.30) of the modified Coulomb gas.

11. Modular Invariance

The result (10.46) for the operator content of the 2D Ising model raises a question: in principle, having three irreducible representations of the Virasoro algebra for $m = 3$, one could combine in 9 possible ways the holomorphic and antiholomorphic parts to obtain the primary operators. However, just the 5 found can be realized. A part of the explanation comes from the locality requirement for the correlation functions discussed in Chaps. 5–7. A finer explanation for this selection comes from the requirement of **modular invariance** for the partition function. The presentation follows the work of Cardy [137, 138, G14].

11.1 The Modular Group

Consider a system defined on a 2D lattice with periodic boundary conditions and in both directions of finite extent which has the topology of a **torus**. Mathematically, a torus is conveniently characterized by a complex number τ with imaginary part Im $\tau > 0$ (not to be confused with the lattice spacing in time direction). The finite periodic lattice can regarded as a rectangle in the complex plane with horizontal extent L_2 and vertical extent iL_1, which is a torus with $\tau = iL_1/L_2$. The partition function $\mathcal{Z}$ depends on τ.

The **modular group** $\mathrm{Sl}(2, \mathbb{Z})$ is defined by the transformations

$$\tau \rightarrow \frac{a\tau + b}{c\tau + d} \ ; \quad ad - bc = 1, \tag{11.1}$$

where a, b, c, d are integers. It is generated by the transformations T and S

$$T: \ \tau \rightarrow \tau + 1 \ , \quad S: \ \tau \rightarrow -\frac{1}{\tau} \tag{11.2}$$

whose effect on a torus characterized by τ in shown in Fig. 11.1. They satisfy the defining relations $S^2 = 1$ and $(ST)^3 = 1$.

It is necessary to have the partition function $\mathcal{Z}(\tau)$ *invariant* under the modular group. To show this, it is sufficient to discuss the practical consequences of the two basic transformations T and S. Due to the periodic boundary conditions, invariance under T means that we could equivalently regard the same periodic lattice as having $iL_1 + L_2$ as "vertical" extent. We

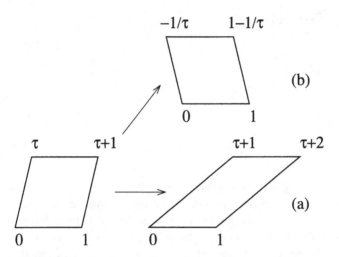

Fig. 11.1. Modular transformations of the torus generated by (a) T and (b) S

can also reverse our image of the lattice and look at the torus from the side, rather from underneath, to have a practical realization of S. The partition $\mathcal{Z}$ is insensitive to such arbitrary choices to describe the periodic lattice and is therefore modular invariant.

11.2 Implementation for Minimal Models

The requirement of modular invariance, although apparently trivial by itself, does become important when combined with the information obtained in the strip geometry. Let $q := \exp(2\pi i \tau)$. In Chap. 8 it was shown that (see (8.35)) the quantum Hamiltonian H and the momentum operator $\hat{P}$ act as infinitesimal translation operators along and around the cylinder. Consequently

$$
\begin{aligned}
\mathcal{Z}(q) &= \operatorname{tr} e^{(-2\pi \operatorname{Im} \tau H + 2\pi i \operatorname{Re} \tau \hat{P})} \\
&= e^{\pi c \operatorname{Im} \tau / 6} \operatorname{tr}\left(e^{2\pi i \tau L_0 - 2\pi i \bar{\tau} \bar{L}_0}\right) \\
&= q^{-c/24} \bar{q}^{-c/24} \operatorname{tr} q^{L_0} \bar{q}^{\bar{L}_0} \\
&= \sum_{r,s;r',s'} N_{r,s;r',s'}\, \chi_{r,s}(q)\chi_{r',s'}(\bar{q}),
\end{aligned}
\tag{11.3}
$$

where the Virasoro algebra character functions (10.30) are used and Re and Im denote the real and imaginary part, respectively. Note that $\chi_{r,s}(\bar{q}) = \bar{\chi}_{r,s}(q)$. The non-negative integers $N_{r,s;r',s'}$ give the number of times a certain primary field together with its conformal tower $(\Delta_{r,s}, \bar{\Delta}_{r',s'})$ appears in $\mathcal{Z}$. One also requires that $N_{1,1;1,1} = 1$, which means that there is a single vacuum.

One needs the transformation of the characters under the modular group. These are given for the minimal models by $q \to \tilde{q} := \exp(-2\pi i/\tau)$ and [137, 138]

$$T: \quad \chi_{r,s}(q) \; \to \; \exp\left[2\pi i \left(\Delta_{r,s} - \frac{c}{24}\right)\right] \chi_{r,s}(q)$$

$$S: \quad \chi_{r,s}(q) \; \to \; \sum_{r',s'} S^{r',s'}_{r,s} \chi_{r',s'}(q), \tag{11.4}$$

where

$$S^{r',s'}_{r,s} = \sqrt{\frac{8}{m(m+1)}} \, (-1)^{(r+s)(r'+s')} \, \sin\left(\frac{\pi r r'}{m}\right) \sin\left(\frac{\pi s s'}{m+1}\right). \tag{11.5}$$

Below, we shall give an example how these modular transformations are done, but let us first examine the consequences of these transformation rules. Equation (11.4) implies that the $\chi_{r,s}$ transform under a finite-dimensional representation $\mathcal{R}_m$ of the modular group and that $\mathcal{Z}$ transforms under the product representation $\mathcal{R}^*_m \otimes \mathcal{R}_m$. Modular invariance is equivalent to the statement that this in general reducible product contains the identity representation.

Invariance under T requires that $N_{r,s;r',s'} = 0$ unless $\Delta_{r,s} - \overline{\Delta}_{r',s'}$ is an integer. This was shown in Chap. 5 to imply the locality or singled-valuedness of the correlation functions. Exactly this same condition was met when discussing the monodromy of the multipoint correlation functions. Considering S shows the existence of at least one solution, since $S^2 = 1$, namely the diagonal one $N_{r,s;r',s'} = \delta_{r,r'}\delta_{s,s'}$. This solution contains all *scalar* operators permitted by the Kac formula. For a second solution, one notes the relations (if m is odd)

$$S^{r',s'}_{r,s} = (-1)^{s'-1} S^{r',m+1-s'}_{r,s} = (-1)^{s-1} S^{r',s'}_{r,m+1-s} \tag{11.6}$$

which implies that the space spanned by $\chi_{r,s} \pm \chi_{r,m+1-s}$ is left invariant. Similar expressions are found for m even. Since the indices r, s are not the same, the corresponding physical scaling operators have a non-vanishing spin $\Delta_{r,s} - \Delta_{r,m+1-s}$. Operator product algebras containing such operators were already mentioned in Chap. 7.4. The consistency of modular invariance with the existence of non-scalar scaling operators follows from the specific modular relationship between $\chi_{r,s}$ and $\chi_{r,m+1-s}$.

The modular invariant partition functions have been classified by Cappelli, Itzykson and Zuber [127] and Kato [394]. They showed that the complete list contains two infinite series of modular invariant partition functions (these are the two solutions mentioned above) and a finite number of known exceptions. There is a deep correspondence with the ADE classification of the simply-laced finite-dimensional Lie algebras. In Table 11.1, all modular invariant critical partition functions are given. It is enough to discuss here the unitary case, since the generalization to non-unitary minimal models is

Table 11.1. Critical point partition functions as given by the A, D and E series of unitary models, of central charge $c = 1 - 6/(m(m+1))$

m	Partition function $\mathcal{Z}$	Class
≥ 3	$\frac{1}{2}\sum_{r=1}^{m-1}\sum_{s=1}^{m}\lvert\chi_{r,s}\rvert^2$	(A_{m-1}, A_m)
$4\ell+1$	$\sum_{r=1}^{m-1}\sum_{a=0}^{\ell}\lvert\chi_{r,2a+1}\rvert^2 + \sum_{r=1}^{m-1}\sum_{a=0}^{\ell-1}\chi_{r,2a+1}\bar{\chi}_{m-r,2a+1}$	(A_{m-1}, D_{2l+2})
$4\ell+2$	$\sum_{a=0}^{\ell}\sum_{s=1}^{m}\lvert\chi_{2a+1,s}\rvert^2 + \sum_{a=0}^{\ell-1}\sum_{s=1}^{m}\chi_{2a+1,s}\bar{\chi}_{2a+1,m+1-s}$	(D_{2l+2}, A_m)
$4\ell+3$	$\sum_{r=1}^{m-1}\sum_{a=0}^{\ell}\lvert\chi_{r,2a+1}\rvert^2 + \sum_{r=1}^{2\ell+1}\lvert\chi_{r,2\ell+2}\rvert^2$ $+\sum_{r=1}^{m-1}\sum_{a=1}^{\ell}\chi_{r,2a}\bar{\chi}_{m-r,2a}$	$(A_{m-1}, D_{2\ell+3})$
$4\ell+4$	$\sum_{a=0}^{\ell}\sum_{s=1}^{m}\lvert\chi_{2a+1,s}\rvert^2 + \sum_{s=1}^{2\ell+2}\lvert\chi_{2\ell+2,s}\rvert^2$ $+\sum_{a=1}^{\ell}\sum_{s=1}^{m}\chi_{2a,s}\bar{\chi}_{2a,m+1-s}$	$(D_{2\ell+4}, A_m)$
11	$\frac{1}{2}\sum_{r=1}^{10}[\lvert\chi_{r,1}+\chi_{r,7}\rvert^2 + \lvert\chi_{r,4}+\chi_{r,8}\rvert^2 + \lvert\chi_{r,5}+\chi_{r,11}\rvert^2]$	(A_{10}, E_6)
12	$\frac{1}{2}\sum_{s=1}^{12}[\lvert\chi_{1,s}+\chi_{7,s}\rvert^2 + \lvert\chi_{4,s}+\chi_{8,s}\rvert^2 + \lvert\chi_{5,s}+\chi_{11,s}\rvert^2]$	(E_6, A_{12})
17	$\frac{1}{2}\sum_{r=1}^{16}[\lvert\chi_{r,1}+\chi_{r,17}\rvert^2 + \lvert\chi_{r,5}+\chi_{r,13}\rvert^2 + \lvert\chi_{r,7}+\chi_{r,11}\rvert^2$ $+\lvert\chi_{r,9}\rvert^2 + (\chi_{r,3}+\chi_{r,15})^*\chi_{r,9} + \chi_{r,9}^*(\chi_{r,3}+\chi_{r,15})]$	(A_{16}, E_7)
18	$\frac{1}{2}\sum_{s=1}^{18}[\lvert\chi_{1,s}+\chi_{17,s}\rvert^2 + \lvert\chi_{5,s}+\chi_{13,s}\rvert^2 + \lvert\chi_{7,s}+\chi_{11,s}\rvert^2$ $+\lvert\chi_{9,s}\rvert^2 + (\chi_{3,s}+\chi_{15,s})^*\chi_{9,s} + \chi_{9,s}^*(\chi_{3,s}+\chi_{15,s})]$	(E_7, A_{18})
29	$\frac{1}{2}\sum_{r=1}^{28}[\lvert\chi_{r,1}+\chi_{r,11}+\chi_{r,19}+\chi_{r,29}\rvert^2$ $+\lvert\chi_{r,7}+\chi_{r,13}+\chi_{r,17}+\chi_{r,23}\rvert^2]$	(A_{28}, E_8)
30	$\frac{1}{2}\sum_{s=1}^{30}[\lvert\chi_{1,s}+\chi_{11,s}+\chi_{19,s}+\chi_{29,s}\rvert^2$ $+\lvert\chi_{7,s}+\chi_{13,s}+\chi_{17,s}+\chi_{23,s}\rvert^2]$	(E_8, A_{30})

straightforward. Consider the E-series, which is the easiest one to describe. The *Coxeter number* of the corresponding Lie algebra is given by either m or $m + 1$. From Table 1.2, these are 12, 18 and 30 for E_6, E_7 and E_8 respectively. Moreover, for the E_6 case, the indices $(1, 7, 4, 8, 5, 11)$ are the *Coxeter exponents*, also listed in Table 1.2. Adding one to the exponents, one obtains the order of the Casimir invariants. In addition, the multiplicities of the corresponding scalar operators also correspond to multiplicities of the Casimir invariants at the corresponding order.

The same pattern holds for all modular invariant partition functions. In the D-series, the indices $2a + 1$ for the scalar operators correspond to the exponents of some D_p Lie algebra. In this case, the doubling of some primary scalar operators corresponds to the existence of two Casimir invariants of the same order in the corresponding Lie algebra. In the A-series, all possible values for r, s are taken. We point out that in the pairing of two Lie algebras, a member of the A-series is always present and that no (D, D), (D, E) or (E, E) pairing is possible [127].

We illustrate in an example the modular transformation generated by S [137]. For simplicity, let $\tau = i\delta$ with δ real, so that the torus becomes a rectangle. Then under S, one has $\delta \rightarrow 1/\delta$ and $q \rightarrow \tilde{q} = e^{-2\pi/\delta}$. Consider the generating function of the partitions of the integers $P(q)$ (see (10.31))

$$P^{-1}(q) = \prod_{n=1}^{\infty} \left(1 - e^{-2\pi\delta n} \right) = \sum_{n=-\infty}^{\infty} \exp\left(-\pi\delta(3n^2 + n) + i\pi n \right) \qquad (11.7)$$

by Euler's pentagonal theorem. One uses Poisson's resummation formula

$$\sum_{n=-\infty}^{\infty} f(n) = \sum_{p=-\infty}^{\infty} \int_{-\infty}^{\infty} dx \, f(x) \, e^{2\pi i p x} \qquad (11.8)$$

and this yields, after having performed the integration over x

$$P^{-1}(q) = \frac{2}{\sqrt{3\delta}} \exp\left[\frac{\pi}{12} \left(\delta - \delta^{-1} \right) \right] \sum_{p=-\infty}^{\infty} \cos\left(\frac{(2p+1)\pi}{6} \right) \exp\left(-\frac{p(p+1)\pi}{3\delta} \right). \qquad (11.9)$$

Now using that

$$\cos\frac{(2p+1)\pi}{6} = \begin{cases} (-1)^n \frac{1}{2}\sqrt{3} & \text{if } p = 3n, 3n - 1 \\ 0 & \text{if } p = 3n + 1 \end{cases} \qquad (11.10)$$

one can replace the sum over p by two sums over n. This readily leads to the relation, written again in terms of τ and recalling the definition of $\tilde{q}$

$$P^{-1}(q) = (-i\tau)^{-1/2} \exp\left[-\frac{i\pi}{12} \left(\tau + \frac{1}{\tau} \right) \right] P^{-1}(\tilde{q}). \qquad (11.11)$$

In particular, this implies that the partition function of a free Gaussian model with central charge $c = 1$

$$\mathcal{Z}_1(q, \bar{q}) := q^{-1/24}\bar{q}^{-1/24} P(q) P(\bar{q}) \, (\text{Im } \tau)^{-1/2} \qquad (11.12)$$

is modular invariant. It does differ from the naive expectation by the last factor, which can be understood [G12] in terms of a zero-mode subtraction from the otherwise ill-defined functional integral.

The discussion so far has ignored the possibility of internal symmetries and its applications are therefore restricted exclusively to lattice systems with *periodic* boundary conditions. The possible discrete symmetries compatible with the modular invariant partition functions have been classified recently [543]. All unitary minimal models listed in Table 11.1 have a maximal discrete $\mathbb{Z}_2$ symmetry, with the exception of six models. These are the models (A_4, D_4) and (D_4, A_6), which correspond to the critical and tricritical three-states Potts models, respectively and have a discrete $\mathbb{S}_3$ symmetry. The four models related to E_7 and E_8 have no discrete symmetry at all.[1] If the global discrete symmetry is $\mathbb{Z}_2$, the model may be decomposed into a charge zero (integer spin) and a charge one (half-integer spin) sector. The charge sectors are exactly the eigenspaces of the lattice charge operator $\hat{Q}$, introduced in Chap. 9. One may then also write down the partition function for antiperiodic boundary condition. For the A-series, we list these in Table 11.2 (and refer to [543] for the D series and E_6). They are not invariant under the full modular group, but only under some subgroup (see Chap. 6).

Table 11.2. Critical point partition function for $\mathbb{Z}_2$ symmetric systems with antiperiodic boundary conditions in the A series (after [652])

m	Charge	
	0	1
$4\ell + 1$	$\sum_{r=1}^{2\ell} \sum_{a=0}^{2\ell} \chi_{r,2a+1} \bar{\chi}_{r,m-2a}$	$\sum_{r=1}^{2\ell} \sum_{a=1}^{2\ell} \chi_{r,2a} \bar{\chi}_{r,m+1-2a}$
$4\ell + 2$	$\sum_{a=0}^{2\ell+1} \sum_{s=1}^{2\ell} \chi_{2a+1,s} \bar{\chi}_{m-1-2a,s}$	$\sum_{a=1}^{2\ell} \sum_{s=1}^{2\ell+1} \chi_{2a,s} \bar{\chi}_{m-2a,s}$
$4\ell + 3$	$\sum_{r=1}^{2\ell+1} \sum_{a=1}^{2\ell+1} \chi_{r,2a} \bar{\chi}_{r,m+1-2a}$	$\sum_{r=1}^{2\ell+1} \sum_{a=0}^{2\ell+1} \chi_{r,2a+1} \bar{\chi}_{r,m-2a}$
$4\ell + 4$	$\sum_{a=1}^{2\ell+1} \sum_{s=1}^{2\ell+2} \chi_{2a,s} \bar{\chi}_{m-2a,s}$	$\sum_{a=0}^{2\ell+1} \sum_{s=1}^{2\ell+2} \chi_{2a+1,s} \bar{\chi}_{m-1-2a,s}$

We compare this with the results found in the Ising model (10.46). Since $m = 3$, one has the A series as the only possibility. If one has periodic boundary conditions $\tilde{Q} = 0$, the partition function summed over both charge sectors is from Table 11.1

$$\mathcal{Z} = (0,0) + \left(\frac{1}{2}, \frac{1}{2}\right) + \left(\frac{1}{16}, \frac{1}{16}\right) \tag{11.13}$$

which reproduces the result found in Chap. 10 for periodic boundary conditions, while the results of Table 11.2 agree with the observations for an-

---

[1] These discrete symmetries coincide with the automorphism groups of the corresponding ADE Dynkin diagrams.

tiperiodic boundary conditions, $\tilde{Q} = 1$. It should be stressed that the results presented in Table 11.1 do give a *classification of the critical point partition functions* of conformally and modularly invariant systems. They do not, however, specify how the primary fields contained in $\mathcal{Z}$ are distributed into the different charge sectors of the system. This leads to yet a further degree of freedom in constructing models. Although this possibility does not occur for the Ising case $m = 3$, it is present already for $m = 4$ (see Chap. 12).[2]

We finally mention the **Verlinde formula** [602], which relates the formal operator product expansion (5.23) of primary operators

$$\phi_i \cdot \phi_j = N_{i,j}^k \, \phi_k \tag{11.14}$$

to the modular transformation S, written as

$$S: \quad \chi_i(q) \quad \rightarrow \quad \sum_j S_i^j \chi_j(q), \tag{11.15}$$

where we use a shorthand to avoid too many indices. E. Verlinde has shown that [602]

$$\sum_k S_k^\ell N_{i,j}^k = \frac{S_i^\ell S_j^\ell}{S_0^\ell}, \tag{11.16}$$

where 0 stands for the identity operator $\phi_0 = \mathbf{1}$. This formula establishes a direct link between the partition function and the OPE, which can be obtained without the knowledge of the correlation functions. We remark that this abstract result is useful when studying the Kondo effect [8].

11.3 Modular Invariance at $c = 1$

Up to now, only systems with a *finite* number of primary fields as described by *minimal* models were studied. However, there exist plenty of models which do not fall into this category. We have already encountered the example of the free Gaussian field in Chap. 7 described by the action

$$S = \int \mathrm{d}^2 \boldsymbol{x} \, (\partial \phi)^2 = \int \mathrm{d}^2 z \, (\partial_{\bar{z}} \phi)(\partial_z \phi). \tag{11.17}$$

From its partition function (11.12), one reads off the central charge $c = 1$. Primary operators for this action are the exponentials $\mathrm{e}^{\mathrm{i}\alpha\phi}$, also called **vertex operators**, which have the conformal weights $\Delta = \overline{\Delta} = \alpha^2$. Since the action is invariant under the transformations

$$\phi \rightarrow \phi + \text{ const. }, \quad \phi \rightarrow -\phi \tag{11.18}$$

---

[2] Explicit lattice models for these theories do not seem to have been constructed yet.

one may use these to restrict the configuration space. This leads to two possible compactifications, the **circle or Coulomb models** and the **orbifold models**.

For a Coulombic model, one identifies two configurations with fields ϕ and $\phi + 2\pi\rho$, where ρ is a constant and sometimes is called the **compactification radius**. Nothing changes for the correlation functions of the vertex operators, if one only takes $\alpha = n/\rho$. One example of these models is the **XY model** as defined in Chap. 1. It is known that this model has a second-order transition described by the formation of *vortices* [415]. Besides the vortex operators one can also introduce operators corresponding to a *frustration line* such that as r circles around one of the end points (a vortex), ϕ increases by $2\pi m\rho$ where m is an integer. One would like to identify these operators within the framework of conformal invariance.

The implementation of modular invariance is done via the two compactification schemes mentioned above which we now describe in detail.

11.3.1 Circle or Coulomb Models

Consider the model on the torus, characterized by the two periods ω_1 and ω_2 [243, 244]. The compactification condition is

$$\phi(z + k\omega_1 + k'\omega_2) - \phi(z) = 2\pi\rho(km + k'm'), \tag{11.19}$$

where the dependence on $\bar{z}$ is suppressed. The fields satisfying this condition can be written as $\phi = \phi_{\text{per}} + \phi_{\text{cl}}$ where ϕ_{per} is a fluctuating periodic field and ϕ_{cl} is the solution of the equations of motion compatible with (11.19)

$$\phi_{\text{cl}}(z) = 2\pi\rho \left(\frac{z}{\omega_1} \frac{m\bar{\tau} - m'}{\bar{\tau} - \tau} \right) + \text{c.c.}. \tag{11.20}$$

The action is then $S = S_{\text{per}} + \pi\rho^2 |m\tau - m'|^2 / \text{Im}\,\tau$. The model is thus decomposed into sectors described by the integers m, m'. The partition function of a sector is

$$\mathcal{Z}_{m',m}(\rho, \tau) := \mathcal{Z}_1(\tau) \exp\left(-\pi\rho^2 \frac{|m\tau - m'|^2}{\text{Im}\,\tau} \right), \tag{11.21}$$

where $\mathcal{Z}_1$ is defined in (11.12). One can show that under a modular transformation, using the techniques of the previous section

$$\mathcal{Z}_{m',m} \left(\rho, \frac{a\tau + b}{c\tau + d} \right) = \mathcal{Z}_{am'+bm, cm'+dm}(\rho, \tau). \tag{11.22}$$

Consequently, there are two obvious modular invariants. One is simply $\mathcal{Z}_{0,0} = \mathcal{Z}_1$. The other one is obtained by summing over all values of m, m'

$$\mathcal{Z}(\rho, \tau) := \rho \sum_{m',m} \mathcal{Z}_{m',m}(\rho, \tau) \tag{11.23}$$

$$= (q\bar{q})^{-1/24} P(q) P(\bar{q}) \sum_{n,m} \exp\left[-2\pi i \text{Re}\,\tau \cdot mn - 2\pi \text{Im}\,\tau \frac{1}{2} \left(\rho^2 m^2 + \frac{n^2}{\rho^2} \right) \right]$$

after applying the Poisson formula (11.8) to m'. Note that in this expression the factor Im τ of (11.12) has disappeared. Now let $q = \exp(2\pi i \tau)$ and recalling (11.3), one recognizes the scaling dimension $x_{n,m}$ and the spin $s_{n,m}$ of the primary operators $\mathcal{O}_{n,m}$ with

$$x_{n,m} = \frac{1}{2}\left(\rho^2 m^2 + \frac{n^2}{\rho^2}\right) \quad , \quad s_{n,m} = nm. \tag{11.24}$$

Finally, the Coulombic partition function becomes

$$\mathcal{Z}(\rho, \tau) = \mathcal{Z}(\rho^{-1}, \tau) = (q\bar{q})^{-1/24} P(q) P(\bar{q}) \sum_{n,m=-\infty}^{\infty} q^{\Delta_{n,m}} \bar{q}^{\overline{\Delta}_{n,m}} \tag{11.25}$$

with the conformal weights

$$\Delta_{n,m} = \frac{1}{4}\left(\frac{n}{\rho} + m\rho\right)^2 \quad , \quad \overline{\Delta}_{n,m} = \frac{1}{4}\left(\frac{n}{\rho} - m\rho\right)^2. \tag{11.26}$$

For $\rho > 1$, the quantum number n is said to describe *electric* fields, while the quantum number m is said to describe *magnetic* fields. It is apparent that all conformal weights are ρ-dependent, but the spins are always integer.

To make the connection with the XY model, one identifies the spin operator with the conformal operator $\mathcal{O}_{\pm 1, 0}$ such that the exponent $\eta = 2x_{\pm 1,0} = 1/4$ is reproduced. This implies the choice $\rho = 2$. The corresponding magnetic or vortex operator is then $\mathcal{O}_{0,\pm 1}$, implying $x_{0,\pm 1} = 2$. This is satisfactory, since a *marginal* operator is needed to account for the temperature-dependent exponent $\eta(T)$ is the low-temperature phase of the XY model. In particular, if this identification is correct, this yields for the central charge of the XY model

$$c_{XY} = 1. \tag{11.27}$$

11.3.2 Orbifold Models

An **orbifold** is defined as follows [G15]. Let $\mathcal{M}$ be a manifold and let $\mathbb{G}$ be a finite group acting on $\mathcal{M}$. Consider the *fixed points* under the action of $\mathbb{G}$: $gx = x$, where g is an element of $\mathbb{G}$. Define the equivalence relation $x \equiv gx$ for all elements g of $\mathbb{G}$. Then the quotient space $\mathcal{M}/\mathbb{G}$ is an orbifold. As an example, take the circle S^1 with points $x = x + 2\pi r$ mod 2π and the group $\mathbb{Z}_2$ with the generator $gx = -x$. The fixed points are $x = 0, x = \pi r$. So topologically the orbifold $S^1/\mathbb{Z}_2$ is a line segment (see Fig. 11.2). It is distinguished from a manifold by the existence of a discrete set of singular points.

In the context of conformal invariance, an orbifold construction on a modular invariant theory with some discrete symmetry group $\mathbb{G}$ amounts to forming a quotient of that theory with respect to $\mathbb{G}$ such that the resulting theory is also modular invariant.

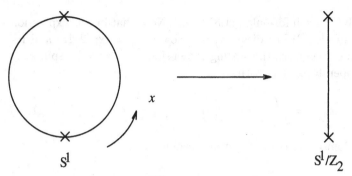

Fig. 11.2. The manifold S^1 and the orbifold $S^1/\mathbb{Z}_2$

Consider the Gaussian model. The symmetry group is $\mathbb{Z}_2$ with the generator $g\phi = -\phi$. Consequently, on the torus one has four possible boundary conditions, obtained as combinations from periodic or antiperiodic boundary conditions $(\pm)$ in either direction. Then the orbifold partition function is [268, G15]

$$\mathcal{Z}_{\mathrm{orb}}(\rho, \tau) := (q\bar{q})^{-1/24} \left[\mathrm{tr}^{(+)} \frac{1}{2}(1+g)q^{L_0}\bar{q}^{\bar{L}_0} + \mathrm{tr}^{(-)} \frac{1}{2}(1+g)q^{L_0}\bar{q}^{\bar{L}_0} \right]$$

$$= \frac{1}{2}\left(\mathcal{Z}(\rho, \tau) + 2\mathcal{Z}(2,\tau) - \mathcal{Z}(1,\tau) \right), \qquad (11.28)$$

where $\mathcal{Z}(\rho, \tau)$ is defined in (11.25).

A few remarks are in order. First, one sees that $\mathcal{Z}_{\mathrm{orb}}(1, \tau) = \mathcal{Z}(2, \tau)$ which is the XY model. Secondly, after some tedious algebra, one has $\mathcal{Z}_{\mathrm{orb}}(\sqrt{2}, \tau) = \mathcal{Z}_{\mathrm{Ising}}^2$ and the model is equivalent to two decoupled Ising models. Third, by writing down the primary fields, some of the critical exponents do depend on the orbifold compactification radius ρ, while others do not. This will be made more explicit when discussing the Ashkin–Teller model in Chap. 12. Finally, one can write the partition functions of the p-state Potts model with $0 < p < 4$ and of the $O(n)$ model with $-2 \leq n \leq 2$ in terms of $\mathcal{Z}$. For the p-state Potts model, take

$$\mathcal{Z}(p; \tau) = \frac{\rho}{2}\mathcal{Z}_1(\tau) \sum_{m', m} \exp\left[-\frac{\pi\rho^2}{4} \frac{|m\tau - m'|^2}{\mathrm{Im}\,\tau} \right] \cos[\pi e_0 < m, m' >]$$

$$+ \frac{1}{2}(p-1)\left[\mathcal{Z}(\rho, \tau) - \mathcal{Z}(\rho/2, \tau) \right], \qquad (11.29)$$

where $< m', m >$ is the largest common positive divisor of m' and m. Put

$$\rho^2 = \frac{4m}{m+1}, \quad p = \left(2\cos\frac{\pi}{m+1} \right)^2, \quad e_0 = \frac{2}{m+1}, \qquad (11.30)$$

where here m is used to calculate the central charge from the unitary Kac formula (10.28). For the $O(n)$ model, put

$$\rho^2 = \frac{m+1}{m} \quad , \quad n = \left(2\cos\frac{\pi}{m}\right)^2 \quad , \quad e_0 = \frac{1}{m} \tag{11.31}$$

and the term proportional to $p - 1$ in (11.29) should be dropped.

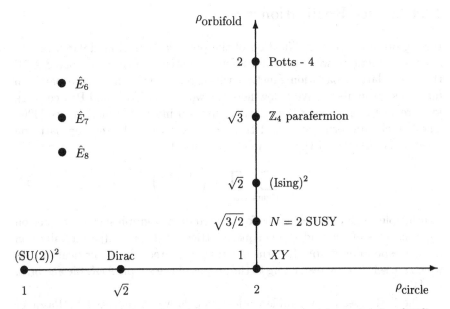

Fig. 11.3. Known conformal modular invariant systems with $c = 1$ (after [268])

In Fig. 11.3 the known conformal systems with central charge $c = 1$ are displayed. It is generally believed that this is the complete list of modular invariant systems with $c = 1$. We remark that the three isolated points correspond to systems which cannot be continuously connected to the Coulombic or orbifold models. Their partition functions were constructed by Pasquier [492] and read

$$\mathcal{Z}(\rho, \tau; \hat{E}_6) = \frac{1}{2}\left(2\mathcal{Z}(3, \tau) + \mathcal{Z}(2, \tau) - \mathcal{Z}(1, \tau)\right)$$

$$\mathcal{Z}(\rho, \tau; \hat{E}_7) = \frac{1}{2}\left(\mathcal{Z}(4, \tau) + \mathcal{Z}(3, \tau) + \mathcal{Z}(2, \tau) - \mathcal{Z}(1, \tau)\right)$$

$$\mathcal{Z}(\rho, \tau; \hat{E}_8) = \frac{1}{2}\left(\mathcal{Z}(5, \tau) + \mathcal{Z}(3, \tau) + \mathcal{Z}(2, \tau) - \mathcal{Z}(1, \tau)\right). \tag{11.32}$$

Since the XY model appears at the meeting point of the two compactifications, one might speculate that this peculiar property might be related to the presence of the Kosterlitz–Thouless transition.

We have seen that the $c = 1$ models studied were classified according the Dynkin diagrams of the simple Lie algebras of the ADE series. It is well known that the subgroups of SU(2) can be also classified according to them. It is then tempting to try to construct new integrable models from the Dynkin diagrams arising from the subgroups of larger algebras, like SU(n). For a review on this line of research, see [245].

11.4 Lattice Realizations

Having obtained the classification of the possible critical modular invariant partition functions, we now list for reference for each of the constructed ADE theories a lattice partition function which reproduces one of the partition functions given above. We follow here the work of O'Brien and Pearce [481], for a review emphasising the relationship with integrable models, see [496]. The models are formulated as IRF models (see Chap. 1) and the partition function is in terms of the Boltzmann weights W

$$\mathcal{Z} = \sum_{\text{heights}} \prod_{\text{faces}} W \begin{pmatrix} d & c \\ a & b \end{pmatrix}. \tag{11.33}$$

In principle, such a model can be formulated on any graph such that spins on adjacent sites of the underlying square lattice must have adjacent values on the corresponding graph. It turns out that it is precisely the critical models which are selected by restricting to the graphs of the simple ADE Lie algebras, see [496].

The first class of integrable critical models was formulated by Pasquier [492, 493]. The Boltzmann weights are

$$W \begin{pmatrix} d & c \\ a & b \end{pmatrix} u \end{pmatrix} = \frac{\sin \lambda - u}{\sin \lambda} \delta_{a,c} G_{a,b} G_{a,d} + \frac{\sin u}{\sin \lambda} \sqrt{\frac{S_a S_c}{S_b S_d}} \delta_{b,d} G_{a,b} G_{b,c}, \tag{11.34}$$

where the local heights a, b, c, d take values on a given ADE graph and u is called **spectral parameter** (here, we only consider the case $0 < u < \lambda$), which controls the spatial anisotropy of the Boltzmann weights. The *adjacency matrix* is

$$G_{a,b} = \begin{cases} 1 & ; a, b \text{ connected on the graph} \\ 0 & ; \text{otherwise} \end{cases}. \tag{11.35}$$

The non-negative components S_a of the Perron-Frobenius eigenvector are determined by[3]

$$\sum_b G_{a,b} S_b = 2 \cos \left(\frac{\pi}{h} \right) S_a, \tag{11.36}$$

where the **crossing parameter** $\lambda = \pi/\hat{h}$ is given in terms of the *Coxeter number* of the associated simple Lie algebra and are listed in Table 1.2. These

---

[3] The same construction can also be performed on the *affine* ADE graphs. Then (11.36) is replaced by $\sum_b G_{a,b} S_b = 2 S_a$, in the Boltzmann weights $\sin x \to x$ and the the central charges in Table 11.3 become $c = 1$ or $c = 3/2$ [496, 608].

models coincide with the critical RSOS models solved by Andrews, Baxter and Forrester [32]. Their partition functions realize the (A, G) series found in [127].

Table 11.3. Physical branches and central charges of the dilute ADE models

branch	crossing parameter λ	inversion point	physical region	central charge c
I	$\pi(1 - 1/\hat{h})/4$	$\gamma = 3\lambda$	$0 < u < \gamma$	$1 - 6/(\hat{h}(\hat{h} + 1))$
II	$\pi(1 + 1/\hat{h})/4$	$\gamma = 3\lambda$	$0 < u < \gamma$	$1 - 6/(\hat{h}(\hat{h} - 1))$
III	$\pi(1 + 1/\hat{h})/4$	$\gamma = 3\lambda - \pi$	$\gamma < u < 0$	$3/2 - 6/(\hat{h}(\hat{h} + 1))$
IV	$\pi(1 - 1/\hat{h})/4$	$\gamma = 3\lambda - \pi$	$\gamma < u < 0$	$3/2 - 6/(\hat{h}(\hat{h} - 1))$

A second class of integrable models on ADE graphs are the so-called **dilute ADE models** [540, 605]. There are four solvable branches of these models, see Table 11.3. The Boltzmann weights are

$$
W \left(\begin{matrix} d & c \\ a & b \end{matrix} \bigg| u \right) = \rho_1(u)\, \delta_{a,b,c,d} + \rho_2(u) \left[\delta_{a,b,c} G_{a,d} + \delta_{a,c,d} G_{a,b} \right]
$$

$$
+ \rho_4(u) \left[\sqrt{\frac{S_a}{S_b}}\, \delta_{b,c,d} G_{a,b} + \sqrt{\frac{S_c}{S_a}}\, \delta_{a,b,d} G_{a,c} \right] + \rho_9(u) \sqrt{\frac{S_a S_c}{S_b S_d}}\, \delta_{b,d} G_{a,b} G_{b,c}
$$

$$
+ \rho_8(u)\, \delta_{a,c} G_{a,b} G_{a,d} + \rho_6(u) \left[\delta_{a,b} \delta_{c,d} G_{a,c} + \delta_{a,d} \delta_{b,c} G_{a,b} \right]. \tag{11.37}
$$

The Perron-Frobenius vector S_a is again given by (11.36), but here $G_{a,b}$ are the elements of the *effective* adjanceny matrix, which is obtained by adding a loop to each node of the ADE graphs. That implies that spins on adjacent sites on the lattice are either the same or adjacent on the ADE graph. Furthermore

$$
\delta_{a,b,c,\dots} := \begin{cases} 1 & ; a = b = c = \dots \\ 0 & ; \text{otherwise} \end{cases} \tag{11.38}
$$

and the trigonometric weight functions are [540, 605]

$$
\rho_1(u) = 1 + \frac{\sin u \sin(3\lambda - u)}{\sin(2\lambda) \sin(3\lambda)} \quad , \quad \rho_2(u) = \frac{\sin(3\lambda - u)}{\sin(3\lambda)}
$$

$$
\rho_4(u) = \epsilon \frac{\sin u}{\sin(3\lambda)} \quad , \quad \rho_6(u) = \epsilon \frac{\sin u \sin(3\lambda - u)}{\sin(2\lambda) \sin(3\lambda)} \tag{11.39}
$$

$$
\rho_8(u) = \frac{\sin(2\lambda - u) \sin(3\lambda - u)}{\sin(2\lambda) \sin(3\lambda)} \quad , \quad \rho_9(u) = -\frac{\sin u \sin(\lambda - u)}{\sin(2\lambda) \sin(3\lambda)},
$$

where the choice $\epsilon = \text{sgn}\, u$ ensures positive Boltzmann weights at the isotropic points $u = \gamma/2$ [481].

The models of branch II give a second realization of the (A, G) series found in [127]. The (G, A) series is realized by branch I [481]. Taken together, one has at least one explicit lattice model for all the conformal field theories listed in Table 11.1. The models of branches III and IV are related to *two-color* models, built on pairs of adjacency graphs, see [481, 607] for details. We leave it as an exercise to write down the first few cases explicitly and to relate them to known statistical systems. For example, the dilute A_3 model corresponds to the adjacency diagram

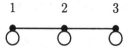

and it can be shown that branch II can be written as a spin-one Ising model [540, 605].

These identifications have been verified numerically. Estimates for the central charges and the critical exponents can be obtained by diagonalising the transfer matrix

$$\langle a|T(u)|b\rangle = \prod_{n=1}^{N} W\left(\begin{array}{cc} b_n & b_{n+1} \\ a_n & a_{n+1} \end{array}\middle| u\right). \tag{11.40}$$

In an extensive study [481], the expected values of the central charges and the scaling dimensions have been fully confirmed for the dilute ADE models. We refer to [481] for detailed tables which show the degree of numerical precision achieved.

For some of these models, a lot more can be said away from the critical point. We shall come back to this in Chap. 14.

12. Further Developments and Applications

Having treated the Ising model in quite some detail, we now study several other systems to give an overview of some of the possible applications. The selected examples are meant to illustrate the techniques developed so far and to introduce a few new concepts. Many of the results to be described were obtained by the numerical techniques presented in Chap. 9 and we shall mainly give a review on existing results. For computational details, we refer to the original papers.

12.1 Three-States Potts Model

The quantum Hamiltonian of the three-states Potts model is, with the correct normalization for conformal invariance applications [258]

$$H = -\frac{2}{3\sqrt{3}} \sum_{n=1}^{N} \left\{ \sigma_n + \sigma_n^+ + \lambda \left(\Gamma_n^+ \Gamma_{n+1} + \Gamma_n \Gamma_{n+1}^+ \right) \right\},$$ (12.1)

where the 3×3 matrices σ and Γ were defined in Chap. 9. The model has a single second-order transition at the self-duality point $\lambda_c = 1$. Consider the case of toroidal boundary conditions

$$\Gamma_{N+1} = \omega^{\tilde{Q}} \Gamma_1 \; ; \quad \tilde{Q} = 0, 1, 2$$ (12.2)

which gives the blocks $H_Q^{(\tilde{Q})} = H_{\tilde{Q}}^{(Q)}$ and also $H_1^{(\tilde{Q})} = H_2^{(\tilde{Q})}$ in the sense that the spectra are the same. One further has to replace $\mathcal{P}_{\text{shift}}$ from (10.34) by $\mathcal{P}_{\text{shift}} = \tilde{Q}/3$.

From finite-size scaling calculations, the central charge $c = 4/5$ was found (for tables of finite-size data, see [258, 261]) and the operator content can thus be written in terms of the $m = 5$ Virasoro characters. The degeneracies following from the character functions of the primary operators are given in Table 12.1. This gives half of the entries of the Kac table. The second half of the conformal operators is obtained from

$$\phi_{r,s} = \phi_{m-r,m+1-s}.$$ (12.3)

The operator content of the five independent sector is [258, 138]

Table 12.1. Virasoro characters $d(\Delta, I)$ of the primary operators $\phi_{r,s}$ for $m = 5$

Δ	0	1	2	3	4	5	6	7	8	9	10	11	12	13	14	15	r	s
							I											
0	1	0	1	1	2	2	4	4	7	8	12	14	21	24	34	41	1	1
1/8	1	1	1	2	3	4	6	8	11	15	20	26	35	45	58	75	1	2
2/3	1	1	2	2	4	5	8	10	15	19	27	34	46	58	77	96	1	3
13/8	1	1	2	3	4	6	9	12	16	22	29	38	50	64	82	105	1	4
2/5	1	1	1	2	3	4	6	8	11	15	20	26	35	45	58	74	2	1
1/40	1	1	2	3	4	6	9	12	17	23	31	41	54	70	91	117	2	2
1/15	1	1	2	3	5	7	10	14	20	26	36	47	63	81	106	135	2	3
21/40	1	1	2	3	5	7	10	14	19	26	35	46	61	79	102	131	2	4
7/5	1	1	2	2	4	5	8	10	15	19	26	33	45	56	74	92	3	1
3	1	1	2	3	4	5	8	10	14	18	24	31	41	51	66	83	4	1

$$H_{0,+}^{(0)} = (0,0) + \left(\frac{2}{5}, \frac{2}{5}\right) + \left(\frac{7}{5}, \frac{7}{5}\right) + (3,3)$$

$$H_{0,-}^{(0)} = \left(\frac{2}{5}, \frac{7}{5}\right) + \left(\frac{7}{5}, \frac{2}{5}\right) + (3,0) + (0,3)$$

$$H_{1}^{(0)} = \left(\frac{1}{15}, \frac{1}{15}\right) + \left(\frac{2}{3}, \frac{2}{3}\right)$$

$$H_{1}^{(1)} = \left(\frac{2}{5}, \frac{1}{15}\right) + \left(0, \frac{2}{3}\right) + \left(\frac{7}{5}, \frac{1}{15}\right) + \left(3, \frac{2}{3}\right)$$

$$H_{1}^{(2)} = \left(\frac{1}{15}, \frac{2}{5}\right) + \left(\frac{2}{3}, 0\right) + \left(\frac{1}{15}, \frac{7}{5}\right) + \left(\frac{2}{3}, 3\right), \tag{12.4}$$

where[1] the index $\pm$ refers to charge-conjugation eigenstates. Comparing with Table 8.1, we note that only a subset of the possible $m = 5$ primary operators is present and that some of the scaling operators occur twice in the full operator content. For periodic boundary conditions, one has two scalar relevant primary operators with exponents

$$x_\varepsilon = \frac{4}{5}, \quad x_\sigma = \frac{2}{15} \tag{12.5}$$

from which all other bulk exponents can be found by the familiar scaling relations (1.41) or (9.21) and in agreement with earlier numerical studies, see [625, 450]. Constructing the partition function for periodic boundary conditions and comparing with Table 11.1, one sees that the model is the first member of the infinite D series. In contrast to the models of the A series, the three-states Potts model does contain also non-scalar operators, even for periodic boundary conditions. These correspond to *conserved currents* of spins 1 and 3, respectively and have negative charge conjugation.

---

[1] We remind the reader of the notation used here, already introduced in Chap. 10. Instead of the character $\chi_{r,s}\bar{\chi}_{\bar{r},\bar{s}}$ we write simply $(\Delta_{r,s}, \bar{\Delta}_{\bar{r},\bar{s}})$, where the $\Delta_{r,s}$ are the conformal weights of $\chi_{r,s}$ from the Kac formula.

These results can also be obtained analytically from a mapping onto the XXZ chain with boundary fields using the Temperley–Lieb algebra [22]. The simplest example of this correspondence occurs for free boundary conditions and will be described in Chap. 15.

Finally, we remark that the conformal operators constructed from the Verma module of the Virasoro algebra are not yet capable of representing the extra degree of freedom encoded in the charge conjugation, which distinguishes, for example, the *two* order parameters σ and $\bar{\sigma}$. In oder to understand this, the so-called $\mathcal{W}$ **symmetry** must be introduced, but this subject is beyond the scope of an introduction, see e.g. [557].

12.2 Tricritical Ising Model

12.2.1 Operator Content

As a further example, we consider the tricritical Ising model. There are quite a few known realizations of this universality class. One can start from a A_4 RSOS model or a ϕ^6 theory. Widely studied is the realization as an Ising model with vacancies, whose concentration is governed by a chemical potential. We have done so in Chap. 1. This leads to the **Blume–Capel model** which has the quantum Hamiltonian [254, 20]

$$H = -\frac{1}{\zeta} \sum_{n=1}^{N} \left\{ S^x(n)S^x(n+1) - \alpha \left(S^x(n) \right)^2 - \beta S^z(n) - \gamma \left(S^z(n) \right)^2 \right\},$$
(12.6)

where

$$S^x = \frac{1}{\sqrt{2}} \begin{pmatrix} 0 & 1 & 0 \\ 1 & 0 & 1 \\ 0 & 1 & 0 \end{pmatrix} \quad , \quad S^z = \begin{pmatrix} 1 & & \\ & 0 & \\ & & -1 \end{pmatrix}$$
(12.7)

and α, β, γ are free parameters and ζ is chosen to permit the application of conformal invariance. The $\mathbb{Z}_2$-charge operator is

$$\hat{Q} = \prod_{n=1}^{N} \left[1 - \left(S^z(n) \right)^2 \right].$$
(12.8)

Another possibility is to consider an **Ising metamagnet** with quantum Hamiltonian

$$H = -\frac{1}{\zeta} \sum_{n=1}^{N} \left\{ t\sigma^z(n) - \sigma^x(n)\sigma^x(n+1) + \kappa \sigma^x(n)\sigma^x(n+2) + h\sigma^x(n) \right\}.$$
(12.9)

Here the non-ordering field h plays the role of the chemical potential in the Blume–Capel model. The presence of the second term with $\kappa > 0$ is necessary to have a tricritical point. The non-ordering magnetic field h plays the role of the chemical potential $e^{-\mu}$ in the Blume–Capel model.

The phase diagram of these models (see Fig. 1.5) contains lines of first-order and second-order transitions. The meeting point of these lines is the tricritical point. At this point, two ordered and one disordered phase(s) become indistinguishable. The tricritical point is thus characterized by the fact that *two* lengths will diverge simultaneously. These are the correlation length ξ and the **persistence length** $\tilde{\xi}$. They can be calculated from the two lowest gaps in the energy spectrum using the notation of (10.34)

$$\xi^{-1} = E_1^{(0)}(0,1;N) - E_0^{(0)} \quad , \quad \tilde{\xi}^{-1} = E_0^{(0)}(0,2;N) - E_0^{(0)} \tag{12.10}$$

and where $E_0^{(0)}$ is the ground state energy. The tricritical point is now found by solving simultaneously [536]

$$N\xi^{-1}(N;\alpha_t,\beta_t) = (N+1)\xi^{-1}(N+1;\alpha_t,\beta_t)$$
$$N\tilde{\xi}^{-1}(N;\alpha_t,\beta_t) = (N+1)\tilde{\xi}^{-1}(N+1;\alpha_t,\beta_t), \tag{12.11}$$

where the parameter notation of the Blume–Capel model was used. Computationally, one has to be aware that the finite-size scaling region for tricritical behaviour, as opposed to ordinary second-order or first-order criticality, might become relatively small and a rather fine mesh of data points (α,β) or (t,h) may be necessary. The numerical results for the tricritical point are for the Blume–Capel quantum spin model with $\gamma = 0$ [262]

$$\alpha_t = 0.910207(4) \quad , \quad \beta_t = 0.415685(6) \quad , \quad \zeta_t = 0.56557(50) \tag{12.12}$$

and for the Ising metamagnet quantum chain [314]

$$\kappa = 1 : t_t = 1.0940(9) \quad , \quad h_t = 1.8794(15) \quad , \quad \zeta_t = 4.36(4)$$
$$\kappa = \frac{1}{2} : t_t = 0.5628(3) \quad , \quad h_t = 1.92776(5) \quad , \quad \zeta_t = 2.154(5). \tag{12.13}$$

For estimates of phase boundaries and the tricritical point of isotropic and quantum systems, see [425, 426, 536, 67, 328, 264].

Table 12.2. Virasoro characters $d(\Delta,I)$ of the primary operators $\phi_{r,s}$ for $m = 4$

Δ	I																	
	0	1	2	3	4	5	6	7	8	9	10	11	12	13	14	15	r	s
0	1	0	1	1	2	3	4	4	7	8	12	14	20	23	32	38	1	1
1/10	1	1	1	2	3	4	6	8	11	14	19	24	32	40	51	64	1	2
3/5	1	1	2	2	4	5	7	9	13	16	22	27	36	45	58	71	1	3
7/16	1	1	1	2	3	4	6	8	10	14	18	23	30	38	48	61	2	1
3/80	1	1	2	3	4	6	8	11	15	20	26	34	44	56	71	90	2	2
3/2	1	1	2	2	3	4	6	7	10	13	17	21	28	34	44	54	3	1

The central charge is found to be $c = 7/10$ [43], which implies $m = 4$. The corresponding degeneracies of the primary operators are given in Table 12.2. The operator content is [43]

$$H_0^{(0)} = (0,0) + \left(\frac{1}{10}, \frac{1}{10}\right) + \left(\frac{3}{5}, \frac{3}{5}\right) + \left(\frac{3}{2}, \frac{3}{2}\right)$$

$$H_0^{(1)} = H_1^{(0)} = \left(\frac{3}{80}, \frac{3}{80}\right) + \left(\frac{7}{16}, \frac{7}{16}\right)$$

$$H_1^{(1)} = \left(0, \frac{3}{2}\right) + \left(\frac{3}{2}, 0\right) + \left(\frac{1}{10}, \frac{3}{5}\right) + \left(\frac{3}{5}, \frac{1}{10}\right) \tag{12.14}$$

and one sees that the partition function for periodic boundary conditions

$$\mathcal{Z}(A_4) := H_0^{(0)} + H_1^{(0)} \tag{12.15}$$

is a member of the modular invariant A series of Table 11.1. In particular, one reads off the tricritical bulk exponents

$$x_\varepsilon = \frac{1}{5} \quad, \quad x_{\varepsilon'} = \frac{6}{5} \quad, \quad x_\sigma = \frac{3}{40} \quad, \quad x_{\sigma'} = \frac{7}{8} \tag{12.16}$$

in excellent agreement with the more traditional approaches, see [430]. The tricritical exponents can be recovered from these through familiar scaling relations, see Chap. 1.

However, this is *not* the only model with a modular invariant partition function and $m = 4$. A second model can be obtained by projecting out the corresponding states from the XXZ quantum chain to be described later. The model obtained has among others the sectors [24]

$$D_0^{(1)} = (0,0) + \left(\frac{7}{16}, \frac{7}{16}\right) + \left(\frac{3}{2}, \frac{3}{2}\right)$$

$$D_0^{(2)} = \left(\frac{3}{80}, \frac{3}{80}\right) + \left(\frac{1}{10}, \frac{1}{10}\right) + \left(\frac{3}{5}, \frac{3}{5}\right) \tag{12.17}$$

and the partition function

$$\tilde{\mathcal{Z}} := D_0^{(1)} + D_0^{(2)} = \mathcal{Z}(A_4) \tag{12.18}$$

but the *assignment of the primary operators to the sectors* of the model is different. In particular, one can also show that the operator product expansions of the sector $D_0^{(1)}$ closes. The exponents of the relevant operators are

$$x_\varepsilon = \frac{7}{8} \quad, \quad x_\sigma = \frac{3}{40} \quad, \quad x_{\sigma'} = \frac{1}{5} \quad, \quad x_{\sigma''} = \frac{6}{5}. \tag{12.19}$$

This identification is based on taking $(0,0)$ as the identity primary operator.

This example shows that the ADE classification of the modular invariant partition functions $\mathcal{Z}$ of unitary minimal models is *not* equivalent to a classification of the universality classes of the 2D conformally and modularly invariant systems. Also, it has become evident that one cannot characterize an universality class just by giving a few critical exponents. The specification of the critical point partition function and the distribution of the primary operators into all the sectors $H_Q^{(\tilde{Q})}$ of the model is required.

12.2.2 Supersymmetry and Superconformal Invariance

Sometimes it is possible to put several irreducible representations of some algebra together into a *single* irreducible representation of a larger algebra. This happens in the tricritical Ising model with **superconformal invariance**. The concept of *extended symmetries* is an important one with arises in many systems and we present superconformal invariance just as one example standing for a wealth of others.

We first recall briefly the notion of **supersymmetry**. For a detailed mathematical introduction, see [554] and for supersymmetric field theory, see e.g. [613]. A **Lie superalgebra** is spanned by so-called *even* generators E_i and so-called *odd* generators V_α, where the indices i and α run over the values $i = 1, \ldots N$ and $\alpha = 1, \ldots M$. The generators satisfy the commutation and anticommutation relations

$$[E_i, E_j] = C_{i,j}^k \, E_k \ , \quad [E_i, V_\alpha] = D_{i,\alpha}^\beta \, V_\beta \ , \quad \{V_\alpha, V_\beta\} = F_{\alpha,\beta}^i \, E_i, \qquad (12.20)$$

where C, D, F are structure constants such that the *Jacobi identities*

$$[E_i, [E_j, E_k]] + [E_j, [E_k, E_i]] + [E_k, [E_i, E_j]] = 0$$
$$[V_\alpha, [E_i, E_j]] + [E_i, [E_j, V_\alpha]] + [E_j, [V_\alpha, E_i]] = 0$$
$$[E_i, \{V_\alpha, V_\beta\}] + \{V_\alpha, [V_\beta, E_i]\} - \{V_\beta, [E_i, V_\alpha]\} = 0$$
$$[V_\alpha, \{V_\beta, V_\gamma\}] + [V_\beta, \{V_\gamma, V_\alpha\}] + [V_\gamma, \{V_\alpha, V_\beta\}] = 0 \qquad (12.21)$$

are satisfied. The commutator $[,]$ and the anticommutator $\{ , \}$ can be defined in terms of products of matrices

$$[A, B] := AB - BA \ , \quad \{A, B\} := AB + BA. \qquad (12.22)$$

To motivate a physicist's interest in this apparently rather abstract definition, we note that supersymmetries already show up for very well-known systems. For example, the quantum mechanic harmonic oscillator as described by the Hamiltonian $H = (\hat{x}^2 + \hat{p}^2)/2$, where $\hat{x}$ and $\hat{p}$ are the canonical coordinate and momentum operators, has a dynamic supersymmetry [174]. The even generators are S^z, S^+, S^- where

$$S^z = H = \frac{1}{2}\{b^-, b^+\} \ , \quad S^- = \frac{1}{2}\{b^-, b^-\} \ , \quad S^+ = \frac{1}{2}\{b^+, b^+\} \qquad (12.23)$$

and the odd generators are the bosonic creation and annihilation operators b^+, b^- built in the familiar way from $\hat{x}, \hat{p}$ via $b^\pm = (\hat{x} \mp i\hat{p})/\sqrt{2}$. These five generators generate a Lie superalgebra which is known as **orthosymplectic** superalgebra $Osp(1|2)$. Since the Hamiltonian appears as one of the generators of the algebra, representation theory can be used to calculate the energy spectrum and transition rates. Since the harmonic oscillator spectrum can be described by a *single* irreducible representation of $Osp(1|2)$, this Lie superalgebra gives the *maximal* dynamical symmetry for that system [174]. The familiar unitary $U(1)$ or symplectic $Sp(2)$ harmonic oscillator dynamical symmetries are Lie subalgebras of the orthosymplectic Lie superalgebra [174].

The **superconformal algebra** [247, 89, 216] contains besides the "even" Virasoro generators L_n also "odd" generators G_n. The commutation and anticommutation relations are

$$[L_n, L_m] = (n - m)L_{n+m} + \frac{c}{12}\left(n^3 - n\right)\delta_{n+m,0}$$

$$[L_n, G_m] = \left(\frac{n}{2} - m\right)G_{n+m}$$

$$\{G_n, G_m\} = 2L_{n+m} + \frac{c}{3}\left(m^2 - \frac{1}{4}\right)\delta_{n+m,0}. \tag{12.24}$$

While the L_n have integer indices, for the G_n one can have a half-integer index (Neveu-Schwarz) or else an integer index (Ramond). This implies that in the strip geometry, the Neveu-Schwarz "fermionic" fields have antiperiodic boundary conditions, while the Ramond "fermionic" fields have periodic ones. In the Ramond sector, G_0 acts as a supersymmetry charge, since $G_0^2 = L_0 - c/24$. Thus in the Ramond sector, the highest weight states come in pairs $|\Delta^\pm\rangle$, with $|\Delta^-\rangle = G_0|\Delta^+\rangle$. Supersymmetry is unbroken if $G_0|\Delta^+\rangle = 0$, which is possible if $\Delta^+ = c/24$.

The unitary irreducible representations are described by a super Kac formula, if $c < 3/2$

$$c = \frac{3}{2}\left(1 - \frac{8}{M(M+2)}\right) \; ; \; M = 2, 3, 4, \ldots$$

$$\Delta_{r,s} = \frac{[(M+2)r - Ms]^2 - 4}{8M(M+2)} + \frac{1}{32}\left(1 - (-1)^{r-s}\right) \tag{12.25}$$

with $1 \le r \le M - 1$ and $1 \le s \le M + 1$. For the Neveu-Schwarz sector, one has $r - s$ even and for the Ramond sector, one has $r - s$ odd. The character functions [271, 465] are for the Neveu-Schwarz sector

$$\chi_{r,s}(q) := q^{-c/24} \prod_{n=1}^{\infty} \left(\frac{1 + q^{n-1/2}}{1 - q^n}\right) \sum_{k=-\infty}^{\infty} \left(q^{\Delta_{2kM+r,s}} - q^{\Delta_{2kM+r,-s}}\right) \tag{12.26}$$

$$\tilde{\chi}_{r,s}(q) := q^{-c/24} \prod_{n=1}^{\infty} \left(\frac{1 - q^{n-1/2}}{1 - q^n}\right) \sum_{k=-\infty}^{\infty} (-1)^{Mk}\left(q^{\Delta_{2kM+r,s}} - q^{\Delta_{2kM+r,-s}}\right)$$

and for the Ramond sector

$$\chi_{r,s}(q) := q^{-c/24+1/16} \prod_{n=1}^{\infty} \left(\frac{1+q^n}{1-q^n}\right) \sum_{k=-\infty}^{\infty} \left(q^{\Delta_{2kM+r,s}} - q^{\Delta_{2kM+r,-s}}\right)$$

(12.27)

and there does exist a classification of the modular invariant partition functions [392, 454, 128]. If M is odd, then $\Delta_{M/2,1+M/2} = c/24$ and supersymmetry can be preserved in the Ramond sector. The techniques to calculate the operator product expansion coefficients and the n-point correlation functions have been extended to superconformal invariance as well [400, 31].

Returning to the tricritical Ising model, the central charge $c = 7/10$ corresponds to $M = 3$ and one has the superconformal representations in the Neveu-Schwarz sector (the index VIR labels the Virasoro characters)

$$(0) = (0)_{\text{VIR}} + \left(\frac{3}{2}\right)_{\text{VIR}} \quad , \quad \left(\frac{1}{10}\right) = \left(\frac{1}{10}\right)_{\text{VIR}} + \left(\frac{3}{5}\right)_{\text{VIR}}$$

(12.28)

and one has the decomposition of the superconformal characters, e.g.

$$\chi_{0,\text{VIR}} + \chi_{3/2,\text{VIR}} = \chi_0 \quad , \quad \chi_{0,\text{VIR}} - \chi_{3/2,\text{VIR}} = \tilde{\chi}_0.$$

(12.29)

For the Ramond sector, one has

$$\left(\frac{3}{80}\right) = \left(\frac{3}{80}\right)_{\text{VIR}} \quad , \quad \left(\frac{7}{16}\right) = \left(\frac{7}{16}\right)_{\text{VIR}}.$$

(12.30)

We remark that also for the second $m = 4$ model (12.17) those Virasoro operators which combine into a super Virasoro operator are in the same sector $D_a^{(b)}$ of the model.

Superconformal invariance is merely the simplest example of the many possible extensions of the conformal algebra. A detailed presentation of this goes beyond the scope of an introduction to conformal invariance. In Table 12.3 the central charges of some further examples of extended conformal algebras as compiled in [272] are listed. These can be obtained from a coset construction and we refer to the literature [270, 272] for further information.

Table 12.3. Discrete series of some extended conformal algebras and their central charges. m is a non-negative integer (after [272])

algebra	c	coset
Virasoro	$1 - \frac{6}{(m+2)(m+3)}$	$(su(2)_m \oplus su(2)_1)/su(2)_{m+1}$
Supervirasoro	$\frac{3}{2}\left(1 - \frac{8}{(m+2)(m+4)}\right)$	$(su(2)_m \oplus su(2)_2)/su(2)_{m+2}$
$\mathbb{S}_3$ parafermion	$2\left(1 - \frac{12}{(m+2)(m+6)}\right)$	$(so(3)_m \oplus su(3)_1)/so(3)_{m+4}$
$\mathbb{D}_N$ parafermion	$(N-1)\left(1 - \frac{N(N-2)}{(m+N)(m+N-2)}\right)$	$(so(N)_m \oplus su(N)_1)/so(N)_{m+2}$
$\mathbb{Z}_3$ parafermion	$2\left(1 - \frac{12}{(m+3)(m+4)}\right)$	$(su(3)_m \oplus su(3)_1)/su(3)_{m+1}$
$\mathbb{Z}_N$ parafermion	$2\frac{m-1}{m+2}$	$su(2)_m/u(1)$

12.3 Yang–Lee Edge Singularity

The Yang-Lee singularity had been introduced in Chap. 1. It may be described by an Ising model in an imaginary magnetic field, for which the quantum Hamiltonian is [263]

$$H = -\frac{1}{2\zeta} \sum_{n=1}^{N} [\sigma^z(n) + \lambda \sigma^x(n)\sigma^x(n+1) + ih\sigma^x(n)]. \tag{12.31}$$

Although the critical behaviour is signaled by the divergence of some correlation length, given by the lowest gap of H, there is no apparent symmetry of the model. A few values for the location of the Yang–Lee edge singularity in the Ising quantum chain (12.31) are given in Table 12.4 [263]. Since the transition mechanism is given by the merging of two non-degenerate levels and the subsequent development of non-vanishing imaginary parts of the eigenvalues, one can show [263] that the lowest energy gap takes the form, for μ not too large

$$\xi^{-1}(\lambda, h; N) = \Delta E(\lambda, h; N) \simeq \frac{2\pi x_1}{N} \sqrt{1 + \frac{\mu}{B(\lambda)}} + \dots, \tag{12.32}$$

where $\mu := (h_c(\lambda) - h)N^{2-x_{\text{inf}}}$ is the scaling variable and x_{inf} the scaling dimension of the lowest primary operator. Some values of $B(\lambda)$ are also included in Table 12.4.

Table 12.4. Location of the Yang–Lee singularity and conformal normalization (after [263])

λ	0.002	0.10	0.20	0.40	0.60	0.80	1.0
$h_c(\lambda)$	0.970715(1)	0.636640(3)	0.458498(1)	0.23202(1)	0.09807(2)	0.02468(2)	0
$\zeta(\lambda)$	0.097	0.376	0.49	0.657	0.773	0.890	1
$B(\lambda)$	0.02975(2)	0.4741(2)	0.8500(3)	1.714(6)	2.97(1)	5.50(6)	$\pi/4$

Since the Hamiltonian is not hermitian, there is no reason to demand that the Virasoro generators L_n come from a unitary representation. In fact, there exists a class of **minimal**, albeit non-unitary theories which contain a *finite* number of primary operators. Their structure is quite similar to the unitary theories. Minimal models, denoted by $\mathcal{M}_{p,p'}$, are characterized by two positive coprime integers p and p' with $p < p'$.[2] One has the Kac formula

$$c = 1 - \frac{6(p-p')^2}{pp'} \tag{12.33}$$

$$\Delta_{r,s} = \Delta_{p-r,p'-s} = \frac{(rp'-sp)^2 - (p-p')^2}{4pp'},$$

---

[2] The convention $p > p'$ is also possible, e.g. [G12, G18].

where $1 \leq r \leq p - 1$ and $1 \leq s \leq p' - 1$ and the unitary case is recovered when $p = p' - 1 = m$ with $m \geq 2$. In particular, the Rocha–Caridi formula (10.30) can be used for the calculation of the characters. The considerations of modular invariance can be repeated and a classification of the critical modular invariant partition functions can be obtained.[3]

The simplest example of the non-unitary minimal models [134] is the one with just one primary operator besides the identity. Since the Kac table will contain each primary operator twice, one needs a table with four entries. This is achieved by choosing $p = 2, p' = 5$. Then one finds for the central charge and the conformal weights

$$c = -\frac{22}{5} , \quad \Delta = 0, -\frac{1}{5}. \tag{12.34}$$

Due to the non-unitarity, both the central charge and some conformal weights are negative in this example. In Table 12.5 we give the first few degeneracies $d(\Delta, I)$.

Table 12.5. Virasoro characters $d(\Delta, I)$ of the primary operators $\phi_{r,s}$ for $p = 2, p' = 5$

Δ	I																r	s
	0	1	2	3	4	5	6	7	8	9	10	11	12	13	14	15		
0	1	0	1	1	1	1	2	2	3	3	4	4	6	6	8	9	1	1
$-1/5$	1	1	1	1	2	2	3	4	5	6	8	10	13	15	20	25	1	2

However, since the lowest Δ is negative, some extra care is needed when interpreting the finite-size data. Usually, one assumes that the lowest eigenstate of H corresponds to the identity operator with zero conformal weight. Here, one has a case where the lowest eigenstate does have a non-vanishing scaling dimension, sometimes referred to as the **anomalous dimension of the vacuum** x^*. The two- and three-point functions are now [232]

$$\langle \phi(r_1)\phi(r_2) \rangle = |r_{12}|^{-2x + x^*} \tag{12.35}$$

$$\langle \phi_1(r_1)\phi_2(r_2)\phi_3(r_3) \rangle = C_{123}\, r_{12}^{-x_1 - x_2 + x_3 + x^*/3}\, r_{23}^{-x_2 - x_3 + x_1 + x^*/3}$$
$$\times\, r_{31}^{-x_3 - x_1 + x_2 + x^*/3}.$$

Consequently, when just calculating the eigenvalues of H, one will find *effective* values for the central charge and the exponents, see (3.49,3.54)

$$f = f_b - \frac{\pi c_{\text{eff}}}{6}\frac{1}{N^2} + \dots , \quad c_{\text{eff}} = c - 12x_{\text{inf}}$$
$$\xi_i^{-1} = 2\pi\, x_{\text{eff},i}\, N^{-1} + \dots , \quad x_{\text{eff},i} = x_i - x_{\text{inf}}, \tag{12.36}$$

---

[3] These systems might appear to be quite artificial, but arise in many applications, some of which are discussed below.

where $x_{\text{inf}} = \Delta_{\text{inf}} + \bar{\Delta}_{\text{inf}} = x^*/2$ is the scaling dimension of the lowest state. For the minimal models $\mathcal{M}_{p,p'}$, one has $\Delta_{\text{inf}} = (1 - (p - p')^2)/(pp')$.

Returning to the Yang–Lee singularity, numerical finite-size calculations have given evidence that [362]

$$c_{\text{eff}} = \frac{2}{5} \ , \quad x_{\text{eff},1} = \frac{2}{5} \tag{12.37}$$

from which (12.36) implies

$$c = -\frac{22}{5} \ , \quad x_1 = -\frac{2}{5} \tag{12.38}$$

which means that the Yang–Lee singularity of the 2D Ising model is described by the minimal model $\mathcal{M}_{2,5}$ [134, 362]. The operator content is for periodic boundary conditions

$$H = (0,0) + \left(-\frac{1}{5}, -\frac{1}{5}\right) \tag{12.39}$$

which is modular invariant. Since $\eta_{\text{YL}} = 2x_1 = -4/5$, this implies, using the scaling relations of Chap. 1 that indeed $\delta_{\text{YL}} = -6$ in agreement with earlier conclusions [229, 134]. A non-unitary *tricritical* point has been observed in the Blume–Capel model with an imaginary magnetic field, which apparently realizes the minimal model $\mathcal{M}_{2,7}$ [264].

One might get the impression that models with a non-vanishing x^* are a peculiarity of non-unitary models. This is not the case. As an example, consider the three-states Potts quantum chain (12.1) with the boundary conditions

$$\Gamma_{N+1} = \omega^{\tilde{Q}} \Gamma_1^+. \tag{12.40}$$

This model has only a global $\mathbb{Z}_2$ symmetry. The Hamiltonian spectrum can be decomposed into the sectors $K_Q^{(\tilde{Q})}$, with $Q = 0, 1$ and $\tilde{Q} = 0, 1, 2$. One has $K_Q^{(0)} = K_Q^{(1)} = K_Q^{(2)}$. The operator content is [138, 24]

$$K_0^{(0)} = \left(\frac{1}{40}, \frac{1}{40}\right) + \left(\frac{1}{8}, \frac{1}{8}\right) + \left(\frac{21}{40}, \frac{21}{40}\right) + \left(\frac{13}{8}, \frac{13}{8}\right)$$

$$K_1^{(0)} = \left(\frac{21}{40}, \frac{1}{40}\right) + \left(\frac{1}{40}, \frac{21}{40}\right) + \left(\frac{1}{8}, \frac{13}{8}\right) + \left(\frac{13}{8}, \frac{1}{8}\right), \tag{12.41}$$

where the energies are measured with respect to the lowest state[4] in the sector $H_{0,+}^{(0)}$, see (12.4). Consequently, already the lowest state of $K_0^{(0)}$ has a non-vanishing scaling dimension. We also remark that the unitary $m = 5$ primary operators present in the K sectors are exactly those absent from the ordinary Potts model.

---

[4] This is less arbitrary than it might appear at first sight, since the degeneracies of the excited states, to be calculated from the Rocha–Caridi formula, allow a much more rigid identification of a representation of the Virasoro algebra than the mere knowledge of the conformal weights of the associated primary operator.

12.4 Ashkin–Teller Model

We now introduce a further spin system not already defined in Chap. 1. This is the Ashkin–Teller model. It describes the coupling of two sets of Ising spin variables placed on the same lattice and is defined by the classical spin Hamiltonian

$$\mathcal{H} = -J_2 \sum_{\langle i,j \rangle} (\sigma_i \sigma_j + \tau_i \tau_j) - J_4 \sum_{\langle i,j \rangle} \sigma_i \sigma_j \tau_i \tau_j, \qquad (12.42)$$

where $\sigma_i = \pm 1$ and $\tau_i = \pm 1$ are the Ising spin variables. The coupling J_4 measures the coupling between the two Ising models described by the σ or τ spins alone.

This example might give some idea on the complexities of the conformal description of more general models than the simple spin systems considered so far. Furthermore, the 2D Ashkin–Teller model turns out to contain a section of the orbifold compactification line of Fig. 11.3. The global symmetry of the model (pending a precise statement of the boundary conditions) is the **dihedral group** $\mathbb{D}_4$, which is generated by the transformations $\sigma \to -\sigma$, $\tau \to -\tau$ and $\sigma \leftrightarrow \tau$. Already at this stage, there is an apparent relationship with the orbifold partition function. $\mathcal{Z}_{\text{orb}}$ contains two spin operators $\hat{\sigma}_{1,2}$ with conformal weights $(1/16, 1/16)$, which arise in the $(-)$ sector of the traces, (11.28). This implies that the statistical model having $\mathcal{Z}_{\text{orb}}$ as partition function should have as a global symmetry at least the group generated by $\hat{\sigma}_1 \to -\hat{\sigma}_1$, $\hat{\sigma}_2 \to -\hat{\sigma}_2$, $\hat{\sigma}_1 \leftrightarrow \hat{\sigma}_2$, which is nothing else than $\mathbb{D}_4$.

To obtain the quantum Hamiltonian, it is convenient to rewrite the model in terms of a four-state variable Θ via

$$2\cos(\Theta_i - \Theta_j) = \sigma_i \sigma_j + \tau_i \tau_j \ , \quad \cos(2\Theta_i - 2\Theta_j) = \sigma_i \sigma_j \tau_i \tau_j. \qquad (12.43)$$

Now one proceeds as in Chap. 8. In the extreme anisotropic limit, where $\tau \to 0$ and

$$\beta J_{2,r} = \tau t \ , \quad \beta J_{4,r} = \tau t \varepsilon \qquad (12.44)$$

$$\beta J_{2,\tau} = \frac{1}{4}\left(\ln \tau^{-1} - \ln t\right) \ , \quad \beta J_{4,\tau} = \frac{1}{4}\left(\ln \tau^{-1} + \ln t\right)$$

the quantum Hamiltonian becomes [412]

$$H = -\frac{1}{\zeta} \sum_{n=1}^{N} \left\{ \left(\Gamma_n + \Gamma_n^3 + \varepsilon \Gamma_n^2 \right) + t \left(\sigma_n \sigma_{n+1}^3 + \sigma_n^3 \sigma_{n+1} + \varepsilon \sigma_n^2 \sigma_{n+1}^2 \right) \right\}, \qquad (12.45)$$

where

$$\zeta^{-1} = \frac{1 - 4g}{4\sqrt{t}g\sin(\pi/4g)} \ , \quad g = \frac{\pi}{4\arccos(-\varepsilon)} \qquad (12.46)$$

and σ and Γ are 4×4 matrices as defined in Chap. 9 and ζ is already chosen in view for conformal invariance applications. In this case, the charge conjugation invariance corresponds to the exchange $\sigma_i \leftrightarrow \tau_i$ in the isotropic Ashkin–Teller model. The self-duality of H is evident.

12.4.1 Relation with the XXZ Quantum Chain

There exists a mapping onto a staggered XXZ chain with $2N$ sites. By making first a duality transformation on the τ spins alone and then a second duality transformation on all spins, one finds, up to boundary terms [412]

$$H = -\frac{1}{\zeta} \sum_{n=1}^{2N} \left(\frac{1+t}{2} + (-1)^n \frac{1-t}{2} \right) \left\{ \sigma_n^x \sigma_{n+1}^x + \sigma_n^y \sigma_{n+1}^y - \varepsilon \, \sigma_n^z \sigma_{n+1}^z \right\},$$

(12.47)

where the $\sigma^x, \sigma^y, \sigma^z$ are Pauli matrices. This quantum Hamiltonian can also be obtained from the *six-vertex model* [G19]. For the critical point $t = 1$, it can be solved via Bethe ansatz techniques [22]. If the boundary conditions are taken into account, the relationship becomes more complicated. In Table 12.6 the correspondence is given for periodic boundary conditions in the Ashkin–Teller model with N sites. For each sector Q of the Ashkin–Teller model, the corresponding boundary conditions of the XXZ chain are given. This result follows [22] since the link variables in both the Ashkin–Teller chain and the XXZ chain can be shown to satisfy the same algebra. This relationship led to an exact solution of the model on a three-dimensional submanifold of the full six-dimensional space of the thermodynamic variables [495].

Table 12.6. Sector equivalence of the XXZ chain with the Ashkin–Teller chain (after [22])

Q	XXZ boundary conditions		
0	$\sigma_{2N+1}^x = \sigma_1^x$	$\sigma_{2N+1}^y = \sigma_1^y$	$\sigma_{2N+1}^z = \sigma_1^z$
1, 3	$\sigma_{2N+1}^x = \sigma_1^x$	$\sigma_{2N+1}^y = -\sigma_1^y$	$\sigma_{2N+1}^z = \sigma_1^z$
2	$\sigma_{2N+1}^x \pm i\sigma_{2N+1}^y = -(\sigma_1^x \pm i\sigma_1^y)$	$\sigma_{2N+1}^z = \sigma_1^z$	

12.4.2 Global Symmetry and Boundary Conditions

The global symmetry of H is given by the transformation [36]

$$(\sigma_n)^a \to \left(\sigma_n' \right)^a := A^{a,b} (\sigma_n)^b,$$

(12.48)

where the 3×3 matrices $\hat{A}$ form a representation of the group $\mathbb{G}$. For the Ashkin–Teller model, $\hat{A}$ can be taken to be one of the matrices $\Sigma^k C^\ell$, with $k = 0, 1, 2, 3$, $\ell = 0, 1$ and

$$\Sigma = \begin{pmatrix} i & & \\ & -1 & \\ & & -i \end{pmatrix}, \quad C = \begin{pmatrix} & & 1 \\ & 1 & \\ 1 & & \end{pmatrix}$$

(12.49)

which form a representation of the finite group $\mathbb{D}_4$. The symmetry sectors are given by the *irreducible representations* of $\mathbb{D}_4$. There is a two-dimensional one

$$\mathcal{R}_Q(\Sigma) = \begin{pmatrix} i^Q & \\ & i^{-Q} \end{pmatrix} , \quad \mathcal{R}_Q(C) = \begin{pmatrix} 0 & 1 \\ 1 & 0 \end{pmatrix} \tag{12.50}$$

and four one-dimensional ones

$$\begin{aligned}
\mathcal{R}_{0,+}(\Sigma) &= 1 , \quad \mathcal{R}_{0,+}(C) = 1 \\
\mathcal{R}_{0,-}(\Sigma) &= 1 , \quad \mathcal{R}_{0,-}(C) = -1 \\
\mathcal{R}_{2,+}(\Sigma) &= -1 , \quad \mathcal{R}_{2,+}(C) = 1 \\
\mathcal{R}_{2,-}(\Sigma) &= -1 , \quad \mathcal{R}_{2,-}(C) = -1.
\end{aligned} \tag{12.51}$$

The presence of the two-dimensional irreducible representation of $\mathbb{D}_4$ implies that in the charge sectors $Q = 1$ and $Q = 3$ one has the same spectrum of H. The one-dimensional irreducible representations provide the decomposition with respect to the charge conjugation C in the sectors $Q = 0$ and $Q = 2$. The *conjugacy classes* of $\mathbb{D}_4$ are [293]

$$\left\{ \Sigma^0 \right\} , \left\{ \Sigma^2 \right\} , \left\{ \Sigma, \Sigma^3 \right\} , \left\{ C, \Sigma^2 C \right\} , \left\{ \Sigma C, \Sigma^3 C \right\} . \tag{12.52}$$

Toroidal boundary conditions are defined by

$$(\sigma_{N+1})^a = B^{a,b} (\sigma_1)^b , \tag{12.53}$$

where $\hat{B}$ is any of the matrices of (12.52). Two Hamiltonians with boundary conditions specified by matrices $\hat{B}_1, \hat{B}_2$ have the same spectrum if $\hat{B}_1$ and $\hat{B}_2$ are in the same conjugacy class. The global symmetry depends on the choice of $\hat{B}$ as illustrated in Table 12.7. For further details and the generalization to n-state quantum chains, we refer to the review [538]. These symmetries are generically valid. For peculiar choices of ε, the global symmetry can be higher. An obvious case is $\varepsilon = 1$, when $\mathbb{G} = \mathbb{S}_4$, the permutation group. This case corresponds to the four-state Potts model.

Table 12.7. Dependence of the symmetries of the Ashkin–Teller model on the boundary conditions [36]

Boundary condition	Group	number of elements	Elements[a]
Σ^0	$\mathbb{D}_4$	8	$\Sigma^k C^\ell$
$\Sigma^{\tilde{Q}}$	$\mathbb{Z}_4$	4	Σ^k
Σ^2	$\mathbb{D}_4$	8	$\Sigma^k C^\ell$
$\Sigma^R C$	$\mathbb{Z}_2 \otimes \mathbb{Z}_2$	4	$\Sigma^0, \Sigma^2, \Sigma^R C, \Sigma^{R+2} C$

[a] The indices are $k = 0, 1, 2, 3$ and $\ell = 0, 1$.

12.4.3 Phase Diagram

The various phases described by the quantum Hamiltonian (12.45) are displayed in Fig. 12.1. One distinguishes five phases.

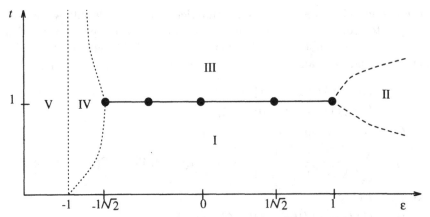

Fig. 12.1. Phase diagram of the Ashkin–Teller quantum chain ([169] after [412])

I A paramagnetic phase, both σ_i and τ_i are disordered.

II A partially ordered phase, with one of the σ_i or τ_i ordered and the other disordered.

III A fully ordered ferromagnetic phase.

IV A *"critical fan"*, where the order parameters of σ_i, τ_i are both zero, but their correlation functions decay algebraically.

V An antiferromagnetically frozen phase.

The transitions between these phase are the following [412, 260]

1. Along the self-dual line $t = 1$, critical exponents change continuously as a function of the *four-spin coupling* ε (full line). The central charge is $c = 1$. The section between $\varepsilon = -1/\sqrt{2}$ and $\varepsilon = 1$ of this line corresponds to the orbifold compactification line of Fig. 11.3 with $1 \le \rho \le 2$. The case $\varepsilon = -1/\sqrt{2}$ corresponds to the meeting point of the two compactification lines and is in the Kosterlitz–Thouless universality class.

2. Two lines which are in the 2D Ising universality class (dashed lines). These transition values of t are dual to each other. It was checked that indeed $c = 1/2$ [260].

3. The borders of the critical fan have a Kosterlitz–Thouless transition (dotted lines).

It should be noted that the *isotropic* Ashkin–Teller model (12.42) does not contain the critical fan,[5] although the antiferromagnetic frozen phase is again present. The critical fan corresponds to a region with negative Boltzmann

[5] See [382] for a recent study on the phase diagram.

weights in the isotropic system. To have positive Boltzmann weights, one needs that $\varepsilon > -1/2$ [412].

12.4.4 Operator Content on the $c = 1$ Line

Since the actual expressions become quite lengthy, we shall only write here the operator content for those boundary conditions which preserve the full $\mathbb{D}_4$ symmetry, see Table 12.7. The operator content is known for all toroidal boundary conditions and is given in [36]. One now needs some notation. Let

$$
\begin{aligned}
A &= (\{0\}, \{0\}) + (\{1\}, \{1\}) + A_1 \\
B &= (\{0\}, \{1\}) + (\{1\}, \{0\}) + A_1 \\
C &= R(4, 2; 4, 2|g) + R(4, 2; -4, -2|g) \\
F(g) &= \sum_{n \geq 0} \left(\frac{(2n+1)^2}{16}, \frac{(2n+1)^2}{16} \right) + R(4, 2; 4, 4|g) + R(4, 2; -4, -4|g) \\
G(g) &= F(1/g) \\
H &= (\{1/16\}, \{1/16\}) + (\{9/16\}, \{9/16\}) \\
K &= (\{1/16\}, \{9/16\}) + (\{9/16\}, \{1/16\})
\end{aligned}
\tag{12.54}
$$

with the abbreviations

$$
R(p, q; r, s|g) = \sum_{m,n \geq 0} \left(\frac{pm + q + (rn+s)g)^2}{16g}, \frac{pm + q - (rn+s)g)^2}{16g} \right)
$$

$$
A_1 = \sum_{n \geq 0} ((n+1)^2 g, (n+1)^2 g) + \left(\frac{(n+1)^2}{g}, \frac{(n+1)^2}{g} \right)
$$
$$
+ R(4, 4; 4, 4|g) + R(4, 4; -4, -4|g)
\tag{12.55}
$$

and

$$
\left\{ \frac{1}{16} \right\} = \sum_{m=-\infty}^{\infty} \left(\frac{(8m+1)^2}{16} \right), \quad \left\{ \frac{9}{16} \right\} = \sum_{m=-\infty}^{\infty} \left(\frac{(8m+3)^2}{16} \right)
$$

$$
\{0\} = \sum_{m=0}^{\infty} (4m^2), \quad \{1\} = \sum_{m=0}^{\infty} ((2m+1)^2),
\tag{12.56}
$$

where (Δ) stands everywhere for a $c = 1$ Virasoro character. These are

$$
\chi_\Delta(q) = q^{-1/24} \begin{cases} q^{n^2/4} \left(1 - q^{n+1}\right) P(q) & \text{if } \Delta = n^2/4 \\ q^\Delta P(q) & \text{if } \Delta \neq n^2/4 \end{cases},
\tag{12.57}
$$

where n is an integer and $P(q)$ is the generating function of the partitions of the integers, see (10.31). The particular form when $\Delta = n^2/4$ is important for the calculation of the secondary states. In Table 12.8 the correspondence between the sectors of the model and the operator content (12.54) is stated.

Table 12.8. Operator content of the Ashkin–Teller model for the $\mathbb{D}_4$-invariant boundary conditions Σ^0, Σ^2 [36]

Sector	Σ^0	Σ^2
$\mathcal{R}_{0,+}$	A	G
$\mathcal{R}_{0,-}$	C	F
$\mathcal{R}_{2,+}$	G	B
$\mathcal{R}_{2,-}$	F	C
$\mathcal{R}_1$	H	K

Note that since $\mathcal{R}_1$ is two-dimensional, the corresponding primary operators appear twice. One can use the charge conjugation quantum number to distinguish between them. We also observe that only the one-dimensional irreducible representations of $\mathbb{D}_4$ are related to g-dependent primary operators while the operators coming from the two-dimensional irreducible representations are g-independent.

In order to illustrate these results, we list as an example the relevant (or marginal) scalar primary operators for periodic boundary conditions as shown in Table 12.9, extending earlier results of [380]. The correspondence

Table 12.9. Scalar primary operators for periodic boundary conditions of the Ashkin–Teller quantum chain

$(0,0)$	$\left(\frac{1}{16},\frac{1}{16}\right)\times 2$	$\left(\frac{g}{4},\frac{g}{4}\right)$	(g,g)	$\left(\frac{9}{16},\frac{9}{16}\right)\times 2$	$\left(\frac{1}{4g},\frac{1}{4g}\right)$	$(1,1)$
1	$\hat{\sigma}\pm\hat{\tau}$	$\hat{\sigma}\hat{\tau}$	$\hat{\varepsilon}_\sigma+\hat{\varepsilon}_\tau$	$\hat{\sigma}\hat{\varepsilon}_\tau\pm\hat{\tau}\hat{\varepsilon}_\sigma$	$\hat{\varepsilon}_\sigma-\hat{\varepsilon}_\tau$	$\hat{\varepsilon}_\sigma\hat{\varepsilon}_\tau$

is given in terms of the primary operators $\hat{\sigma}, \hat{\varepsilon}_\sigma$ and $\hat{\tau}, \hat{\varepsilon}_\tau$ of the two Ising models into which the Ashkin–Teller chain decouples at $\varepsilon = 0$. The $\pm$ refers to the charge conjugation eigenvalue, or C-parity. For $\varepsilon \neq 0$, this simple decomposition no longer applies, but the operators are continuously deformed when changing ε. Since there are no level crossings, the operators retain their identity, however. The operators in the first three columns are often referred to as *magnetization*, *polarization* and *energy density*, respectively. Their bulk critical exponents are

$$x_\sigma = \frac{1}{8} \ , \quad x_P = \frac{g}{2} \ , \quad x_\varepsilon = 2g. \tag{12.58}$$

Examining the operator content, one notes a few special values of ε where the model has a particular symmetry. A few cases are listed below, where the first column refers to the coupling, the second is the symmetry and the third identifies the model

$$
\begin{array}{lll}
\varepsilon = & 1 & \mathbb{S}_4 \quad \text{four-states Potts model} \\
\varepsilon = & 1/\sqrt{2} & \mathbb{Z}_4 \quad \text{parafermion} \\
\varepsilon = & 0 & \mathbb{Z}_2\otimes\mathbb{Z}_2 \quad \text{two decoupled Ising models} \\
\varepsilon = & -\sqrt{2-\sqrt{2}}\big/2 & N=2 \text{ twisted SUSY} \\
\varepsilon = & -1/\sqrt{2} & \text{Kosterlitz–Thouless}
\end{array}
\tag{12.59}
$$

These findings, some obtained long before the construction of the modular invariant partition functions, are in agreement with the conformal invariance results of Fig. 11.3.

One aspect completely left out here is the reorganisation of the operator content into irreducible representations of larger algebraic structures which are not necessarily Lie algebras any more. This is treated in detail in the references [36, 243, 631, 632].

12.5 XY Model

Previously, the XY model (1.57) was tentatively identified within the framework of modular invariant systems with central charge $c = 1$. We now describe, following the original work of Allton and Hamer [30], a finite lattice study to confirm this identification. The quantum Hamiltonian of the XY model is

$$H = \frac{1}{\zeta} \sum_{n=1}^{N} \left[J_3(n) - \frac{\lambda}{2} \left(J_+(n) J_-(n+1) + J_-(n) J_+(n+1) \right) \right.$$
$$\left. - \frac{h}{2} \left(J_+(n) + J_-(n) \right) \right], \qquad (12.60)$$

where the $J_\pm, J_3$ are angular momentum operators with commutators

$$[J_3(n), J_\pm(n')] = \pm J_\pm(n) \delta_{n,n'} \qquad (12.61)$$

while λ is the thermal variable and h a magnetic field. The total spin

$$\hat{S} = \sum_{n=1}^{N} J_3(n) \qquad (12.62)$$

commutes with H. Since the model is expected to have not a power-law divergence of the correlation length and other physical quantities but rather a singularity of the form

$$\xi \sim \exp \left[b \left(\frac{T_K}{T - T_K} \right)^\sigma \right], \qquad (12.63)$$

finite-size techniques must be adapted accordingly. Consider the Callan–Symanzik **beta function**

$$\beta(\lambda) := \frac{\xi^{-1}(\lambda)}{\xi^{-1}(\lambda) - 2\lambda(\xi^{-1})'(\lambda)}, \qquad (12.64)$$

where the prime denotes the derivative. For a singularity of the form (12.63) one has $\beta(\lambda) \sim (\lambda - \lambda_c)^{1+\sigma}$, while for an ordinary second-order transition, the beta function would be linear in the vicinity of λ_c. This allows to distinguish these two situations and to find the exponent σ. A very useful tool to compute the beta-function from finite-lattice data is provided by the **Roomany–Wyld approximant** [541]

$$
\beta_N^{RW} := \left[1 + \ln \left(\frac{\xi^{-1}(\lambda; N)}{\xi^{-1}(\lambda; N-1)} \right) \middle/ \ln \frac{N}{N-1} \right]
$$
$$
\times \left[1 - \lambda \left(\frac{\partial \xi^{-1}(\lambda; N)/\partial \lambda}{\xi^{-1}(\lambda; N)} + \frac{\partial \xi^{-1}(\lambda; N-1)/\partial \lambda}{\xi^{-1}(\lambda; N-1)} \right) \right]^{-1} . \quad (12.65)
$$

Then finite-size scaling yields

$$
\lambda_c = 2.06(4) \;,\;\; \sigma = 0.501(5), \quad (12.66)
$$

in excellent agreement with the Kosterlitz–Thouless prediction $\sigma = 1/2$. On the other hand, various determinations of λ_c using different methods have yielded results which scatter between 1.9 and 2.1.

Due to the conservation of $\hat{S}$, one can consider its sectors separately. Allton and Hamer [30] considered the two lowest states in the sector $S = 0$, denoted by $|0\rangle$ and $|2\rangle$, and in the sector $S = 1$ giving $|1\rangle$ and $|3\rangle$. Now, using (3.49), the energy gaps are at the critical point

$$
E_i - E_j = 2\pi x_{i,j} N^{-1} \quad (12.67)
$$

with $i, j = 0, 1, 2, 3$. One expects that $x_{0,1} = x_{2,3} = \eta/2$ and $x_{0,2} = x_{1,3} = 2$. In writing this, the normalization factor ζ was determined from the ground state energy and (3.54) under the assumption that $c = 1$. The determination of the exponent η was checked by also computing it from the magnetization and the susceptibility per spin. These are computed as outlined in Chap. 9

$$
\chi = \frac{1}{N^2} \sum_{n,m=1}^{N} \langle 0 | J_+(n) J_-(m) + J_-(n) J_+(m) | 0 \rangle
$$
$$
M = \frac{1}{N} \sum_{n=1}^{N} \langle 1 | J_+(n) + J_-(n) | 0 \rangle. \quad (12.68)
$$

Since finite-size scaling was formulated in terms of the correlation length, see Chap. 3, one can calculate the exponent η from

$$
\chi_N \sim \xi^{2-\eta} \sim N^{2-\eta} \;,\;\; M_N \sim \xi^{-\eta/2} \sim N^{-\eta/2}. \quad (12.69)
$$

In Table 12.10, some of the results of [30] for η as a function of λ are displayed. The indices refer to the results obtained either from the susceptibility χ, the magnetization M, the gap $2x_{0,1}$ or the leading weak-coupling result $\eta_{\text{weak}} = 1/(\pi\sqrt{2\lambda})$ for $\lambda \to \infty$. The inferred conformal normalization ζ is also given. The results are seen to be in good agreement with each other. This is consistent with the central charge being equal to unity. On the other hand, the data do not reproduce very precisely the expected value $\eta = 1/4$

Table 12.10. Exponent $\eta(\lambda)$ for the (1+1)D XY model (after [30])

λ	1.8	2.0	2.3	2.6	2.9	3.2	3.5	3.8	4.0
η_χ	0.229	0.205	0.183	0.168	0.157	0.148	0.141	0.135	0.132
η_M	0.229	0.204	0.182	0.168	0.157	0.148	0.140	0.135	0.132
$2x_{0,1}$	0.231	0.205	0.183	0.169	0.157	0.148	0.140	0.133	0.129
η_{weak}						0.126	0.120	0.115	0.113
ζ	1.52	1.62	1.77	1.91	2.04	2.16	2.28	2.40	2.47

at the critical point. This is ascribed to the presence of additional logarithmic factors, see (1.59), which in other models are known to severely affect the convergence of finite-lattice data. This phenomenon was also observed in Monte Carlo simulations using lattices up to 1200×1200 sites, which yield $\sigma = 0.48(10)$ [366]. The deviation of η from the value $1/4$ at $T = T_K$ is explained in terms of logarithmic corrections, but the value of the associated coefficient does not agree with predictions for the Kosterlitz–Thouless transition [367]. The same difficulty also appears in high-temperature series expansions. In fact, the precise determination of the critical behaviour in the two-dimensional XY model remains a challenging problem.

12.6 XXZ Quantum Chain

It was already mentioned above that the partition functions for the minimal models are contained in those of the compactified Coulomb gas partition functions, provided that certain frustrations lines at the boundaries were introduced. In the language of quantum Hamiltonians, the same can be achieved by projecting out the spectra of the minimal models from the one of the spin-1/2 XXZ quantum chain. The presentation follows the original work of Alcaraz, Grimm and Rittenberg [24]. The XXZ Hamiltonian is

$$H = -\frac{\gamma}{2\pi \sin \gamma} \sum_{n=1}^{N} [\sigma^x(n)\sigma^x(n+1) + \sigma^y(n)\sigma^y(n+1) - \cos \gamma \, \sigma^z(n)\sigma^z(n+1)] \tag{12.70}$$

where the σ are Pauli matrices. (This Hamiltonian can also be obtained by taking the Hamiltonian limit in the eight-vertex model [G19].) One writes

$$g = \frac{1}{4}(1 - \gamma/\pi)^{-1}. \tag{12.71}$$

$g = \rho^2/2$ is related to the bosonic compactification radius. Consider the toroidal boundary conditions, characterized by the parameter ℓ

$$\sigma^x(N+1) + i\sigma^y(N+1) = e^{2\pi i\ell} (\sigma^x(1) + i\sigma^y(1)) \tag{12.72}$$

with $0 \leq \ell < 1$. The charge operator is

$$\hat{Q} = \frac{1}{2} \sum_{n=1}^{N} \sigma^z(n). \tag{12.73}$$

Denote by $E_{Q,i}^{\ell}(P;N)$ the ith level in the sector Q, with lattice momentum P, boundary condition ℓ and for N sites. Define the **scaled gaps**

$$\mathcal{E}_{Q,i}^{\ell} := \frac{N}{2\pi} \left(E_{Q,i}^{\ell}(P;N) - E_{0,0}^{0}(0;N) \right). \tag{12.74}$$

The operator content [23] is given by the $N \to \infty$ limit of the partition functions

$$\mathcal{Z}_Q^{\ell}(q,\bar{q}) := \sum_{i,P} q^{(\mathcal{E}_{Q,i}^{\ell}(P;N)+P)/2} \bar{q}^{(\mathcal{E}_{Q,i}^{\ell}(P;N)-P)/2} \tag{12.75}$$

$$= P(q)P(\bar{q}) \sum_{k=-\infty}^{\infty} q^{[Q+4g(\ell+k)]^2/16g} \bar{q}^{[Q-4g(\ell+k)]^2/16g}.$$

In order to give the projection rules for the unitary minimal models, a few preparations are necessary. First, discretize the boundary conditions

$$\ell = \frac{k}{n} \; ; \; k = 0, 1, \ldots, n-1, \tag{12.76}$$

where the integer n will be fixed later. Second, take $E_{0,0}^{1/n}(0;N)$ as the new ground state and define the new scaled gaps

$$\tilde{\mathcal{F}}_{Q,i}^{k}(P;N) := \frac{N}{2\pi} \left(E_{Q,i}^{k/n}(P;N) - E_{0,0}^{1/n}(0;N) \right). \tag{12.77}$$

Since the boundary conditions are in the finite group $\mathbb{Z}_n$, take the charges Q to be in $\mathbb{Z}_n$ by summing up all charges $Q = R \bmod n$ and obtain

$$\mathcal{F}_{R,i}^{k}(P;N) := \frac{N}{2\pi} \left(E_{R,i}^{k/n}(P;N) - E_{0,0}^{1/n}(0;N) \right), \tag{12.78}$$

where $R = 0, 1, \ldots, n-1$ for N even and $2R = 1, 3, \ldots, 2n-1$ for N odd. Finally, choose

$$g = \frac{n^2}{4m(m+1)}, \tag{12.79}$$

where m gives the central charge via the Kac formula (4.42). Then the partition functions are, where again the leading factor $(q\bar{q})^{-1/24}$ is suppressed

$$\mathcal{Y}_R^k(q,\bar{q}) := \mathcal{Y}_{n-R}^{n-k}(q,\bar{q}) = \sum_{i,P} q^{(\mathcal{F}_{R,i}^{k}(P;N)+P)/2} \bar{q}^{(\mathcal{F}_{R,i}^{k}(P;N)-P)/2}. \tag{12.80}$$

One now specifies two classes of systems, labeled by a positive integer M. The systems are called M_L and M_R and are defined through the angles $\varphi_{M,L}$ and $\varphi_{M,R}$

$$\varphi_{M,L} = \pi \left(\frac{1}{M^2 g} - 1 \right) \; ; \; 0 \le \varphi_{M,L} \le \frac{\pi}{M}$$

$$\varphi_{M,R} = \pi \left(1 - \frac{1}{M^2 g} \right) \; ; \; 0 \le \varphi_{M,R} \le \frac{\pi}{M}. \tag{12.81}$$

One can now recognise a few known models. For $M = 1, 2$, these are [24]

1. *The low-temperature* $O(p)$ *model with* $p = 2 \cos \varphi_{1,R}$ *and* $g \ge 1$.
2. *The* $O(p)$ *model with* $p = 2 \cos \varphi_{1,L}$ *and* $\frac{1}{2} \le g \le 1$.
3. *The p-states Potts model with* $\sqrt{p} = 2 \cos \varphi_{2,R}$ *and* $1/4 \le g \le 1/2$.
4. *The tricritical p-states Potts model with* $\sqrt{p} = 2 \cos \varphi_{2,L}$ *and* $1/6 \le g \le 1/4$.

The values of n and g are chosen as

$$M_R : \quad n = \frac{2(m+1)}{M} \; , \quad g = \frac{m+1}{mM^2} \; ; \; m \ge M - 1$$

$$M_L : \quad n = \frac{2m}{M} \; , \quad g = \frac{m+1}{(m+1)M^2} \; ; \; m \ge M. \tag{12.82}$$

Together with (12.71) and (12.79), the models are completely specified. Now the projection to the unitary minimal models can be stated. We do so here for the case $M = 2$ and refer to [24] for the other cases. For the 2_R system, consider the quantities

$$\mathcal{X}_R^k = \mathcal{X}_{m+1-R}^{m+1-k} := \mathcal{Y}_R^k - \mathcal{Y}_k^R \tag{12.83}$$

$$= \sum_{r=1}^{m-1} \chi_{r,k-R} \bar{\chi}_{r,k+R} = \sum_{r=1}^{m-1} \chi_{r,R-k} \bar{\chi}_{r,-k-R},$$

where the $\chi_{r,s}$ are the Virasoro character functions of the unitary model with central charge $c = 1 - 6/(m(m+1))$. As an example, consider $m = 3$. The sectors are

$$\mathcal{X}_0^1 = \mathcal{X}_0^3 = (0,0) + \left(\frac{1}{2}, \frac{1}{2} \right)$$

$$\mathcal{X}_0^2 = 2 \left(\frac{1}{16}, \frac{1}{16} \right)$$

$$\mathcal{X}_1^2 = \mathcal{X}_3^2 = \left(0, \frac{1}{2} \right) + \left(\frac{1}{2}, 0 \right). \tag{12.84}$$

Combining all sectors, one obtains the partition function $\mathcal{Z}_T = 2Z$, where Z is called the **half-partition function**. One can check that the sectors of Z are identical to the ones of the Ising model given in (10.46). In general, it is the half-partition function which should be considered. As a second example, we note that for $m = 4$ one finds the system with exponents given in (12.17). In general, the thermal sector is always $\mathcal{X}_0^1$ and the magnetic sectors are $\mathcal{X}_0^k$ with $k > 1$. The corresponding scaling dimensions are

$$x_{\varepsilon,v} = \frac{v^2(m+1)+2v}{2m} \quad ; \quad 1 \le v \le m-2$$

$$x_{\sigma,k} = \frac{k^2-1}{2m(m+1)} \quad ; \quad k > 1. \tag{12.85}$$

For the 2_L system, one has

$$\mathcal{X}_R^k = \mathcal{X}_{m-R}^{m-k} := \mathcal{Y}_R^k - \mathcal{Y}_k^R \tag{12.86}$$

$$= \sum_{s=1}^m \mathcal{X}_{k+R,s}\bar{\mathcal{X}}_{k-R,s} = \sum_{s=1}^m \mathcal{X}_{-k-R,s}\bar{\mathcal{X}}_{R-k,s}.$$

For $m = 4$, one recovers the tricritical Ising model. The thermal and magnetic scaling dimensions are

$$x_{\varepsilon,v} = \frac{v^2m-2v}{2(m+1)} \quad ; \quad 1 \le v \le m-1$$

$$x_{\sigma,k} = \frac{k^2-1}{2m(m+1)} \quad ; \quad k > 1. \tag{12.87}$$

We leave it as an exercise to find the operator contents for $m = 5$.

Also, it was observed that the projection mechanism outlined in these cases and valid for $N \to \infty$, is still working even for finite systems [24]. The energy levels remaining in $\mathcal{X}_R^k$ will be those which do *not* occur simultaneously in both $\mathcal{Y}_R^k$ and $\mathcal{Y}_{R'}^{k'}$, where R' and k' depend on the class of models studied.

Let us finish the Heisenberg model with an experimental application [7, E4]. Consider the isotropic antiferromagnetic spin-s chain

$$H = \sum_n \left(S_n^x S_{n+1}^x + S_n^y S_{n+1}^y + S_n^z S_{n+1}^z \right), \tag{12.88}$$

where the $S^{x,y,z}$ are spin-s matrices. Here, the total spin $S^z = \sum_n S_n^z$ is conserved. In the continuum formulation of the model, this will turn into a conserved current which, for the fully isotropic model, will satisfy a SU(2) Kac–Moody algebra with central charge k, see Chap. 7. The corresponding continuum theory is a SU(2) Wess-Zumino-Witten non-linear σ model [7, G16] and conformal invariance combined with the Kac–Moody symmetry completely determine all properties of these models. In particular, the Kac–Moody central charge $k = 2s$ and the conformal central charge $c = 3k/(2+k)$.

On the other hand, the Kac–Moody central charge k can be measured experimentally in 1D quantum chains. It is related to the zero-temperature susceptibility [7, E4]

$$\chi = \frac{1}{2\pi v_s}k, \tag{12.89}$$

where v_s is the spin-wave velocity. In Chap. 3, a similar result had been derived in (3.55), relating the specific heat to the conformal central charge c. Substances such as $CuCl_2 \cdot 2NC_5H_5$ (with $s = 1/2$) and $(CH_3)_4NMnCl_3$

(with $s = 5/2$) behave to a very good approximation as 1D magnets and it has been possible to check the values predicted from the WZW model for the low-temperature susceptibility and specific heats [E4].

12.7 Ising Correlation Functions on Cylinders

The applications discussed so far focussed on the operator content, obtained from the critical point spectrum of the quantum Hamiltonian (or the transfer matrix). We now describe a confirmation of the full correlation function in the infinite-cylinder geometry, following Abraham *et al.* [4]. Since *all* secondary operators enter into it, this calculation is a more demanding test of conformal invariance than merely calculating a finite number of the low-lying energies of the quantum Hamiltonian.

Consider the two-point function of the energy density operator ε in the classical Ising spin model (1.1). Lattice representations of ε may be defined in several ways, for example [4]

$$\varepsilon_\perp(r_\|, r_\perp) = \sigma(r_\|, r_\perp)\sigma(r_\|, r_\perp + 1) \ , \quad \varepsilon_\|(r_\|, r_\perp) = \sigma(r_\|, r_\perp)\sigma(r_\| + 1, r_\perp),$$
(12.90)

where $r_\perp$ and $r_\|$ are the directions perpendicular and parallel to the axis of the infinitely long cylinder of finite circumference L. One expects from (3.45) at $T = T_c$

$$\langle \varepsilon_\perp(0,0)\varepsilon_\perp(r_\|, r_\perp)\rangle \sim \langle \varepsilon_\|(0,0)\varepsilon_\|(r_\|, r_\perp)\rangle$$
(12.91)

$$\sim \frac{1}{L^2}\left(\sinh^2\left(\frac{\pi r_\|}{L}\right) + \sin^2\left(\frac{\pi r_\perp}{L}\right)\right)^{-1}$$

up to normalization, since $x_\varepsilon = 1$ in the 2D Ising model. These energy correlators have been obtained exactly (for all temperatures) in [4]. Precisely *at* the critical point, they consider the case $r_\perp = 0$, $r_\| = r$ and find in the limit $r \gg 1$ as the leading term

$$\langle \varepsilon_\perp(0,0)\varepsilon_\perp(r,0)\rangle \sim \langle \varepsilon_\|(0,0)\varepsilon_\|(r,0)\rangle \sim \left(\frac{r}{\pi}\sum_{n=-\infty}^{\infty}\frac{(-1)^n}{r^2 + n^2 L^2}\right)^2 .$$
(12.92)

Taking r complex allows to identify (12.91) and (12.92) by the Mittag-Leffler theorem (since $1/\sinh(\pi r/L)$ has simple poles at $r = nLi$ for integer n). It is possible to extend this agreement for the general case $r_\perp \neq 0$, see [4].

12.8 Alternative Realizations of the Conformal Algebra

So far, it had always been implicitly assumed that primary operators transform according to (2.30) under the action of the conformal generators. Here we briefly mention two alternatives which might be useful in some applications.

12.8.1 Logarithmic Conformal Theories

If in a certain model one has two local operators with the same conformal weights, then in their operator product expansion logarithmic operators may appear [288]. Here, we shall illustrate what happens in the simplest possible case. Generalizations can be found in the literature, see [288, 528, 236]. Consider the transformation (2.30) of a set of primary operators ϕ_i under a conformal transformation. Formally, one might view the conformal weights Δ_i as the elements of a diagonal matrix $\hat{D}$. However, even if $\hat{D}$ is taken to be non-diagonal, the associated generators still satisfy the commutation relations of the conformal algebra [288, 528]. Non-trivial situations arise when $\hat{D}$ is assumed to take a Jordan form. In the simplest case, one has just two operators ϕ and ψ which for infinitesimal $\epsilon(z)$ transform as

$$\delta_\epsilon \phi(z) = (\Delta \partial_z \epsilon + \epsilon \partial_z) \phi(z) \tag{12.93}$$
$$\delta_\epsilon \psi(z) = (\Delta \partial_z \epsilon + \epsilon \partial_z) \psi(z) + (\partial_z \epsilon) \phi(z).$$

The finite transformation under any conformal transformation $w(z)$ is[6]

$$\phi(z) \rightarrow \left(\frac{\partial w}{\partial z}\right)^\Delta \phi(w(z)) \tag{12.94}$$

$$\psi(z) \rightarrow \left(\frac{\partial w}{\partial z}\right)^\Delta \left[\psi(w(z)) + \ln\left(\frac{\partial w}{\partial z}\right) \phi(w(z))\right].$$

Although ϕ and ψ have the same conformal weight, only ϕ is a primary operator. The projective Ward identities then yield the two-point functions [288, 156, 528]

$$\langle \phi(z_1)\phi(z_2)\rangle = 0 \ , \ \langle \phi(z_1)\psi(z_2)\rangle = a(z_1 - z_2)^{-2\Delta} \ ,$$
$$\langle \psi(z_1)\psi(z_2)\rangle = [b - 2a \ln(z_1 - z_2)](z_1 - z_2)^{-2\Delta}, \tag{12.95}$$

where a, b are constants. Similar results can be derived for the higher order correlators. However, a simple short-cut is to set [528]

$$\psi(z) = \frac{\partial}{\partial \Delta} \phi(z) \tag{12.96}$$

and to derive the known n−point functions $\langle \phi \ldots \phi \rangle$ (with $n \geq 3$) of the primary operator ϕ formally with respect to Δ. The same technique also allows to reduce the calculation of the operator product expansion coefficients to the OPE coefficients of the conventional primary operator ϕ [528]. Modular invariant partition functions have been studied as well [236].

Logarithmic operators may arise naturally in various situations, such as the dynamics of polymers, percolation, the quantum Hall effect or 2D turbulence, where non-unitary, non-minimal conformal models must be considered.

---

[6] The $\bar{z}$ dependence is suppressed. Full operators can be recovered via $z^\Delta \rightarrow z^\Delta \bar{z}^{\bar{\Delta}}$ and $\ln z \rightarrow \ln z\bar{z}$.

12.8.2 Lattice Two-Point Functions

By definition, conformal invariance does require the absence of any length scale and is manifestly broken on the scale of the lattice constant a of the underlying physical model. However, it is possible to introduce new realizations of the conformal algebra which mimic the effect of a discrete lattice constant a. To see this, recall that the infinitesimal generators $l_n = -\hat{X}^{n+1}\hat{P}$ will automatically satisfy the conformal algebra commutation relations (2.19), provided that $[\hat{P}, \hat{X}] = 1$. Usually, one takes $\hat{X} = z$ and $\hat{P} = \partial_z$, recovering (2.17). Here, we consider the choice [323, 326]

$$\hat{X} = \frac{1}{\cosh \frac{a}{2}\partial_z} z \ , \quad \hat{P} = \frac{2}{a} \sinh\left(\frac{a}{2}\partial_z\right), \tag{12.97}$$

where a is a free parameter which might be interpreted as a lattice constant. In particular, the usual infinitesimal translation operator $l_{-1}\phi(z) = -\partial_z\phi(z)$ is replaced by the symmetric discrete difference $l_{-1}^{(a)}\phi(z) = -a^{-1}[\phi(z+a/2) - \phi(z-a/2)]$. In particular, the usual translations and dilatations are still finite transformations generated by this realization.

In order to illustrate the calculation of the critical two-point function of a quasi-primary (scalar) operator $\Phi(r,t)$, we work again in the Cartesian coordinates r, t (referred to as "space" and "time", $z = r + it$) and assume that "space" is discretized while the "time" remains continuous. Consider the two-point function $F(r_1, r_2; t_1, t_2) = \langle \Phi(r_1, t_1)\Phi(r_2, t_2) \rangle$. The projective Ward identities (only those which for $a \to 0$ reduce to the translations, rotation and dilatation) needed for the determination of F are

$$\left(\frac{\partial}{\partial t_1} + \frac{\partial}{\partial t_2}\right) F(r_1, r_2; t_1, t_2) = 0$$

$$\frac{2}{a}\left[\sinh\left(\frac{a}{2}\frac{\partial}{\partial r_1}\right) + \sinh\left(\frac{a}{2}\frac{\partial}{\partial r_2}\right)\right] F(r_1, r_2; t_1, t_2) = 0$$

$$\sum_{i=1}^{2}\left[\frac{2}{a} t_i \sinh\left(\frac{a}{2}\frac{\partial}{\partial r_i}\right) - \frac{1}{\cosh(\frac{a}{2}\partial/\partial r_i)} r_i \frac{\partial}{\partial t_i}\right] F(r_1, r_2; t_1, t_2) = 0$$

$$\left\{\sum_{i=1}^{2}\left[t_i\frac{\partial}{\partial t_i} + \frac{2}{a}\frac{1}{\cosh(\frac{a}{2}\partial/\partial r_i)} r_i \sinh\left(\frac{a}{2}\frac{\partial}{\partial r_i}\right)\right] + 2x\right\} F(r_1, r_2; t_1, t_2) = 0 \tag{12.98}$$

where x is the scaling dimension of Φ. The physical assumptions behind this are shown through the Casimir operator, which leads to the dispersion relation $E = (2/a)|\sin(ak/2)|$. In the $a \to 0$ limit, one recovers the usual dispersion relation $E = |k|$ of continuum conformal field theory, while for a finite, one has the dispersion relation of a free relativistic particle in a Brillouin zone. The particular form of the dispersion relation depends on the explicit choice of the operators $\hat{X}, \hat{P}$.

The solution of the projective Ward identities is, up to a normalization constant and with $\rho := (r_1 - r_2)/a$ an integer [326]

$$F \sim a^{-2x} \int_0^\infty dv\, v^{-x-1/2} \exp\left(-\frac{(t_1 - t_2)^2}{2a^2 v} - v\right) I_{(r_1-r_2)/a}(v) \qquad (12.99)$$

where I_ρ is a modified Bessel function. For $t_1 = t_2$, the result simplifies considerably

$$F\big|_{t_1=t_2} = a^{-2x} \frac{\Gamma(\rho + \frac{1}{2} - x)}{\Gamma(\rho + \frac{1}{2} + x)} \qquad (12.100)$$

which reduces to the conventional form (2.36) for $\rho \gg 1$. For $x > 3/2$, the two-point function (12.100) becomes singular even before the distance between the two local operators has been reduced to a single lattice spacing a and the correlator behaves as if the operators Φ would contain a hard core sphere of diameter $(x - 1/2)a$.

The result (12.99) should apply exactly to lattice models with an underlying free field theory and obeying the dispersion relation $E = (2/a)|\sin(ak/2)|$. Indeed, it is easy to check that the spin-spin correlation function of the spherical model in $2 < d < 4$ dimensions is reproduced from (12.99) by letting $x = x_\sigma = \beta/\nu = (d-2)/2$. Also, considering the energy-energy correlation of the quantum Ising chain (which implies the setting $t_1 = t_2$ and thus a check on (12.100)), the result $\langle \varepsilon\varepsilon \rangle \sim (4\rho^2 - 1)^{-1}$ [503] is reproduced from (12.100), setting $x_\varepsilon = 1$.

12.9 Percolation

Previously, we had already encountered the Yang–Lee singularity as an example of a non-unitary statistical model. Further, and apparently more natural examples, are provided by geometric phase transitions such as percolation.

From Chap. 1, we recall the relationship of the percolation problem to the $q \to 1$ limit of the q-states Potts model. Calculating that limit naively, one arrives at $\mathcal{Z} = 1$ for the Potts partition function and thus the central charge is expected to be $c = 0$. As we shall see, however, this does not mean that the problem is trivial. The scaling dimensions of the physical operators can be obtained from the critical point eigenvalues of the transfer matrix $\mathcal{T}$. In particular, it can be shown [479] that $\mathcal{T} = \mathcal{T}_e \otimes \mathcal{T}_s$ factorizes into two submatrices, where $\mathcal{T}_e$ describes correlations of 'energy' type, and $\mathcal{T}_s$ describes correlations of 'spin' type, quite analogously to the charge sectors found for example in the Ising model. In Table 12.11 we collect some finite-lattice estimates for the scaling dimensions, as obtained from (3.47).

Since the central charge vanishes, the model cannot correspond to a unitary minimal model, since in that case, $\Delta = 0$ were the only possibility. However, the exponent values found can be explained in the context of a non-unitary minimal model. The simplest possibility with a central charge

Table 12.11. Some scaling dimensions in 2D percolation as estimated from strips of width up to $N = 8$ ('spin' type) and $N = 9$ ('energy' type) with total momentum $\mathcal{P} = 0$, together with the expected exact value (after [551])

	'spin' type				'energy' type			
numerical	0.1042	1.635	2.112	3.756	1.254	3.303	4.161	4.113
expected	5/48	77/48	2 + 5/48	2 + 77/48	5/4	2 + 5/4	4	4

$c = 0$ is the model $\mathcal{M}_{4,6}$. The Kac table can then be found from (12.33) and we list in Table 12.12 the Virasoro characters of the primary operators.[7] We point out that there are *two* distinct operators with $\Delta = 0$, although there is still a unique ground state.

Table 12.12. Virasoro characters $d(\Delta, I)$ of the primary operators $\phi_{r,s}$ for $p = 4, p' = 6$

Δ	I																r	s
	0	1	2	3	4	5	6	7	8	9	10	11	12	13	14	15		
0	1	0	1	1	2	2	4	4	7	8	8	14	19	24	34	41	1	1
5/8	1	1	1	2	3	4	6	8	11	15	19	25	33	42	53	68	1	2
2	1	1	2	2	4	4	7	8	12	14	21	24	34	40	54	64	1	3
0	1	1	1	2	3	4	6	8	11	15	20	26	34	44	56	72	2	1
1/8	1	1	2	3	4	6	9	11	16	21	28	36	48	60	88	98	2	2
1	1	1	2	3	4	6	8	11	15	20	26	34	44	56	72	91	2	3
1/3	1	1	2	2	4	5	8	10	15	18	26	32	44	54	72	88	3	1
−1/24	1	1	2	3	5	7	9	13	18	24	32	42	55	71	91	116	3	2

The conformal weights of the scalar operators in the 'energy' sector are [551]

$$\Delta^{(\varepsilon)} = \frac{(6s - 4)^2 - 4}{96} = \Delta_{1,s}. \tag{12.101}$$

On the other hand, the conformal weights of the scalar operators in the 'spin' sector are given by [551]

$$\Delta^{(\sigma)} = \frac{(6t - 3)^2 - 4}{96} = \Delta_{3/2,t+1/2} = \Delta_{3/4,t}, \tag{12.102}$$

where *fractional* values of the indices r, s must be used in the Kac formula. While this certainly reproduces the numerical results for the exponents, the procedure appears at first sight as completely arbitrary. However, the conformal field theory which correctly describes this geometric phase transition is not $\mathcal{M}_{4,6}$, but a twisted $N = 2$ superconformal field theory [552]. In that

---

[7] In some papers, the Kac table of the unitary minimal model $\mathcal{M}_{2,3}$ is formally extended beyond its borders $1 \leq r \leq 1, 1 \leq s \leq 2$. But consideration of the well-defined non-unitary minimal model $\mathcal{M}_{4,6}$ allows to use the Rocha–Caridi formula $\chi_{r,s} = \sum_{k=-\infty}^{\infty} \chi(\Delta_{r+2kp,s}) - \chi(\Delta_{r+p(2k-1),p'-s})$ to find the characters. The OPE from Chap. 7 can be generalized similarly.

context, the above procedure can be completely justified. In particular, it can be shown that the critical point partition function is [551, 552]

$$\mathcal{Z} = \mathcal{Z}_c \left[\frac{2}{3}, 6\right] = \mathcal{Z}_c \left[\frac{3}{2}, 4\right], \tag{12.103}$$

where $\mathcal{Z}_c$ is a Coulombic partition function, using the notation of equations (11.25,11.26), which contains ρ implicitly

$$\mathcal{Z}_c\left[\rho, \sigma\right] := (q\bar{q})^{-1/24} P(q) P(\bar{q}) \sum_{n,m=-\infty}^{\infty} q^{\Delta_{n/\sigma,m\sigma}} \bar{q}^{\overline{\Delta}_{n/\sigma,m\sigma}}. \tag{12.104}$$

From this, the full operator content can be extracted. In particular [475, 476]

$$x_\sigma = \frac{5}{48} \ , \ \ x_\varepsilon = \frac{5}{4}. \tag{12.105}$$

Scaling relations from Chap. 1 then imply the well-known results $\alpha = -2/3$, $\beta = 5/36$, $\gamma = 43/18$, $\nu = 4/3$ and $\eta = 5/24$, e.g. [625, 577, 428], which have been confirmed numerically many times. For example, phenomenological renormalisation studies in strips of finite width which give from both square and triangular lattices $\lambda = 1/\nu = 0.750(2)$ [526], while the amplitude-exponent relation (3.47) for the transfer matrix eigenvalues gives $\eta = 0.2088(8)$ [192] or $\eta = 0.20834(2)$ [526], see Table 15.3. We are not aware of any numerical study testing the full operator content (12.103).

12.10 Polymers

12.10.1 Linear Polymers

We now briefly review the results of the application of Coulomb gas techniques to the calculation of the exponents describing infinitely long polymers, e.g. [476, 214]. We make use of the setting defined in Chap. 1 for the discussion of the scaling properties of (interacting) self-avoiding walks. It is useful to consider the **watermelon topology** made from Q self-avoiding walks (SAW) tied together at two vertices $\mathbf{R}_{1,2}$, see Fig. 12.2 for an example. To a Q-leg vertex, one associates a scaling operator ϕ_Q and it can be shown that there is a critical value $\lambda = \lambda_c$ of the fugacity for which the two-point function $G_Q(\mathbf{R}_1 - \mathbf{R}_2) = \langle \phi_Q(\mathbf{R}_1)\phi_Q(\mathbf{R}_2)\rangle \sim |\mathbf{R}_1 - \mathbf{R}_2|^{-2x_Q}$, which defines the scaling dimension x_Q. It is easy to see that $x_{1,2}$ are related to the conventional exponents as follows

$$\eta = 2x_1 \ , \ \ \frac{1}{\nu} = 2 - x_2 \ , \ \ \gamma = (2 - \eta)\nu. \tag{12.106}$$

Specifying a lattice model, it is convenient to formulate it on a hexagonal lattice (as usual, this does not influence the exponent values). Then the SAWs

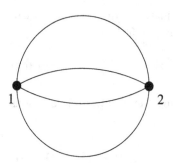

Fig. 12.2. Watermelon topology with $Q = 4$ and the fixed vertices at the points $R_{1,2}$

are domain walls between regions of constant values of the spins in the associated $O(n)$ model. In the continuum limit, this becomes a Gaussian model[8] such that

$$n = -2\cos\left(\pi\rho^2\right), \qquad (12.107)$$

where ρ is the compactification radius (Chap. 11). Since the vertices at $R_{1,2}$ act as dislocations and the correlator $G_Q(R_1 - R_2) = \langle \mathcal{O}_{e_0,m} \mathcal{O}_{e_0,-m} \rangle$ where the 'magnetic' contribution comes from the defect lines, viz. $m = Q/2$, and the 'electric' charge $e_0 = \rho^2 - 1$ in order to give contracting and winding $O(n)$ loops the same weight on the torus [476, 547, 214] (the central charge $c = 1 - 6e_0^2/\rho^2$). Consequently,

$$x_Q = \frac{1}{8}\rho^2 Q^2 - \frac{(1-\rho^2)^2}{2\rho^2}. \qquad (12.108)$$

For the SAW, we need $n = 0$, corresponding[9] to $\rho^2 = 3/2$. Thus $x_Q = (9Q^2 - 4)/48$.[10] As expected from the arguments in Chap. 1, $c = 0$. It follows that for the SAW

$$\eta = \frac{5}{24} \; , \quad \nu = \frac{3}{4} \; , \quad \gamma = \frac{43}{32}. \qquad (12.109)$$

Let us check this numerically. Consider the two-point function $G_1^{(L)}$ on an infinitely long lattice of finite width L. Then

$$G_1^{(L)}(R_1 - R_2) = \sum_{\text{SAW};1\to2} v^N \qquad (12.110)$$

---

[8] The number of star polymers $c_N^{(Q)} \sim N^{\gamma_Q - 1}\mu^N$ is described by the exponent $\gamma_Q = \nu(2 - x_Q - x_1)$, with $\gamma = \gamma_1$. In addition, since the associated Gaussian field theory has a $U(1)$ symmetry with a conserved current J, Q can be interpreted as the value of this conserved charge.

[9] A second solution is $\rho^2 = 1/2$, leading to $x_Q = (Q^2 - 4)/16$. Physically, this corresponds to a dense polymer phase, with central charge $c = -2$.

[10] Formally, these scaling exponents can be obtained from the Kac formula as $x_Q = 2\Delta_{Q/2,0}$ with $m = 2$ ($c = 0$) for the SAW and $x_Q = 2\Delta_{0,Q/2}$ with $m = 1$ ($c = -2$) for the dense phase. See [552] for the justification of the non-integer indices from $N = 2$ superconformal theory.

where N is the number of monomers and the sum is over all SAWs linking the vertex $\mathbf{R}_1$ with $\mathbf{R}_2$. For $R = |\mathbf{R}_1 - \mathbf{R}_2|$ large, we expect $G_1^{(L)} \sim \exp(-R/\xi_L(v))$. The correlation length $\xi_L(v) = -1/\ln \Lambda_{\max}$ can in turn be obtained from the largest *nontrivial* eigenvalue of the transfer matrix [190, 191, 192]. The transfer matrix acts parallel to the infinite direction and maps a configuration $\mathcal{V}$ of L occupied or empty sites onto a new configuration $\mathcal{V}'$. The calculation of its elements is not completely straightforward, because one has to do the necessary bookkeeping to ensure that the occupied sites form a single connected cluster such that *all* sites can be visited through a single self-avoiding walk. Therefore, it is not enough to simply store the occupied or empty sites (as it is for the Ising model) but the connectivity has to be stored as well. This constraint introduces an effective long-range interaction which prevents the formulation of an appropriate Hamiltonian limit and the full non-sparse matrix must be treated. The critical fugacity v_c can be found from the condition that the largest eigenvalue of the transfer matrix is equal to unity [190, 471] or else from phenomenological renormalization (Chap. 3). We illustrate this in Table 12.13, where estimates for the critical point v_c and the exponents ν and η are shown for several values of the transverse size L. The first two parameters were obtained from the phenomenological renormalization procedure for the correlation length described in Chap. 3 and the third one is calculated using (3.47) precisely *at* $v = v_{c,\infty} = 0.3790523$, obtained from exact enumeration studies. The extrapolation towards $L \to \infty$ was done with the BST algorithm, but taking into account small oscillations in the estimates, the sequences for L even and odd, respectively, were extrapolated separately [588, 565]. The extrapolated results are close to the predictions in (12.109).

Table 12.13. Finite-lattice estimates for the critical parameters of a self-avoiding walk (after [588])

L	v_c	ν	η	comments
1	0.347810385	0.6684732	0.308786	
2	0.365304779	0.7244766	0.258368	
3	0.373399472	0.7391245	0.233014	
4	0.376632894	0.7450054	0.221516	
5	0.377909540	0.7476797	0.216198	
6	0.378447688	0.7489294	0.213518	
7	0.378698393	0.7495269	0.212020	
8	0.378827984	0.7498262	0.211102	
9	0.378901312	0.7499830	0.210494	
10	0.378945913	0.7500675	0.210068	
11	0.378974611	0.7501136	0.209758	
12			0.209524	
13			0.209344	
∞	0.379055(30)	0.75017(15)	0.208327(15)	L even
∞	0.379052(7)	0.75018(8)	0.208325(10)	L odd

We now turn to *interacting* polymers, following [579]. The interactions may be defined in terms of contact fugacities as will be described in the context of lattice animals below. Exact results for the exponents, however, are easier to obtain by using annealed vacancies as mediators of an effective interaction of self-avoiding walks [170, 211]. Consider a hexagonal lattice. Each hexagon is occupied or vacant with probability p or $1 - p$, respectively and one has thus defined a percolation problem on the dual triangular lattice. Now, for each percolative configuration C, define a self-avoiding walk which is only allowed to visit the occupied hexagons.[11] The polymer generating function is

$$Z = \sum_{C} P(C) \sum_{\text{SAW}}{}' v^{N}, \qquad (12.111)$$

where $P(C)$ is the probability of the configuration C, v the walk fugacity and N is the number of monomers. The second sum goes over the SAWs compatible with C. The sum over C can be carried out

$$Z = \sum_{\text{SAW}} p^{\bar{H}} v^{N}, \qquad (12.112)$$

where the sum is now over *all* SAWs and $\bar{H}$ (which depends on the SAW) is the number of distinct lattice hexagons whose edges are visited by the walk. One can show that $\bar{H} = N + 1 - N_2 - 2N_3$, where $N_{2,3}$ are the numbers of hexagons visited, not consecutively, two and three times, respectively, by a given walk [211]. Thus, through the percolation vacancies, one has introduced local attractive interactions, which may mediate the collapse of the polymer to a compact phase. In distinction to more common models (referred to as Θ-models), one has here both nearest-neighbour and next-nearest neighbour interactions and the collapse point is called a Θ'-point [170]. One expects, however, that the universality class is not influenced through the extra interactions, which conveniently reduce the problem to a much simpler SAW.[12] At the Θ'-point, it can be shown that a self-avoiding ring at the critical fugacity $v_c = 1$ has the same statistics as the hull of percolation clusters [211], of fractal dimension $d_{\text{hull}} = 7/4$ [548] and $\nu = 1/d_{\text{hull}}$. As before for the SAW, the exponent η can be obtained from the two-point function, which involves only walks between two fixed points $\boldsymbol{R}_{1,2}$. It turns out that this can be identified with the spin-spin correlator $\langle \sigma_{\boldsymbol{R}_1} \sigma_{\boldsymbol{R}_2} \rangle$ of an Ising model at zero temperature,

---

[11] Or alternatively, which must not visit the occupied hexagons, which amounts to $p \to 1 - p$.

[12] While universality apparently holds for the *bulk* exponents, this appeared for a while to be no longer so when *surface* critical exponents (Chap. 15) are considered. A long debate arose, e.g. [211, 510, 562, 563, 212, 463, 447, 600, 606, 239, 159] It took quite a while until this point could be completely clarified [579]. It turned out that the surface exponents calculated in [211] belong to the special transition. The correct exponents for the ordinary transition were given in [598, 579] and are in agreement with the numerical simulations and universality.

thus $\eta = 0$ due to the existing long-range order in the Ising model [211, 579]. Summarising, the exponents at the Θ'-point are

$$\eta = 0 \ , \ \ \nu = \frac{4}{7} \ , \ \ \gamma = \frac{8}{7}. \tag{12.113}$$

The exponent ν in (12.109,12.113) has also been measured experimentally, see Table 1.7.

12.10.2 Lattice Animals

A related, but intrinsically different, problem is the description of randomly **branched polymers**, which are also often referred to as **lattice animals**. By definition, a lattice animal is a connected graph of occupied sites on a (hypercubic) lattice, see Fig. 12.3 for examples. Two neighboring occupied sites may or may not be immediately connected. If they are, one says that there is a **bond** between them, and if not, one says that there is a **contact** between them. Between an occupied site and an empty site one says there is a **solvent** contact. One is interested in the generating function [234, 564]

$$Z = \sum_{n,s,k} a_{n,s,k} \exp\left(\beta_0 n + \beta_1 s + \beta_2 k\right), \tag{12.114}$$

where $a_{n,s,k}$ is the number of animals with n sites, s solvents and k contacts and the β_i are coupling constants. When restricting attention to infinite lattice animals ($n \to \infty$), one of the couplings is fixed in terms of the other two, e.g. $\beta_0 = \beta_0(\beta_1, \beta_2)$, which we shall use from now on. An important special case is the $\beta_2 \to -\infty$ limit, called **strong embedding** model, where two neighboring occupied sites are always connected by a bond.

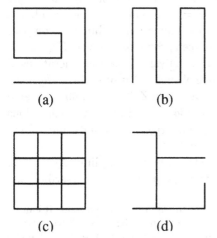

(a)   (b)

(c)   (d)

Fig. 12.3. Some examples of 2D lattice animals

The lattice animal is related to a generalized q-state Potts model, with Hamiltonian (analogously to percolation) [300]

$$\mathcal{H} = -J \sum_{\langle i,i'\rangle} \delta(\Theta_i - \Theta_{i'}) - J' \sum_{\langle i,i'\rangle} \delta(\Theta_i)\delta(\Theta_{i'}) - TB \sum_i \delta(\Theta_i). \quad (12.115)$$

Its partition function $\mathcal{Z} = \sum_{\{\Theta\}} e^{-\mathcal{H}/T}$ is related to the lattice animal generating function Z, with $K = J/T$, via $K' = J'/T$

$$Z = \lim_{q \to 1} \frac{\partial}{\partial q} \ln \mathcal{Z}, \quad (12.116)$$

where, for a square lattice

$$\beta_0 = -B + 2\ln\left(e^K - 1\right) - 2(K + K'), \quad \beta_1 = -\frac{1}{2}\left[\ln\left(e^K - 1\right) + K + K'\right]$$

$$\beta_2 = -\ln\left(e^K - 1\right). \quad (12.117)$$

In particular, along the line $J' = 0$ the model reduces to percolation [234, 564]. Transitions between different states of the infinite lattice animal correspond to multicritical points within the critical manifold of the generalized Potts model, see [182].

The phase diagram is shown in Fig. 12.4. There is a general agreement about the existence (and the location) of a second-order Θ-transition from a compact phase, with animal configurations as in Fig. 12.3abc, to an extended phase (Fig. 12.3d). Physically, these are distinguished by the scaling of the mean radius of gyration, $\bar{R}_G$, with the number of monomers, viz. $\bar{R}_G \sim N^\nu$. In the compact phase, $\nu = 1/2$, and in the extended phase, $\nu = 0.64075(15)$ [192], both in 2D. At $\beta_1 = -\frac{1}{2}\ln 2$, $\beta_2 = 0$, the transition maps onto the percolation critical point. For $\beta_2 > 0$, numerical results suggest that $\nu = 8/15$ [564]. The transition line for $\beta_2 < 0$ is controlled by the strong embedding fixed point [191, 192, 564], leading to $\nu = 1/2$. On the other hand, the proposed existence of a further continuous transition within the compact phase (the horizontal line in Fig. 12.4), between ground states as in Fig. 12.3ab and in Fig. 12.3c, is controversial. While the analysis of exact enumerations of finite animal graphs suggested the presence of such a transition because of the presence of a maximum in the 'specific heat' $\partial^2 \ln Z / \partial \beta_2^2$ [234, 497], no sign of it was found in transfer matrix studies [564], closed-form approximations and the renormalization group [324, 373] and rigorous bounds and Monte Carlo simulations in the $\beta_1 \to -\infty$ limit [580].[13]

From now on, we concentrate on the strong embedding limit $\beta_2 \to -\infty$. It has been known for quite a while that this system *cannot* be described in a

---

[13] For *directed* lattice animals, ground states as in Fig. 12.3ab are absent. It can be shown exactly that there is only a single compact phase [324, 580]. Furthermore, the 'specific heat' shows a finite maximum along a curve in the (β_1, β_2) plane which is roughly horizontal [580].

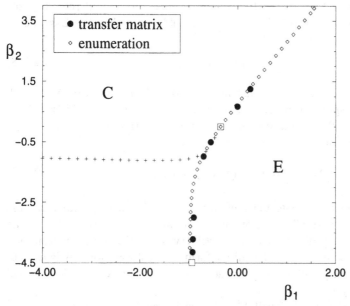

Fig. 12.4. Phase diagram of the infinite lattice animal. The collapse transition between the extended phase (E) and the compact phase (C) was found from the transfer matrix [564] and exact enumeration [234, 497]. The open squares mark the percolation critical point $\beta_1 = -\frac{1}{2}\ln 2, \beta_2 = 0$ and the strong embedding limit $\beta_{1,c} \simeq -0.934$ [191]. The crosses indicate the position of the (controversial) compact-compact transition proposed in [234, 497] (after [324])

simple way using conformal invariance. For example, it has not been possible to fit the precise result $\nu = 0.64075(15)$ into a Kac table of some minimal model. Furthermore, the amplitude-exponent relation (3.47) is apparently *not* satisfied [192, G11], although the correlation length on a cylinder does satisfy the usual scaling $\xi \sim L$ with the system size L.[14] On the other hand, there is a field theory formulation of the model available, which was shown by Parisi and Sourlas [490] to be supersymmetric. As a consequence of the supersymmetry, certain correlators of the lattice animal problem in d dimensions are equal to correlators of the field theory associated to the Yang–Lee in $d-2$ dimensions, at least near to the upper critical dimension $d^* = 8$. It appears plausible, however, that the supersymmetry and the operator content of the Parisi–Sourlas field theory persist down to 2D, see [466] for details.

If that is indeed the case, it can be shown that the lowest dimension field ϕ in the Parisi–Sourlas field theory cannot be a conformal primary operator, in contrast to what is the case for the conformal field theories related to, for example, the Ising or Yang–Lee models [466]. We limit ourselves here to the

---

[14] There are further tests on conformal invariance for systems near to the tip of a cone, which are not satisfied by this model, see Chap. 15.

particularly simple case of the free Parisi–Sourlas theory and refer to [466] for the full theory. The free Parisi–Sourlas field theory has the Lagrangian [466]

$$\mathcal{L} = \int \mathrm{d}^2 r \, \left(\nabla^2 \phi\right)^2 . \tag{12.118}$$

For $\mathcal{L}$ to be scale invariant, ϕ must have conformal weights $\Delta = \bar{\Delta} = -1/2$. If ϕ would correspond to a conformal primary operator, it should transform according to (2.30) under the conformal transformation $z \to z + \epsilon(z)$. The variation in $\mathcal{L}$ is then

$$\delta\mathcal{L} = \int \mathrm{d}^2 r \, \left[\epsilon''(z)(\bar{\partial}\phi)(\partial\bar{\partial}\phi) + \text{c.c.}\right] . \tag{12.119}$$

For special conformal transformations, $\epsilon(z) = 1, z, z^2$, the integrand is at most a total derivative and $\delta\mathcal{L} = 0$. For more general transformations, that is no longer the case. In addition, one might still try to modify the transformation law (2.30) to keep $\delta\mathcal{L} = 0$ under all conformal transformations. It is easy to see that this is not possible [466]. Consequently, $\mathcal{L}$ is not even at the classical level conformally invariant.

In conclusion, while the available information clearly suggests that lattice animals in the strong embedding limit are *not* conformally invariant, it has not yet been understood in a physical way how this comes about.

12.11 A Sketch of Conformal Turbulence

In this section, we shall present a sketch on some recent ideas relating conformal invariance with two-dimensional turbulence. It is obvious that it is impossible to do justice to a field so old and enormous as turbulence on a few pages. We shall limit ourselves to a brief presentation of a beautiful idea of Polyakov [508] and shall provide only the barest of background. A nice guided tour into the many roads of turbulence together with an entry into the literature is provided in the book by Frisch [248]. This section has served its purpose if the curiosity of the reader is piqued to look further into this fascinating field for himself.

The basic equation describing an incompressible turbulent fluid is the Navier–Stokes equation for the velocity field $\boldsymbol{v}(\boldsymbol{r}, t)$

$$\partial_t \boldsymbol{v} + \boldsymbol{v} \cdot \nabla \boldsymbol{v} = -\nabla p + \nu \nabla^2 \boldsymbol{v} \ , \quad \nabla \cdot \boldsymbol{v} = 0 \tag{12.120}$$

supplemented by convenient initial and boundary conditions. Here p is the pressure and ν is the kinematic viscosity. The pressure can be eliminated by defining the **vorticity** $\omega = \nabla \times \boldsymbol{v}$. Its equation of motion does not contain p explicitly. An important physical parameter is the **Reynolds number**

$$\mathrm{R} = \frac{LV}{\nu}, \tag{12.121}$$

where L, V are a typical length scale and velocity of the flow, respectively. For a given geometrical shape of the boundaries, R is the only control parameter of the problem.

In general, due to the high non-linearity of the Navier–Stokes equation, obtaining exact solutions in a very difficult problem. For very small Reynolds numbers, linear approximations work well (laminar regime) and the solutions have the same space-time symmetries as (12.120). That situation changes when R is increasing [248]. Through a series of bifurcations, the individual solutions loose the symmetries of the Navier–Stokes equation, while the flow patterns become increasingly complicated. However, for very large Reynolds numbers (e.g. R $\gtrsim 10^3 - 10^4$), the regime of *homogeneous isotropic turbulence* (Kelvin 1887) is reached, characterized by eddy motion on all scales, suggesting that on average some form of *scale invariance* might be present. Formally, that is only reached in the R $\to \infty$ limit, which makes the Reynolds number play the same role as the quantity $t = (T - T_c)/T_c$ in magnetic systems. Roughly, turbulence might be viewed to be analogous to conventional critical phenomena, provided that long and short scales are interchanged. If that idea were correct and precise, one should expect long-distance universal behaviour in turbulence [252]. In a turbulent system, energy is injected into the system, to be transferred between eddies of different size and finally dissipated. Kolmogorov (1941) has shown that for large Reynolds numbers it is typical to have energy injection confined to largest scales and the energy dissipation confined to the smallest scales [248]. Physical quantities of interest include the **cumulative energy** $\mathcal{E}(K)$, the **cumulative enstrophy** $\Omega(K)$

$$\mathcal{E}(K) := \frac{1}{2} \sum_{k \leq K} |v_k|^2 \ , \quad \Omega(K) := \frac{1}{2} \sum_{k \leq K} k^2 |v_k|^2 \tag{12.122}$$

(where v_k is the Fourier component of the velocity with wave vector k) and the **energy spectrum**

$$E(K) := \frac{d\mathcal{E}(K)}{dK} = \frac{1}{2} \frac{d}{dK} \left(\int_{k \leq K} dk \int dr \ \langle v(0) \cdot v(r) \rangle \ e^{ik \cdot r} \right) \tag{12.123}$$

which is related to the equal-time velocity correlation function. For large enough Reynolds numbers, $E(k)$ is observed to show a power law behaviour in k and one of the important questions is to predict the value of the associated exponent.

Now we specialise to two-dimensional situations. Then only a single component of the vorticity remains, conveniently written $\omega = \nabla^2 \psi$, where ψ is the **stream function**. Following Polyakov [508, 509], it is assumed that for large values of R, there exists an inertial range in which both viscosity and the stirring force are negligible. Then the Navier–Stokes equation goes over into the inviscid Hopf equation

$$\sum_{i=1}^{n} \langle \omega(r_1) \dots \dot{\omega}(r_i) \dots \omega(r_n) \rangle = 0 \ , \quad \dot{\omega} = -\varepsilon^{ab} \partial_a \psi \partial_b \nabla^2 \psi, \tag{12.124}$$

where ε^{ab} is the totally antisymmetric tensor. However, care is needed in defining the correlators in coinciding points, since the Hopf equation can only be expected to be satisfied at intermediate scales, well separated from the scales of energy injection and energy dissipation. More precisely, a point-splitting procedure must be invoked, viz.

$$\dot{\omega}(r) = -\lim_{a\to 0} \varepsilon^{ab} \partial_a \psi(r+a) \partial_b \nabla^2 \psi(r), \qquad (12.125)$$

where the limit procedure includes an averaging over all directions before a is taken to zero. The key assumption is now that ψ is a primary operator of some conformal field theory. Then the r.h.s. of (12.125) can be found from the operator product expansion $\psi \cdot \psi = \phi$, where ϕ is the operator with minimal conformal weight which arises in the operator product. Furthermore, the result in (12.125), to be formed out of ϕ and its secondaries, must be a pseudoscalar. That leads to the important result [508]

$$\dot{\omega}(r) \sim |a|^{\Delta_\phi - 2\Delta_\psi} \left[L_{-2}\bar{L}_{-1}^2 - L_{-1}^2 \bar{L}_{-2} \right] \phi(r) \qquad (12.126)$$

and it follows directly from (12.123) that the energy spectrum takes a power-law form[15]

$$E(k) \sim k^{4\Delta_\psi + 1}. \qquad (12.127)$$

Previous predictions in 2D have produced[16] $E \sim k^{-3}$ [416, 58], $E \sim k^{-11/3}$ [467] and $E \sim k^{-4}$ [546], all of them roughly in agreement with experiment. Consistency of (12.127) with any of these results can be only achieved for *non-unitary* conformal theories.

Equation (12.126) gives a precise meaning to the $a \to 0$ limit.[17] Now, if $\Delta_\phi \leq 2\Delta_\psi$, the r.h.s. has to vanish. The simplest way of realizing this is by requiring that ϕ is a null operator at level two,[18] see (4.16). On the other hand, if $\Delta_\phi \geq 2\Delta_\psi$, no further condition is required in order to satisfy the Hopf equation in the $a \to 0$ limit.

In addition, Polyakov assumes that enstrophy is a conserved quantity in the inertial range. Since $\Omega \sim \int dk \, \langle \dot{\omega}(k)\omega(-k) \rangle$, its conservation implies [508] $\Delta_\omega + \Delta_{\dot{\omega}} = 0$ and

$$\Delta_\psi + \Delta_\phi + 3 = 0. \qquad (12.128)$$

---

[15] Because $v_a = \varepsilon_a{}^b \partial_b \psi$ and the energy takes contributions from the full scaling operators, which are assumed to be scalars, thus $\Delta_\psi = \bar{\Delta}_\psi$.

[16] Possible logarithmic factors are suppressed.

[17] It is not at all straightforward to show that the theory is also well-defined at large distances, since the physical correlators and those calculable from conformal invariance differ by δ-function terms in momentum space. In this section, it is taken for granted that this point does not spoil the solutions to the Hopf equation, but a detailed study [508, 509] is necessary.

[18] The simplest model satisfying this is the Yang–Lee singularity, with $\Delta_\psi = \Delta_\phi = -1/5$.

That would be consistent with taking $\Delta_\psi = -1, \Delta_\phi = -2$, thereby reproducing[19] the Kraichnan-Batchelor result for the spectrum [416, 58].

The so-called *'minimal' minimal model* is the minimal conformal theory with the smallest number of primaries which satisfies the above constraints. Convenient candidates are the minimal models $\mathcal{M}_{2,2N+1}$, which have N primary operators ψ_s, with $s = 1, \ldots, N$. The OPE is (see Chap. 7) $\psi_{s_1} \cdot \psi_{s_2} = \psi_{s_3} + \psi_{s_3-2} + \ldots$, where $s_3 = \min(s_1 + s_2 - 1, 2N + 2 - s_1 - s_2)$. The conformal weights are $\Delta_s = -(2N - s)(s - 1)/(4N + 2)$, see also (12.33). Identifying $\psi = \psi_s$, it follows from (12.128) and the OPE that

$$\Delta_s + \Delta_{2s-1} + 3 = 0. \tag{12.129}$$

The unique solution to this is $s = 4, N = 10$, and consequently $\Delta_\psi = \Delta_4 = -8/7$ and $\Delta_\phi = \Delta_7 = -13/7$. Finally, the energy spectrum [508]

$$E(k) \sim k^{-25/7} \tag{12.130}$$

is well inside the range of the earlier theoretical predictions [416, 58, 467, 546]. Certainly, this is nothing but one particular solution and many others can easily be constructed, e.g. [527]. Also, the restriction to a conformal minimal model is completely formal and has been critically reexamined in [220].

The condition (12.128), resulting from enstrophy conservation, has been tested using a direct numerical simulation of the vorticity equation. If $P(k)$ is the power spectrum of $\dot{\omega}$, it can be shown that $P(k) \sim k^{7+4\Delta_\phi}$ [73]. With (12.128), this leads to $E(k)P(k) \sim k^{-4}$, also in agreement with the prediction of [416, 58], although the scaling for $E(k)$ is in general different. The simulations were found to agree with the expected scaling of the product $E(k)P(k)$, at least when no large-scale coherent vortices dominate the flow. Presently, it appears likely that different initial conditions and/or forcing mechanisms might be described by different conformal theories with different operator contents [73]. The experimental status of 2D turbulence goes beyond the scope of this short discussion and will not be discussed here, but see [446, 103].

Finally, we mention that the presence of logarithmic factors in the energy spectrum can be explained in the context of logarithmic conformal theories, see [237, 529].

Closing this overview of some applications of conformal invariance, we remark that the examples presented here were selected as illustrations for the main lines usually occurring in applications and are meant to stimulate the reader to look for more. Certainly our sample is not exhaustive.

---

[19] Here it is always assumed that the expectation value of a single operator vanishes. This condition can be relaxed [508].

12.12 Some Remarks on 3D Systems

Given the success of combined conformal and modular invariance to describe the operator content of a given model in detail, one might ask whether at least some of the techniques could be extended into higher dimensions as well. It is an open question whether such an extension is possible. Here we merely sketch proposals made towards this possibility.

Cardy [135, 143] considered a 3D space of the topology $\Sigma^3 = S^2 \otimes S^1$ with the metric

$$ds^2 = R^2 \left(d\theta^2 + \sin^2 \theta \, d\phi^2 \right) + R'^2 d\tau^2, \qquad (12.131)$$

where R and R' are the radii of the 2-sphere S^2 and the 1-sphere S^1, respectively. Interpreting τ as imaginary time, the quantum Hamiltonian H in this formulation will be defined on the 2-sphere S^2 and by the transformation $\tau \to (R/R') \ln r$, H is the generator of scale transformations $r \to \lambda r$ for the space $\mathbb{R}^3 \backslash \{0\}$ with the flat metric. Then the partition function is, up to anomaly terms [143]

$$\mathcal{Z} = \sum_x q^x, \qquad (12.132)$$

where x runs over the list of scaling dimensions in the model and $q = e^{-2\pi R'/R}$ is the modular parameter. If the eigenvalues of H are taken to be in units of R' one has for the critical point correlation length [135]

$$\xi_i^{-1} = R^{-1} 2\pi x_i \qquad (12.133)$$

in close analogy to the 2D situation (see (3.47)). Unfortunately, there is at present no confirmation from any model beyond free field theory. For an attempted numerical study in the Ising model, see [21]. Cardy has further shown that for 3D free fields, modular invariance of the partition function (12.132) can be obtained if *antiperiodic* boundary conditions on the 1-sphere are taken [143]. At present, it is not known whether this result remains true for interacting fields, at least in some asymptotic sense.

On the other hand, some model-dependent information is available. Consider the quantum Hamiltonian of the (2+1)D Ising model [303]

$$H = -t \sum_n \sigma^z(n) - \frac{1}{2} \sum_{\langle n,n' \rangle} [(1 + \eta)\sigma^x(n)\sigma^x(n') + (1 - \eta)\sigma^y(n)\sigma^y(n')] \qquad (12.134)$$

with nearest neighbor interactions and on a square $N \times N$ lattice, with the topology $S^1 \otimes S^1$. As in the (1+1)D case (10.1), η is an irrelevant parameter introduced to test universality explicitly. The inverse correlation lengths are given by the energy spectrum $\xi_i^{-1} = E_i - E_0$. In Table 12.14, we give

Table 12.14. Critical point t_c and amplitude ratio Ξ in the square lattice (2+1)D Ising model, for periodic (P) and antiperiodic (A) boundary conditions (after [304])

η	1.0	0.9	0.7	0.5	0.3
t_c	3.047(1)	2.938(1)	2.720(1)	2.500(2)	2.30(1)
Ξ_P	3.64	3.64	3.66	3.67	3.5
Ξ_A	2.76	2.78	2.70	2.80	2.77

the location[20] of the Ising transition $t_c(\eta)$ and for both periodic (P) and antiperiodic (A) boundary conditions the amplitude ratio

$$\Xi := S_\varepsilon(0,0)/S_\sigma(0,0) \tag{12.135}$$

as obtained from a finite-size scaling calculation. We observe that Ξ is universal for both periodic (P) and antiperiodic (A) boundary conditions, as expected from finite-size scaling. The apparent variation of Ξ with η gives a realistic idea on the numerical precision of the data. Motivated from the simple 2D result (3.47) for $S_i(0,0)$ we compare with the exponent ratio $x_\varepsilon/x_\sigma \simeq 2.73\ldots$ for the (2+1)D Ising model (see [155, 285] for numerical values of 3D Ising exponents). This yields evidence for a *linear amplitude-exponent relation* in the critical (2+1)D Ising model with *antiperiodic* boundary conditions (in *both* finite directions) on a square lattice

$$\xi_i = N\mathcal{A}/x_i \tag{12.136}$$

as checked for $i = \sigma, \varepsilon$. The constant $\mathcal{A}$ cannot be determined from the quantum Hamiltonian (12.134), since its correct normalization is unknown.[21]

Table 12.15. Critical point K_c, amplitude ratio Ξ and the ratio x_ε/x_σ of scaling dimensions in the 3D O(n) model (after [611])

model	n	K_c	Ξ_P	Ξ_A	x_ε/x_σ
Ising	1	0.2216544(3)	3.67(3)	2.736(13)	2.7326(16)
XY	2	0.454167(3)	3.97(3)	2.93(5)	2.923(7)
Heisenberg	3	0.693004(7)	4.248(9)	3.08(8)	3.091(8)

Recently, the universal amplitude ratios $\Xi_{P,A}$ have also been found through a careful Monte Carlo study in the O(n) model [611]. A column geometry, of topology $S^1 \otimes S^1 \otimes \mathbb{R}$, with one 'infinite' direction and a square cross-section of $L \times L$ sites was considered. This is the analogue of the geometry studied through the quantum Hamiltonian (12.134). In Table 12.15, we

---

[20] The phase diagram has the same structure as in Fig. 10.1, but the disorder line is now given by $(t/2)^2 + \eta^2 = 1$, see [421]. A recent estimate is $t_c = 3.0444(3)$ for $\eta = 1$ [95].

[21] The universality of a large class of universal finite-size scaling amplitudes has been checked in the 3D Ising model [636].

show the results for the universal ratios Ξ_P and Ξ_A. For the Ising model, they agree with Table 12.14, as expected from universality. However, the agreement of the *anti*periodic amplitude ratio Ξ_A with the exponent ratio x_ε/x_σ is made more precise. Furthermore, the same relation apparently holds for the 3D XY and Heisenberg models as well. Finally, the correlation lengths were found exactly in the 3D spherical model. The result $\Xi_A = x_\varepsilon/x_\sigma = 2$ is again consistent with (12.136) [318].[22]

While Ξ_A appears to related to the scaling dimension through (12.136), there is no obvious link of Ξ_P with any exponents.

In conclusion, the result $\Xi_A = x_\varepsilon/x_\sigma$ has been established for the $O(n)$ model in 3D and for $n = 1, 2, 3$ and $n \to \infty$. Nevertheless, this remains a purely empirical remark. It is presently not understood why antiperiodic boundary conditions should be important for this kind of relationship.[23]

The question of an amplitude-exponent relation has also been studied in the context of a 3D dimer model [102]. Although the model does reproduce the linear relation (12.136) in 2D with periodic boundary conditions, as expected, it fails to do so in 3D with antiperiodic boundaries. That negative result might indicate that some other "twisting" boundary conditions must be used to recover (12.136).[24]

Conformal field theory in $2 < d < 4$ dimensions (with and without a boundary) was studied in detail in [219, 456, 486], in connection with conserved currents and applications to ϕ^4 theories. Another attempt tries to calculate the entire set of scaling dimensions for some three-dimensional system. This program has been carried through for the $O(n)$ vector model via a $1/n$ expansion up to second order [427]. The available results might point towards the existence of analogues of conformal families.

Exercises

1. Find the Kac tables for minimal models with one or two distinct primary operators.
2. From the Kac tables, calculate the conventional critical exponents of the spin models treated in this chapter and compare with the experimental results quoted in Chaps. 1 and 3.
3. Prove (12.35).
4. Find the Kac table and the Virasoro characters of the minimal model $\mathcal{M}_{6,8}$. Compare with the unitary minimal model $\mathcal{M}_{3,4}$.

---

[22] In the spherical model, the scaling dimension x_ε is read off directly from the decay of the two-point function and cannot be obtained simply through a scaling relation.

[23] Monte Carlo simulations allow to determine the constant $\mathcal{A}$. Early results [614, 318] lead to speculations that $\mathcal{A}$ might be model-independent, but the precise data of [611] for the $O(n)$ model display a clear variation of $\mathcal{A}$ with n.

[24] For example, if there *were* a second-order transition in the 3D three-states Potts model (there is not!), (12.136) might be recovered not for antiperiodic, but rather for the toroidal boundary conditions (12.2) with $\tilde{Q} = 1, 2$.

13. Conformal Perturbation Theory

In the last chapters we have obtained numerous results for statistical systems exactly *at* the critical point. Finite-size corrections were eliminated by the finite-lattice extrapolation procedures described and the finite-size scaling variables were set to zero. We now turn towards an investigation of these and shall show how finite-size corrections and finite-size scaling functions can be derived from the known operator content of a given model. These techniques do not require the integrability of the system under consideration.

13.1 Correlation Functions in the Strip Geometry

We recall the notation used with the logarithmic transformation Fig. 3.4. The complex coordinate on the strip is $w = u + iv = (N/2\pi)\ln z$, where the lattice constant a was put to one. The critical two-point correlation function of a primary operator is on the strip (see (3.45))

$$
\begin{aligned}
&\langle \varphi(w_1, \bar{w}_1)\varphi(w_2, \bar{w}_2)\rangle_w \\
&= \left(\frac{\pi}{N}\right)^{2x}\left(\sinh\frac{\pi}{N}(w_1 - w_2)\right)^{-2\Delta}\left(\sinh\frac{\pi}{N}(\bar{w}_1 - \bar{w}_2)\right)^{-2\overline{\Delta}} \\
&= \left(\frac{2\pi}{N}\right)^{2x}\sum_{m=0}^{\infty}\sum_{\bar{m}=0}^{\infty}a_m\bar{a}_{\bar{m}}\exp\left[-\frac{2\pi}{N}(x + m + \bar{m})(u_1 - u_2)\right] \\
&\quad \times \exp\left[-\frac{2\pi i}{N}(s + m - \bar{m})(v_1 - v_2)\right],
\end{aligned}
\tag{13.1}
$$

where $x = \Delta + \overline{\Delta}$, $s = \Delta - \overline{\Delta}$ are the scaling dimension and the spin, respectively and

$$
a_m = \frac{\Gamma(2\Delta + m)}{\Gamma(2\Delta)\,m!}\ , \quad \bar{a}_{\bar{m}} = \frac{\Gamma(2\overline{\Delta} + \bar{m})}{\Gamma(2\overline{\Delta})\,\bar{m}!}.
\tag{13.2}
$$

This follows from the binomial theorem and the identities

$$
\binom{-2\Delta}{k} = (-1)^k\frac{\Gamma(2\Delta + k)}{\Gamma(2\Delta)k!}
$$

$$
(\sinh W)^{-2\Delta} = 2^{2\Delta}\sum_{k=0}^{\infty}\frac{\Gamma(2\Delta + k)}{\Gamma(2\Delta)k!}(-1)^{-2\Delta}e^{W(2\Delta + 2k)}.
\tag{13.3}
$$

On the other hand, rewrite the two-point function while going from the "Heisenberg picture" with fields $\varphi(u,v)$ to the "Schrödinger picture" with fields $\varphi(v)$

$$\langle \varphi(u_1,v_1)\varphi(u_2,v_2)\rangle = \sum_{n,k}\langle 0|\varphi(v_1)|n,k\rangle \, e^{-(E_n-E_0)(u_1-u_2)}\langle n,k|\varphi(v_2)|0\rangle,$$

(13.4)

where $|n,k\rangle$ is a complete set of intermediate states with energy E_n and momentum k, the lowset one of which is just the primary state $|\varphi\rangle$. Comparison with (13.1) yields, see [137, G11]

$$\langle 0|\varphi(v)|\varphi\rangle = \left(\frac{2\pi}{N}\right)^x.$$

(13.5)

When considering a lattice operator, the coordinate v will correspond to the sites along the strip. This gives a means to fix the normalization of some lattice operator.

Similarly, the critical three-point function of primary operators is in the plane

$$\langle \varphi_1(z_1,\bar{z}_1)\varphi_2(z_2,\bar{z}_2)\varphi_3(z_3,\bar{z}_3)\rangle = C_{123}\, z_{12}^{-\Delta_{12}} z_{23}^{-\Delta_{23}} z_{31}^{-\Delta_{31}} \bar{z}_{12}^{-\bar{\Delta}_{12}} \bar{z}_{23}^{-\bar{\Delta}_{23}} \bar{z}_{31}^{-\bar{\Delta}_{31}},$$

(13.6)

where $C_{123} = \mathbf{C}_{123}$ is the **operator product expansion coefficient** (see Chaps. 2, 5), $\Delta_{12} := \Delta_1 + \Delta_2 - \Delta_3$ and $z_{12} := z_1 - z_2$ and so on. In the strip geometry, for $u_1 \gg u_2 \gg u_3$

$$\langle \varphi_1\varphi_2\varphi_3\rangle \simeq \left(\frac{2\pi}{N}\right)^{x_1+x_2+x_3} \exp\left[-\frac{2\pi}{N}(x_1(u_1-u_2)+x_3(u_2-u_3))\right]$$
$$\times \exp\left[\frac{2\pi i}{N}(s_1(v_1-v_2)+s_3(v_2-v_3))\right]\mathbf{C}_{123}.$$

(13.7)

Rewriting in the Schrödinger picture

$$\langle \varphi_1\varphi_2\varphi_3\rangle = \langle 0|\varphi_1(v_1)|\varphi_1\rangle e^{-\frac{2\pi}{N}x_1(u_1-u_2)}\langle \varphi_1|\varphi_2(v_2)|\varphi_3\rangle$$
$$\times e^{\frac{2\pi}{N}x_3(u_2-u_3)}\langle \varphi_3|\varphi_3(v_3)|0\rangle$$

(13.8)

one obtains [137, G11]

$$\langle \varphi_1|\varphi_2(v)|\varphi_3\rangle = \left(\frac{2\pi}{N}\right)^{x_2}\mathbf{C}_{123}\exp\left[-\frac{2\pi i}{N}(s_1-s_3)v\right].$$

(13.9)

This can be used either to calculate some $\mathbf{C}_{123}$ from finite-lattice data, see Table 7.1, or else to obtain some matrix elements from a known operator product. As shown in Chap. 7, the knowledge of x_1, x_2, x_3 and s_1, s_2, s_3 is enough to find $\mathbf{C}_{123}$. The explicit expression for $\mathbf{C}_{123}$ was given in (7.52). If the lattice operator happens to be not yet normalized, (13.5) can be used to fix the correct conformal normalization.

13.2 General Remarks on Corrections to the Critical Behaviour

In general, any lattice Hamiltonian will contain correction terms to the critical Hamiltonian H_c

$$H = H_c + \sum_v \sum_k a_k \varphi_k(v), \qquad (13.10)$$

where the a_k are (non-universal) parameters. In the sequel, we assume that the critical spectrum is non-degenerate for technical simplicity. So the following discussion only applies to the few lowest eigenvalues of H, but the generalization, if needed, is straightforward. The eigenvalues of H are then, if $E_{n,c}$ are the critical eigenvalues

$$E_n = E_{n,c} + \sum_k a_k \sum_v \langle n|\varphi_k(v)|n\rangle$$

$$+ \sum_{k,\ell} a_k a_\ell \sum_{v,v'} \sum_{i \neq n} \frac{\langle n|\varphi_k(v)|i\rangle \langle i|\varphi_\ell(v')|n\rangle}{E_{n,c} - E_{i,c}}$$

$$+ \text{ higher order terms.} \qquad (13.11)$$

Consider the first-order corrections for the correlation lengths. One has, taking for simplicity just one perturbing operator

$$\xi_n^{-1} = E_n - E_0 \simeq \frac{2\pi}{N} x_n + \sum_v a_k \left(\langle n|\varphi_k(v)|n\rangle - \langle 0|\varphi_k(v)|0\rangle\right) + \cdots$$

$$= \frac{2\pi}{N} x_n + a_k \left(\frac{2\pi}{N}\right)^{x_k} (\mathbf{C}_{nkn} - \mathbf{C}_{0k0}) N$$

$$= \frac{2\pi}{N} \left[x_n + 2\pi a_k (\mathbf{C}_{nkn} - \mathbf{C}_{0k0}) \left(\frac{2\pi}{N}\right)^{x_k-2} \right]. \qquad (13.12)$$

One now has to distinguish two cases.

1. **Irrelevant fields.** One has $y_k = 2 - x_k < 0$. These fields give rise to **finite-size corrections** to the critical behaviour. The possible finite-size correction exponents, as defined in Chap. 2, can be read off from the operator content of H_c. We illustrate this in a few examples. One source of corrections is the conformal tower of the identity operator. These operators are present in *any* conformal invariant system. Possible correction terms appearing in the Hamiltonian must satisfy a few requirements. They should be quasiprimary and if they shall contribute in first order, they should be invariant under the global symmetry of the model. In principle, the following operators appear in the conformal tower of the identity

$$L_{-2},\ L_{-3},\ L_{-4},\ L_{-2}^2,\ \ldots \qquad (13.13)$$

and their combinations with the $\bar{L}_n$. It is usually sufficient to consider symmetric combinations like $L_{-2} + \bar{L}_{-2}$. This particular operator only changes the normalization ζ. The antisymmetric combination $L_{-2} - \bar{L}_{-2}$ is not consistent with parity conservation $E(\mathcal{P}) = E(-\mathcal{P})$. Terms involving L_{-3} alone are not quasiprimary. The first terms allowed are $L_{-2}\bar{L}_{-2}$ and $L_{-2}^2 - \frac{3}{5}L_{-4} + \bar{L}_{-2}^2 - \frac{3}{5}\bar{L}_{-4}$. The second one describes the breaking of rotational symmetry by a square lattice and is a tensor operator under the dihedral group $\mathbb{D}_4$. We shall see in the examples of the Ising and the three-states Potts model that only this operator has a non-vanishing coupling. These two operators have a scaling dimension $x = 4$. They thus produce finite-size corrections of order $1/N^2$. Corrections of this size will therefore be always present (unless the coefficient vanishes). For the Ising model, corrections of this size are in fact the only ones present (see below). This is not so in other systems. For the tricritical Ising model with periodic boundary conditions, the operator content (12.14) shows an irrelevant field with $\Delta = \bar{\Delta} = 3/2$ in the even sector. This field generates correction terms $1/N$ in leading order. Finally, for the three-states Potts model, (12.4) has in the sector $Q = 0$ a field with $\Delta = \bar{\Delta} = 7/5$. The leading finite-size correction due to this field is of order $1/N^{0.8}$.

In Chap. 12, we remarked that for the tricritical Ising model the correction to scaling terms might be larger than for the Ising model. This numerical observation is confirmed by the above result on the leading correction terms.

Finally, it is common to characterize the leading corrections by Wegner's **correction-to-scaling exponent** ω, as already defined in Chap. 3. The above discussion may be summarized by stating that $\omega = 2, 1, \frac{4}{5}$ for the Ising, tricritical Ising and Potts models, respectively.

2. **Relevant fields.** For these, $y_k > 0$. Relevant fields describe the crossover to another universality class. We shall describe this in more detail in the next chapter. Here we shall limit ourselves to a study of the **finite-size scaling limit** and define **finite-size scaling variables** $z_k := a_k N^{y_k}$ which are kept finite in the **finite-size scaling limit** $a_k \to 0$ and $N \to \infty$ (see Chap. 3). We then obtain a power series expansion of the finite-size scaling functions. We stress that this can be obtained without having to know how to solve a given model exactly on the lattice!

After these general remarks about the order of the leading finite-size corrections, we now turn towards the calculation of the correction amplitudes, following [533]. For an alternative approach, see [550].

13.3 Finite-Size Corrections

The Hamiltonian on the strip of width N is

$$H = H_c + \sum_k a_k \int_{-N/2}^{N/2} \mathrm{d}v \, \varphi_k(0, v), \qquad (13.14)$$

where a_k is a non-universal constant and φ_k an irrelevant conformal operator with scaling dimension

$$x_k = \Delta_k + r_k + \overline{\Delta}_k + \bar{r}_k, \qquad (13.15)$$

where $\Delta_k, \overline{\Delta}_k$ are the conformal weights of the corresponding primary operator. Note that N does not appear as a coupling constant. To calculate the matrix element of φ_k consider the two-point function

$$\langle \varphi_{\Delta,\overline{\Delta}}(u_1, v_1) \varphi_{\Delta,\overline{\Delta}}(u_2, v_2) \rangle$$
$$= \sum_{r,\bar{r};\alpha} \langle 0, 0 | \varphi_{\Delta,\overline{\Delta}}(0,0) | \Delta + r, \overline{\Delta} + \bar{r}; \alpha \rangle \langle \Delta + r, \overline{\Delta} + \bar{r}; \alpha | \varphi_{\Delta,\overline{\Delta}}(0,0) | 0, 0 \rangle$$
$$\times (p_1 p_2)^{\Delta+r} (\bar{p}_1 \bar{p}_2)^{\overline{\Delta}+\bar{r}}$$
$$= \sum_{r,\bar{r}} b(r, \bar{r}) \, (p_1 p_2)^{\Delta+r} (\bar{p}_1 \bar{p}_2)^{\overline{\Delta}+\bar{r}} \qquad (13.16)$$

with

$$p_k := \exp \left\{ \frac{2\pi}{N} \left[(u_k - u_{k+1}) + \mathrm{i}(v_k - v_{k+1}) \right] \right\} \qquad (13.17)$$

which defines $b(r, \bar{r})$. The three-point function is

$$\langle \varphi_{\Delta,\overline{\Delta}}(u_1, v_1) \varphi_k(u_2, v_2) \varphi_{\Delta,\overline{\Delta}}(u_3, v_3) \rangle$$
$$= \sum_{r_1, r_2, \bar{r}_1, \bar{r}_2} \sum_{\alpha, \beta} \langle 0, 0 | \varphi_{\Delta,\overline{\Delta}}(0,0) | \Delta + r_1, \overline{\Delta} + \bar{r}_1; \alpha \rangle$$
$$\times \langle \Delta + r_1, \overline{\Delta} + \bar{r}_1; \alpha | \varphi_k(0,0) | \Delta + r_2, \overline{\Delta} + \bar{r}_2; \beta \rangle$$
$$\times \langle \Delta + r_2, \overline{\Delta} + \bar{r}_2; \beta | \varphi_{\Delta,\overline{\Delta}} | 0, 0 \rangle \, p_1^{\Delta+r_1} p_2^{\Delta+r_2} \bar{p}_1^{\overline{\Delta}+\bar{r}_1} \bar{p}_2^{\overline{\Delta}+\bar{r}_2}$$
$$=: \sum_{r_1, r_2, \bar{r}_1, \bar{r}_2} a(r_1, r_2; \bar{r}_1, \bar{r}_2) \, p_1^{\Delta+r_1} p_2^{\Delta+r_2} \bar{p}_1^{\overline{\Delta}+\bar{r}_1} \bar{p}_2^{\overline{\Delta}+\bar{r}_2} \qquad (13.18)$$

defining $a(r_1, r_2; \bar{r}_1, \bar{r}_2)$. Comparing, one has the desired matrix element, taking translation invariance into account

$$\langle \Delta + r, \overline{\Delta} + \bar{r}; \alpha | \varphi_k(0,0) | \Delta + r, \overline{\Delta} + \bar{r}; \alpha \rangle = \frac{a(r, r; \bar{r}, \bar{r})}{b(r, \bar{r})}. \qquad (13.19)$$

13.3.1 Tower of the Identity

This is always present and thus studied first. As shown in [533], the operators $\int_{-N/2}^{N/2} dv L_{-k}(w)$ with $k \geq 3$ do not give a correction in any order of perturbation theory since the corresponding matrix elements vanish. To leading order there remain two possibilities

$$\varphi_1(w, \bar{w}) := L_{-2}(w)\bar{L}_{-2}(\bar{w})$$
$$\varphi_2(w, \bar{w}) := L^2_{-2}(w) + \bar{L}^2_{-2}(\bar{w}). \tag{13.20}$$

One now has to calculate the correlation functions of these non-primary operators. In the plane, these are known and can be obtained by applying the differential operator $\mathcal{L}_k$ of (5.5) to correlation functions of primary operators. These in turn were shown in Chap. 4 to satisfy linear partial differential equations which can be solved. This was demonstrated for the Ising model example in Chap. 6. Here, it is sufficient to recall that under the conformal transformation $z = \exp(2\pi w/N)$ the energy–momentum tensor transforms as

$$T(w) \rightarrow \left(\frac{2\pi}{N}\right)^2 \left(z^2 T(z) - \frac{c}{24}\right) \tag{13.21}$$

and that

$$L_{-k}(w) = \oint dw' \frac{T(w')}{(w' - w)^{k-1}}. \tag{13.22}$$

It then follows that

$$\langle L_{-2}(w_1)L_{-2}(w_2)\rangle = \left(\frac{2\pi}{N}\right)^4 \left\{\frac{c}{2}\frac{(p_1 p_2)^2}{(1 - p_1 p_2)^4} + \left(\frac{c}{24}\right)^2\right\} \tag{13.23}$$

and with a similar expression for the three-point function one obtains [533]

$$\langle \Delta + r|L_{-2}(0,0)|\Delta + r\rangle = \left(\frac{2\pi}{N}\right)^2 \left(\Delta + r - \frac{c}{24}\right) \tag{13.24}$$

$$\langle \Delta + r|L^2_{-2}(0,0)|\Delta + r\rangle = \left(\frac{2\pi}{N}\right)^4 \left[\left(\frac{c}{24}\right)^2 + \frac{11}{1440}c + A(\Delta, r)\right],$$

where $r = 0, 1, 2, 3$ for $\Delta \neq 0$ and $r = 0, 2, 3$ for $\Delta = 0$ and

$$A(0, r) = \left(\frac{11}{30} + \frac{c}{12}\right) r(2r^2 - 3) \tag{13.25}$$

$$A(\Delta, r) = (\Delta + r)\left[\left(\Delta - \frac{2+c}{12}\right) + \frac{r(2\Delta + r)(5\Delta + 1)}{(\Delta + 1)(2\Delta + 1)}\right] ; \quad \Delta \neq 0.$$

Summarising, the leading finite-size corrections coming from the tower of the identity are described by

$$H = H_c + a_1 \int_{-N/2}^{N/2} dv \, (L_{-2}\bar{L}_{-2}(0, v)) + a_2 \int_{-N/2}^{N/2} dv \, (L^2_{-2}(0, v) + \bar{L}^2_{-2}(0, \bar{v}))$$

$$(13.26)$$

and lead to the following results for the singular free energy density f^{sin} and the scaled and non-degenerate gaps $\mathcal{E}$ [533] (see (10.34))

$$N^2(2\pi)^{-1} f^{\mathrm{sin}} = -\frac{c}{12} + \frac{\pi^3}{N^2} \left[a_1 \frac{c^2}{72} + a_2 \left(\frac{c^2}{36} + \frac{11}{90} c \right) \right] + \mathcal{O}(N^{-4})$$

$$\mathcal{E}(\Delta + r, \bar{\Delta} + \bar{r}) = \Delta + r + \bar{\Delta} + \bar{r}$$

$$+ \frac{8\pi^3}{N^2} \left\{ a_1 \left[\left(\Delta + r - \frac{c}{24} \right) \left(\bar{\Delta} + \bar{r} - \frac{c}{24} \right) - \left(\frac{c}{24} \right)^2 \right] \right.$$

$$\left. + a_2 \left[A(\Delta, r) + A(\bar{\Delta}, \bar{r}) \right] \right\} + \mathcal{O}(N^{-4}).$$

$$(13.27)$$

The only model-dependent parameters in these expressions are the values of the central charge c and the conformal weights $\Delta, \bar{\Delta}$ of the primary operators. For the values of r one has the same restrictions as above.

We now compare these results with explicit model calculations. Besides illustrating the general expressions with some specific numbers, these applications also point towards some further subtleties involved in calculating finite-size corrections.

13.3.2 Application to the Ising Model

Finite-size corrections for the Ising model were already calculated for the entire spectrum and the lowest levels are given in (10.41) and (10.47). Comparison with the general results of (13.27) gives complete consistency if [533]

$$a_1 = 0 \ , \ a_2 = \frac{3}{56\pi} \left(\frac{1}{\eta^2} - \frac{4}{3} \right). \tag{13.28}$$

This results identifies the ubiquitous non-universal correction amplitude a_2 in terms of the Ising model parameters. This is one of the still very few examples where scaling and conformal predictions can be confirmed in full detail.

However, the 2D Ising model has the peculiarity that there is a further irrelevant operator in the conformal tower of the energy density ε, namely $\varphi_3 := L_{-1}\bar{L}_{-1}\varepsilon$ which also gives corrections of order $1/N^2$. Since its scaling dimension $x_3 = 3$, the first order, if present, would give rise to terms of order $1/N$. These terms vanish since φ_3 is a derivative operator, however. Similarly, the operator $L_{-2}\varepsilon = 3/4L^2_{-1}\varepsilon$ does not contribute in first order ($\sim 1/N^2$). Second-order terms involving φ_3 do contribute, however. If a_1 and a_2 were both vanishing, consistency were achieved if

$$a_3^2 = \frac{1}{16\pi^2} \left(\frac{1}{\eta^2} - \frac{4}{3} \right). \tag{13.29}$$

To have H hermitian, one needs the a_k to be real. For $a_1 = a_2 = 0$, this is not possible if $\sqrt{3}/2 < \eta \leq 1$. Consequently, it is not enough to have a_3 alone non-vanishing. The leading finite-size corrections of the Ising model are thus described by the Hamiltonian [533]

$$H = H_c + a \int_{-N/2}^{N/2} dv \, (L_{-1}\bar{L}_{-1}\varepsilon) \tag{13.30}$$

$$+ \frac{3}{56\pi} \left[\frac{1}{\eta^2} - \frac{4}{3} - (4\pi a)^2 \right] \int_{-N/2}^{N/2} dv \, (L_{-2}^2(0, v) + \bar{L}_{-2}^2(0, \bar{v})) ,$$

where a is a free parameter. There is no known criterion to fix its value.

13.3.3 Application to the Three-States Potts Model

The three-states Potts model serves to illustrate the case of *non-analytic* corrections arising from an additional irrelevant primary operator. The finite-size corrections of this model were studied in great detail in [261]. The numerical techniques needed were described in Chap. 9. The numerical results are

$$N^2 (2\pi)^{-1} f^{\mathrm{sin}} = -\frac{1}{15} + 0.0280(5)N^{-2} + \dots$$

$$\mathcal{E}(\Delta + r, \overline{\Delta} + \bar{r}) = x + c_1 N^{-0.8} + c_2 N^{-2} + \dots, \tag{13.31}$$

where c_1 and c_2 are level-dependent constants. Their values are given in Table 13.1.

Table 13.1. Finite-size correction coefficients of (13.31) for the three-states Potts model (after [261])

$(\Delta + r, \overline{\Delta} + \bar{r})$	c_1	c_2
$(0 + 2, 0)$		-5.810(1)
$(1/15, 1/15)$	0.00657(2)	0.03238(2)
$(1/15 + 1, 1/15)$	0.034(2)	-1.6(1)
$(2/5, 2/5)$	0.2364(2)	-0.328(2)
$(2/5, 1/15)$	-0.0395(3)	-0.15(3)

We now discuss how these data can be explained in terms of conformal finite-size corrections.

1. For terms of order $N^{-0.8}$, write the perturbed Hamiltonian

$$H = H_c + a \int_{-N/2}^{N/2} dv \, \varphi_{7/5,7/5}(0, v) \tag{13.32}$$

and take $a = 0.009237(7)$ to reproduce the values of the c_1 as given in Table 13.1. Since the corresponding operator product expansion coefficient vanishes, there is no contribution to f^{sin} in this order.

2. For the order N^{-2}, take the constants a_1, a_2 as in (13.27). A fit to the numerical data yields $a_1 = -0.002(2)$ and $a_2 = -0.0056(3)$. Since this is consistent with $a_1 = 0$, the fit was repeated with a_2 alone and yields [533]

$$a_1 = 0 \ , \quad a_2 = -0.0056(6). \tag{13.33}$$

From the operator content of the Potts model it is clear that there are no further contributions at this order. It is non-trivial that a single parameter is enough to reproduce all values of c_2 in Table 13.1.

3. There are second-order contributions from $\varphi_{7/5,7/5}$, of order $N^{-1.6}$. Generally, the contribution to f^{sin} from a second-order term of an irrelevant scalar primary operator is

$$N^2(2\pi)^{-1} f^{\text{sin}} + \frac{c}{12} = -\left[2\pi a \left(\frac{2\pi}{N}\right)^{x-2} \right]^2 \frac{1}{2} \int_0^1 dv \, v^{x/2-1} {}_2F_1(x,x,1;v) \tag{13.34}$$

with $x = 2\Delta > 2$ and ${}_2F_1$ is the hypergeometric function. The integral is only convergent for $0 < \text{Re } x < 1$, but is needed for $x > 2$. It can be made finite by analytical continuation which is achieved by $2k$-fold partial integration [533]

$$\int_0^1 dv \, v^{x/2-1} {}_2F_1(x,x,1;v) \tag{13.35}$$

$$= -\sum_{m=1}^k \frac{\Gamma^2(1-x)\Gamma^2(m-x/2)\Gamma(2m-2x)}{\Gamma^2(1-x/2)\Gamma^3(1+m-x)\Gamma(m-x)} (3m-2x)$$

$$+ \frac{\Gamma^2(k+1-x/2)\Gamma^2(1-x)}{\Gamma^2(1-x/2)\Gamma^2(k+1-x)} \int_0^1 dv \, v^{x/2-1} {}_2F_1(x-k, x-k, 1; v),$$

where the right-hand side is now well-defined for $0 < \text{Re } x < k+1$ and x non-integer. One then finds for the Potts model

$$N^2(2\pi)^{-1} f^{\text{sin}} = -\frac{1}{15} - 0.001721(3)N^{-1.6} + \ldots$$

$$\mathcal{E}(2,0) = 2 - 0.01093(2)N^{-1.6} + \ldots. \tag{13.36}$$

Comparing these with the results of Table 13.1, one sees that these second order contributions are at least an order of magnitude smaller than those terms of order N^{-2}, at least for the largest lattice of size $N = 12$ used in the numerical calculations reported. This finding nicely illustrates a general remark that for small lattices the effective correction exponent can be substantially affected by relatively large or small values of the correction amplitudes, see [514].

The same techniques were also applied to study the finite-size corrections of the order parameter [534], where the finite-lattice order parameter is defined in (9.17). Finite-size corrections have been calculated in a huge variety of integrable systems. For a review, see [583].

13.3.4 Checking the Operator Content from Finite-Size Corrections

We have seen how the known operator content was used to predict the leading finite-size corrections of critical conformal invariant systems. Blöte and den Nijs [94] have asked conversely whether these corrections are always the *only* ones appearing. This question is well motivated, since, as we have seen in Chap. 6, we came across solutions for the values of the conformal weights Δ which did not fit into the Kac table of *unitary minimal* models. In particular, in the context of the Ising model, there is an operator which would correspond to the conformal operator $\phi_{3,1}$, see (6.22), but which is not in the Kac table for $m = 3$. This operator, if it *were* present, would represent a **vacancy operator** with scaling dimension $x_{\text{vac}} = 10/3$, producing finite-size corrections of order $N^{-4/3}$. The presence of such an operator would still be consistent with conformal invariance but *not* with the closed operator product algebra used to classify the possible primary operators of unitary models. This vacancy operator is absent in the integrable realisation of the 2D Ising universality class, as it is evident from the exact results for the finite-size correction terms of Chap. 10. One might argue, however, that in an antiferromagnetic Ising model with next to nearest neighbor interactions an eventual coupling constant of the vacancy operator to the rest of the model might be enhanced. Numerical calculations [94] of the correlation length, however, clearly indicate the leading finite-size corrections to be of order N^{-2}, which implies that the vacancy operator cannot be present.

It does occur, however, in the random-cluster representation of the $p = 2 + \varepsilon$ state Potts model. The amplitude of this operator vanishes in the Ising case. For a discussion of the weakly disordered $2 + \epsilon$ state Potts model, see [443].

13.4 Finite-Size Scaling Functions

We now turn to the case of *relevant* perturbations. We write the finite-size scaling variable $z = \vartheta N^{2-x}$, where ϑ measures the distance from the critical point, like e.g. $\vartheta = T - T_c$. The Hamiltonian is

$$H = H_c + \bar{a} z N^{x-2} \int_{-N/2}^{N/2} dv\, \varphi(0,v) = H_c + V, \qquad (13.37)$$

where $\bar{a}$ is a non-universal constant, related to the metric factors introduced in Chap. 3.

For the first order, one just needs the matrix elements of V. For the second order, one has, with $\omega_{i,n} = E_{i,c} - E_{n,c}$ and k denoting the momentum eigenvalue [533]

$$E_n^{(2)}(k) = -\sum_{i \neq n} \omega_{i,n}^{-1} \langle k, n|V|k, i\rangle \langle k, i|V|k, n\rangle$$

$$= -\left(\bar{a}zN^{x-2}\right)^2 \int_0^\infty d\tau \int\int_{-N/2}^{N/2} dv_1 dv_2 \cdot$$

$$\left\{ \sum_i e^{-\omega_{i,n}\tau} \langle k, n|\varphi(0, v_1)|k, i\rangle \langle k, i|\varphi(0, v_2)|k, n\rangle \right.$$

$$- \langle k, n|\varphi(0, v_1)|k, n\rangle \langle k, n|\varphi(0, v_2)|k, n\rangle$$

$$\left. - \sum_{i<n} \left(e^{-\omega_{i,n}\tau} + e^{\omega_{i,n}\tau}\right) \langle k, n|\varphi(0, v_1)|k, i\rangle \langle k, i|\varphi(0, v_2)|k, n\rangle \right\}$$

$$= -\left(\bar{a}zN^{x-2}\right)^2 \int_0^\infty d\tau \int\int_{-N/2}^{N/2} dv_1 dv_2 \cdot$$

$$\left\{ \langle k, n|\varphi(0, v_1)\varphi(0, v_2)|k, n\rangle - \langle k, n|\varphi(0, v_1)|k, n\rangle \langle k, n|\varphi(0, v_2)|k, n\rangle \right.$$

$$\left. - \sum_{i<n} \left(e^{-\omega_{i,n}\tau} + e^{\omega_{i,n}\tau}\right) \langle k, n|\varphi(0, v_1)|k, i\rangle \langle k, i|\varphi(0, v_2)|k, n\rangle \right\}$$

$$=: -\left(\bar{a}zN^{x-2}\right)^2 \int_0^\infty d\tau B_{k,n;2}\left(\frac{2\pi}{N}\tau\right) \tag{13.38}$$

using the spectral decomposition. For the scaled gaps one finds

$$\mathcal{E}(\Delta + r, \overline{\Delta} + \bar{r}) = -\left((2\pi)^{x-1}\bar{a}z\right)^2 \int_0^\infty d\tau \left(B_{k,n;2}(\tau) - B_{0,0;2}(\tau)\right) \tag{13.39}$$

and similar expressions can be written down for the higher orders. In general, for the nth order perturbation theory some $(n+2)$-point correlation functions are needed. The n-point correlation functions are known in the infinite plane and must be rewritten into the strip via the standard logarithmic transformation. We shall not give the details of the rather tedious calculation and refer for this to the original work [533].

13.4.1 Ising Model: Thermal Perturbation

The perturbing operator $\varphi = \varepsilon$ is the energy density with $x_\varepsilon = 1$. The finite-size scaling variable is $z := (t-1)N$. The results for the gaps are, including the fifth order contributions [533]

$$\mathcal{E}(r + 1/16, \bar{r} + 1/16) = r + \bar{r} + \frac{1}{8} + \bar{a}z\frac{1}{2}(2\delta_{r,0} - 1)(2\delta_{\bar{r},0} - 1)$$

$$+ (\bar{a}z)^2 \left(\ln 2 + \frac{1 - \delta_{r,0}}{2} + \frac{1 - \delta_{\bar{r},0}}{2}\right)$$

$$+ (\bar{a}z)^4 \left(\frac{3}{4}\zeta(3) + \frac{1 - \delta_{r,0}}{8r^3} + \frac{1 - \delta_{\bar{r},0}}{8\bar{r}^3}\right) + \mathcal{O}(z^6), \tag{13.40}$$

where ζ is the Riemann zeta function and, if $\Delta = 0, 1/2$

$$\mathcal{E}(r + \Delta, \bar{r} + \overline{\Delta}) = \Delta + r + \overline{\Delta} + \bar{r} + (\bar{a}z)^2 \left(\mathcal{A}(r, \Delta) + \mathcal{A}(\bar{r}, \overline{\Delta})\right)$$
$$+ (\bar{a}z)^4 \left(\mathcal{B}(r, \Delta) + \mathcal{B}(\bar{r}, \overline{\Delta})\right) + \mathcal{O}(z^6), \qquad (13.41)$$

$$\mathcal{A}(r, 0) = 1 + \frac{1}{2r - 1} \ , \ \mathcal{B}(r, 0) = 1 + \frac{1}{(2r - 1)^3} \ ; \ r = 0, 2, 3$$

$$\mathcal{A}(r, 1/2) = \frac{1}{2r - 1} \ , \ \mathcal{B}(r, 1/2) = \frac{1}{(2r - 1)^3} \ ; \ r = 0, 1, 2, 3. \quad (13.42)$$

Comparing with the Ising finite-size scaling functions (10.48), the non-universal constant $\bar{a}$ is fixed[1]

$$\bar{a} = (-1)^{\tilde{Q}} \ \text{sgn}\,(z) \frac{1}{2\pi\eta}. \qquad (13.43)$$

We see that conformal perturbation theory is capable to find the *exact* expansion of the finite-size scaling functions in terms of the finite-size scaling variable z. The only model-dependent input needed are the values of the conformal weights.

One may now ask whether the series of which the first few terms were derived is convergent and if so, for which values of z it converges. It can be shown that for *scalar* perturbations with $\Delta = \bar{\Delta} < 1/2$, the conformal perturbation series has an infinite radius of convergence [171]. On the other hand, the Ising model example shows that the condition $x < 1$ cannot be relaxed further. The correlation lengths were calculated exactly in Chap. 10, see (10.48). The remnant function $R_{1\frac{1}{2},0}(w)$ involved is analytic at the origin [228], where ζ is the Riemann zeta function,

$$R_{1\frac{1}{2},0}(w) = \Gamma(-1/2) \sum_{n=2}^{\infty} \frac{\zeta(2n - 1)}{\Gamma(3/2 - n)} \frac{w^n}{n!} \qquad (13.44)$$

and has a *finite* radius of convergence $|w| < 1$. Reinterpreting this in terms of the Ising quantum chain parameters, the conformal perturbation expansion for *all* correlation lengths converges for $z/(\pi\eta) < 1$ and we see that for $\eta = 0$, this representation ceases to converge. This is not surprising, since for $\eta = 0$ the global symmetry and the critical behaviour of the model changes completely.

Summarising, conformal invariance allows to calculate the universal finite-size scaling functions explicitly. The universality of the finite-size scaling function has been confirmed in quite a few studies. For the Ising model, it was shown analytically that the finite-size scaling functions (10.48) obtained from

---

[1] Comparing the exact results (10.48,13.40,13.41) with (13.12), one may read off from the linear term in z the OPE coefficients $\mathbf{C}_{\varepsilon,\varepsilon}^\varepsilon = 0$ and $\mathbf{C}_{\sigma,\varepsilon}^\sigma = 1/2$, in agreement with the explicit calculation in Chap. 6 and the numerical estimates in Table 7.1.

the Ising quantum chain are identical to those obtained from the isotropic classical 2D Ising model [117]. Numerically, the universality with respect to different lattices (including the quasi-periodic Penrose lattice) had been confirmed for correlation lengths, [181] the free energy, the order parameter and the susceptibility [592, 181, 483], as well as for the spin-one Ising model [592, 181, 338]. Universality of the finite-size scaling function for the same quantities has been also confirmed for the three-states Potts model [592, 181, 432]. The universality of the correlation length scaling functions for the tricritical Ising model was checked in [314]. For percolation, universal finite-size scaling functions have been calculated for the existence probability of percolation and for the percolation probability [342, 344, 345]. These references contain extensive tables of the values of the non-universal metric factors.

13.4.2 Ising Model: Magnetic Perturbation

Now consider the case $\varphi = \sigma$ with $x_\sigma = 1/8$. Since the order parameter breaks the global symmetry, only even orders will contribute. The finite-size scaling variable is $\mu := hN^{15/8}$ where h is the magnetic field. The second-order results are [533]

$$\mathcal{E}(r, \bar{r}) = r + \bar{r} + \left(\frac{\bar{a}\mu}{(2\pi)^{7/8}} \right)^2 (1 - \gamma_r \gamma_{\bar{r}}) \cdot \delta \; ; \quad r, \bar{r} = 0, 2, 3$$

$$\mathcal{E}(1/2 + r, 1/2 + \bar{r}) = 1 + r + \bar{r} + \left(\frac{\bar{a}\mu}{(2\pi)^{7/8}} \right)^2 \cdot \delta \; ; \quad r, \bar{r} = 0, 1, 2, 3$$

$$\mathcal{E}(1/16 + r, 1/16 + \bar{r}) = \frac{1}{8} + r + \bar{r} + \left(\frac{\bar{a}\mu}{(2\pi)^{7/8}} \right)^2 (\delta - g_{r,\bar{r}}) \; ; \quad r, \bar{r} = 0, 1, 2$$

$$(13.45)$$

up to terms of order $\mathcal{O}(\mu^4)$ and with the constants

$$\gamma_0 = 1 \; , \quad \gamma_2 = \frac{19}{124} \; , \quad \gamma_3 = \frac{512}{1457}$$

$$\delta = \sum_{m=0}^{\infty} \left(\frac{\Gamma(1/8 + m)}{\Gamma(1/8)m!} \right)^2 \frac{1}{2m + 1/8} = 8.00949725\ldots \qquad (13.46)$$

$$g_{00} = -7.70684920(1) \; , \quad g_{10} = 0.22667203(2) \; , \quad g_{20} = 4.2823790(1)$$

$$g_{11} = -0.00666683(2) \; , \quad g_{21} = 0.1508220(2) \; , \quad g_{22} = 0.6177484(6)$$

and in the last two lines the given errors are exact upper and lower bounds.

We now compare this with numerical data in the $(1+1)$D Ising model. We take $\eta = 1$. In view of the thermal result for the non-universal metric factor $\bar{a}$ in (13.43), the choice

$$\bar{a} = (2\pi)^{-1} \tag{13.47}$$

is suggestive. For the comparison with numerical data, some care is needed, however. In principle, there are two different sources for finite-size corrections:

1. The finiteness of the system for $\mu = 0$. These purely finite-size corrections were calculated above.
2. Keeping a finite value of μ on a finite lattice also implies that h is non-zero. This leads to the appearance of *bulk* corrections.

Both correction terms, if present, will break finite-size scaling. A scaling plot is therefore necessary to show that finite-lattice data obtained really yield information on the scaling behaviour. In fact, using the Ising quantum chain with moderately large lattices (up to $N = 15$), finite-size corrections are still apparent. In order to check the conformal predictions, we calculate the **reduced finite-size scaling functions** $H_i(\mu)$ for the two lowest gaps

$$\xi_\sigma^{-1} = \frac{\pi}{4} N^{-1} + \frac{\pi^3}{192} N^{-3} + h^{8/15} \frac{1}{2} H_1(\mu)$$

$$\xi_\varepsilon^{-1} = 2\pi N^{-1} - \frac{\pi^3}{12} N^{-3} + h^{8/15} \frac{1}{2} H_2(\mu), \tag{13.48}$$

where the first terms are just the conformal critical point results and the second terms are the leading finite-size corrections for $\mu = 0$. These reduced scaling functions are shown in Fig. 13.1. The different symbols correspond to different values of h. Clearly the corrected data display a scaling behaviour. The full curves were obtained by using the second order results together with (13.47). There is excellent agreement for $\mu \lesssim 0.3$. For larger values of μ, deviations appear which are probably due to the higher order corrections. These results represent an impressive confirmation of the theoretical framework of conformal invariance for the precise quantitative description of critical two-dimensional systems. Extending these results to higher orders appears feasible only with the DMRG algorithm. However, we shall see in Chap. 14 that there is a class of relevant perturbations which preserve the integrability even of the non-critical system. Then the thermodynamic Bethe ansatz can be used to render the computations manageable. However, the requirement of integrability is a very strong one, and is restricted in the case of unitary models to the relevant operators ϕ_{12}, ϕ_{21} and ϕ_{13}. In all other cases, more general techniques as the one presented in this chapter have to be used.

We have seen that conformal perturbation theory only contains contributions analytic in h. It had been suggested [208] that a more general scheme might apply, which argues that because of the operator product expansion in the Ising model $\sigma \cdot \sigma = 1 + \varepsilon$, the effective perturbed Hamiltonian should read $H = H_c + h\sigma + Ah^\psi \varepsilon$ with $\psi = 8/15$ and A being a free parameter. However, the presence of a free parameter is not in agreement with the conventional scaling theory (see Chaps. 1 and 3) and also in contradiction to the numerically confirmed universality [181] of the finite-size scaling functions of both

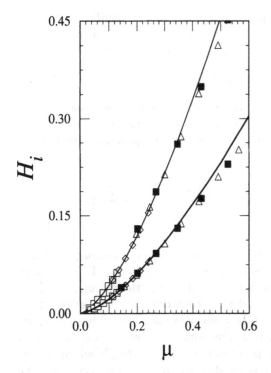

Fig. 13.1. Reduced finite-size scaling functions H_1 (upper curve) and H_2 (lower curve) for the quantum Ising chain in a magnetic field (after [314])

the Ising and Potts-3 models when varying t or h. Quantitatively, the idea of [208] would imply for the scaling function of the spin-spin correlation length of the Ising model the form $m_\sigma = h^{8/15}(\pi/(4\mu^{8/15}) + A + \mathcal{O}(A^2\mu^{8/15}))$, leading to $H_1(0) = 2A.$[2] Comparison with Fig. 13.1 shows that A must indeed be very small and an upper bound $A \lesssim 3 \cdot 10^{-4}$ [314] is obtained. This is consistent with $A = 0$ and with conventional finite-size scaling.

13.5 Truncation Method

We now present a different technique for the approximate calculation of the spectrum of the perturbed Hamiltonian H, following the original work of Yurov and Al. Zamolodchikov [634].

One wants to calculate the spectrum of the perturbed Hamiltonian (13.37). This could be done in an approximate way by selecting a convenient small subspace of the full Hilbert space and diagonalize the interaction matrix V only in that subspace. The matrix elements of V are completely specified in terms of the operator product expansion coefficients $\mathbf{C}_{123}$. What is needed is to find a sensible way to select a truncated space. For simplicity, the perturbing operator ϕ is assumed to be scalar.

---

[2] Similarly, for the energy-energy correlation length one has $m_\varepsilon = h^{8/15}(2\pi\mu^{-8/15} + \pi^{-1}A^2\mu^{8/15} + \mathcal{O}(\mu^{22/15}))$.

The conformal operators appearing in a model are classified as follows. Either they are *quasiprimary operators*, that is, its heighest weight state $|\Delta\rangle$, generated by $|\Delta\rangle = \phi(0)|0\rangle$, is annihilated by L_{-1}

$$L_{-1}|\Delta\rangle = 0 \qquad (13.49)$$

but it may or may not be annihilated by L_{-2}. Alternatively, they can be written in terms of derivatives of quasiprimary operators. Matrix elements of derivatives can be reduced to matrix elements of quasiprimary operators via

$$\langle\Delta_1|L_1^k\phi(0,0)L_{-1}^n|\Delta_3\rangle = n!k!$$
$$\times \sum_\ell \frac{(\Delta_1 + \Delta_3 - \Delta)_\ell(\Delta + \Delta_3 - \Delta_1)_{n-\ell}(\Delta + \Delta_1 - \Delta_3)_{k-\ell}}{\ell!(n-\ell)!(k-\ell)!}\langle\Delta_1|\phi(0,0)|\Delta_3\rangle,$$
$$(13.50)$$

where Δ is the conformal weight of the scaling operator ϕ and $(a)_n := \Gamma(a + n)/\Gamma(a)$ is the Pochhammer symbol. The matrix elements of the primary operators among themselves are given by (13.9). For the calculation of the matrix elements of V, both left and right conformal algebras as generated by the L_n and the $\bar{L}_n$ must be taken into account. The Hilbert subspace to be considered is taken to include the primary operators appearing in a given sector of the model, if the perturbation respects the global symmetry. Otherwise, all sectors must be considered simultaneously. One then adds the quasiprimary operators and derivatives up to a certain level, which in most applications so far has been the level 5, which leads to matrices of size of about 200×200 at most. The matrix elements of V are then calculated and V is diagonalized. A Mathematica^{TM} program to do these computations is available [423].

To illustrate the calculations, consider the example of the Yang–Lee singularity. In Chap. 12 it was shown that at the critical point, the model contains as primary operators the identity $\mathbf{1}$ and a single relevant operator φ with $\Delta = -1/5$. The only non-trivial OPE coefficient is

$$C_{\varphi\varphi\varphi} = i\kappa \ , \quad \kappa := \frac{1}{5}\left(\gamma\left(\frac{1}{5}\right)\right)^{3/2}\gamma\left(\frac{2}{5}\right) = 1.91131\ldots, \qquad (13.51)$$

where $\gamma(x) = \Gamma(x)/\Gamma(1-x)$. Calculating the elements of V, it is enough to consider the scaling operator $\varphi(0,0)$ because of translation invariance. The interaction matrix has the form

$$\langle\Delta_1|V|\Delta_3\rangle = -\frac{2\pi}{N}G\,B_{1,3}, \qquad (13.52)$$

where $G = (2\pi)^{-7/5}hN^{12/5}$ is the dimensionless coupling and corresponds to the finite-size scaling variable $\bar{a}z$ in (13.37), where h is the imaginary magnetic field coupled to φ and N the width of the strip. Since G is kept fixed in these calculations, one is *automatically inside the finite-size scaling region*.

Table 13.2. Quasiprimary states up to level four for the Yang–Lee singularity [634]

State	Norm
φ	1
$(L_{-4} - \frac{625}{624}L^4_{-1})\varphi$	57/130
1	1
$T = L_{-2}\mathbf{1}$	−11/5

The symmetric matrix $\hat{B}$ is N-independent. In Table 13.2, the quasiprimary states are listed together with their norms, up to level 4. The matrix $\hat{B}$ between these four states is

$$\hat{B} = \begin{pmatrix} \kappa & \kappa/130 & -i & i/5 \\ \kappa/130 & 1083\kappa/8450 & 0 & -57i/130 \\ -i & 0 & 0 & 0 \\ i/5 & -57i/130 & 0 & 0 \end{pmatrix}. \tag{13.53}$$

This information is enough to calculate any matrix element of V for any state up to level 5. For illustration, we merely give here the results up to level 3 for scalar states. To shorten notation, we write $\partial = L_{-1}$ and $T = L_{-2}\mathbf{1}$. For the complete table of all non-vanishing matrix elements and eigenvalue calculations, see [634]. The scalar states up to level 3 are given with their scaling dimensions in Table 13.3 and the symmetric matrix $\hat{B}$ is then, taking into account both the left and the right conformal operatorss

$$\hat{B} = \begin{pmatrix} \kappa & \kappa/10 & 4\kappa/75 & i & i/55 \\ & 4\kappa/25 & 49\kappa/750 & 2i/5 & 72i/275 \\ & & 289\kappa/625 & 3i/25 & 588i/1375 \\ & & & 0 & 0 \\ & & & & 0 \end{pmatrix}, \tag{13.54}$$

where we wrote only the upper part explicitly, since the lower part is given by the symmetry of $\hat{B}$.

This technique does provide accurate numerical values for the eigenvalues of H as a function of the finite-size scaling variable G, in particular for

Table 13.3. Normalized scalar states and their scaling dimensions for the truncated space up to level 3 (after [634])

number	State	scaling dimension
1	φ	−2/5
2	$\frac{5}{2}\partial\bar{\partial}\varphi$	8/5
3	$\frac{25}{12}\partial^2\bar{\partial}^2\varphi$	18/5
4	$\mathbf{1}$	0
5	$\frac{5}{11}T\bar{T}$	4

the lower ones, provided that the perturbing coupling G does not become too large. It is remarkable that this can be achieved by treating a fairly small matrix. This should be compared to the huge matrices required for the conventional finite-size scaling calculations. In a sense, this technique is a nice and efficient way to implement an "importance sampling" on quantum Hamiltonians, which is normally done by Monte Carlo methods and it is this property which made it very popular in the investigation of a given conformal field theory. On the other hand, the method requires quite some knowledge on the system it is to be applied to, in particular the conformal operator content must be known beforehand. This is not always immediately available for a given statistical lattice system. Further, there is at present no systematic theory on how the truncations affect the energy levels and the truncation normally done at level 5 at present is just empirical. In particular when two levels of the system become very close or cross, then in the context of the truncation method there will be a spurious level repulsion be generated. The first applications of the method were studies of models with large values of the coupling G, where its disadvantages become more apparent. We postpone until the next chapter a detailed numerical comparison with conventional finite-size scaling results.

Summarising, the truncation method provides sometimes a useful alternative to conventional numerical techniques.

Exercises

1. Work out the values of the correction-to-scaling exponent ω from the operator contents given in Chap. 12. Compare with (8.67).
2. Consider a 2D classical spin model on a square lattice, which breaks rotational invariance down to $\mathbb{D}_4$. This may be described through the quasi-primary operator $\Phi_4 := \left(L_{-2}^2 - \frac{3}{5}L_{-4} + \bar{L}_{-2}^2 - \frac{3}{5}\bar{L}_{-4}\right)\mathbf{1}$. Show that the leading correction to the two-point function of a scalar primary operator ϕ with conformal weight Δ is proportional to, with $z = re^{i\theta}$
$$\int \mathrm{d}^2 z_1 \, \langle \phi(z,\bar{z})\phi(0,0)\Phi_4(z_1,\bar{z}_1)\rangle \sim \frac{1}{r^{4\Delta}}\frac{\cos 4\theta}{r^2}.$$
3. Find the finite-size scaling function of the Ising model with the truncation method, e.g. [635].
4. Estimate numerically the OPE coefficients of the three-states Potts model. Compare with Table 7.1 and with [181].
5. Work out the details in the derivation of (3.60) [E40].
6. Consider a 1D quantum chain at *finite* temperature T and with low-energy dispersion relation $E = v_s k$. Show that for a scalar scaling operator ϕ with scaling dimension x the T-dependent Green's function is
$$G_T(r) = \left[(\pi T/v_s)^{-1}\sinh\left(r\pi T/v_s\right)\right]^{-2x}. \tag{13.55}$$

Hint: consider (13.1) and recall from Chaps. 1 and 3 the link between T and the strip width N. Explain the periodic boundary conditions.

14. The Vicinity of the Critical Point

Up to now, we have concentrated on describing the results of conformal invariance for the universal finite-size scaling functions Y and S_i of the singular free energy density and the correlation lengths

$$f^{\text{sin}} = L^{-2} Y(C_1 t L^{1/\nu}, C_2 h L^{(\beta+\gamma)/\nu})$$

$$\xi_i^{-1} = L^{-1} S_i(C_1 t L^{1/\nu}, C_2 h L^{(\beta+\gamma)/\nu}) \tag{14.1}$$

for the case of *small* values of the finite-size scaling variables $z = tL^{1/\nu}$ and $\mu = hL^{(\beta+\gamma)/\nu}$. We now ask the converse question on their behaviour for z, μ becoming large.

To be specific, take $z \to \infty$, but let $\mu = 0$. Generalizations are obvious. We then have

$$f^{\text{sin}} \sim t^{2\nu} C_1^{2\nu} \tilde{Y}(0) = At^{2-\alpha} \ , \quad \xi_i^{-1} \sim t^{\nu} C_1^{\nu} \tilde{S}_i(0) = B_i t^{\nu}, \tag{14.2}$$

where $2 - \alpha = 2\nu$ and

$$Y(z, 0) = z^{2\nu} \tilde{Y}(1/z) \ , \quad S(z, 0) = z^{\nu} \tilde{S}_i(1/z). \tag{14.3}$$

Although the *bulk amplitudes* A, B_i are non-universal[1] one readily identifies the universal amplitude ratios (see (3.32))

$$R_{i,j} := \frac{\tilde{S}_i(0)}{\tilde{S}_j(0)} = \left(\frac{\xi_i}{\xi_j}\right)^{-1} \ , \quad Q_i := \frac{\tilde{Y}(0)}{\tilde{S}_i^2(0)} = f^{\text{sin}} \xi_i^2. \tag{14.4}$$

What are the predictions of conformal invariance for these ?

To answer this, one has to consider systems with a Hamiltonian perturbed away[2] from the critical point

$$H = H_c + \lambda \int d^2 r \, \phi(r), \tag{14.5}$$

---

[1] They are independent of the boundary conditions however, at least if these do not change the global symmetry.

[2] In this context might occur applications to supercritical fluids as solvents for chemical and materials processing [E26].

where the critical point Hamiltonian H_c is described by a conformal field theory and ϕ is some perturbing scaling operator. If ϕ is relevant, the dimensionful coupling λ is related to a new, physical, length scale $\mathcal{R}$ in the problem (such that $\mathcal{R} \to \infty$ for $\lambda \to 0$). The perturbation thus describes the crossover from the ultraviolet (UV) fixed point described by H_c, which governs the system on scales $\ll \mathcal{R}$, and the new infrared (IR) fixed point, which governs the scales $\gg \mathcal{R}$. The IR fixed point may be trivial, which means that the system has become non-critical, or else describe a new scale-invariant critical point in a different universality class than the one started from.

14.1 The c-Theorem

This theorem, due to A. Zamolodchikov [640], asserts that, *under the assumptions of rotation invariance, positivity and energy–momentum conservation in a two-dimensional theory, there exists a quantity C which is non-increasing along a renormalization group trajectory, stationary at a RG fixed point and takes as its values at these fixed points the corresponding central charge* c. As a corollary, Cardy [141] obtained an expression for the Q_i of (14.4).

We begin with energy–momentum conservation,

$$\partial_\mu T^{\mu\nu} = 0, \tag{14.6}$$

where $T^{\mu\nu}$ is the energy–momentum tensor and rewrite this in complex coordinates $z, \bar{z}$ for the 2D case

$$\partial_{\bar{z}} T + \frac{1}{4} \partial_z \Theta = 0 \ , \quad \partial_z \bar{T} + \frac{1}{4} \partial_{\bar{z}} \Theta = 0, \tag{14.7}$$

where $\Theta := T^\mu_\mu$ vanishes at the critical point. At criticality, these relations were seen in Chap. 2 to imply analyticity for $T = T(z)$. This does no longer hold for the situation under consideration now and $T = T(z, \bar{z})$. Rotation invariance implies for the correlation functions of T and Θ

$$\langle T(z, \bar{z}) T(0, 0) \rangle = \frac{F(z\bar{z})}{z^4}$$

$$\langle \Theta(z, \bar{z}) T(0, 0) \rangle = \langle T(z, \bar{z}) \Theta(0, 0) \rangle = \frac{G(z\bar{z})}{z^3 \bar{z}}$$

$$\langle \Theta(z, \bar{z}) \Theta(0, 0) \rangle = \frac{H(z\bar{z})}{z^2 \bar{z}^2} \tag{14.8}$$

and similarly for $\bar{T}$. Combining the first of these with (14.7)

$$\langle \partial_{\bar{z}} T(z, \bar{z}) T(0, 0) \rangle + \frac{1}{4} \langle \partial_z \Theta(z, \bar{z}) T(0, 0) \rangle = 0. \tag{14.9}$$

We adopt the notation $\dot{F} := z\bar{z} F'(z\bar{z})$ where the prime stands for the derivative of $F(x)$ with respect to its argument. One finds

$$\dot{F} + \frac{1}{4}\dot{G} - \frac{3}{4}G = 0 \qquad (14.10)$$

and similarly by combining with Θ

$$\dot{G} - G + \frac{1}{4}\dot{H} - \frac{1}{2}H = 0. \qquad (14.11)$$

Eliminating G from (14.10,14.11), and defining

$$C := 2F - G - \frac{3}{8}H \qquad (14.12)$$

one finds [640]

$$\dot{C} = -\frac{3}{4}H \leq 0 \qquad (14.13)$$

due to positivity. This last condition is for example always satisfied for *unitary* models. C as defined is a non-increasing function of $R := (z\bar{z})^{1/2}$, for fixed values $\{g\}$ of the couplings (like temperature or magnetic field). The R-dependence is related to the dependence on the $\{g\}$ via the renormalization group (Callan–Symanzik equation)

$$\left(R\frac{\partial}{\partial R} + \sum_i \beta(\{g_i\})\frac{\partial}{\partial g_i} \right) C(R, \{g\}) = 0, \qquad (14.14)$$

where $\beta(\{g_i\})$ is the beta-function. Consequently, the quantity $C(\{g\}) := C(1, \{g\})$ is non-increasing along the RG trajectories. C is stationary if and only if $H = 0$, corresponding to a RG fixed point. At a fixed point, we have seen earlier from conformal invariance that $F = c/2$, $G = H = 0$ and consequently $C = c$ at a RG fixed point. This proves the stated c-**theorem**[3].

In particular, this implies that under the assumptions stated, the RG flow is always going "downhill" and is *irreversible*. Limit cycles and more complicated behaviour is ruled out.[4] This is not surprising in view of the usual real-space renormalization procedure. Each step of coarse-graining implies a "loss" of information on the ultraviolet behaviour of the theory. One may interpret C as some measure of this. Integrating (14.13), one finds the change in the central charge $\Delta c = c^{(\mathrm{UV})} - c^{(\mathrm{IR})}$ between two fixed points

$$\Delta c = \frac{3}{4\pi} \int_0^\infty \mathrm{d}^2 R\, R^2 \langle \Theta(R)\Theta(0)\rangle. \qquad (14.15)$$

Sometimes this formula is also referred to as c-**theorem**. Furthermore,

$$\Theta(z, \bar{z}) = -4\pi\lambda(1 - \Delta)\phi(z, \bar{z}), \qquad (14.16)$$

---

[3] One might contemplate the existence of other quantities with a similar behaviour, but this question goes beyond the scope of this book. See [455] for the generalization to logarithmic conformal theories.

[4] Non-monotonous flow may arise, however, if the assumptions of the c-theorem are not met. This occurs for example in disordered systems [152, 524, 569].

where ϕ is the perturbing relevant scaling operator with conformal weights (Δ, Δ) and λ the corresponding coupling. To see this, one notes first from the operator product expansion (2.51) by expanding around z

$$T(z)\phi(z_1, \bar{z}_1) = \frac{\Delta}{(z - z_1)^2}\phi(z, \bar{z}) + \frac{1 - \Delta}{z - z_1}\partial_z\phi(z, \bar{z}) + \cdots. \qquad (14.17)$$

To leading order in λ, $\langle T(z)\mathcal{X}\rangle \simeq \langle T(z)\mathcal{X}\rangle_* + \lambda \int d^2z_1 \langle T(z)\phi(z_1, \bar{z}_1)\mathcal{X}\rangle_*$, where $*$ refers to the fixed point averages. Through (14.17), this generates an effective dependence on $\bar{z}$ for T. However, the second term diverges when $z \simeq z_1$. One may regularize this by cutting out a small section $|z - z_1|^2 \le a^2$, where a is some microscopic length scale, e.g. a lattice constant. Then

$$\begin{aligned}
\partial_{\bar{z}}T &= \lambda \int d^2z_1 \frac{1 - \Delta}{z - z_1}\partial_z\phi(z, \bar{z}) \cdot (z - z_1)\,\delta(|z - z_1|^2 - a^2) \\
&= \lambda(1 - \Delta)2\pi\,\partial_z\phi(z, \bar{z})\int_0^\infty d\rho\,\rho\,\delta(\rho^2 - a^2) \\
&= \lambda(1 - \Delta)2\pi\,\partial_z\phi(z, \bar{z})\frac{1}{2a}a \qquad (14.18)
\end{aligned}$$

which is independent of a and δ was the Dirac delta function. Comparison with energy–momentum conservation (14.7) then yields the result (14.16).

On the other hand, one can define a correlation length $\tilde{\xi}_i$ via the second moment of the scaling operator ϕ_i

$$\tilde{\xi}_i^2 := \frac{\int d^2r\, r^2 \langle\phi_i(r)\phi_i(0)\rangle_c}{4\int d^2r\,\langle\phi_i(r)\phi_i(0)\rangle_c}. \qquad (14.19)$$

To simplify the notation, consider the example where $\phi_i = \varepsilon$, the energy density. Then by the fluctuation-dissipation theorem

$$\int d^2r\,\langle\varepsilon(r)\varepsilon(0)\rangle_c = -\frac{\partial^2 f^{\text{sin}}}{\partial t^2} \simeq -(2 - \alpha)(1 - \alpha)t^{-2}f^{\text{sin}}, \qquad (14.20)$$

where α is the specific heat critical exponent, which implies that [141]

$$f^{\text{sin}}\tilde{\xi}_\varepsilon^2 \simeq -[(2 - \alpha)(1 - \alpha)]^{-1}(t/2)^2\int d^2r\, r^2 \langle\varepsilon(r)\varepsilon(0)\rangle. \qquad (14.21)$$

Since a thermal perturbation leads to an off-critical system with $c = 0$, one has the following **sum rule**[5] expression for c

$$c = 6\pi^2 t^2 (2 - x_\varepsilon)^2 \int_0^\infty dR\, R^3 \langle\varepsilon(R)\varepsilon(0)\rangle. \qquad (14.22)$$

This expression relates the central charge, characteristic of the critical point in the 'ultraviolet' regime, to the two-point function on *all* length scales.

---

[5] The terminology comes from the analogy with results in strong-interaction particle physics.

Combining (14.21,14.22) yields for $\tilde{Q}_\varepsilon$ and similarly, considering the case $\phi_i = \sigma$, also for $\tilde{Q}_\sigma$ of (14.4) [141]

$$\tilde{Q}_\varepsilon = -\frac{c}{48\pi}\frac{2-\alpha}{1-\alpha} \ , \quad \tilde{Q}_\sigma = -\frac{c}{48\pi}\frac{2-\gamma}{1-\gamma} \ , \tag{14.23}$$

where γ is the susceptibility exponent. We reemphasize that $\tilde{Q}_\varepsilon$ applies to a thermal and $\tilde{Q}_\sigma$ to a magnetic perturbation. They are found from $\tilde{\xi}_i$ and *not* from ξ_i.

We stress again that the $\tilde{\xi}_i$ used here are only proportional to the correlation lengths ξ_i defined from the exponential decrease of the correlation functions. Their relationship is quite complicated. At the critical point, for example, when $\tilde{\xi}$ is defined by

$$4\tilde{\xi}^2\chi^{(2)} = \sum_{u,v} u^2\langle\phi(u,v)\phi(0,0)\rangle, \tag{14.24}$$

where $\chi^{(2)}$ is the corresponding susceptibility, then $\tilde{\xi} = \tilde{A}L$ where [518]

$$\tilde{A} = \frac{1}{4\pi}\left[\psi'\left(\frac{x}{2}\right) - \frac{\pi^2}{2\sin^2(\pi x/2)}\right]^{1/2} \tag{14.25}$$

and ψ is the digamma function [1] and x the scaling dimension of ϕ. Only in the limit $x \to 0$, one recovers $\tilde{\xi}/\xi \to 1/\sqrt{2}$, which one would also obtain if only the purely exponential large-distance behaviour of the two-point correlation function had to be considered.

Remarkably, a result similar to (14.15) also holds for the conformal weights. This is referred to as Δ-**theorem** [184]. To see this, let again ϕ be the scaling operator perturbing the system away from criticality and consider some other scaling operator Φ. One has

$$\langle T(z,\bar{z})\Phi(0,0)\rangle = \frac{F(z\bar{z})}{z^2} \ , \quad \langle\Theta(z,\bar{z})\Phi(0,0)\rangle_c = \frac{G(z\bar{z})}{z\bar{z}}. \tag{14.26}$$

As before, energy–momentum conservation (14.7) leads to

$$\dot{D} = G/4 \tag{14.27}$$

with the same notation as before and $D := F + G/4$. Because of (14.16), the short-distance behaviour of G is determined by the OPE expansion $\phi(r)\Phi(0)$, see (5.16), which leads for $r \to 0$ to

$$G(r) \simeq 2\pi\lambda(2-2\Delta)\mathcal{C}_{\phi\Phi A_0}|r|^{2\gamma_0}\langle A_0\rangle, \tag{14.28}$$

where A_0 is the most relevant (scalar) scaling operator, with conformal weight Δ_0, arising in the OPE $\phi \cdot \Phi$ and $\gamma_0 = \Delta_0 - \Delta_\Phi - \Delta + 1$. Now, if $\gamma_0 > 0$, G vanishes as $r \to 0$ and thus D is stationary and equal to F at the fixed point. In this case, $F(0) = \Delta_\Phi\langle\Phi\rangle$ can be deduced from the OPE of $T(z)$ and Φ. Integrating (14.27), one obtains [184]

$$\Delta_{\Phi}^{(\text{UV})} - \Delta_{\Phi}^{(\text{IR})} = -\frac{1}{4\pi\langle\Phi\rangle} \int \mathrm{d}^2 r \, \langle\Theta(r)\Phi(0)\rangle_c, \tag{14.29}$$

where the indices UV and IR refer to the values of Δ_{Φ} at the ultraviolet and infrared fixed points, respectively. For a trivial fixed point, of course $\Delta_{\Phi}^{(\text{IR})} = 0$. For $\gamma_0 \leq 0$, the full Ward identities from scale invariance must be employed and one finally ends up with a similar formula, see [184] for the details.

14.1.1 Application to Polymers

The above results can be fruitfully employed to derive non-trivial expressions for universal amplitudes. As an example, we review here an application to a gas of loops (not necessarily self-avoiding), following [149]. Consider a loop of (large, but finite) length ℓ and its radius of gyration $R_{G,\ell}$. Such loops may be self-avoiding loops or may arise as the boundaries between domains with a fixed value of a spin variable, e.g. in the Ising model.

One needs a generalization of the $O(n)$ model introduced in Chap. 1, where now the indices a run from 1 to $n + n'$

$$
\begin{aligned}
\mathcal{Z} &= \mathbf{Tr} \prod_{\langle i,i' \rangle} \left(1 + \lambda \sum_{a=1}^{n} S_i^a S_{i'}^a + \lambda' \sum_{a=n+1}^{n+n'} S_i^a S_{i'}^a \right) \\
&= \mathcal{Z}_{O(n)} \left(1 + n' \mathcal{A} \sum_{\ell} p_\ell \lambda'^\ell + \mathcal{O}\left(n'^2\right) \right)
\end{aligned}
\tag{14.30}
$$

which defines the coefficient p_ℓ, $\mathcal{A}$ is the total area of the lattice and $\mathcal{Z}_{O(n)}$ is the partition function already defined in Chap. 1. The model has a critical point at $\lambda = \lambda' = \lambda_c$. Switching on the perturbation $\lambda' - \lambda_c$ breaks the symmetry $O(n+n')$ down to $O(n) \otimes O(n')$ and the degrees of freedom with $n < a \leq n'$ are expected to become non-critical. Since $\sum_\ell p_\ell \lambda'^\ell$ is analogous to a free energy and, using hyperscaling, its singular part should scale with the correlation length $\xi'^{-2} \sim (\lambda' - \lambda_c)^{2\nu'}$, where $\nu' = 2/(2 - x')$ where x' is the scaling dimension of the perturbation to (14.30) generated by changing λ'. It follows that $p_\ell \sim \ell^{-1-2\nu'}$.

Turning now to R_G, one has to count all loops which pass through two chosen points $\boldsymbol{R}_{1,2}$. The energy density of the perturbation of the $O(n+n')$ model on the bond at i is $E_i' = \sum_{a=n+1}^{n+n'} S_i^a S_i^a$ and $E_{\boldsymbol{R}_1}' E_{\boldsymbol{R}_2}'$ has only contributions from the loops one is interested in. In the limit $n' \to 0$, only one loop remains and $\boldsymbol{R}_{1,2}$ will be on it. Thus $\sum_{\boldsymbol{R}_1,\boldsymbol{R}_2} (\boldsymbol{R}_1 - \boldsymbol{R}_2)^2 E_{\boldsymbol{R}_1}' E_{\boldsymbol{R}_2}'$, inserted into a correlator, gives R_G^2. Thus, averaging

$$
2n'\mathcal{A} \sum_\ell p_\ell \ell^2 \bar{R}_{G,\ell}^2 \lambda'^{\ell-2} = \sum_{\boldsymbol{R}_1,\boldsymbol{R}_2} \langle (\boldsymbol{R}_1 - \boldsymbol{R}_2)^2 E_{\boldsymbol{R}_1}' E_{\boldsymbol{R}_2}' \rangle \tag{14.31}
$$

$$
\sim \mathcal{A} \int \mathrm{d}^2 \boldsymbol{R} \, R^2 \langle E'(\boldsymbol{R}) E'(0) \rangle,
$$

where the last correlator is calculated in the *perturbed* $O(n+n')$ model.

This can be done by the c-theorem (14.15). The trace of the energy–momentum tensor is $\Theta = 2\pi\nu'^{-1}(\lambda' - \lambda_c)E'$. The UV theory is the $O(n+n')$ model, thus $\Delta c = c(n+n') - c(n) \simeq n'\frac{dc}{dn}$. Inserting into (14.31), one has

$$2\sum_\ell p_\ell \ell^2 \bar{R}^2_{G,\ell} \lambda'^{\ell-2} \simeq \left(\frac{2\pi(\lambda' - \lambda_c)}{\nu'}\right)^{-2} \frac{4\pi}{3}\frac{dc}{dn}. \tag{14.32}$$

Finally, one may extract from this the large ℓ behaviour $\bar{R}^2_{G,\ell} \sim \ell^{2\nu'}$ [149].

We mention that the mean area $\bar{A}_\ell$ of loops of length ℓ has also been calculated. It turns out that the ratio [149]

$$\frac{\bar{A}_\ell}{\bar{R}^2_{G,\ell}} = \frac{\pi}{2}\frac{1 + 2\rho^2}{1 + \rho^2} \tag{14.33}$$

is a universal constant, where ρ is the compactification radius of the free Gaussian model. Equation (14.33) describes the compactness of loops. If all loops are circular, the above ratio would be π. Some other values are [148, 149, 150]

$$\frac{\bar{A}_\ell}{\bar{R}^2_{G,\ell}} = \begin{cases} 7\pi/10 & \text{Ising clusters, } T > T_c & ; \rho^2 = 2/3 \\ 11\pi/14 & \text{Ising clusters, } T = T_c & ; \rho^2 = 4/3 \\ 4\pi/5 & \text{single self-avoiding loop} & ; \rho^2 = 3/2 \end{cases}. \tag{14.34}$$

The last case has been checked through exact enumerations for finite ℓ, leading to $2.515\ldots$, in agreement with (14.34) [148].

14.2 Conserved Currents Close to Criticality

Having found some of the Q_i, we now turn to the R_{ij}. Their calculation is done using a theory developed by A. Zamolodchikov [643, 644]. The perturbed Hamiltonian is

$$H = H_c + \lambda \int d^2r\, \phi(r), \tag{14.35}$$

where ϕ is a scalar relevant operator with $x = 2\Delta$. The coupling λ has the scaling dimension $2 - x > 0$. Consider the conformal tower of the identity generated by the left Virasoro generators L_{-n}. Denote the operators in this tower by T_s^α where s refers to the spin and α is a label for further quantum numbers which might be needed. In particular, one has $T_2 = T$. In view of the energy–momentum conservation (14.7), one forms the derivative

$$\partial_{\bar{z}}T_s^\alpha = \lambda R_{s-1}^{(\alpha,1)} + \ldots + \lambda^n R_{s-1}^{(\alpha,n)} + \ldots, \tag{14.36}$$

where the $R_{s-1}^{(\alpha,n)}$ are constructed from the conformal operators of the model described by H_c. The left side has conformal weights $(s,1)$. The conformal weights of the $R_{s-1}^{(\alpha,n)}$ are $(s - n(1 - \Delta), 1 - n(1 - \Delta))$. They become *negative* if n is large enough. This implies that the perturbation series above *must* terminate after a *finite* number of terms, up to irrelevant operators, at least when a unitary, or more generally a minimal model, describes the critical Hamiltonian H_c. In fact, for almost always just the first term survives

$$\partial_{\bar{z}} T_s^\alpha = \lambda R_{s-1}^{(\alpha,1)} \tag{14.37}$$

one of the few exceptions being the Ising model with a thermal perturbation. The explicit calculation of $R_s^{(\alpha,1)}$ does not pose any problem. As shown by Zamolodchikov, it follows [643] from the Ward identities that there is a set of operators D_n such that

$$\partial_{\bar{z}} = -\pi \lambda D_0$$
$$[L_n, D_m] = -((1 - \Delta)(n + 1) + m) D_{n+m}$$
$$D_{-n-1}\mathbf{1} = \frac{1}{n!} L_{-1}^n \phi(z, \bar{z}). \tag{14.38}$$

For example

$$\partial_{\bar{z}} T = -\pi \lambda D_0 L_{-2}\mathbf{1} = -\pi \lambda (\Delta - 1) D_{-2}\mathbf{1} = -\pi \lambda (\Delta - 1) L_{-1}\phi \tag{14.39}$$

and since $L_{-1} = \partial_z$ one has indeed (14.16). In the case $\Delta = 1$, corresponding to *marginal* perturbations, all orders in λ will contribute, although the lowest order contribution vanishes. Now, consider $T_4 =: T^2 := L_{-2}^2 \mathbf{1}$

$$\partial_{\bar{z}} T_4 = -\pi \lambda D_0 L_{-2} L_{-2}\mathbf{1}$$
$$= -\pi \lambda (\Delta - 1) (D_{-2} L_{-2} + L_{-2} D_{-2})\mathbf{1}$$
$$= -\pi \lambda (\Delta - 1) \left(2 L_{-2} L_{-1} + \frac{\Delta - 3}{6} L_{-1}^3\right) \phi \tag{14.40}$$

but since the right-hand side cannot be rewritten as a derivative, this does not imply the conservation of anything. To make further progress, one can take the perturbing operator to be one of the operators $\phi_{r,s}$ occurring in the Kac table (10.28). Recall from Chap. 4 the null vector relations

$$\left(L_{-2} - \frac{3}{2(2\Delta + 1)} L_{-1}^2\right) \phi_{1,2} = 0$$

$$\left(L_{-3} - \frac{2}{\Delta + 2} L_{-1} L_{-2} + \frac{1}{(\Delta + 1)(\Delta + 2)} L_{-1}^3\right) \phi_{1,3} = 0$$

$$\left(L_{-4} - \frac{4\Delta}{9} L_{-2}^2 - \frac{4\Delta + 15}{6\Delta + 18} L_{-1} L_{-3}\right. \tag{14.41}$$

$$\left. + \frac{2\Delta + 3}{3\Delta + 9} L_{-1}^2 L_{-2} - \frac{1}{4\Delta + 12} L_{-1}^4\right) \phi_{2,2} = 0$$

and in these equations, $\Delta = \Delta_{r,s}$. In fact, for those operators at the boundary of the Kac table, a relatively compact expression for the null vector can be given [71, 63, 396]

$$\left(\sum_{p_1+\ldots+p_k=s} \frac{[(s-1)!]^2(-\tilde{t})^{s-k}}{\prod_{i=1}^{k-1}[(p_1+\ldots+p_i)(s-p_1-\ldots-p_i)]} L_{-p_1}\cdots L_{-p_k} \right) \phi_{1,s} = 0 \tag{14.42}$$

with all $p_i \geq 1$ and where $\tilde{t}$ is related to the conformal weight $\Delta_{1,s}$ through

$$\tilde{t} = \frac{2}{s+1}\left(\frac{2\Delta_{1,s}}{s-1}+1\right). \tag{14.43}$$

An analogous condition is satisfied by the $\phi_{r,1}$ and see [63, G18] for the general case.

If one takes $\phi_{1,3}$ in (14.40), one finds indeed a conservation law $\partial_{\bar{z}}T_4 = \partial_z \Theta_2$ with

$$\Theta_2 = \frac{\pi\lambda(1-\Delta)}{\Delta+2}\left[2\Delta L_{-2} + \frac{(\Delta-2)(\Delta-1)(\Delta+3)}{6(\Delta+1)}L_{-1}^2\right]\phi_{1,3}. \tag{14.44}$$

Thus nontrivial conservation laws hold if only the perturbation of the critical Hamiltonian H_c is suitably chosen! To study this more systematically, one now looks for conservation laws of the form

$$\partial_{\bar{z}}T_s = \partial_z \Theta_{s-2}, \tag{14.45}$$

where T_s is in the conformal tower of the identity and Θ_s in the conformal tower of the perturbing operator ϕ. The existence of a conservation law is equivalent to that the linear operator $\partial_{\bar{z}}$, acting between these two conformal towers has a non-vanishing kernel, up to derivative fields. In principle, one can calculate its matrix elements in the way indicated, but calculations rapidly become tedious. Very often, however, the first few conserved quantities of this type can be found from a simple **counting criterion** [643], which goes as follows.

Define the numbers $\hat{\Lambda}_s$ and $\hat{\Phi}_s$ by the generating functions

$$\sum_{s=0}^{\infty} q^s \hat{\Lambda}_s = (1-q)q^{c/24}\chi_{1,1}(q) + q$$

$$q^{\Delta_{r,s}}\sum_{s=0}^{\infty} q^s \hat{\Phi}_s = (1-q)q^{c/24}\chi_{r,s}(q), \tag{14.46}$$

where $\chi_{1,1}(q)$ and $\chi_{r,s}(q)$ are the Virasoro character functions (see (10.30)) of the identity operator $\phi_{1,1}$ and the relevant perturbing operator $\phi = \phi_{r,s}$ of conformal weight $\Delta_{r,s}$. *If one has for some s that*

$$\boxed{\hat{\Lambda}_{s+1} \geq \hat{\Phi}_s + 1}, \tag{14.47}$$

then there does exist at least one conservation law of the above form. This is seen as follows. The $\hat{\Lambda}_s$ and $\hat{\Phi}_s$ defined above are just the dimensions of the spaces of quasi-primary fields constructed in the conformal towers of either the identity or the perturbing field ϕ at the level s. The operator $\partial_{\bar{z}}$ is a linear operator acting between them. Thus, if the condition (14.47) is satisfied, it must have a non-vanishing kernel. This is equivalent to the existence of a conservation law of the type (14.45) [643].

As an example, consider the Ising model perturbed by a magnetic field, corresponding to $\phi_{1,2}$. The characters are taken from Table 10.2. In Table 14.1, we illustrate the application of A. Zamolodchikov's counting criterion (14.47). One recognizes that T_s is indeed conserved if [643]

$$s = 1, 7, 11, 13, 17, 19. \tag{14.48}$$

One notes that all spins appearing in the conservation laws are odd. This is a consequence of the conservation of charge conjugation in a two-dimensional system. In general, a 2D conserved quantity with odd spin has a charge conjugation eigenvalue $C = +1$, while for even spin, it has $C = -1$. Since the Ising model does not have a non-trivial charge conjugation, all its states must be considered to have $C = +1$ [637].

Table 14.1. Counting criterion for the Ising model perturbed with $\phi_{1,2}$ for odd spin. A $*$ labels a conservation law

s	1^*	3	5	7^*	9	11^*	13^*	15	17^*	19^*	21
$\hat{\Lambda}_{s+1}$	1	1	1	2	2	3	4	5	7	9	11
$\hat{\Phi}_s$	0	1	1	1	2	2	3	5	6	8	12

Similarly, the counting criterion may be used to show the existence[6] of conserved quantities T_s with the following spins [643]

$$\text{for } \phi_{1,3}: \quad s = 1, 3, 5, 7 \tag{14.49}$$
$$\text{for } \phi_{1,2}, \phi_{2,1}: \quad s = 1, 5, 7, 11 \ ; \ m \geq 5$$

while the cases with $m < 5$ must considered separately. The conservation law for $s = 1$ is always present, since it corresponds to the conservation of energy–momentum. To find any further conserved quantities with $s \neq 1$ is highly nontrivial. We leave it as an exercise to find the first conserved T_s for $m = 3, 4$. It is generally believed that these first few conserved T_s are just the first members in an infinite series of conservation laws. If true, this would be strong evidence for the complete **integrability** of these non-critical systems. We stress that the fact that having the central charge $c < 1$ is essential in finding these conservation laws.[7] Also, non-trivial integrals of motion have only been found for the perturbations $\phi_{1,s}$ and $\phi_{r,1}$.

---

[6] Very roughly, the T_s may be identified with the powers $:T^s:$.

[7] For $c \geq 1$, a simple counting criterion can only work if some extended conformal symmetry, such as superconformal invariance, is realized.

14.3 Exact S-Matrix Approach

What is the meaning of the extra conservation laws for the calculation of quantities like the amplitude ratios R_{ij} in (14.4)? To see this, some information about factorizable scattering and integrable systems is needed. We shall do this in an as elementary way as possible and refer to the specialised literature for details. This field owes a lot to A. and Al. Zamolodchikov [637, 643].

To be specific, suppose that the perturbation of the critical theory is such that *all* correlation lengths are finite. Provided that a continuum description still applies and that in the continuum limit rotation invariance holds, any correlation function will decay exponentially fast. The underlying continuum theory will be a relativistic massive quantum field theory. Therefore, the inverse correlation lengths $(1/\xi_j)$ can be interpreted as the masses m_j of the relativistic particles. The short-distance (ultraviolet) behaviour of the underlying massive field theory is given by the critical conformal theory, because the masses of the particles become negligible in this limit. In the long-distance (infrared) regime, all particles are far away from each other and essentially free. A physical state is characterized by their energies and momenta, which are referred to as *on-shell asymptotic particle momenta*[8] $\boldsymbol{p}(j) = (p^0(j), p^1(j))$, with $(p^0(j))^2 - (p^1(j))^2 = m_j^2$. In scattering theory, the main object of interest is the scattering amplitude, which relates *incoming* asymptotic particles to *outgoing* ones. It is defined as

$$|\boldsymbol{p}_1(j_1), \ldots, \boldsymbol{p}_n(j_n)\rangle_{\text{in}}$$
$$= \mathcal{S}(\boldsymbol{p}_1(j_1), \ldots, \boldsymbol{p}_n(j_n)\,;\, \boldsymbol{p}'_1(j'_1), \ldots, \boldsymbol{p}'_m(j'_m))\,|\boldsymbol{p}'_1(j'_1), \ldots, \boldsymbol{p}'_m(j'_m)\rangle_{\text{out}}.$$
$$(14.50)$$

The function $\mathcal{S}$ of the momenta is called the S-**matrix**. The profound and remarkable relationship between off-critical conservation laws, established in perturbed conformal theories, and the infrared limit of the underlying field theory was first realized by A. Zamolodchikov [643]. This section intends to present a brief overview of this beautiful theory. We shall only do so to the extent needed for understanding the applications to be presented in the next sections and refer to the rich literature for more detailed and more precise information, e.g. [470]. The presentation given here follows [169].

The consequences of the existence of extra conserved quantities, besides the energy–momentum tensor, in the context of *integrable quantum field theories* in $(1+1)$ dimensions are well known [637]. Although in principle a single new conservation law is sufficient, in practise one usually encounters infinitely many of them. That implies that *only elastic scattering can take place in an integrable quantum field theory. The scattering process only exchanges the momenta and the internal quantum numbers. The scattering matrix $\mathcal{S}$*

---

[8] In this section, $\boldsymbol{p}(j)$ denotes the relativistic four-momentum of the particle j, specialised to $(1+1)$D.

factorizes into the product of the $n(n-1)/2$ possible two-particle scattering matrices [348].[9]

We now reformulate this in a more practical way. Recall the definition of the **rapidity** variable θ

$$p^0(j) = m_j \cosh\theta_j \ , \ \ p^1(j) = m_j \sinh\theta_j \tag{14.51}$$

which is useful to parametrize the particle four-momenta in a simple way. Lorentz invariance implies that the scattering matrix only depends on the rapidity differences $\theta_{ij} := \theta_i - \theta_j$. For a collision process between two particles i and j with masses $m_{i,j}$, the **Mandelstam variable**

$$s(i,j) := (\boldsymbol{p}(i) + \boldsymbol{p}(j))^2 = m_i^2 + m_j^2 + 2m_i m_j \cosh\theta_{ij} \tag{14.52}$$

is often used. One is interested in the analyticity properties of the S-matrix as a function of s, where of course real values of s correspond to physical processes (referred as *physical sheet* of s). The physical sheet of the s-plane is mapped onto the strip $0 < \mathrm{Im}\,\theta < \pi$, see Fig. 14.1. The two-particle scattering

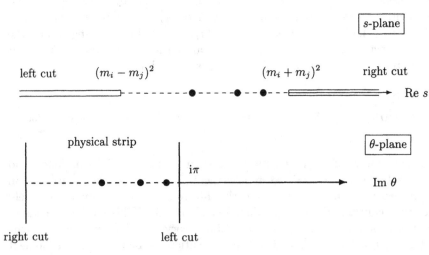

Fig. 14.1. Mapping of the s-plane to the θ-plane. The positions of the cuts are shown and the bound states are indicated as dots [169]

matrix $\mathcal{S}(s)$ has cuts along $s \geq (m_i + m_j)^2$ and $s \leq (m_i - m_j)^2$. These cuts come from the requirement of two-particle unitarity, see any text on quantum field theory. Furthermore, it can be shown that the two-particle scattering of

---

[9] To understand this in an elementary way, one may consider a collision process of identical particles, admitting only the conservation laws $\sum_i p_i^s = \sum_j p_j'^s$, of some powers s of the particle momenta. It is easy to show that a single additional conservation law with $s \neq 1$ is enough to forbid the processes 1 particle $\rightarrow$ 2 particles and 2 particles $\rightarrow$ 3 particles.

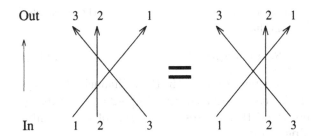

Out

In

Fig. 14.2. Factorizable scattering and the Yang–Baxter equation. The particles move up

an integrable theory is a pure elastic phase shift which defines the two-particle S-matrix $S(\theta)$

$$|\theta_1(i), \theta_2(j)\rangle_{\text{in}} = S_{ij}(\theta_{12}) |\theta_1(i), \theta_2(j)\rangle_{\text{out}}. \tag{14.53}$$

If all masses are distinct, the S-matrix is simply a scalar function of θ. If several masses should coincide, permutations of the momenta among the particles with the same masses but different internal quantum numbers are the only allowed scattering processes, because of elasticity. Here, we shall restrict ourselves to the simplest case of a non-degenerate mass spectrum and of particles which are their own antiparticles. Generalizations of this pose no problem.

The two-particle scattering matrix, which depends on the Mandelstam variable s, is related to $S(\theta)$ as follows by the relation

$$\mathcal{S}(s(i,j)) = 4m_i m_j \sinh\theta_{12}\, S_{ij}(\theta_{12}) \tag{14.54}$$

because of the Jacobian due to the overall energy–momentum δ-function because of Lorentz invariance. The remarkable property of integrable field theories is that the n-particle S-matrix simply factorizes into a product of the possible two-particle processes. In Fig. 14.2, this is illustrated for three particles $ij\ell$ (or 123) and one has

$$S_{ij\ell}(\theta_1, \theta_2, \theta_3) = S_{ij}(\theta_{12})\, S_{i\ell}(\theta_{13})\, S_{j\ell}(\theta_{23}) = S_{j\ell}(\theta_{23})\, S_{i\ell}(\theta_{13})\, S_{ij}(\theta_{12}). \tag{14.55}$$

Whenever two particle world lines cross, there is a scattering process completely given in terms of the corresponding two-particle S-matrix. In fact, factorization implies that the three-particle S-matrix may be written in two alternative ways, as sketched in Fig. 14.2. This gives a simple way to visualize the celebrated **Yang–Baxter equation**.[10]

Because of factorization, *unitarity* of the scattering process need be imposed only on the two-particle S-matrix

$$S_{ij}(\theta)\, S_{ij}(-\theta) = 1. \tag{14.56}$$

---

[10] Since in (14.55) the sequence of collisions is different, there may have been permutations of the rapidities and (14.55) is indeed non-trivial.

The transformation $\theta \to -\theta$ reverses the direction of motion of the particles. Because of parity invariance, one has

$$S_{ij}(\theta) = S_{ji}(\theta). \tag{14.57}$$

So far, we have discussed the scattering process in the s-channel. One may also use the t-channel, where the basic variable is $t := 2m_i^2 + 2m_j^2 - s$. Writing the S-matrix in the t-channel and then expressed in terms of the rapidity variable, implies the **crossing condition**

$$S_{ij}(\theta) = S_{ij}(i\pi - \theta). \tag{14.58}$$

It follows from these equations that $S_{ij}(\theta)$ is a $2\pi i$-periodic function of θ.

The conditions written down so far a very general and do not allow to determine the S-matrices explicitly. This can be done, however, through **bound-state structure** of the theory. Since particle production is disallowed, it is only there where the interactions can manifest themselves. The S-matrix is a meromorphic function in θ which may have regular poles in $0 \leq \operatorname{Im} \theta \leq i\pi$. The poles correspond to physical processes taking place in the intermediate region between the asymptotic regimes and at purely imaginary values of the rapidities. Such poles are shown as dots in Fig. 14.1. They are called **anomalous thresholds**. The *particles corresponding to these poles are stable and have also to be considered as asymptotic states*. Their existence leads to further conditions, fixing the underlying field theory to a large extent.

The kinematic constraints on the poles give rise to relations between the different masses of the theory. Let $S_{ij}(\theta)$ have a simple pole at[11] $\theta = iu_{ij}^k$ and let m_k be the mass of a particle k created by particles i and j with masses $m_{i,j}$ and the total square momentum of the system. Therefore

$$m_k^2 = m_i^2 + m_j^2 + 2m_i m_j \cos(u_{ij}^k). \tag{14.59}$$

Such a bound-state is diagrammatically illustrated in Fig. 14.3a. One has the identities

$$u_{ij}^k + u_{ik}^j + u_{jk}^i = 2\pi \ , \quad u_{ij}^k = u_{ji}^k. \tag{14.60}$$

One now has to localize these poles systematically. The residue of the S-matrix at a physical pole u in the s-channel is positive in a unitary theory. Such a pole is graphically represented in Fig. 14.3b. Because of the crossing condition, there is a second pole at $\bar{u}_{ij}^k := \pi - u_{ij}^k$, with negative residue. However, it is a physical pole in the t-channel. These poles are present any n-particle scattering which involves the particles m_i and m_j, in particular in the three-particle scattering sketched in Fig. 14.2. The pole structure of the two-particle S-matrices must be chosen such that it is consistent with the Yang–Baxter equation. This can be expressed in a simple way by the **bootstrap equation**

---

[11] No confusion should arise between the index i labeling a particle and the complex number i.

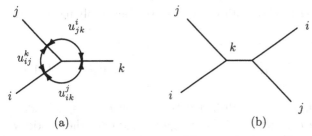

Fig. 14.3. Anomalous threshold (a) giving in the s-channel (b) of S_{ij} (after [169])

$$S_{k\ell}(\theta) = S_{i\ell}(\theta - i\bar{u}_{ik}^j)\, S_{j\ell}(\theta + i\bar{u}_{jk}^i) \qquad (14.61)$$

of the pole $\theta_{12} = iu_{ij}^k$ of the S-matrix (14.55). It is an easy exercise to check that the bootstrap equation is indeed consistent will all conditions introduced so far [348, 637].

This beautiful equation essentially fixes the S-matrices. More precisely, their pole structure inside the *physical strip* $0 < \mathrm{Im}\,\theta < \pi$ is completely determined. A S-matrix without further singularities is called a **minimal S-matrix**. The bootstrap allows to find the minimal S-matrices recursively. In general, there are infinitely many solutions of (14.61), since the knowledge of poles implies (14.61), but the converse is not true. The minimal S-matrix is the simplest solution to (14.61) with the proper bound-state structure.

Now, we have to combine the two structures discussed separately so far, namely the new conservation laws and the bound-state structure of the S-matrix and see what happens. The counting criterion gave in particular the spin s of the conserved charges $\mathcal{P}_s$:

$$\mathcal{P}_s = \oint d\bar{z}\, T_{s+1} + \oint dz\, \Theta_{s-1}. \qquad (14.62)$$

In general, there is a complex conjugate set of conserved charges $\overline{\mathcal{P}}_s$ with spin $-s$, generated by $\overline{T}$. The charges act on the asymptotic states and under a Lorentz transformation $\theta \to \theta + v$, they transform covariantly as[12]

$$\mathcal{P}_s \longrightarrow e^{sv}\mathcal{P}_s. \qquad (14.63)$$

Because of (14.63), when applying a conserved charge $\mathcal{P}_s$ to a single-particle state, one has

$$\mathcal{P}_s|\theta(j)\rangle = q(s,j)(m_j)^s e^{s\theta}\,|\theta(j)\rangle \qquad (14.64)$$

where the $q(s,j)$ are dimensionless coefficients, in principle to be determined but not always explicitly needed. Since the conserved charges, because of

---

[12] $p(j) := p^0(j) + p^1(j) = m_j e^{\theta}$ transforms like a spin-one tensor, to be identified with $\mathcal{P}_1$, since $T_2 = T$ and $\Theta_0 = \Theta/4$ are the components of the energy–momentum tensor. Similarly, $\bar{p}(j) = p^0(j) - p^1(j) = m_j e^{-\theta}$, related to $\overline{\mathcal{P}}_1$, has spin -1.

(14.62), are integrals over local densities, one expects when applying them to n-particle states

$$\mathcal{P}_s|\theta_1(j_1),\ldots,\theta_n(j_n)\rangle = \sum_{r=1}^{n} q(s,j_r)(m_{j_r})^s e^{s\theta_r} |\theta_1(j_1),\ldots,\theta_n(j_n)\rangle. \quad (14.65)$$

At this stage, one can formally establish that the existence of a single additional conserved charge $\mathcal{P}_s$, with $s > 1$, is enough to guarantee the elasticity of the scattering process. To see this, consider matrix elements of $\mathcal{P}_{1,s}$ between different asymptotic states. From $\mathcal{P}_1$ one recovers the conservation of the sum of the ingoing and outgoing momenta, familiar from translation invariance. However, from $\mathcal{P}_s$ one gets an additional relation between powers of the momenta. This is incompatible with particle production/annihilation and only allows permutations of the individual momenta.

We can now connect the conserved charges with the bootstrap equation (14.61) and the bound-state structure. Equation (14.61) suggests to *identify a bound-state as an asymptotic state, defined in the limit where a two-particle state has rapidity variables corresponding to a pole of the S-matrix*

$$\lim_{\epsilon \to 0} \epsilon \left| \left(\theta - i\bar{u}_{ik}^j + \frac{\epsilon}{2}\right)(i) \left(\theta + i\bar{u}_{jk}^i - \frac{\epsilon}{2}\right)(j) \right\rangle = |\theta(k)\rangle. \quad (14.66)$$

Acting with $\mathcal{P}_s$ on both sides of (14.66) leads to a system of linear equations for the parameters $q(s,j)$

$$q(s,i)(m_i)^s e^{-is\bar{u}_{ik}^j} + q(s,j)(m_j)^s e^{+is\bar{u}_{jk}^i} = q(s,k)(m_k)^s. \quad (14.67)$$

Some of the $q(s,i)$ may be zero for a given value of s. However, $\mathcal{P}_s = 0$ only if $q(s,i) = 0$ for all values i.

All this may appear to be quite abstract, but we now outline how the ratios

$$R_{ij} = \left(\frac{\xi_i}{\xi_j}\right)^{-1} = \frac{m_i}{m_j} \quad (14.68)$$

may be explicitly found. Along the way, one builds up the simplest possible form for the S-matrix. The procedure was developed by A. Zamolodchikov for the example of the Ising model in a magnetic field, but at $T = T_c$ [643]. From the counting criterion discussed in the previous section, the existence of conserved charges with spins $s = 1, 7, 11, 13, 19$ was established. Then a few heuristic assumptions are made. First, there should be several stable particles with masses $m_1 < m_2 < \ldots$. Second, it appears sensible that the lightest particle with mass m_1 should be fundamental and generates by "fusion" all other particles of the theory. For example, the two-particle S-matrix of two fundamental particles has a fundamental bound-state. From (14.59) with $m_i = m_j = m_k = m_1$, it follows that $u_{11}^1 = 2\pi/3$. Equation (14.67) then simplifies

$$e^{-i\frac{\pi}{3}s} + e^{+i\frac{\pi}{3}s} = 1. \quad (14.69)$$

This means that possible values for s must be odd and no multiples of three, which is the set $s = 1, 5, 7, 11, 13, 17, 19, \ldots$. The values of s obtained for the magnetic perturbation of the Ising model are contained in this set and to this point, the chosen procedure appears to be applicable to this model.

If the same two-particle scattering also has a particle with mass m_2 as a bound-state, one gets the equations

$$2m_1^2(1 + \cos(u_{11}^2)) = \left[2m_1 \cos\left(\frac{u_{11}^2}{2}\right)\right]^2 = m_2^2$$

$$q(s,1)(m_1)^s \left(e^{-is\bar{u}_{12}^1} + e^{is\bar{u}_{12}^1}\right) = q(s,2)(m_2)^s \quad (14.70)$$

considering $s = 1, 7, 11, 13, 19$ and (14.60). This simplifies to

$$2\cos\left(s\frac{u_{11}^2}{2}\right) = \left[2\cos\left(\frac{u_{11}^2}{2}\right)\right]^s \left(\frac{q(s,2)}{q(s,1)}\right). \quad (14.71)$$

Now, at some stage, yet one more hypothesis is needed to make the equations written down compatible with one and only one model. For the Ising model in a magnetic field, A. Zamolodchikov [643] found that the correct condition is to require that the two-particle process with two masses m_2 has a bound state with mass m_1. One finds similar equations as before, with the indices 1 and 2 interchanged

$$\frac{m_2}{m_1} = 2\cos\left(\frac{u_{11}^2}{2}\right) = \left[2\cos\left(\frac{u_{22}^1}{2}\right)\right]^{-1}$$

$$2\cos\left(s\frac{u_{22}^1}{2}\right) = \left[2\cos\left(\frac{u_{22}^1}{2}\right)\right]^s \left(\frac{q(s,1)}{q(s,2)}\right) \quad (14.72)$$

with the solution

$$u_{11}^2 = \frac{2\pi}{5} \quad, \quad u_{22}^1 = \frac{4\pi}{5} \quad (14.73)$$

leading to the mass ratio [643, 644]

$$R_{21} = \frac{m_2}{m_1} = 2\cos\left(\frac{\pi}{5}\right) = 1.6180339\ldots \quad (14.74)$$

Besides this remarkable result, A. Zamolodchikov determined the minimal solution to the bootstrap equation and found for example

$$S_{11}(\theta) = \frac{\tanh\left(\frac{\theta}{2} + i\frac{\pi}{3}\right)}{\tanh\left(\frac{\theta}{2} - i\frac{\pi}{3}\right)} \frac{\tanh\left(\frac{\theta}{2} + i\frac{\pi}{5}\right)}{\tanh\left(\frac{\theta}{2} - i\frac{\pi}{5}\right)} \frac{\tanh\left(\frac{\theta}{2} + i\frac{\pi}{30}\right)}{\tanh\left(\frac{\theta}{2} - i\frac{\pi}{30}\right)}. \quad (14.75)$$

Each of the three factors of S_{11} gives rise to a physical bound-state. The presence of hyperbolic functions reflects the $2\pi i$-periodicity of the S-matrix. The crossing condition (14.58) doubles the physical poles between 0 and $i\pi$ such that there is one with positive residue in the s-channel and another one with negative residue in the crossed t-channel. The unitarity condition (14.56) implies the existence of one zero in the S-matrix for each physical pole. Due to the negative $-\theta$ in (14.56), this zero is located on the imaginary segment $-i\pi \leq \mathrm{Im}\,\theta \leq 0$. These arguments are sufficient to explain the two hyperbolic tangents in each part of the solution (14.75). The first two parts on the r.h.s. of (14.75) correspond to the bound-states with masses m_1 and m_2, already discussed. The last part is a natural consequence of the particular bootstrap equation (14.61)

$$S_{11}(\theta) = S_{11}(\theta - i\pi/3)\, S_{11}(\theta + i\pi/3). \tag{14.76}$$

This enforces the presence of a further bound-state at $\theta = iu_{11}^3 = i\pi/15$ corresponding to another stable particle with mass

$$\frac{m_3}{m_1} = 2\cos\left(\frac{\pi}{30}\right) = 1.9890437\ldots. \tag{14.77}$$

The S-matrix S_{11} is minimal because all poles and zeroes have a specific meaning, necessary for the physical interpretation.

The bootstrap program can be performed step by step, from the lightest particle to the heaviest one, until this last one is reached.[13] This can be seen when the bootstrap program is closed, that is when all physical poles of the S-matrices are explained with the knowledge of the existing particles and when all possible bootstrap equations have been shown to be consistent. We shall not present all the steps, but simply quote [643], where it is shown that for the Ising model, at $T = T_{\mathrm{c}}$, but in a magnetic field, that is indeed the case. It turns out that in this model, *eight stable particles* must be introduced. *A posteriori*, this number may not be too surprising for a mathematical reason: the spins of the conserved current are

$$s = 1, 7, 11, 13, 17, 19, 23, 29 \bmod(30). \tag{14.78}$$

These numbers coincide with the *Coxeter exponents* of the Lie algebra E_8, and 30 is its *Coxeter number*, see Chap. 1. This example illustrates the observed relation between integrable perturbed systems and quantum integrable field theories such as the **affine Toda** field theories [215, 166, 167, 99, 339, 188, 470]. These describe an r-component real scalar field $\Phi^a, a = 1, \ldots r$ with Lagrangian

$$\mathcal{L} = \frac{1}{2}\partial_\mu \Phi^a \partial^\mu \Phi^a - \frac{g}{\beta^2}\sum_{i=1}^{r+1} n_i \left[e^{\beta\alpha_i \cdot \Phi} - 1 \right], \tag{14.79}$$

where the r-dimensional $(r + 1)$ vectors α_i are the simple roots of some **affine** simple Lie algebra $\hat{g}$ and r is the rank of the classical Lie algebra g. The integers $\{n_i\}$ verifies the identity $\sum_i n_i\alpha_i = 0$, and $\sum_i n_i = \hat{h}$ is the Coxeter number of $\hat{g}$. For example, the Ising model in a magnetic field

---

[13] There is a very compact notation which allows to make the calculations very quickly [99].

has the same mass-spectrum, the same bound-state structure, and the same conserved spins as the $\hat{E}_8$ affine Toda theory. However, the S-matrices of the Toda models differ from those of the Ising model by (non-minimal) so-called CDD-terms which take into account the coupling constant dependence of the Toda model. As $\beta \to 0$, the scattering in the Toda model must become trivial. In this limit, the theory reduces to a set of uncoupled free massive bosons, and the two-particle S-matrix simplifies to

$$S_{ij}(\theta) = \delta_{i,j}. \tag{14.80}$$

We refer to [99, 167, 339, 470] for further information.

For later references, we compile the complete list of mass ratios for the Ising model in a magnetic field [643, 644].

$$
\begin{aligned}
m_2/m_1 &= & 2\cos\pi/5 &= 1.6180339\ldots \\
m_3/m_1 &= & 2\cos\pi/30 &= 1.9890437\ldots \\
m_4/m_1 &= & 2m_2/m_1\cos 7\pi/30 &= 2.4048671\ldots \\
m_5/m_1 &= & 2m_2/m_1\cos 2\pi/15 &= 2.9562952\ldots \\
m_6/m_1 &= & 2m_2/m_1\cos\pi/30 &= 3.2183404\ldots \\
m_7/m_1 &= & 4m_2/m_1\cos\pi/5\cos 7\pi/30 &= 3.8911568\ldots \\
m_8/m_1 &= & 4m_2/m_1\cos\pi/5\cos 2\pi/15 &= 4.7833861\ldots
\end{aligned}
\tag{14.81}
$$

One can also show that these mass ratios do not depend on the fact that only the minimal S-matrix was taken. Explicit expressions for the masses, the three-point couplings, and the minimal factorizable S-matrices for all affine Toda theories are given in [99].

From a *physical* point of view, this result is highly unexpected. Previously, we had described the Ising model either by a spin variable or alternatively by a scalar ϕ^4 field theory. From none of these formulations is it obvious that there might be several local conserved quantities leading to stable, soliton-like excitations. These are completely absent if one considers the Ising model in a vanishing magnetic field and just changes the temperature. It would be desirable to have some intuitive understanding on the origin of these stable states.

The Ising model in a magnetic field has turned out to be one of the most complicated examples of integrable massive field theories related to perturbed conformal theories. Much simpler is the Ising model with a thermal perturbation. The energy-density operator, which couples to the temperature, introduces a single correlation length in the system. It is known that this off-critical system is, like the critical one, described by the scalar combinations of Majorana fermions, which are massive in this case. The mass is proportional to the inverse correlation length. Zamolodchikov's counting criterion shows the existence of conserved quantities for all odd spins. In this case, the integrability of the system has been understood a long time ago. The off-critical system is essentially free and therefore the S-matrix is free from any pole structure. However, the underlying fermionic structure of this scalar theory appears through the minus sign of the two-particle S-matrix given by [384, 385, 84]

$$S(\theta) = -1. \tag{14.82}$$

It is easy to verify that (14.82) is compatible with all constraints discussed in this section.

The procedure presented in this section, being potentially extremely powerful, has the disadvantage that the physical implications of the various assumptions which had to be made along the way are not obvious. When the theory was first presented, it was not even clear whether it was applicable to statistical mechanics models at all and it required precise numerical checks on the mass spectrum to remove this doubt. The reader should retain from this section the message that techniques coming from factorizable scattering may sometimes permit enormous progress not within the reach of different approaches. Rather than developing the abstract theory further, we prefer to review in the next section numerical tests performed on the predictions made so far on mass ratios the whole S-matrix in order to show to what extent the Zamolodchikov theory has been confirmed. Besides, there are new applications coming from this, some of which will be briefly described.

14.4 Phenomenological Consequences

14.4.1 Integrable Perturbations

The first check of the predictions for the R_{ij} given in the last section for the Ising model *exactly at* the critical temperature but with a magnetic field was made using the quantum Hamiltonian is [308]

$$H = -\frac{1}{\zeta} \sum_{n=1}^{N} (\sigma^z(n) + \sigma^x(n)\sigma^x(n+1) + h\sigma^x(n)) \tag{14.83}$$

with the conformal normalization $\zeta = 2$ and periodic boundary conditions. It is enough to restrict attention to the ratios $r_i := R_{i+1,1}$. In Fig. 14.4 we display the $r_i, i = 1, \ldots 5$ as functions of the finite-size scaling variable $\mu = hN^{15/8}$. The symbols refer to different values of the magnetic field h. We first note that the data do indeed collapse into a finite-size scaling form. Secondly, for the critical point $\mu = 0$, we see that the mass ratios do reproduce the values expected from critical point conformal invariance

$$r_1 = 8, r_2 = 17, r_3 = 24, r_4 = 32, r_5 = 33, \ldots \tag{14.84}$$

since $R_{ij} = x_i/x_j$ at $\mu = 0$. If μ increases, the r_i start to decrease until they pass through some minimum. This finding is at least qualitatively accounted for by conformal perturbation theory. Define the finite-size scaling functions G_i and the reduced scaling functions H_i for the inverse correlation lengths

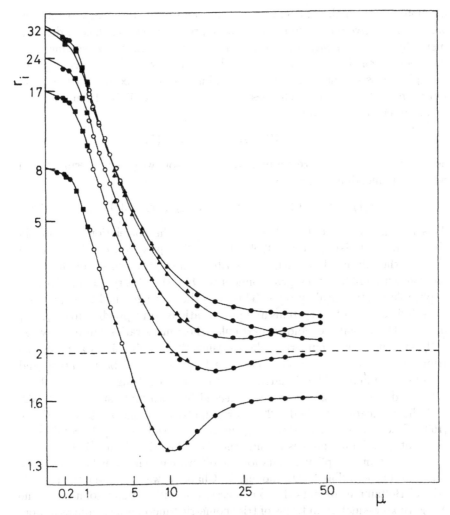

Fig. 14.4. Mass ratios $r_i = (\xi_{i+1}/\xi_1)^{-1}$ for the Ising model in a magnetic field as a function of μ [169]

$$\xi_i^{-1} = h^{1/y}G_i(\mu) \ , \quad G_i(\mu) = 2\pi x_i \mu^{-1/y} + H_i(\mu), \qquad (14.85)$$

where $y = 15/8$ for the Ising model with a magnetic field. The first term in G_i is nothing else than the conformal invariance prediction (3.49). For a symmetry-breaking field, the lowest-order term in H_i is of order $\mu^{2-1/y}$. For a magnetic perturbation, one usually has $y \simeq 2$. The minimum is thus produced by the combined effect of the leading singularity and the perturbative corrections. Finally, the r_i tend towards some limit value when $\mu \to \infty$. Already from the figure one recognizes agreement with the prediction (14.81). We see that two mass ratios are in the discrete spectrum, while above $m/m_1 = 2$ there is a continuum of states. This is generated by the multiparticle excitations in which the particles inside the composite state can have a non-vanishing relative momentum. For a more precise check, one has to deal numerically with the *double limit* that $\mu \to \infty$ after the finite-size scaling limit has been taken. In fact, computationally it has proved to be advantageous to replace this by the double limit of taking first $N \to \infty$ with h fixed and letting $h \to 0$ afterwards. The best numerical data available from standard quantum chains with up to $N = 21$ sites give [317]

$$r_1 = 1.6181(5) \ , \ \ r_2 = 1.994(5) \tag{14.86}$$

as obtained from finite-size scaling. For comparison, we give the results found with the truncation method [635]

$$r_1 = 1.61(1) \ , \ \ r_2 = 1.98(2) \ , \ \ r_3 = 2.43(4) \ , \ \ r_4 = 3.03(7). \tag{14.87}$$

This has also been confirmed in a study a of the correlation function in the isotropic 2D Ising model [429]. Much higher precision can be achieved through the diluted A_3 model, as described below. The success of this comparison with (14.81) does give confidence that the theory leading to it is applicable to statistical systems. This is further strengthened by similar results derivable for other systems and perturbations corresponding to $\phi_{1,2}, \phi_{2,1}$ or $\phi_{1,3}$. The S-matrices and in particular their mass ratios are by now exactly known. Precise numerical checks using either finite-size scaling or the truncation method have been done on a huge variety of both unitary and non-unitary models. The S-matrices for the tricritical Ising model have been calculated [166, 221, 284] and been checked by both finite-size scaling [262] and the truncation method [422]. For S-matrices of the Potts model, see [163]. External, symmetry-breaking fields were studied in [29]. Studies on non-unitary minimal models include the Yang–Lee singularity [142, 648].

More recently, explicit results for the off-critical behaviour have been obtained from the (dilute) ADE models. In Chap. 11, we have given the explicit critical Boltzmann weights. In some cases, the critical-point solution of the Yang–Baxter equation in terms of trigonometric functions also admits a more general solution in terms of elliptic functions, see [496] for a classification. In these cases, the elliptic nome plays the role of a coupling constant describing the deviation away from the critical point. If the elliptic nome vanishes, the explicit expressions for the Boltzmann weights of the ADE models and the diluted ADE models quoted in Chap. 11 are recovered.

Without entering into the details, we merely quote two results. First, for the simple A_m model, the elliptic nome plays the role of a temperature-like variable t. On can then define a class of order parameters and derive the scaling behaviour [32, 347]

$$M^{(k)} \sim t^{\beta_k} \ , \ \ \beta_k = \frac{(k+1)^2 - 1}{8m} \ , \ \ k = 1, 2, \ldots, m - 2. \tag{14.88}$$

In Chap. 1, we had argued that the A_3 model is in the 2D Ising universality class. This is in agreement with $\beta_1 = 1/8$ for $m = 3$. A remarkable difference, at least when m is odd, is found for the *dilute* A_m models (from now on, we only consider branch II, see Chap. 11, but the other branches are also treated in [608]). Explicit, but lengthy, expressions for the Boltzmann weights in terms of Jacobi elliptic functions are known, e.g. [540, 605, 283]. It turns out that in these cases, the elliptic nome plays the role of a *symmetry-breaking* magnetic field h. The scaling behaviour of the order parameters is, for m odd [608][14]

$$M^{(k)} \sim h^{1/\delta_k}, \ \delta_k = \frac{3m(m+2)}{(k+1)^2 - 1}, \ k = 1, 2, \ldots, m-2; \ \delta_{m-1} = \frac{3m(m+2)}{(m+2)^2 - 1}.$$

$$(14.89)$$

Consider the case $m = 3$. For the dilute model, the effective adjacency diagram is

$$\begin{array}{ccc} 1 & 2 & 3 \end{array}$$

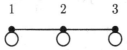

One can make the identification of the three states $\{1, 2, 3\}$ of the diluted A_3 model with the states $\{+, 0, -\}$ of a spin-1 Ising model, with the restriction that a '+' and a '−' spin must not be adjacent on the lattice [605, 608]. Because the elliptic nome breaks the global symmetry, it plays the role of the magnetic field. In other words, *the dilute A_3 model is an exactly solvable lattice model which is in the same universality class as the 2D Ising model at the critical point but in a magnetic field!* In this sense, the old problem of solving the 2D Ising model in a *finite* magnetic field has been completed,[15] although the original problem as specified in Chap. 1 will probably remain intractable. In particular, from (14.89) one can read off the Ising magnetic exponent $\delta = 15$ [605, 608], a result so far only obtained through scaling relations. In addition, the free energy per site $f = -\ln \mathcal{Z}/\mathcal{N}$ has been calculated, at $T = T_c$ [608]

$$f = -2 \sum_{k=-\infty}^{\infty} \int_{-\infty}^{\infty} dx \, \exp(32i\epsilon kx)$$
$$\times \frac{\cosh(9\pi x) \cosh(5\pi x) \sinh(16\pi u x) \sinh((15\pi - 16u)x)}{x \sinh(16\pi x) \cosh(15\pi x)}.$$

$$(14.90)$$

Here, the spectral parameter u controls the spatial anisotropy of the Boltzmann weights (the isotropic point is at $u = 15\pi/32$) and $e^{-\epsilon} \sim h$, the symmetry breaking field. In a similar fashion, explicit results for the symmetry-breaking order parameters can be derived and the scaling operator which drives the system away from the critical point can be identified in the Kac table. This can be done for general m [608].

---

[14] For branch I, $\delta_{m-1} = 3m(m+2)/[(m+3)^2 - 1]$.

[15] The thermodynamic Bethe ansatz for the dilute A_3 model can be recast in terms of the root system of the Lie algebra E_8, which allows to make the last link in the relation with the corresponding S-matrix [66].

From a phenomenological point of view, it is interesting to see how the numerical results obtained *before* the exact solution compare with it. The transfer matrix eigenvalues $\Lambda(u)$ for a system of size N with periodic boundary conditions can be found from the Bethe ansatz for the diluted A_m model [66, 283]

$$
\Lambda(u) = \omega^\ell \left[-\frac{\vartheta_1(u - 2\lambda)\vartheta_1(u - 3\lambda)}{\vartheta_1(2\lambda)\vartheta(3\lambda)} \right]^N \prod_{j=1}^N \frac{\vartheta_1(u - u_j + \lambda)}{\vartheta_1(u - u_j - \lambda)}
$$

$$
+ \left[-\frac{\vartheta_1(u)\vartheta_1(u - 3\lambda)}{\vartheta_1(2\lambda)\vartheta(3\lambda)} \right]^N \prod_{j=1}^N \frac{\vartheta_1(u - u_j)\vartheta_1(u - u_j - 3\lambda)}{\vartheta_1(u - u_j - \lambda)\vartheta_1(u - u_j - 2\lambda)}
$$

$$
+ \omega^{-\ell} \left[-\frac{\vartheta_1(u)\vartheta_1(u - \lambda)}{\vartheta_1(2\lambda)\vartheta_1(3\lambda)} \right]^N \prod_{j=1}^N \frac{\vartheta_1(u - u_j - 4\lambda)}{\vartheta_1(u - u_j - 2\lambda)}, \tag{14.91}
$$

where $\omega = \exp[i\pi/(m + 1)]$ and the u_j, $j = 1, \ldots, N$, are solutions of

$$
\omega^\ell \left[\frac{\vartheta_1(u_j - \lambda)}{\vartheta_1(u + \lambda)} \right]^N = -\prod_{k=1}^N \frac{\vartheta_1(u_j - u_k - 2\lambda)\vartheta_1(u_j - u_k + \lambda)}{\vartheta_1(u_j - u_k + 2\lambda)\vartheta_1(u_j - u_k - \lambda)} \tag{14.92}
$$

and $\ell = 1, \ldots, m$ labels the sectors of the model. The elliptic nome occuring in the theta functions ϑ_1 [1] has not been written explicitly. For the A_3 model, $\lambda = 5\pi/16$.

The advantage of this formulation of the problem is that for N sites, one has merely N coupled equations to solve, which *a priori* makes much larger system sizes available than in direct diagonalization approaches using the Lanczos algorithm, where $\mathcal{O}(2^N)$ coupled equations arise. On the other hand, the Bethe ansatz equations (14.92) are *non*linear.[16] Also, it is not immediately obvious which solutions of (14.92) will give the states with the lowest conformal weights at criticality or the lowest masses away from criticality. To find this relationship, one may first diagonalize the transfer matrix via the Lanczos algorithm for small systems (i.e. $N \leq 10$) and compare with the solutions of (14.92). Second, these solutions are extended to larger values of N and can be followed varying the nome [283].

For the critical point, this is checked in Table 14.2 and from the agreement found with the exactly known operator content[17] (see Chap. 10) it is clear that the solutions of the Bethe ansatz equations have been correctly identified. Extensive tables with the solutions u_j of (14.92) are provided in [283]. Mass ratios for off-critical case are shown in Table 14.3 for several values of of the scaling variable μ. This illustrates the convergence of the finite-size data for

---

[16] Solving (14.92) numerically, e.g. via the Newton method, may require the use of extended precision (16 bytes) arithmetics [283]. Equations (14.92) have also been solved analytically and the masses and universal amplitude ratios were derived [61].

[17] Only scalar states were considered.

Table 14.2. Finite-size approximants for central charge c and smallest eight scalar scaling dimensions x_j for the diluted A_3 model (after [283])

N	c	x_1	x_2	x_3	x_4	x_5	x_6	x_7	x_8
10	0.499 681	0.125 080	1.014 498	2.196	3.148	4.272	4.417	5.449	6.608
20	0.499 920	0.125 020	1.003 543	2.142	3.034	4.062	4.191	5.098	6.142
50	0.499 987	0.125 003	1.000 563	2.128	3.005	4.010	4.135	5.015	6.022
75	0.499 994	0.125 001	1.000 250	2.126	3.002	4.004	4.130	5.007	6.010
100	0.499 997	0.125 001	1.000 141	2.126	3.001	4.002	4.128	5.004	6.005
∞	$\frac{1}{2}$	$\frac{1}{8}$	1	$2\frac{1}{8}$	3	4	$4\frac{1}{8}$	5	6

Table 14.3. Mass ratios $R_i = m_{i+1}/m_1$ for three values of the scaling variable μ in the diluted A_3 model (after [283])

μ	N	R_1	R_2	R_3	R_4	R_5	R_6	R_7
	10	1.618 749	1.983 943	2.403 558	2.083 461	2.616 910	2.910 107	2.317 908
	20	1.618 037	1.983 923	2.403 649	2.081 017	2.617 335	2.930 942	2.320 478
10	50	1.618 024	1.983 977	2.403 797	2.079 804	2.617 468	2.934 096	2.314 580
	75	1.618 024	1.983 984	2.403 815	2.079 550	2.617 482	2.934 583	2.313 631
	100	1.618 025	1.983 990	2.403 834	2.079 551	2.617 497	2.934 617	2.313 523
	10	1.623 692	1.989 446	2.412 441	2.017 536	2.623 691	2.958 322	2.085 027
	20	1.618 144	1.987 688	2.405 009	2.025 316	2.618 144	2.954 218	2.129 306
20	50	1.618 034	1.987 597	2.404 866	2.026 212	2.618 034	2.954 289	2.132 554
	75	1.618 034	1.987 591	2.404 865	2.026 286	2.618 034	2.954 306	2.132 780
	100	1.618 034	1.987 589	2.404 865	2.026 315	2.618 034	2.954 311	2.132 871
	10	1.634 823	1.992 979	2.428 402	2.003 678	2.634 823	2.970 263	2.021 663
	20	1.618 424	1.988 671	2.405 376	2.012 146	2.618 424	2.956 329	2.071 838
30	50	1.618 036	1.988 497	2.404 869	2.013 483	2.618 036	2.956 009	2.078 248
	75	1.618 034	1.988 489	2.404 867	2.013 604	2.618 034	2.956 007	2.078 742
	100	1.618 034	1.988 486	2.404 867	2.013 644	2.618 034	2.956 007	2.078 950
∞	∞	1.618 034	1.989 044	2.404 867	2.000 000	2.618 034	2.956 295	2.000 000

μ fixed and also shows the approach towards the mass ratios (14.81) obtained from the S-matrix for $\mu \to \infty$. Comparison with (14.86,14.87) shows the huge increase in precision.

Although the considerations of this section are restricted to the models with central charge $c < 1$, there are special cases where the integrability exists even if $c \geq 1$. One example is the Ashkin–Teller model with a thermal perturbation which can be expressed by the integrable (1+1)D sine-Gordon model. Mass ratios computed this way were confirmed numerically [316]. Another example is given by a set of $\mathbb{Z}_p$-invariant systems [638] and the mass ratios have been calculated [26]. The XXZ Heisenberg chain perturbed by homogeneous and staggered magnetic field was investigated in [28, 484]. We do not attempt to provide an exhaustive review here and refer to [470] instead.

14.4.2 Universal Critical Amplitude Ratios

While all this is very impressive and beautiful, the practically-minded reader might wonder whether there is more to this than merely bright intellectual firework? To illustrate that this may be not so, we now quote from recent work [186, 187] which shows how the information contained in the S-matrix can be translated into specific predictions on statistical systems. Consider the following **critical amplitudes**, defined by [521]

$$C \simeq \frac{A_\pm}{\alpha} |t|^{-\alpha} \ , \quad M \simeq B(-t)^\beta \ , \quad \chi \simeq \Gamma_\pm |t|^{-\gamma} \ , \quad \tilde\xi \simeq \xi_0^\pm |t|^{-\nu} \qquad (14.93)$$

and the the $\pm$ refers to $t \gtrless 0$, respectively. This generalizes the definition of exponents given in Chap. 1. From scaling (including hyperscaling) it follows that, although the individual amplitudes contain metric factors and are not universal, there exist quite a few **universal amplitude ratios**, for example (in d dimensions)

$$\frac{A_+}{A_-} \ , \quad \frac{\Gamma_+}{\Gamma_-} \ , \quad \frac{\xi_0^+}{\xi_0^-} \ , \quad R_C := \frac{A_+ \Gamma_+}{B^2} \ , \quad R_\xi^+ := A_+^{1/d} \xi_0^+ . \qquad (14.94)$$

For the q-states Potts model, these quantities can be found from the S-matrix [187]. To see this, one first relates the physical observables to off-critical (connected) correlators

$$C = \int d^2r \, \langle 0|\varepsilon(r)\varepsilon(0)|0\rangle_c \ , \quad M = \langle 0|\sigma|0\rangle \qquad (14.95)$$

$$\chi = \int d^2r \, \langle 0|\sigma(r)\sigma(0)|0\rangle_c \ , \quad \tilde\xi^2 = \frac{1}{4} \frac{\int d^2r \, |r|^2 \langle 0|\sigma(r)\sigma(0)|0\rangle_c}{\int d^2r \langle 0|\sigma(r)\sigma(0)|0\rangle_c} .$$

Next, these correlators are found from the spectral representation, e.g.

$$\langle \phi_1(r)\phi_2(0)\rangle = \sum_{n=0}^{\infty} \int_{\theta_1 > ... > \theta_n} \frac{d\theta_1}{2\pi} \cdots \frac{d\theta_n}{2\pi} \langle 0|\phi_1(0)|n\rangle \langle n|\phi_2(0)|0\rangle \, e^{-|r|E_n},$$
$$(14.96)$$

where E_n is the energy of the n-particle state $|n\rangle$. The matrix elements arising in this expansion are called **form factors**. For integrable theories with factorizable S-matrices, unitarity and crossing symmetry relate the form factors to the S-matrix (form factors are explained in detail in a book by Smirnov [571]). Since it is in general not feasible to sum the entire series in (14.96), one often restricts one-self to merely the two-particle states $|2\rangle$. Empirically, this can be justified in some cases through the very rapid convergence of the spectral representation series. Another possibility is to use the c-theorem and the Δ-theorem since the central charges and the conformal weights of the critical (unperturbed) theory are known.

For the q-states Potts model perturbed with the energy density $\varepsilon = \phi_{2,1}$, the factorizable S-matrix has been explicitly constructed in [163]. Using this

Table 14.4. Universal amplitude ratios (14.94) for the q-states Potts model as determined from the two-particle form factors. Here $q = 1$ stands for percolation (after [187])

q	1	2	3	4
A_+/A_-	1	1	1	1
Γ_+/Γ_-	74.2*	37.699	13.848	4.013
R_C	3.0*	0.3183	0.1041	0.0204
ξ_0^+/ξ_0^-	3.76*	3.162	2.657	1.935
R_ξ^+	0.926	0.3989	0.3262	0.2052

as input, the two-particle form factors have been calculated for the $q = 1, 2, 3, 4$-states Potts model and the universal amplitudes determined [187]. To check the accuracy of the truncation to two-particle form factors, the central charge c and the scaling dimension x_σ of the magnetization were calculated and found to agree with the known exact values up to a few percent. The results for the amplitude ratios are displayed in Table 14.4.[18] The result $A_+/A_- = 1$ is expected from self-duality.[19] For the Ising case $q = 2$, the accuracy can be further checked by comparing with the exact results [G22, 627]

$$\frac{\Gamma_+}{\Gamma_-} = 37.69365\ldots \; , \quad R_C = 0.318569\ldots \; , \quad \frac{\xi_0^+}{\xi_0^-} = 3.16\ldots . \tag{14.97}$$

For percolation, lattice studies still disagree considerably on the correct values of some of these amplitude ratios, for example estimates for Γ_+/Γ_- are in the range $14 - 200$. We refer to [521] for an extensive compilation of theoretical and experimental results on universal critical amplitude ratios.

For the Ising model, also the magnetic perturbation is integrable and the same program can be carried out. This allows to determine the amplitude ratios involving a non-zero magnetic field [186]. The results agree well with the exact solution (when existing) and with numerical studies.

An alternative technique to extract the physical contents of the S-matrix is provided by the thermodynamic Bethe ansatz. This technique goes back to the work of Yang and Yang [630] and was related to perturbed conformal theories by Al. Zamolodchikov [647]. For example, one may check the

---

[18] For the $q \to 1$ (percolation) case, amplitudes must be defined with respect to the site or bond probability p and the $q \to 1$ limit must be carefully taken, see Chap. 1 and [187] for details. In addition, the formal solution of the functional equation fixing the form factors, while known for $q = 2, 3, 4$, has not yet been achieved for $q = 1$. In that case, values were obtained from an *ad hoc* extrapolation from the other q-values. The accuracy of this procedure was checked for $q = 2$, with an estimated error $\lesssim 10\%$. The amplitudes so determined are marked by a * in Table 14.4.

[19] See [521] for a collection of experimental results on A_+/A_- for several Ising antiferromagnets. For the CO/graphite reorientation transition, which is also in the 2D Ising universality class, $A_+/A_- = 1.03(3)$ [E95].

consistency of the S-matrix by recalculating the central charge which govern the ultraviolet regime. It turns out that the *minimal* S-matrix is required [647, 645, 649, 168, 452, 401, 169]. Recent applications of the technique for the calculation of the mass spectrum (and more) can be found in [66, 222, 461]. Of course, these results may also be recovered by solving the Bethe ansatz equations for the diluted A_m models. For example, for $m = 3, 4, 6$, one recovers the mass spectra of the affine E_6, E_7, E_8 theories [62].

14.4.3 Chiral Potts Model

In Chap. 1, the p-states chiral Potts model was defined and it was stated that in the so-called superintegrable case, the thermal exponents of the specific heat and the order parameters retain the values found for the *non*-chiral model and described by the $\mathbb{Z}_p$-invariant conformal field theories first studied in [638], but the correlation length exponents depend on the space direction, see [488, 346, 343, 256, 65, 16, 18, 35, 340] and references therein. A simple argument explaining this observation goes as follows [147].

For simplicity, we restrict to the three-states chiral Potts model. Consider the vector $\delta = (\delta_x, \delta_y)$ of couplings which, at least on the lattice, breaks rotation invariance. Its conjugate operator transforms covariantly under rotations of 90° and should therefore contain a scaling operator with spin 1 (higher spin operators are irrelevant). Form the operator content of the three-states Potts model (12.4), the corresponding relevant scaling operators must be $\varphi = (7/5, 2/5)$ and $\bar{\varphi} = (2/5, 7/5)$ such that φ couples to $\delta = \delta_x + i\delta_y$ and $\bar{\varphi}$ couples to $\bar{\delta} = \delta_x - i\delta_y$. It can be shown [147] that under a duality transformation, $\varphi \to \varphi$ and $\bar{\varphi} \to -\bar{\varphi}$. This is consistent with the OPE $\varphi \cdot \bar{\varphi} = \varepsilon$, since the energy density $\varepsilon \to -\varepsilon$ under duality. One is led to consider

$$H = H_c + \int d^2 r \left(\delta\varphi + \bar{\delta}\bar{\varphi} + \tau\varepsilon \right), \qquad (14.98)$$

where H_c describes the critical three-states Potts model and H is self-dual if $\bar{\delta} = \tau = 0$. The chiral Potts model is obtained by perturbing H_c with the spin-one operator φ.

Consider the renormalization group equation of the dimensionless coupling $g = \delta a^{1/5}$, where a is the lattice constant. The beta function can be found in terms of connected (lattice regularized) correlators of φ evaluated in the unperturbed theory, which has rotational symmetry. Thus [147]

$$\dot{g} = \frac{1}{5}g \ , \quad \dot{\tau} = \frac{6}{5}\tau + \mathcal{O}\left(\tau^3\right) \qquad (14.99)$$

since higher-order terms in g would not transform correctly under rotations. The second relation is found when considering a *scalar* perturbation of H, since the renormalization group equation cannot contain terms depending on g, again by rotational symmetry. It follows that there are only two fixed points, $g^* = 0, \infty$ and that the renormalization group eigenvalues of *all scalar* *scaling operators* are the same for both. This explains the independence of the thermal critical exponents on the chiral perturbation.

Now consider the strongly anisotropic scaling of the correlation functions. It is enough examine only the leading correction in δ. Since $\varphi = (7/5, 2/5)$ is primary, one has the operator product expansion (14.17)

$$T(z)\varphi(z_1, \bar{z}_1) = \frac{7/5}{(z - z_1)^2}\varphi(z, \bar{z}) + \frac{1 - 7/5}{z - z_1}\partial_z\varphi(z, \bar{z}) + \cdots. \tag{14.100}$$

Because of $\partial_{\bar{z}}(z - z_1)^{-1} = \frac{\pi}{2}\delta^{(2)}(z - z_1)$, one finds

$$\partial_{\bar{z}}T = \frac{\pi}{2}\left(1 - \frac{7}{5}\right)\delta\,\partial_z\varphi. \tag{14.101}$$

In analogy with the dimensional arguments described above for scalar perturbations, one sees that there are no higher-order terms present, in agreement with the renormalization group equation (14.99). Comparison with energy–momentum conservation $\partial_{\bar{z}}T + \partial_z T_{\bar{z}z} = 0$ yields (and similarly for $\overline{T}$) [147]

$$T_{\bar{z}z} = -\frac{\pi}{2}\left(1 - \frac{7}{5}\right)\delta\varphi \;, \quad T_{z\bar{z}} = -\frac{\pi}{2}\left(1 - \frac{2}{5}\right)\delta\varphi \tag{14.102}$$

and one has $T_{\bar{z}z} = -\frac{2}{3}T_{z\bar{z}} \neq T_{z\bar{z}}$. This shows that rotation invariance is broken. Rather, the system is invariant under the transformation

$$z \to \lambda^3 z \;, \quad \bar{z} \to \lambda^2 \bar{z} \tag{14.103}$$

which implies for the ratio of the correlation length exponents $\nu_\perp/\nu_\parallel = 2/3$.[20] Scaling arguments may be invoked to yield $\nu_{\parallel,\perp}$ separately.

All this is readily extended to the p-states chiral Potts model [147]. Conservation laws and the generalization of the Zamolodchikov counting criterion are discussed in [609].

14.4.4 Oriented Interacting Polymers

An interesting generalization of the polymer models discussed so far is provided by the interacting **oriented self-avoiding walk**. An example is shown in Fig. 14.5. As in Chap. 12, interactions are modeled in terms of contacts. Parallel and antiparallel contact are distinguished through different values of the couplings $\beta_{p,a}$. The generating function is

$$Z = \sum_{\text{OSAW}} \exp\left(k_p\beta_p + k_a\beta_a\right), \tag{14.104}$$

---

[20] As it stands, the principal axes with respect to which the anisotropic scaling takes places are the light-cone coordinates $r^\pm = r_\parallel \pm r_\perp$ and not $r_{\parallel,\perp}$, as found from the exact solution, but it can be shown that irrelevant operators may rotate these axes without affecting the thermal exponents [147].

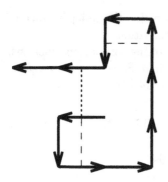

Fig. 14.5. Interacting oriented self-avoiding walk on a square lattice. The dashed lines represent antiparallel contacts and the dotted line a parallel one

where $k_{p,a}$ are the number of parallel and antiparallel contacts for a given oriented self-avoiding walk (OSAW) and the sum is over all those walks. As before, critical exponents $\gamma = \gamma(\beta_p, \beta_a)$ and ν can be defined. The values of these exponents have been calculated [151]. For linear oriented polymers, the result is that for $\beta_a = 0$, the exponent $\nu = 3/4$ is independent of β_p while, at least for $\beta_p \leq 0$

$$\gamma(\beta_p, 0) = \nu \left(\frac{43}{24} - 4\pi\Lambda(\beta_p) \right), \qquad (14.105)$$

where $\Lambda(\beta_p)$ is an undetermined continuous and monotonous function of β_p such that $\Lambda(0) = 0$. We shall first describe the argument leading to (14.105) and shall afterwards review attempts to verify this prediction numerically.

In general, exponents depending continuously on a parameter are only possible for central charges $c \geq 1$, at least for unitary models, as is evident from the explicit operator contents given in previous chapters. But it was realized that continuously varying exponents can also occur in *non-unitary* theories with $c < 1$, of which the O(2n)-model with $-1 \leq n \leq 1$ is an example [151]. A sufficient condition for continuously varying exponents is the presence of a conserved current, with complex components $J, \bar{J}$, which generates a U(1) symmetry of the theory. For a perturbation of the form

$$H = H_c + \Lambda \int \mathrm{d}^2 z \; :J\bar{J}:, \qquad (14.106)$$

Λ is a *marginal* scaling field.[21] The scaling dimensions x_Q of the scaling operators of the unperturbed theory with a U(1) charge Q then become [151]

$$x_Q(\Lambda) = x_Q(0) + \frac{2\pi\Lambda Q^2}{1 - 2\pi k\Lambda}, \qquad (14.107)$$

---

[21] Λ remains marginal to all orders. This is only true in 2D, for $d > 2$, Λ is irrelevant from naive dimensional counting. The same mechanism also accounts for the non-universality in the 2D XY and Ashkin–Teller models.

where k is the Kac–Moody central charge, to be read off from the correlator $\langle J(z)J(0)\rangle = k/z^2$.[22]

To understand this in a simple way, consider the Gaussian model, a free scalar field, compactified on the circle with action $S_G = g_0/(4\pi)\int d^2z\,(\partial\phi)^2$, see Chap. 7. This model has an U(1) invariance corresponding to $\phi \to \phi +$ cste., with the conserved current $J_\mu = 2ig_0\partial_\mu\phi$. The effect of the marginal perturbation on a two-point function of two vertex operators $V_Q(z) =:e^{iQ\phi}:$, see (7.21), can be worked out explicitly and it turns out that the ultraviolet divergencies of the relevant integrals can be absorbed into a multiplicative renormalization of the operators $V_{\pm Q}(z)$. This leads to the scaling dimension $x_Q = Q^2/(2g_0^2) + 2\pi\Lambda Q^2 + \mathcal{O}(\Lambda^2)$. Alternatively, one might have observed that the perturbation corresponds to a shift in the coupling constant [151]

$$g_0 \to g = g_0 - 4\pi\Lambda g_0^2. \tag{14.108}$$

Since in the Gaussian model $x_Q = Q^2/(2g)$, (14.107) is recovered. The same type of argument can now be applied to a general theory with a conserved current, by inserting the operator $J\bar{J}$ into the correlators. The corresponding integrals are completely fixed by the U(1) Ward identities and the OPE of the current J and it turns out that all correlations are identical to those of the Gaussian model, provided that the constant $k_G := 2g_0$ is replaced by the Kac–Moody central charge k.[23]

To see the relationship of this with polymers, consider

$$\mathcal{Z} = \mathbf{Tr} \prod_{\langle i,i'\rangle} (1 + \lambda\, \boldsymbol{S}_i \cdot \boldsymbol{S}_{i'}^* + \text{c.c.}), \tag{14.109}$$

where $\boldsymbol{S}_i$ is a *complex* n-component spin at site i, normalized such that $\mathbf{Tr}S_i^a S_{i'}^{b*} = \delta^{a,b}\delta_{i,i'}$. As in Chap. 1, expansion in λ shows that $\mathcal{Z}$ may be rewritten as a sum over configurations of closed self-avoiding loops and in the $n \to 0$ limit (assumed from now on), a single self-avoiding loop remains. As in Chap. 12, a source at which Q distinct walks originate is related to an operator $\phi_Q^{(a_1,...,a_Q)} = S^{a_1}...S^{a_Q}$ with U(1) charge Q and (12.108) is recovered. The exponent describing the number of star oriented polymers is then, again, $\gamma_Q = \nu(2 - x_Q - Qx_1)$, with $\gamma = \gamma_1$. Now, an orientation-dependent interaction energy is added, which for simplicity we take here to be of the form $\Lambda\sum_{\langle i,i'\rangle}\boldsymbol{J}_i\cdot\boldsymbol{J}_{i'}$. Performing a formal continuum limit, the OPE $\boldsymbol{J}_i\cdot\boldsymbol{J}_{i'} = $ singular term$+ : J\bar{J}: +\ldots$ is used. The singular term merely yields an overall constant in the energy and the neglected terms are irrelevant. One has thus a perturbation of the type (14.106). For the model

---

[22] Any conserved current such that the OPE $J(z) \cdot J(0)$ does *not* contain a term proportional to z^{-1} will generate through (14.106) a marginal perturbation [161].

[23] It may also be shown that the energy–momentum tensor ($T(z) = -g_0 : (\partial\phi)^2 :$ for the Gaussian model) is modified such that the OPE $T \cdot V_Q$ gives the correct scaling dimension $x_Q(\lambda)$ and that the central charge c remains unchanged, as required from the c-theorem [151].

of the interacting oriented self-avoiding walks defined above, with $\beta_a = 0$, the critical point is not shifted by this perturbation [151, 69]. [24] The relation $\gamma_Q = \nu(2 - x_Q(\Lambda) - Qx_1(\Lambda))$ should thus remain valid for all Λ and one finally arrives[25] at (14.105). Since the energy operator $\varepsilon \sim S^a S^{a*}$ has $Q = 0$, $\nu = 3/4$ remains independent of Λ. Another remarkable consequence is the relation, valid for N large (see exercises)

$$\overline{k_{p,N}} = \frac{\sum_M M c_{N,M}}{\sum_M c_{N,M}} \sim Q^2 \ln N \qquad (14.110)$$

for the average number of parallel contacts of an oriented star polymer of N steps [151].

We now review numerical studies attempting to verify the predictions (14.105,14.110) [69, 70, 413, 589, 55, 57, 235, 512].[26] This has turned out to be surprisingly difficult:

1. The exponent shift should be maximal in the case when there are no parallel contacts. While for SAWs on the **Manhattan lattice** (Fig. 14.6), parallel contacts are forbidden by construction, they are eliminated on the square lattice in the $\beta_p \to -\infty$ limit. From exact enumeration, the connnectivity constant is found to be the same as for the non-oriented polymer, whereas [70] (updating earlier results [69])

$$\gamma = \begin{cases} 1.3385 \pm 0.003 & \text{; SAW on Manhattan lattice, 53 steps} \\ 1.3395 \pm 0.002 & \text{; polygons on square lattice, 84 steps, } \beta_p \to -\infty \end{cases}$$

$$(14.111)$$

to be compared with the exact value $\gamma = 43/32 = 1.34375$ for the non-oriented polymer.

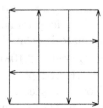

Fig. 14.6. Manhattan lattice. Interactions are only permissible in the direction of the arrows

---

[24] To see this, let $c_{N,M}$ be the number of these walks of N steps and M parallel contacts and consider $c_N(\beta_p) := \sum_M c_{N,M} e^{\beta_p M}$. If $\beta_p \le 0$, clearly $c_{N,0} \le c_N(\beta_p) \le c_N(0)$. Both bounds scale as μ^N, where $\mu = \lambda^{-1}$ is the connectivity constant of the non-oriented polymer, with $\beta_p = 0$. Now, $c_{N,0} \ge c_{\text{return}}$, the number of walks which return to the vicinity of the origin. In the continuum limit, this should become the same as the number of closed loops of length N, known to scale as μ^N. Similar arguments hold for star polymers.

[25] For $n = 0$, the central charge k of the O(2n) model vanishes [148].

[26] I thank J.L. Cardy for kindly sending a copy of [70] before publication.

2. A thorough transfer matrix study on the dependence of the exponent $\eta = 2x_1$ on Λ was presented in [589]. Besides the bond model defined above, two alternative formulations are also considered, namely (i) interactions between sites rather than bonds are considered (site model) and (ii) next-nearest neighbour interactions across an elementary square are included (plaquette model). For $\beta_p \to -\infty$, lattices of width up to[27] $L = 11$ were considered. This leads[28] to [589]

$$\eta = \begin{cases} 0.2094(32) & ; \text{bond model} \\ 0.2096(10) & ; \text{site model} \\ 0.2085(14) & ; \text{plaquette model} \end{cases} , \qquad (14.112)$$

where the quoted uncertainties are estimates of the extrapolation error. Since all three estimates are very close to each other, (expected) universality is confirmed and the apparent scatter of the data gives a further indication of the reliability of the results. An upper bound $|\Lambda(-\infty)| \lesssim 10^{-4}$ was obtained [589]. In addition, data for two-leg walks are also consistent with a vanishing of $\Lambda(-\infty)$ [589, Fig. 9].

3. Several Monte Carlo simulations [55, 57, 235, 512] tried to verify the logarithmic growth (14.110) of the mean number of parallel contacts with N, but instead found saturation (for $N \lesssim \mathcal{O}(5 \cdot 10^3)$), consistent with no change in the exponents x_Q[29] (see the note on page 320).

Summarizing, it is probably just fair to leave it to the reader to make up his mind for himself. Why does $|\Lambda(-\infty)|$ come out so small?

14.4.5 Non-integrable Perturbations

The perturbing fields studied so far all have in common that the existence of several conservation laws involving some T_s was established. However, there are many physical fields, for example the order parameter (with the only exception of the Ising model), which in unitary models is given by $\phi_{2,2}$, for which no conservation law involving T_s with $s \neq 1$ is presently known. These models are commonly referred to as *non-integrable* in the literature.

The first example studied to be in this class was the tricritical Ising model coupled to a magnetic field. The quantum Hamiltonian is [314]

$$H = -\frac{1}{\zeta} \sum_{n=1}^{N} [\, \{t_t \sigma^z(n) - \sigma^x(n)\sigma^x(n+1) + \sigma^x(n)\sigma^x(n+2) + h_t \sigma^x(n)\}$$
$$+ h_s(-1)^n \sigma^x(n)\,], \qquad (14.113)$$

---

[27] An earlier study [413], on lattices up to $L = 7$, did not produce results accurate enough for this problem.

[28] Recall that $\eta = 5/24 \simeq 0.20833\ldots$ in the non-oriented case.

[29] It is argued in [70] that polymer lengths of order $N \sim \mathcal{O}(10^9)$ were needed for a check of (14.110), relating parallel contacts to winding angles.

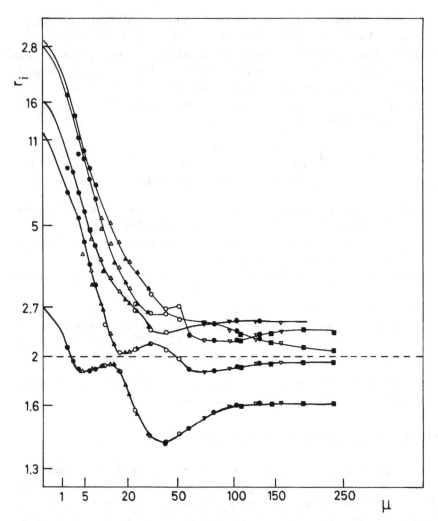

Fig. 14.7. Mass ratios $r_i = (\xi_{i+1}/\xi_1)^{-1}$ as functions of μ for the tricritical Ising model [314]

where h_s is the symmetry-breaking staggered field coupled to the order parameter which is described by the primary field $\phi_{2,2}$. The tricritical point (t_t, h_t) was given in Chap. 12. The magnetic scaling variable is $\mu = h_s N^{77/40}$. In Fig. 14.7, we show the mass ratios r_i defined analogously to the Ising model case as a function of μ. Comparing with the corresponding plot for the Ising model case (Fig. 14.4), we note that they are quite similar in many but not all aspects. We observe that the data do display finite-size scaling. One reproduces the conformal expectation in the $\mu \to 0$ limit

$$r_1 = 2\frac{2}{3}, r_2 = 11\frac{2}{3}, r_3 = 16, r_4 = 27\frac{2}{3}, \ldots . \qquad (14.114)$$

However, attempts to find a conservation law with some T_s and $s \neq 1$ have failed so far.[30] Nevertheless, finite-size scaling calculations yield for the lowest two mass ratios in the $\mu \to \infty$ limit [314]

$$r_1 = 1.62(1) \ , \ r_2 = 1.98(2). \tag{14.115}$$

Surprisingly, these are found to agree, within numerical errors, with the Ising model result! This appears to be the first example where two systems being in *different* universality classes yet show the same values for some universal amplitudes, within numerical errors. Could this indicate a deep relation between these two systems? From the truncation method [422]

$$r_1 = 1.6(2) \ , \ r_2 = 1.9(2). \tag{14.116}$$

Martins [453] has extended these calculations, using the truncation method, to the tetracritical ($m = 5$) and pentacritical ($m = 6$) Ising models, perturbed with their order parameter $\phi_{2,2}$. He finds for the lowest mass ratio [453]

$$r_1 = 1.63(2) \text{ if } m = 5 \ , \ r_1 = 1.62(2) \text{ if } m = 6, \tag{14.117}$$

again in agreement with the Ising model result. Finally, the very same mass ratios are also reproduced within the Ashkin–Teller model [315] coupled to a magnetic field, which has central charge $c = 1$. The Hamiltonian is

$$H = -\frac{1}{\zeta} \sum_{n=1}^{N} \left[\{ (\Gamma_n + \Gamma_n^3 + \varepsilon \Gamma_n^2) + (\sigma_n \sigma_{n+1}^3 + \sigma_n^3 \sigma_{n+1} + \varepsilon \sigma_n^2 \sigma_{n+1}^2) \} \right.$$
$$\left. + h(\sigma_n + \sigma_n^3) \right]. \tag{14.118}$$

The order parameter perturbation also breaks a $\mathbb{Z}_2$ symmetry. The finite-size scaling variable is $\mu = hN^{15/8}$. The mass ratios r_i are given as a function of ε in the $\mu \to \infty$ limit. We refer to Chap. 12 for the variable definitions and to Figs. 11.3 and 12.1 for the phase diagram. In contrast to the models considered so far in this section, the Ashkin–Teller model does have a charge conjugation and states are labeled by a C-parity. When calculating the mass spectrum in the $\mu \to \infty$ limit, one observes that in the *entire interval* $0 \leq \varepsilon < 1/\sqrt{2}$, the mass ratios in the $C = +1$ sector do again reproduce the Ising model result. This is shown in Table 14.5, where the $*$ give those $C = +1$ states which are close to the Ising model values.[31]

---

[30] It is known that up to $s = 17$, no conservation law exists, but for three exotic values of the central charge which do not correspond to any minimal model: (i) $c = -11 \Rightarrow \Delta_{22} = 3/2$ and the conserved spins 1, 7, 9, 11, 13 (twice), 15 (twice), 17 (3 times). (ii) $c = 33 \Rightarrow \Delta_{22} = -4$ and the conserved spins 1, 7, 11, 13, 17. (iii) $c = -35 \Rightarrow \Delta_{22} = 9/2$ and the conserved spins $1, 7, \ldots$. This follows from the explicit construction of the matrix of $\partial_{\bar{z}}$ in the corresponding subspaces of fixed spin and looking for zeroes of its determinant [169].

[31] Since for the Ising model, the integrals of motion all have odd spin and therefore $C = +1$, it is reasonable to find the Ising mass ratios in the $C = +1$ sector of the Ashkin–Teller model.

Table 14.5. Mass ratios $r_i = m_{i+1}/m_1$ for the Ashkin–Teller model in the limit $\mu \to \infty$ for several values of ε [315]

ε	r_1	r_2	r_3	r_4	r_5	r_6
$-1/\sqrt{2}$	1.934(6)	2.0	-	-	-	-
-0.6	1.652(4)	2.0	-	-	-	-
-0.3	1.221(4)	1.837(6)	1.94(1)	2.0	-	-
0.0	1	1.618*	1.618	1.989*	1.989	2
0.1	1.050(3)	1.618(4)*	1.651(5)	1.99(1)*	<2.0(1)	2.0
0.25	1.125(5)	1.621(5)*	1.71(1)	1.98(2)*	<2.0(1)	2.0
0.5	1.245(5)	1.623(5)*	1.785(10)	1.99(1)*	<2.0(1)	2.0
$1/\sqrt{2}$	1.330(5)	1.615(10)*	1.79(2)	1.91(2)*	>1.95	2.0
0.875	1.370(5)	1.57(1)	1.86(1)	1.85(2)	2.00(5)	2.0
1.0	1.395(5)	1.52(5)	1.75(5)	1.8(1)	?	

This finding is quite surprising. Previously, we had understood the appearance of stable, soliton-like excitations (one can check that they all have momentum zero) as a consequence of the existence of at least one non-trivial integral of motion T_s. For all the models discussed here, however, no integral of motion besides energy–momentum is known. One might wonder whether some important aspect of critical systems perturbed with a relevant, symmetry-breaking scaling operator has yet escaped attention.

Returning to Fig. 14.7, we observe that some of the eigenvalues of H appear to show some kind of level repulsion. Effects of this kind are absent in the Ising model, see Fig. 14.4, but are also seen in the higher multicritical Ising models [453] and in the Ashkin–Teller model, see Fig. 14.8. If this interpretation of the finite-size scaling functions is correct, this indicates the non-integrability of the $\phi_{2,2}$ perturbation.

We conclude that at least for perturbations which break a global $\mathbb{Z}_2$ symmetry the mass ratios predicted for the integrable Ising model in a magnetic field seem to be quite a general feature.[32] We tentatively illustrate this diagrammatically in Fig. 14.9. One can show that by perturbing a multicritical Ising model, with a central charge given by m via the Kac formula, by $\phi_{1,3}$ one arrives at the end of the renormalization group trajectory at a multicritical point given by $m - 1$. It has been determined how the primary fields change when m goes to $m - 1$ [403, 530]. For a generalization to non-unitary minimal models, see [424]. It remains at present an open question in what sense the apparent relationship between these different universality classes can be understood. Because of the non-integrability, most of the powerful machinery developed recently is not available for studying this problem further, but application of the density-matrix renormalization group algorithm (see Chap. 9) may allow to make progress.

---

[32] Perturbing the non-unitary minimal model $\mathcal{M}_{2,11}$ with ϕ_{14}, the same values of the spins of the conserved quantities and the same mass ratios as for the Ising model in a magnetic field are obtained [531].

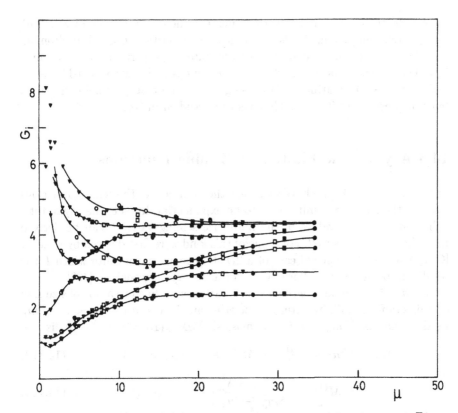

Fig. 14.8. Scaled masses $G_i(\mu)$ for the Ashkin–Teller model with $\epsilon = 1/\sqrt{2}$ perturbed with the order parameter. The scaling variable is $\mu = hN^{15/8}$. The different symbols correspond to different values of h, thus showing that the scaling regime is reached. In the $\mu \to \infty$ limit, the first, third, fifth and sixth levels have C-parity $C = +1$, while the second and forth levels have $C = -1$

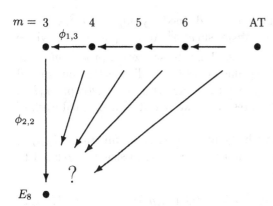

Fig. 14.9. Relation between $\mathbb{Z}_2$ breaking perturbations $\phi_{2,2}$ in multicritical Ising models with $m = 3, 4, 5, 6$ and the Ashkin–Teller model (AT) via renormalization group trajectories. The existence of a common fixed point E_8 is tentative. The role of the perturbation $\phi_{1,3}$ is also indicated (after [169])

A different type of non-integrable massive theory arises when two *simultaneous* relevant scaling fields are added to the critical point Hamiltonian. If at least one of them alone leads to an integrable perturbation, one may include the second one as a perturbation of an integrable massive field theory. This is a possible treatment of the Ising model in a magnetic field and close, but not precisely at $T = T_c$. This has been studied in [185].

14.5 Asymptotic Finite-Size Scaling Functions

In Figs 14.4 and 14.7 we have seen that the approach of the scaling functions towards their $\mu \to \infty$ limit seems to be exponential. In fact, the asymptotic behaviour for large values of the finite-size scaling variable can be derived from the S-matrix as well. Formulae of this kind were first derived by Lüscher [G17, 444]. For the special case of a 2D strip geometry of finite width L the result can be stated as follows, valid for arbitrary Lagrangians [402].

Let $a = 1, \dots, n$ denote the n stable particles confined to a box of size L in one direction and infinite extent in the other and let $0 < m_1 < m_2 < \dots < m_n$ be their masses. If $m_a < 2m_1$, the mass shift $\Delta m_a(L) = m_a(L) - m_a$ is

$$\Delta m_a(L)/m_1 = M_a^{(1)} + M_a^{(2)} + \mathcal{O}(\exp(-\sigma_a L)), \qquad (14.119)$$

$$M_a^{(1)} = -\frac{1}{8m_a^2} \sum \frac{\lambda_{abc}^2}{\mu_{abc}} \exp(-\mu_{abc} L), \qquad (14.120)$$

where the sum is over all pairs (b, c) of stable particles with $|m_b^2 - m_c^2| < m_a^2$,

$$M_a^{(2)} = -\frac{1}{2\pi} \sum_b{}' \int_{-\infty}^{\infty} d\theta \, \exp(-\cosh\theta \, m_b L) \, m_b \cosh\theta \left[S_{ab}\left(\theta + \frac{i\pi}{2}\right) - 1 \right],$$

$$(14.121)$$

where $\mu_{abc} = m_b m_c m_a^{-1} \sin u_{bc}^a$, u_{bc}^a is defined by $m_a^2 = m_b^2 + m_c^2 + 2m_b m_c \cos u_{bc}^a$ and $\lambda_{abc}^2 = -8i \, (m_b m_c \sin u_{bc}^a)^2 \, \mathrm{Res}_{\theta = iu_{bc}^a} S_{bc}(\theta)$, where $S_{ab}(\theta)$ is the two-particle S-matrix describing the scattering process $ab \to ab$ and θ is the rapidity variable. The prime in (14.121) indicates that one should only sum over terms which are larger than the error term $\mathcal{O}(\exp(-\sigma_a L))$, and $\sigma_a \leq 2\mu_{a11}$. For precise estimates of σ_a, we refer to the original work of Klassen and Melzer [402].

The correspondence of this with the finite-size scaling functions $G_i(\mu)$ of the correlation lengths (14.85) is a follows. The system size L is related to the finite-size scaling variable μ via

$$m_1 L = \zeta^{-1} \xi_1^{-1} N = \zeta^{-1} G_1(\infty) \mu^{1/y} \qquad (14.122)$$

(recall that $y = 15/8$ in the Ising model). The mass shifts are then calculated with $i = a$ for all states below the continuum threshold

$$\delta G_i(\mu) = (G_i(\infty) - G_i(\mu))/G_1(\infty) = -\Delta m_a/m_1(\infty). \qquad (14.123)$$

This results can be used in a two-fold way. First, one can explicitly test the S-matrix beyond the mass ratios using finite-lattice data. We shall do so below. Secondly, one can also predict the asymptotic finite-size scaling function from the knowledge of the two-particle S-matrix. This approach is complementary to the conformal perturbation theory approach described in Chap. 13.

Table 14.6. Numerical values for μ_{abc} and $\rho_{abc} := \lambda_{abc}^2/(8m_a^2 \mu_{abc})$ in the 2D Ising model [402]

	abc	μ_{abc}	ρ_{abc}
G_1	111	$0.866025 \cdot m_1$	$1.046154 \cdot 10^2$
	122	$1.538842 \cdot m_1$	$1.463532 \cdot 10^4$
G_2	211	$0.587785 \cdot m_1$	$1.205778 \cdot 10^2$
	212	$0.951057 \cdot m_1$	$9.045125 \cdot 10^3$
	221	$0.951057 \cdot m_1$	$9.045125 \cdot 10^3$
G_3	311	$0.104528 \cdot m_1$	$1.130876 \cdot 10^{-1}$

In Table 14.6, we list the contributions from the simple poles of the S-matrix which contribute in the example of the 2D Ising model in a magnetic field, where the *minimal* S-matrix was used. For a comparison with the numerical data, we need the value of $G_1(\infty)$. Since finite-size scaling techniques alone require a double extrapolation, its value will normally be known to somewhat lesser precision than the finite-lattice data for $G_i(\mu)$ for μ finite. Further, the comparison with the asymptotic expression $M_a^{(1)}$ requires that the values of μ must be large enough so that the corrections to it are still negligible. If on the other hand μ becomes too large, the numerical errors in the determination of of the $G_i(\mu)$ will become larger than the value of $\delta G_i(\mu)$ itself. The calculation of the $G_i(\mu)$ is improved by knowing the size of the leading finite-size corrections as obtained from conformal perturbation theory, here in the Ising model example

$$\xi_i^{-1} = h^{8/15} \left(G_i(\mu) + N^{-2}R_i(\mu) + \mathcal{O}(N^{-4})\right). \qquad (14.124)$$

In Fig. 14.10, we compare the mass shift predicted from the S-matrix with the *minimal* number of poles, using the contribution of $M_a^{(1)}$ only (full curves), with the finite-lattice data. We first note that the mass shifts are surprisingly large. Still, we find an excellent agreement with the asymptotic predictions over several orders of magnitude for G_1 and G_2, while for G_3, due to the smallness of μ_{311}, the available data just reach the beginning of the asymptotic regime. This means that indeed the available evidence is consistent with the *minimal* solution for the S-matrix. This result has also been obtained from the calculation of the central charge using the S-matrix

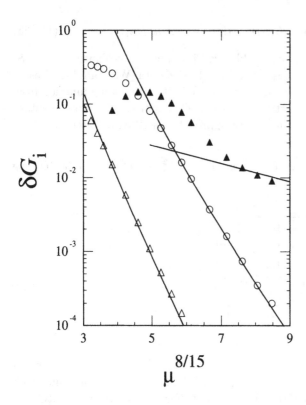

Fig. 14.10. Mass shift δG_i, $i = 1, 2, 3$ in the (1+1)D Ising model with a magnetic field. The lowest curve corresponds to δG_1, the uppermost to δG_3 (after [317])

through the thermodynamic Bethe ansatz [647, 645, 649, 168, 452, 401], and not only for the Ising model. In any case, the observed agreement does provide an impressive confirmation of the minimal solution to the bootstrap program. However, there are a few more lessons to be drawn from this plot. The data for $\delta G_i(\mu)$ were obtained in using the finite-size estimate $G_1(\infty) = 5.4156(3)$ [317]. Fig. 14.10 shows clearly for μ large enough a *linear* relation between $\delta G_i(\mu)$ and $\mu^{1/y}$ as expected from the asymptotic prediction. This is even independent of any particular S-matrix and can be used by itself as a check on the quality of the data. Further, the slope only depends on the masses but not on the couplings λ_{abc}, which offers an additional test. However, for very small values of δG_1, the finite-lattice data fall systematically above the asymptotic form. Eventually the data would appear to be independent of μ. This means that the value of $G_1(\infty)$ used was a little overestimated. The order of magnitude of this systematic error can be read off to be $\sim 10^{-4}$. We stress that this way one has a means at hand to tell about systematic extrapolation errors and their sizes *a posteriori*.

These remarks are fully confirmed by an extended study which combines the finite-size data with the truncation method. This yields the improved estimate $G_1(\infty) = 5.41550(1)$ [402]. In Table 14.7 we compare the results obtained from the truncation method (TM) and from the finite-size scaling

Table 14.7. Mass shifts δG_i for the Ising model in a magnetic field (after [402])

$m_1 L$	$\delta G_1 \cdot 10^3$			$\delta G_2 \cdot 10^3$		
	TM	FSS	ASY	TM	FSS	ASY
9.24575	40.56(1)	40.556(6)	57.29			
10.4142	15.02(2)	15.018(3)	17.11			
11.4778	5.85(1)	5.845(3)	6.158			
12.4613	2.44(3)	2.440(3)	2.486	128.97(3)	128.988(6)	208.4
13.3811	1.07(3)	1.084(3)	1.085	80.49(5)	80.53(1)	100.1
14.2486	0.48(4)	0.508(6)	0.5014	47.45(8)	47.54(3)	51.36
15.0722	0.20(6)	0.253(4)	0.2422	27.3(1)	27.5(2)	27.90
15.8582	0.1(1)	0.13(2)	0.1213	15.8(3)	16.08(3)	15.89
16.6114		0.07(4)	0.06268	9.1(5)	9.61(6)	9.422
18.0348				3(1)	3.7(1)	3.646
19.3660					1.6(1)	1.554
20.6216					0.71(9)	0.7113
21.8135					0.33(12)	0.3434

data (FSS), using the improved value of $G_1(\infty)$, with the asymptotic formula (14.119) (ASY). Comparing these results and recalling that the gaps are originally of order unity, the agreement shown is truly impressive. We also see that the truncation method is really pushed beyond its limits in this particular application. This is not yet yet so apparent for the lowest level, but already quite visible for the second one. The finite-size scaling method reproduces the large μ behaviour to at least one order of magnitude better, provided the improved value of $G_1(\mu)$ is used.

In conclusion, while these examples examples presented a major advance in numerical precision when first proposed, the data obtained from the Bethe ansatz in the context of diluted ADE models are much more precise.

Exercises

1. Consider the unitary minimal models $\mathcal{M}_{m,m+1}$, perturbed by one of the relevant scalar operators $\phi_{1,2}$, $\phi_{2,1}$ and $\phi_{1,3}$. For what values of s is there a conserved quantity $T_s \sim :T^s:$?
2. Extend the previous exercise to the non-unitary minimal models $\mathcal{M}_{p,p'}$. Which relevant scaling operators $\phi_{r,s}$ lead to integrable perturbations?
3. Show that unitarity (14.56) of the two-particle S-matrix implies unitarity of the three-particle (n-particle) S-matrix, if the S-matrix is factorizable.
4. Find the factorizable S-matrix for the Yang–Lee model $\mathcal{M}_{2,5}$ perturbed by its relevant operator [142].
5. Prove the universality of the amplitude ratios in (14.94).
6. Consider the number $c_N^{(Q)}(\beta_p, 0)$ of Q-star oriented self-avoiding walks of N steps. Derive (14.110) and estimate also the prefactor.

Note added: A recent high-statistics Monte Carlo Simulation [657] for oriented self-awoiding walks on the Manhattan lattice yields the estimate $\gamma \gtrsim 1.3425(3)$, again consistent with a constant value $\gamma = 43/32$ and in disagreement with the estimates (14.111) found from exact enumeration.

15. Surface Critical Phenomena

Up to now we have been always considering a statistical system defined in the infinite two-dimensional plane. However, conformal invariance can also be applied to systems with boundaries present. It is impossible to give on just a few pages a full description on the rich field of surface effects and we will only consider some of the problems for which conformal invariance has proved to be useful. For more background on surface critical phenomena, we refer to the detailed reviews [91, 195, 198, 351]. Wetting and related questions are reviewed in [200].

In this chapter, we shall again follow the same sequence of logical steps done before for the case of bulk critical phenomena. After defining the relevant thermodynamics, we introduce surface critical exponents and discuss scaling. Conformal invariance will then be used to determine completely the two-point correlation function - a problem of the same degree of difficulty as finding the four-point correlation function in the bulk. Finite-size scaling techniques will be described and we shall use them to give the surface operator content of the models discussed before. Defect lines, which interpolate between a free surface and the bulk, will be studied and we mention applications of this to aperiodic systems and persistent currents.

15.1 Systems with a Boundary

In general, real experimentally considered systems are always finite and possess a boundary. The presence of boundaries gives rise to new important effects neglected so far. We shall restrict attention to a particularly simple case: a flat surface with free boundary conditions. We are therefore totally neglecting the fluctuation properties of boundaries or interfaces, these being interesting topics in their own right (see e.g. [522] for an introduction and further references). While in itself this is a very crude approximation, this does not affect the universal properties of surface phase transitions. We consider exclusively hypercubic systems within a **slab geometry**, that is, just one direction is of finite length L with free boundary conditions, while the other direction(s) are taken to be infinite in extent.

We further assume that the thermodynamic parameters of the system are such that bulk and surface properties can be well separated. For example, we write for the free energy density

$$f = f_b(T, B, V, \mathcal{N}) + \frac{2}{L} f_s(T, B, B_1, V, \mathcal{N}) + \ldots, \qquad (15.1)$$

where f_b is the **bulk free energy density** studied so far and f_s is the **surface free energy density**. While B is the magnetic field as measured for the bulk system, we have also included a **surface magnetic field** B_1 to allow for a variation of thermodynamic variables close to the free surface. The factor 2 takes the presence of two free interfaces into account.

Here, we want to limit ourselves to just one rather restricted question, namely about the types of singularities present in f_s and the quantities derived from it close to the bulk critical point. To get an overview over the generic situation, consider the space-dependent variation of the order parameter $m(r)$, where r measures the distance to the free surface, assumed to be at $r = 0$. Taking Landau theory [91] as a guide, there are in general four possible situations, as illustrated in Fig. 15.1 (see [91]). The critical temperature $T_{c,b}$ is the **bulk critical temperature** (called critical temperature up to now).

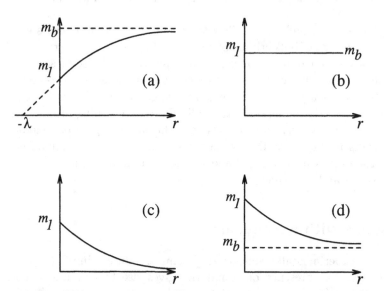

Fig. 15.1. Schematic local order parameter profiles $m(r)$ in dependence of the temperature and the extrapolation length λ ([169] after [91])

1. Let $T < T_{c,b}$ and take a positive **extrapolation length** λ, as shown in Fig. 15.1a. The transition from the disordered state to this state is called the **ordinary transition** O where m_b is the magnetization deep inside the bulk and m_1 the magnetization at the surface.
2. Let $T = T_{c,b}$. The extrapolation length is infinite. The transition from the disordered state to this state is called the **special transition** SP. It is a surface tricritical point (Fig. 15.1b).

3. Let $T_{c,b} < T < T_{c,s}$. The extrapolation length is negative and $T_{c,s}$ is some surface critical temperature. The transition from the disordered state into this state is called the **surface transition** S. (Fig. 15.1c)
4. Let $T < T_{c,b}$ and take a negative extrapolation length. The transition from the state of Fig. 15.1c to this state is called the **extraordinary transition** E. (Fig. 15.1d)

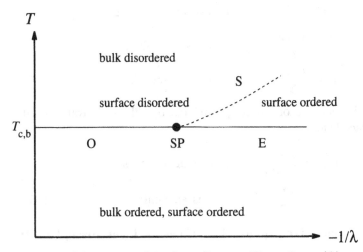

Fig. 15.2. Schematic surface phase diagram. The ordinary (O), extraordinary (E), special (SP) and surface (S) transitions are indicated

It becomes apparent that the local behaviour of thermodynamic quantities can be quite distinct from those deep inside the bulk. In particular, we will have to allow for the possibility that the critical exponents close to the surface may be different from their bulk values. This can be already verified in the context of Landau theory [91]. We summarize the generic surface phase diagram in Fig. 15.2.[1] Although the results sketched in Figs 15.1 and 15.2 are merely a consequence of mean-field theory, they remain qualitatively correct in three and higher dimensions, at least in the context of the Ising model. For systems like the 2D Ising model however, a free surface cannot order by itself at a non-zero temperature in the absence of ordering fields (for short-ranged interactions). Therefore for such a system, only the ordinary surface transition exists.

We now describe the calculation of the surface thermodynamical quantities, see [91, 195]. This proceeds in quite an analogous manner with those of the bulk as defined in Chap. 1. There we had introduced the reduced temperature t and the reduced magnetic field h. Here, we introduce in addition a

---

[1] The value of the extrapolation length λ may for example be controlled by couplings close to the surface. See (15.134) for a 2D example.

reduced **surface magnetic field** h_1, which only acts on the boundary spins. For a large class of models, this is the only new relevant scaling field which arises. The new surface quantities can now be defined by taking derivatives of the surface free energy density (15.1). First, we define **excess quantities** by taking derivatives of f_s with respect to bulk scaling fields

$$m_s := -\frac{\partial f_s}{\partial h} \qquad \text{excess magnetization}$$

$$e_s := \frac{\partial f_s}{\partial t} \qquad \text{excess internal energy}$$

$$\chi_s := \frac{\partial m_s}{\partial h} \qquad \text{excess susceptibility}$$

$$C_s := \frac{\partial e_s}{\partial t} \qquad \text{excess specific heat.} \qquad (15.2)$$

Second, we define surface or **local quantities** by taking derivatives with respect to surface fields

$$m_1 := -\frac{\partial f_s}{\partial h_1} \qquad \text{surface magnetization}$$

$$\chi_1 := \frac{\partial m_1}{\partial h_1} \qquad \text{local susceptibility.} \qquad (15.3)$$

Finally, we can define **mixed quantities** such as the **layer susceptibility**

$$\chi_{11} := \frac{\partial m_1}{\partial h}. \qquad (15.4)$$

In these definitions, it is understood that all those thermodynamic variables not explicitly involved in the derivatives are kept constant.

The distinct nature of these quantities becomes apparent when expressing a space-dependent profile such as $m(r)$ (see Fig. 15.1) in terms of them. For example, for the order parameter one has

$$m_b = m(r = \infty)$$
$$m_1 = m(r = 0)$$
$$m_s = \int_0^\infty dr\, [m(r) - m_b]. \qquad (15.5)$$

Next, we turn to the definition of the critical exponents. Their definition is again quite analogous to the bulk exponents in (1.11). We assume that those of the parameters t, h, h_1 not explicitly mentioned are zero. The **surface exponents** are then defined by[2]

---

[2] The surface exponent Δ_1 should not be confused with the correction-to-scaling exponent $\Delta_1 = \omega\nu$ [G20] defined in Chap. 3 and related to the Wegner exponent ω.

$$C_s \sim |t|^{-\alpha s}$$

$$m_s \sim (-t)^{\beta s}$$

$$m_1 \sim (-t)^{\beta_1}$$

$$\chi_s \sim |t|^{-\gamma s}$$

$$\chi_1 \sim |t|^{-\gamma_1}$$

$$\chi_{11} \sim |t|^{-\gamma_{11}}$$

$$m_s \sim h^{1/\delta s}$$

$$m_1 \sim h^{1/\delta_1}$$

$$m_1 \sim h_1^{1/\delta_{11}}$$

$$\frac{\partial^{2\ell} f_s}{\partial h_1^{2\ell}} \sim |t|^{-2\Delta_1} \frac{\partial^{2\ell-2} f_s}{\partial h_1^{2\ell-2}}. \tag{15.6}$$

For the correlation function, surface critical exponents are defined as follows. Denote $\boldsymbol{r}_s = (r_\parallel, 0)$ and $\boldsymbol{r} = (r_\parallel, r_\perp)$. Then one has correlations parallel and perpendicular to the free surface at $r_\perp = 0$

$$G_\parallel(r_\parallel - r'_\parallel) := G(\boldsymbol{r}_s, \boldsymbol{r}'_s) \ , \quad G_\perp(r_\perp, r'_\perp) := G(\boldsymbol{r}, \boldsymbol{r}')|_{r_\parallel = r'_\parallel} , \tag{15.7}$$

where $G(\boldsymbol{r}, \boldsymbol{r}')$ is the two-point order parameter correlation function defined in (1.7). The surface exponents are then defined exactly *at* the critical point $t = h = h_1 = 0$

$$G_\parallel(r_\parallel) \sim |r_\parallel|^{-d+2-\eta_\parallel}$$

$$G_\perp(r_\perp, r'_\perp) \sim |r_\perp - r'_\perp|^{-d+2-\eta_\perp}. \tag{15.8}$$

These definitions apply to the ordinary, special and extraordinary transitions. Exponents belonging to the surface transition can be defined by replacing t by $t_s := (T - T_{c,s})/T_{c,s}$, but since the bulk is not critical at a surface transition, critical exponents only exist for the local quantities of (15.3).

As already seen for the bulk critical exponents, there exist many scaling relations, relating different exponents to each other. The renormalization group treatment can be extended to surface critical phenomena and the notions of relevant, marginal and irrelevant scaling fields are introduced as in the bulk, see Chap. 1. For notational simplicity, we only consider here the case of the ordinary transition and refer to [91, 195] for the other cases. This is sufficient for us, since later we shall only deal with two-dimensional systems in detail, where there is only an ordinary surface transition. Since for an ordinary transition, there is only a relevant surface magnetic field h_1, we can write the scaling form for the singular part of the surface free energy density

$$f_s^{\text{sin}} = |t|^{2-\alpha_s} W_s^\pm \left(h|t|^{-\beta-\gamma}, h_1|t|^{-\Delta_1} \right). \tag{15.9}$$

For the bulk system, *hyperscaling* was seen to be equivalent to the dimensional argument that $f_b^{\text{sin}} \sim \xi_b^{-d}$, valid below the upper critical dimension $d < d^*$, since one assumes that there is only one relevant length scale. For the free surface, the analogous statement is $f_s^{\text{sin}} \sim \xi_b^{-(d-1)}$ which implies

$$\alpha_s = \alpha + \nu. \tag{15.10}$$

Since α_s is related to bulk exponents, it follows from scaling (15.9) that there is only one new independent surface critical exponent. Knowing one of them, for example $\eta_\|$, all the others can be readily found by scaling relations. These read, see [91, 195]

$$\beta_s = \beta - \nu \qquad\qquad \gamma_s = \gamma - \nu \qquad\qquad \delta_s = (\beta + \gamma)/\beta_s$$
$$\beta_1 = (\nu/2)(d - 2 + \eta_\|) \qquad \gamma_1 = \nu(2 - \eta_\perp) \qquad \gamma_{11} = \nu(1 - \eta_\|)$$
$$\Delta_1 = (\nu/2)(d - \eta_\|) \qquad \delta_1 = (\beta + \gamma)/\beta_1 \qquad \delta_{11} = \Delta_1/\beta_1$$
$$\eta_\perp = (\eta + \eta_\|)/2$$

$$\tag{15.11}$$

There exists a plethora of ϵ-expansion results for the critical exponents. We merely mention as an example the result for the exponent $\eta_\|$ of the $O(n)$ vector model at the ordinary transition in $d = 4 - \epsilon$ dimensions [532, 194, 195]

$$\eta_\| = 2 - \frac{n + 2}{n + 8}\epsilon - \frac{(n + 2)(17n + 76)}{2(n + 8)^3}\epsilon^2 + \mathcal{O}(\epsilon^3). \tag{15.12}$$

Comparison with the results quoted for 3D bulk exponents in Chap. 1 show that surface exponents are indeed different from their bulk counterparts.

Finally, we quote some experimental results. To our best knowledge, there are no data on the surface critical behaviour of 2D systems available. For 3D systems, the first clear experimental evidence for a surface magnetization exponent $\beta_1 = 0.825^{+0.025}_{-0.040}$ different from the bulk value $\beta \sim 0.36$ was obtained studying m_1 on the (100) surface of Ni [E5], which is in the bulk $O(3)$ (Heisenberg) universality class. The theoretical estimate is $\beta_1 = 0.84(1)$ [196]. The binary alloy FeAl shows a transition which is believed to be in the bulk 3D XY universality class ($\beta \simeq 0.345$) and the surface exponent is $\beta_1 = 0.75 \pm 0.02$ [E55], close to estimates $\beta_1 \simeq 0.8$ found from the ϵ-expansion [194, 532].

15.2 Conformal Invariance Close to a Free Surface

Repeating the program followed in the bulk, we now ask for the consequences of local, space-dependent scaling close to a free surface. This was done first by Cardy [133]. As we did in Chap. 2, we first examine the projective conformal transformations. Consider scaling operators $\phi_i(r)$ and the two-point correlation function

$$G(r_1, r_2) = \langle \phi_1(r_1)\phi_2(r_2) \rangle. \tag{15.13}$$

The coordinate space is restricted to a semi-infinite geometry, namely the half-plane $r = (u, v)$ with $v \geq 0$, as shown in Fig. 15.3. It is convenient to

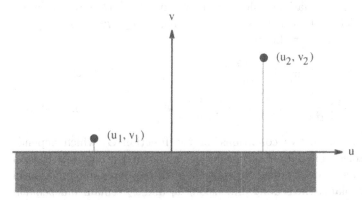

Fig. 15.3. Semi-infinite geometry. The physically inaccessible lower half-plane is shaded

rewrite the projective conformal generators $l_n = -z^{n+1}\partial_z$ and $\bar{l}_n = -\bar{z}^{n+1}\partial_{\bar{z}}$ which act on the complex coordinates $z = u + iv$, in terms of the real coordinates u and v. This is done in Table 15.1. While the infinite complex plane is invariant under the full set of projective transformations generated by $\{l_n\} \otimes \{\bar{l}_n\}$, the half-plane is merely invariant under those projective transformations generated by the *diagonal subalgebra* $\{l_n + \bar{l}_n\}$.

Table 15.1. Real projective conformal generators, for $z = u + iv$

$l_{-1} + \bar{l}_{-1} = -\partial_u$	$i\left(l_{-1} - \bar{l}_{-1}\right) = -\partial_v$
$l_0 + \bar{l}_0 = -(u\partial_u + v\partial_v)$	$i\left(l_0 - \bar{l}_0\right) = -(u\partial_v - v\partial_u)$
$l_1 + \bar{l}_1 = -\left(\left(u^2 - v^2\right)\partial_u + 2uv\partial_v\right)$	$i\left(l_1 - \bar{l}_1\right) = -\left(\left(u^2 - v^2\right)\partial_v - 2uv\partial_u\right)$

We want to find the general form of the two-point function G from the requirement that it transforms covariantly under those conformal transformations which keep the free surface $v = 0$ invariant (Fig. 15.3). Since these transformations are just those of the diagonal subalgebra, the projective Ward identities for $G = G(u_1, u_2; v_1, v_2)$, built from quasi-primary scaling operators ϕ_i with scaling dimensions x_i, can be read from (2.37)

$$\sum_{i=1}^{2}\left[\frac{\partial}{\partial u_i}\right] G = 0$$

$$\sum_{i=1}^{2}\left[u_i\frac{\partial}{\partial u_i} + v_i\frac{\partial}{\partial v_i} + x_i\right] G = 0 \qquad (15.14)$$

$$\sum_{i=1}^{2}\left[\left(u_i^2 - v_i^2\right)\frac{\partial}{\partial u_i} + 2u_i v_i\frac{\partial}{\partial v_i} + 2x_i u_i\right] G = 0.$$

The solution of these equations follows standard lines [383]. Due to translation invariance parallel to the surface, we have $G = G(u; v_1, v_2)$, where $u = u_1 - u_2$, and the other two relations become

$$u\frac{\partial G}{\partial u} + v_1\frac{\partial G}{\partial v_1} + v_2\frac{\partial G}{\partial v_2} + (x_1 + x_2)G = 0$$

$$\left(v_2^2 - v_1^2\right)\frac{\partial G}{\partial u} + uv_1\frac{\partial G}{\partial v_1} - uv_2\frac{\partial G}{\partial v_2} + u(x_1 - x_2)G = 0. \quad (15.15)$$

Consequently, we have two conditions for the function G, which depends on three variables. Therefore, the projective conformal group will leave a scaling function Φ undetermined. In contrast with the two-point function in the unbounded plane, the scaling dimensions x_1 and x_2 remain *independent* of each other. This implies that correlators which vanish in the bulk may yet become non-zero close to a boundary (this depends of course on the presence of boundary fields). It is easy to check that the solution may be written as

$$G(\boldsymbol{r}_1, \boldsymbol{r}_2) = v_1^{-x_1}\,v_2^{-x_2}\,\Phi\left(\frac{(u_1 - u_2)^2 + v_1^2 + v_2^2}{v_1 v_2}\right), \quad (15.16)$$

Although this result was only derived for a two-dimensional system here, the form (15.16) is readily generalized to higher dimensions as well. The special case $x_1 = x_2$ was obtained before by Cardy [133].

From now on, we always take $x_1 = x_2 = x$. Although $\Phi(V)$ is not yet determined, its form can be used to relate different limiting cases. Comparing with the scaling predictions of the last section, the form $G \sim |u_1 - u_2|^{-2x_s}$ is expected, if $|u_1 - u_2| \to \infty$ with v_1 and v_2 fixed. For the order parameter

$$\boxed{x_{\sigma,s} := (d - 2 + \eta_\parallel)/2 = \beta_1/\nu} \;. \quad (15.17)$$

This implies that $\Phi(V) \sim V^{-x_s}$ as $v \to \infty$. However, this limit is also reached by $v_2 \to 0$ with v_1 and $u_1 - u_2$ fixed, describing a correlation of a density close to the surface with one in the bulk. With $v_1 = R\cos\vartheta, u_1 - u_2 = R\sin\vartheta$,

$$G \sim R^{-x-x_s}\,(\cos\vartheta)^{x_s - x} \quad (15.18)$$

so that the form of G is completely determined one the exponent x_s is known. Note, however, that the projective conformal transformation do not determine the value of x_s.

To find both x_s and Φ in two dimensions, we have to impose covariance of the two-point function under the the diagonal subalgebra of the full conformal algebra. In doing so, we have to restrict to systems which are two-dimensional in the bulk. Unless stated otherwise, we shall assume the absence of surface magnetic fields and thus all what follows only applies to the ordinary transition. In order to keep the free surface geometry invariant, we have to restrict to those conformal transformations $z \to w(z) = z + \epsilon(z)$ where $\epsilon(z)$ is real analytic $(\epsilon(z) = \epsilon(\bar{z}))$. As we have seen in Chap. 2, the central relation from

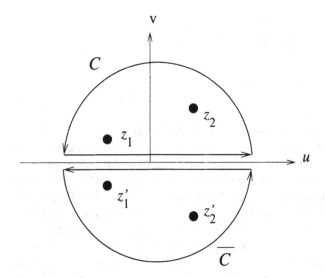

Fig. 15.4. Contours for the Ward identity with a free surface. Only the upper half-plane $v \geq 0$ is physically accessible

which the whole theory was built was the conformal Ward identity (2.50). While this remains valid if the contour C is restricted to lie in the upper half-plane, the subsequent separation of $T(z)$ and $\overline{T}(\bar{z})$ of the bulk system no longer applies.

Rather, following [133], one can extend the domain of definition of $T(z)$ into the lower half-plane via the *definition*

$$T(z) := \overline{T}(z) \; ; \quad \text{if } \operatorname{Im} z < 0 \tag{15.19}$$

and relabel $\bar{z}_i = z'_i$. Using this, the Ward identity (2.50) becomes [133]

$$\frac{1}{2\pi i} \int_C dz \, \hat{\epsilon}(z) \langle T(z)\phi_1(z_1, z'_1) \cdots \rangle + \frac{1}{2\pi i} \int_{\overline{C}} dz \, \hat{\epsilon}(z) \langle T(\bar{z})\phi_1(z_1, z'_1) \cdots \rangle$$

$$= \sum_i \left[\hat{\epsilon}'(z_i)\Delta_i + \hat{\epsilon}(z_i)\partial_{z_i} + \hat{\epsilon}'(z'_i)\Delta_i + \hat{\epsilon}(z'_i)\partial_{z'_i} \right] \langle \phi(z_1, z'_1) \cdots \rangle, \tag{15.20}$$

where $\overline{C}$ is a contour lying in the lower half-plane which is oriented oppositely to the contour C and encloses the points z'_i, see Fig. 15.4. Now consider the physical meaning of the definition (15.19). It states that $T = \overline{T}$ when $\operatorname{Im} z = 0$. Reexpressed in cartesian coordinates, this implies that $T_{uv} = 0$ across the boundary, which means that there is *no* flux of energy across the surface. Since only then the straight parts of $C, \overline{C}$ cancel, this is obviously the correct boundary condition.

As done in Chap. 2, the l.h.s. of (15.20) can be rewritten as an integral along a contour enclosing all the z_i and z'_i. The r.h.s. may be written the same way using Cauchy's theorem. The result is, for *all* values of $\operatorname{Im} z$

$$\langle T(z)\phi_1(z_1, z_1') \cdots \rangle = \left\{ \sum_i \left[\frac{\Delta_i}{(z - z_i)^2} + \frac{1}{z - z_i} \frac{\partial}{\partial z_i} \right] \right.$$

$$\left. + \sum_i \left[\frac{\Delta_i}{(z - z_i')^2} + \frac{1}{z - z_i'} \frac{\partial}{\partial z_i'} \right] \right\} \langle \phi_1(z_1, z_1') \cdots \rangle.$$

$$(15.21)$$

Let us compare this to the corresponding result for the bulk case, (2.51). In particular, the short-distance expansion and operator product algebra only use local properties of the theory and should therefore be independent of the presence of the boundary. So the analysis presented in Chaps. 4 and 5 can be taken over and the correlation functions can be calculated along the same lines. So we arrive at the important conclusion that the correlation function $\langle \phi(z_1, z_1') \cdots \phi(z_n, z_n') \rangle$ in the semi-infinite geometry, regarded as a function of $z_1, \ldots, z_n, \bar{z}_1, \ldots, \bar{z}_n$, satisfies the same differential equation as the bulk correlation function $\langle \phi(z_1, \bar{z}_1) \cdots \phi(z_{2n}, \bar{z}_{2n}) \rangle$ taken as a function of $z_1, \ldots, z_{2n}$. This means that besides the points $z_1, \ldots, z_n$ at which the conformal fields are taken, their *mirror images* obtained by reflecting them at the free boundary lying on the real axis must also be taken into account. All this is of course highly reminiscent of the *method of images*, familiar from electrostatics.

For example, the two-point function will have the form (with $x_1 = x_2$)

$$G_s = [(z_1 - \bar{z}_1)(z_2 - \bar{z}_2) / (z_1 - z_2)(\bar{z}_1 - \bar{z}_2)(z_1 - \bar{z}_2)(\bar{z}_1 - z_2)]^{2\Delta} F_s(V),$$

$$(15.22)$$

where $V := (z_1 - z_2)(\bar{z}_1 - \bar{z}_2)/(z_1 - \bar{z}_1)(z_2 - \bar{z}_2)$ and $F_s(V)$ satisfies the same ordinary holomorhphic differential equation as the corresponding bulk four-point correlation function, derived in Chap. 4 and with V replaced by η. Below, we shall give an example how the free surface boundary conditions are taken into account. This will select the appropriate solution of the holomorphic differential equation.

15.3 Finite-Size Scaling with Free Boundary Conditions

In analogy to what we have seen in Chap. 3, finite-size effects can also be introduced and studied for the case of free boundary conditions. The main distinction with respect to periodic boundary conditions is the presence of a non-universal surface free energy term for the free energy density

$$f = A + \frac{2}{L} B + \frac{1}{L^d} U + \cdots,$$

$$(15.23)$$

where A and B are non-universal amplitudes. The amplitude A is identical to the corresponding one for periodic boundary conditions, while B is due to the presence of the free surface. The leading universal contribution is described by U, which comes from the singular part of the free energy density. The considerations of universality for the *singular* free energy density can be repeated almost literally from Chap. 3. So we can write for free boundary conditions

$$f_{\text{fr}}^{\text{sin}}(t, h, h_1) = L^{-d} Y_{\text{fr}}(C_1 t L^{1/\nu}, C_2 h L^{(\beta+\gamma)/\nu}, C_s h_1 L^{\Delta_1/\nu})$$

$$\xi_{i,\text{fr}}^{-1}(t, h, h_1) = L^{-1} S_{i,\text{fr}}(C_1 t L^{1/\nu}, C_2 h L^{(\beta+\gamma)/\nu}, C_s h_1 L^{\Delta_1/\nu}) \quad (15.24)$$

for both the free energy density and the correlation lengths as measured *parallel* to the free surface, where C_1, C_2, C_s are non-universal metric factors and $Y_{\text{fr}}(x_1, x_2, x_3)$ and $S_{i,\text{fr}}(x_1, x_2, x_3)$ are universal finite-size scaling functions.

Here, we want to consider how finite-size effects can be calculated from conformal invariance. We need a conformal transformation mapping the upper half-plane onto the strip geometry. This is achieved by the following transformation

$$w = \frac{L}{\pi} \ln z \quad (15.25)$$

quite similar to the one considered in Fig. 3.4 which implies that on the infinitely long strip of width L one has indeed free boundary conditions. Now, since the covariance of the correlation functions and the partition function is unchanged, the calculation of Chap. 3 can be almost literally repeated. One finds the analogues of the universal finite-size scaling amplitudes[3] of (3.49) and (3.54) [131, 93]

$$\boxed{S_{i,\text{fr}}(0,0,0) = \pi x_{i,s} \quad , \quad Y_{\text{fr}}(0,0,0) = -\pi c/24} \quad , \quad (15.26)$$

where $x_{i,s}$ is the surface scaling dimension of the scaling operator ϕ_i and c is the central charge. We can also rephrase this into a statement for the critical point spectrum of the quantum Hamiltonian H in the strip geometry with finite width $L = Na$ and *free* boundary conditions

$$H = \frac{\pi}{N} L_0 - \frac{\pi c}{24} \frac{1}{N} + \text{const.}, \quad (15.27)$$

where L_0 is a generator of the Virasoro algebra. We stress that due to the presence of the free boundary conditions, the spectrum is given here by the irreducible representations of a *single* Virasoro algebra.

These results allow to calculate surface critical exponents directly from finite-size scaling data. For the order parameter, $x_s := x_{\sigma,s}$ allows to recover the exponent $\eta_{\parallel}$ (and consequently also the other surface exponents) from (15.17). It is also possible to verify explicitly that the values of the central charge agree with those calculated from systems with periodic boundary conditions. Finally (15.26) gives a further means to fix the correct conformal

---

[3] An universal amplitude-exponent relation can be established for the transverse correlation length of an interface confined in a strip geometry at a second-order wetting transition [491].

normalization of a given quantum Hamiltonian with free boundary conditions. This relies on the result that for the energy density, the surface critical exponent is

$$x_{\varepsilon,s} = d. \tag{15.28}$$

This has been proven to all orders in ε-expansion [199] for the $O(n)$ vector model. We shall also see below that it is satisfied for all two-dimensional spin systems we consider.

15.4 Surface Operator Content

By now we have everything at hand to be able to give the surface operator content for the models we have considered before. So we are not going to repeat the model definitions, but refer the reader for this and the notations used to the previous chapters.

15.4.1 Ising Model

Since we have already discussed the Ising model correlation functions in great detail, we shall use this model again here to illustrate the calculation of surface correlation functions. We shall discuss as an example here the spin-spin correlation function [133]. We have seen above that this amounts (i) to study the four-point function $\langle \sigma\sigma\sigma\sigma \rangle$ in the bulk and (ii) take into account the surface boundary conditions. The order parameter σ is a primary operator of level 2, with operator product algebra $\sigma \cdot \sigma = 1 + \varepsilon$. Recall the form of the two-point correlation function (15.22). The function F_s then satisfies the differential equation (5.12). For the correlator at hand, that equation has already been solved explicitly in Chap. 6 where F_s was shown to be a linear combination of the elementary functions $\sqrt{\sqrt{1-V} \pm 1}$, where, using the notation of the previous section,

$$V = -\left[(u_1 - u_2)^2 + (v_1 - v_2)^2\right] / 4v_1 v_2 \tag{15.29}$$

so that the physical region corresponds to $V < 0$. The correct solution is determined by the boundary conditions.

As $V \to 0$, one expects that $G_s \sim |z_1 - z_2|^{-1/4}$, which is satisfied by both solutions. Further information must come from the asymptotic behaviour of the two-point function near the surface (i.e. the real axis). We know that the order parameter vanishes at the surface for an *ordinary* transition. This is the boundary condition which selects the physical solution. In terms of V, this means that one has to consider the limit $V \to -\infty$. Thus, one requires $F_s(V) \to 0$ in that limit and the solution for F_s satisfying this boundary condition is

$$F_s(V) = \sqrt{\sqrt{1-V}+1} - \sqrt{\sqrt{1-V}-1} \tag{15.30}$$

and consequently $F_s(V) \sim |V|^{-1/4} \sim u^{-1/2}$. However, since the prefactor in (15.22) scales as $u^{-1/2}$ for u large and v_1, v_2 fixed, one finds that $G_s \sim u^{-1}$ close to the surface. In terms of critical exponents, this implies

$$\eta_\| = 1 \tag{15.31}$$

for the ordinary transition of the 2D Ising model.[4] Collecting all this, the final result for the two-point spin-spin correlation function is [133]

$$G_s(u_1 - u_2; v_1, v_2) = (v_1 v_2)^{-1/8} \, \Psi(\rho)$$
$$\rho = \left[(u_1 - u_2)^2 + (v_1 + v_2)^2\right] / \left[(u_1 - u_2)^2 + (v_1 - v_2)^2\right]$$
$$\Psi(\rho) = \sqrt{\rho^{1/4} - \rho^{-1/4}} \tag{15.32}$$

in agreement with earlier exact Ising model calculations of McCoy and Wu [457].

We now consider the quantum Ising chain with free boundary conditions. As for the periodic ones, this provides a further check on the conformal techniques and adds another point of view. We first have to recall how the Ising quantum chain can be diagonalized for free boundary conditions. This is very simple: almost all the calculations of Chap. 10 can be taken over. The quantum Hamiltonian is, see (10.1)

$$H = -\frac{1}{2\eta} \left[\sum_{n=1}^{N} t\sigma^z(n) + \sum_{n=1}^{N-1} \left[\frac{1+\eta}{2} \sigma^x(n)\sigma^x(n+1) + \frac{1-\eta}{2} \sigma^y(n)\sigma^y(n+1) \right] \right] \tag{15.33}$$

Here the coupling constant η should not be confused with a critical exponent. Both the Jordan-Wigner transformation and the diagonalization of the resulting quadratic form are independent of the boundary conditions. The diagonalized Hamiltonian is still given by (10.13) where the eigenvalues Λ_k are obtained from (10.19). The only new feature arises in the form of the matrix $\widehat{\mathcal{M}}$ to be diagonalized, which for free boundary conditions is

$$\widehat{\mathcal{M}} = \begin{pmatrix} t^2 + \frac{(1+\eta)^2}{4} & -t & \frac{1-\eta^2}{4} & & & & \\ -t & t^2 + \frac{1+\eta^2}{2} & -t & \ddots & & & \\ \frac{1-\eta^2}{4} & -t & \ddots & \ddots & & & \\ & \ddots & \ddots & \ddots & \ddots & & \\ & & & \ddots & \ddots & -t & \frac{1-\eta^2}{4} \\ & & & \ddots & -t & t^2 + \frac{1+\eta^2}{2} & -t \\ & & & & \frac{1-\eta^2}{4} & -t & t^2 + \frac{(1-\eta)^2}{4} \end{pmatrix} \tag{15.34}$$

---

[4] For the extraordinary transition, the surface has ordered before the bulk and one expects $G_s = \langle\sigma(z_1)\sigma(z_2)\rangle - \langle\sigma(z_1)\rangle\langle\sigma(z_2)\rangle \sim (v_1 v_2)^{-1/8}$. Then $F_s(V) = \sqrt{\sqrt{1-V}+1} + \sqrt{\sqrt{1-V}-1}$, leading to $\eta_\|^{ex} = 4$.

The solution of this eigenvalue problem can be done is close analogy to solving the one-dimensional Schrödinger equation with a step-like potential. In fact, the technique readily generalizes to other boundary conditions or even interactions which are no longer fully translation invariant. It is enough to illustrate the idea for the special case $\eta = 1$, to which we restrict from now on. For the eigenfunctions, we make the ansatz, with $r = 1, 2, \ldots, N$

$$(\boldsymbol{\Phi}_k)_r = A e^{ikr} + B e^{-ikr} \tag{15.35}$$

and the constants A and B must be conveniently chosen. For $2 \le r \le N - 1$, this ansatz leads again to the explicit form of the lattice dispersion relation, see (10.21)

$$\Lambda_k^2 = (t - \cos k)^2 + \sin^2 k. \tag{15.36}$$

The quantization condition for the k is obtained by considering the eigenvalue equations for $r = 1$ and $r = N$. This leads to the following linear system of equations

$$\begin{pmatrix} 2te^{ik}\cos k - te^{2ik} & 2te^{-ik}\cos k - te^{-2ik} \\ (1 - 2t\cos k)e^{ikN} + te^{ik(N-1)} & (1 - 2t\cos k)e^{-ikN} + te^{-ik(N-1)} \end{pmatrix} \begin{pmatrix} A \\ B \end{pmatrix} = 0 \tag{15.37}$$

and it follows that the determinant of this has to vanish. This condition determines the discretization of the k. Calculating the determinant yields [96, 115]

$$\sin kN = t \sin (k(N + 1)). \tag{15.38}$$

For general values of t, this equation cannot be solved explicitly for k. At the critical point $t = 1$, however, the solution is

$$k = \frac{2m + 1}{2N + 1}\pi \;,\quad m = 0, 1, \ldots, N - 1. \tag{15.39}$$

Then the ground state energy and the first two correlation lengths are [115]

$$E_0^{(F)} = -\frac{1}{2} \sum_{m=0}^{N-1} 2\sin\left(\frac{2m + 1}{2(2N + 1)}\pi\right)$$
$$\simeq -\frac{2}{\pi}N + \frac{1}{2} - \frac{1}{\pi} - \frac{\pi}{48}\frac{1}{N} + \mathcal{O}(N^{-2}), \tag{15.40}$$

$$\xi_{\sigma,s}^{-1} = 2\sin\frac{\pi}{2(2N + 1)} \simeq \frac{\pi}{2N} + \mathcal{O}(N^{-2}) \tag{15.41}$$

$$\xi_{\varepsilon,s}^{-1} = 2\sin\frac{\pi}{2(2N + 1)} + 2\sin\frac{3\pi}{2(2N + 1)}$$
$$\simeq \frac{2\pi}{N} + \mathcal{O}(N^{-2}). \tag{15.42}$$

Let us discuss these results. For the ground state energy, the leading term in the number of sites N is the non-universal bulk free energy density. It is the same as obtained for $\eta = 1$ for both periodic and antiperiodic boundary conditions. The next term is the surface free energy density, which is non-universal as well. The term of order $1/N$ is universal and can be used to extract the central charge. Comparing with (15.26), we read off

$$c = 1/2 \qquad (15.43)$$

as expected from the periodic case. Similarly, the surface critical exponents of the order parameter and the energy density are

$$x_{\sigma,s} = 1/2 \ , \ \ x_{\varepsilon,s} = 2. \qquad (15.44)$$

The first of these reproduces the result $\eta_\parallel = 1$ via (15.17), in agreement with earlier numerical results [210] and the second one reconfirms beyond perturbation theory the result (15.28). In a similar way, the higher energy gaps can be calculated. Comparing with the Virasoro characters of Table 10.2, one finds the complete operator content for the Ising model with free boundary conditions [259]

$$H_0^{(F)} = (0) \ , \ \ H_1^{(F)} = \left(\frac{1}{2}\right) \qquad (15.45)$$

in terms of the Virasoro characters for $c = 1/2$. From this, we can understand why $x_{\varepsilon,s} = 2$ always. The corresponding state should have the same quantum numbers of charge, parity and C-parity as the ground state. The first excited state in this sector is $L_{-2}\mathbf{1}$, the energy–momentum tensor, which has $x = 2$.

We digress briefly towards yet further boundary conditions. We consider the cases of fixed boundary conditions, where the boundary spins are either both "up" $(++)$ or else "up" and "down" $(+-)$, and mixed (M) boundary conditions, which are free boundary conditions on one side and fixed on the other. Fixing a boundary spin can be said to come from an infinite surface magnetic field. It is an easy exercise to show, using (9.12), that the quantum chain with free boundary condition is dual to the chains with fixed boundary conditions and that these fixed boundary conditions correspond to the *extraordinary* transition.[5] The operator content can be worked out following the same lines as above and is found to be [138, 112]

$$H^{(++)} = (0) \ , \ \ H^{(+-)} = \left(\frac{1}{2}\right) \ , \ \ H^{(M)} = \left(\frac{1}{16}\right). \qquad (15.46)$$

---

[5] More precisely, extraordinary transitions only exist if the surface can spontaneously order without the presence of an external field, which is not possible for 2D systems with short-range interactions. However, in these cases ordered boundaries can be obtained by magnetic fields and the corresponding surface transition is called a **normal transition** [197]. It is usually in the same universality class as the extraordinary transition and for that reason, we shall not explicitly distinguish them here (but see [60] for extraordinary vs. normal transitions the O(n) model with $n < 1$). An interesting problem is the crossover between the ordinary and normal transitions for a weak surface field B_1, leading to non-monotonous profiles of the order parameter $m(z)$ [537].

The finite-size scaling functions for all these boundary conditions can be worked out fairly easily [117]. We remark that all these operator contents can also be found from modular invariance considerations [138]. This completes the conformal description of this system.[6]

An interesting application of these results occurs when considering the surface critical behaviour of either binary alloys undergoing a second-order phase transition or else Ising *anti*ferromagnets in the presence of a (non-ordering) magnetic field. While the bulk critical behaviour of ferromagnetic and antiferromagnetic Ising models is the same, it can be shown that *for antiferromagnets, the surface critical exponents depend on the orientation of the surface with respect to the crystal axes* [209]. Here, we restrict entirely to the 2D situation, as illustrated in Fig. 15.5.

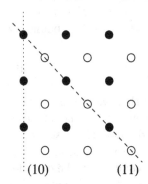

Fig. 15.5. Surfaces (10) and (11) for a square lattice, as indicated by dotted and dashed lines

Consider an antiferromagnetic Ising model, with classical Hamiltonian

$$\mathcal{H} = +J \sum_{\langle i,j \rangle} \sigma_i \sigma_j - B \sum_i \sigma_i - B_1 \sum_{i,\text{surf}}' \sigma_i, \qquad (15.47)$$

where B, B_1 are the bulk and surface magnetic field and the last sum only extends over the sites on the surface. In the direction perpendicular to the surface, periodic boundary conditions are assumed. The ground state of $\mathcal{H}$ is antiferromagnetic, as indicated by black and white sites in Fig. 15.5 and thus, both bulk and surface magnetic fields are non-ordering. The physical order parameter is proportional to the difference of the sublattice magnetizations.

However, the (11) surface respects the symmetry of the model under exchange of the 'black' and 'white' sublattices. The problem can then be mapped back to the usual ferromagnetic Ising model and one expects ordinary surface critical behaviour. On the other hand, the (10) surface breaks the sublattice symmetry. The order parameter profile is symmetric and antisymmetric around the center of a strip with N layers for N odd and even,

---

[6] $\beta_1 = 1/2$ can also be found directly from the surface magnetization, see (15.141).

respectively. If B or B_1 are non-vanishing, this implies that the spins at the boundaries are fixed in a relative $++$ or $+-$ orientation, for N odd or even, respectively. One thus expects an extraordinary surface transition.[7] From the operator contents (15.45,15.46), the exponent $\eta_\parallel$ may be easily read off. These predictions have been checked numerically through transfer matrix calculations. From the largest two eigenvalues of $\mathcal{T}$, the spin-spin correlation length is obtained in the usual way (see Chap. 1), but since the lattice constant a for the (10) surface is twice the one for the (11) surface, this factor must be taken into account when considering scaling amplitudes. On a strip of $N \times \tau$ layers (with $\tau = 1$ for (11) and $\tau = 2$ for (10)), the amplitude $A := \pi \lim_{N\to\infty} N^{-1}\xi_N$ is found numerically to converge rapidly to the values [209]

$$A^{\mathrm{ord}} = 2 \ , \quad A^{++} = \frac{1}{2} \ , \quad A^{+-} = 1 \tag{15.48}$$

in agreement with (15.45,15.46). The order parameter profiles have also been obtained and show the expected (anti)symmetry with respect to the center of the strip [209].[8]

15.4.2 Three-States Potts Model

The quantum Hamiltonian of the three states Potts model was given in (12.1). Free boundary conditions imply $\Gamma_{N+1} = 0$. The charge sectors are $H_Q^{(F)}$, with $Q = 0, 1, 2$. These can be further decomposed by classifying the states according to their C-parity (see Chap. 9). The model is therefore decomposed into the sectors $H_{0,\pm}^{(F)}$ and $H_{1,\pm}^{(F)} = H_{2,\pm}^{(F)}$, where $\pm$ labels the C-parity. The central charge can be obtained by the numerical techniques described in Chap. 9 and one finds that $c = 4/5$, as expected. The surface operator content reads [259, 138]

$$H_{0,+}^{(F)} = (0) \ , \quad H_{0,--}^{(F)} = (3) \ , \quad H_1^{(F)} = H_{1,+}^{(F)} + H_{1,-}^{(F)} = \left(\frac{2}{3}\right) \tag{15.49}$$

in terms of $c = 4/5$ Virasoro characters. In particular we have for the leading surface exponents

$$x_{\sigma,s} = \frac{2}{3} \ , \quad x_{\varepsilon,s} = 2 \tag{15.50}$$

and thus $\eta_\parallel = 4/3$, as was also confirmed numerically [210].

We also mention the results for some further boundary conditions. For fixed boundary conditions, the boundary spins can be the same, like (00), or different, like (01). Mixed boundary conditions (M) correspond to free

---

[7] For $B = B_1 = 0$, the critical magnetization vanishes on both sublattices, leading to an ordinary transition.

[8] The generalization to 3D is obvious, a surface with Miller indices (ijk) is symmetry-preserving iff $i + j + k$ is even. Surface transitions and coarse-grained continuum models are extensively studied on the mean-field level in [434].

boundary conditions on one side and fixed on the other. For the cases (00) and (M), the system has a global $\mathbb{Z}_2$ symmetry and the spectrum can be decomposed into charge sectors $Q = 0, 1$. The operator content is then [138]

$$H_0^{(00)} = (0) \ , \ H_1^{(00)} = (3) \ , \ H^{(01)} = \begin{pmatrix} 2 \\ 3 \end{pmatrix}$$

$$H_0^{(M)} = \begin{pmatrix} 1 \\ 8 \end{pmatrix} \ , \ H_1^{(M)} = \begin{pmatrix} 13 \\ 8 \end{pmatrix}. \tag{15.51}$$

Non-toroidal boundary conditions[9] were already discussed in Chap. 12.

15.4.3 Temperley–Lieb Algebra and Relation with the XXZ Chain

The Temperley–Lieb algebra introduced in Chap. 8 can be used to yield an interesting relationship of the p-state Potts model with the XXZ quantum chain [22]. Consider the (unnormalized) quantum Hamiltonian of the p-state Potts model with free boundary conditions

$$H_{\text{Potts}} = -\sum_{n=1}^{N}\sum_{k=1}^{p-1}\Gamma_n^k - \sum_{n=1}^{N-1}\sum_{k=1}^{p-1}\sigma_n^k\sigma_{n+1}^{p-k} = 2N - 1 - \sqrt{p}\sum_{\ell=1}^{2N-1} e_\ell, \tag{15.52}$$

where the operators e_ℓ are defined by [87]

$$e_{2\ell-1} = \frac{1}{\sqrt{p}}\sum_{k=0}^{p-1}\Gamma_\ell^k \ ; \ \ell = 1, 2, \ldots, N$$

$$e_{2\ell} = \frac{1}{\sqrt{p}}\sum_{k=0}^{p-1}\sigma_\ell^k\sigma_{\ell+1}^{p-k} \ ; \ \ell = 1, 2, \ldots, N-1. \tag{15.53}$$

These satisfy the Temperley–Lieb algebra (8.59). An alternative realization of (8.59) in terms of Pauli matrices is

$$e_\ell = \frac{1}{2}\left(\sigma_\ell^x\sigma_{\ell+1}^x + \sigma_\ell^y\sigma_{\ell+1}^y\right) + \frac{1}{2}\cosh\theta\left(1 - \sigma_\ell^z\sigma_{\ell+1}^z\right) + \frac{1}{2}\sinh\theta\left(\sigma_{\ell+1}^z - \sigma_\ell^z\right), \tag{15.54}$$

where

$$\cosh\theta = \frac{1}{2}\sqrt{p}. \tag{15.55}$$

---

[9] Recently, a new type of boundary condition was discovered [10]. Consider the Hamiltonian $H = H_c - (\lambda_T\sigma_0 + \lambda_T^*\sigma_0^+)$, where H_c is given by (12.1) at $\lambda = \lambda_c = 1$ and with open end at site N, which is defined on $N+1$ sites. The new boundary condition corresponds to $\lambda_T < 0$. The surface operator content (even for $\lambda_T \in \mathbb{C}$) is listed in [10]. It can be shown that this one completes the list of conformally invariant boundary conditions for the three-states Potts model [251]. See [655] for the relation with the classification of conformal field theories.

It then follows [22]

$$H_{\text{Potts}} = (2N - 1)\left(1 - \frac{p}{4}\right) + \sqrt{p}\, H_{\text{XXZ}}, \tag{15.56}$$

where

$$H_{\text{XXZ}} = -\frac{1}{2}\sum_{n=1}^{2N-1}\left(\sigma_n^x\sigma_{n+1}^x + \sigma_n^y\sigma_{n+1}^y - \frac{1}{2}\cosh\theta\,\sigma_n^z\sigma_{n+1}^z\right) \tag{15.57}$$
$$+\frac{1}{2}\sinh\theta\left(\sigma_1^z - \sigma_{2N}^z\right)$$

describes the XXZ quantum chain with a surface field. Equation (15.56) is an operator equivalence, both H_{Potts} and H_{XXZ} are, up to trivial constants, just the sum of Temperley–Lieb operators. Consequently, the possible eigenvalues of H_{Potts} and H_{XXZ} are determined by the algebra (8.59). However, the degeneracy of a given eigenvalue is not determined from the algebra but rather from the representation chosen for the e_ℓ. The XXZ chain does contain eigenvalues which do not appear in the spectrum of the Potts quantum chain. For example, the XXZ eigenstates with $\sigma_\ell^z|0\rangle = \pm|0\rangle$ uniformly for all ℓ have no analogue in the Potts model. The same correspondence can also be obtained by comparing the isotropic partition functions, see [G19].

The systematic construction of quantum systems for Temperley–Lieb algebras has been studied in detail [451], but is beyond the scope of this book.

Since the XXZ chain can be diagonalized exactly using the Bethe ansatz, many of the numerical results in the models described have been confirmed analytically by mapping them onto a XXZ chain. As an example, consider the correct normalization of a quantum Hamiltonian such that conformal invariance can be applied. The correct normalization of the XXZ chain is known analytically. Then the correct normalization of H_{Potts} follows from (15.57) and is given by [22]

$$\zeta = \frac{\pi\sqrt{p(4 - p)}}{4\arccos(\sqrt{p}/2)}. \tag{15.58}$$

For $p = 3$, we recover the normalization already applied in (12.1). In addition, the known results for the (non-universal) bulk and surface free energies [22, 59] of the XXZ chain may be translated into the corresponding quantities of the p–states Potts model. For free boundary conditions, one expects for the ground state energy of the Hamiltonian $H = \zeta^{-1}H_{\text{Potts}}$

$$E_0 = A_0 N + A_1 - \frac{\pi c}{24}\frac{1}{N} + \dots. \tag{15.59}$$

Some values of A_0 and A_1 are collected in Table 15.2. They are non-universal, but A_0 is independent of the boundary conditions. For the Ising model, we have checked this explicitly in Chap. 10.

Similar relationships of H with the XXZ chain via the Temperley–Lieb algebra apply for periodic or toroidal boundary conditions as described in detail in the literature [22].

Table 15.2. Bulk and surface free energy amplitudes A_0, A_1, as defined in (15.59), for the p−states Potts quantum chain $H = \zeta^{-1} H_{\text{Potts}}$, (15.52). After [59]

p	A_0	A_1
1	0	0
2	$-2/\pi$	$1/2 - 1/\pi$
3	$-4/(3\pi) - 8/(9\sqrt{3})$	$1 - 4/(3\sqrt{3})$
4	$2(1 - 4\ln 2)/\pi$	$1 - (1 + 2\ln 2)/\pi$

15.4.4 Tricritical Ising Model

The surface exponents of the tricritical Ising model were obtained from the Blume–Capel quantum chain (12.6) with free boundary conditions $S_{N+1}^{x,z} = 0$ [43]. The central charge is again found to be $c = 7/10$, in agreement with the bulk value. The surface operator content is [43]

$$H_0^{(F)} = (0) \ , \ H_1^{(F)} = \left(\frac{3}{2}\right) \tag{15.60}$$

from which we read off the surface exponents

$$x_{\sigma,s} = \frac{3}{2} \ , \ x_{\epsilon,s} = 2 \tag{15.61}$$

and therefore $\eta_\parallel = 3$.

As in the Ising model, one can introduce fixed and mixed boundary conditions. The operator content is [280]

$$H^{(++)} = (0) \ , \ H^{(+-)} = \left(\frac{3}{2}\right) \ , \ H^{(M)} = \left(\frac{7}{16}\right). \tag{15.62}$$

15.4.5 Yang–Lee Edge Singularity

The Yang–Lee singularity of the Ising model was described by the quantum chain (12.31) with $\sigma_{N+1}^{x,z} = 0$. The surface operator content is simply [263]

$$H^{(F)} = (0) \tag{15.63}$$

and thus $x_{\sigma,s} = \eta_\parallel/2 = 2$.

15.4.6 Ashkin–Teller Model

The quantum Hamiltonian of the Ashkin–Teller model was given in (12.45). We have described how its spectrum can be decomposed into the sectors of the model as given by the irreducible representation of the dihedral group $\mathbb{D}_4$, see (12.51). Free boundary conditions imply $\sigma_{N+1} = 0$. We have the sectors $H_{Q,\pm}^{(F)}$, where $\pm$ labels the C-parity. The surface operator content is then [36]

$$H_{0,+}^{(F)} = \sum_{m=0}^{\infty} (4m^2) + \sum_{m=1}^{\infty} (4m^2/g)$$

$$H_{0,-}^{(F)} = \sum_{m=0}^{\infty} ((2m+1)^2/g)$$

$$H_{2,+}^{(F)} = \sum_{m=0}^{\infty} ((2m+1)^2) + \sum_{m=1}^{\infty} (4m^2/g)$$

$$H_{2,-}^{(F)} = \sum_{m=0}^{\infty} ((2m+1)^2/g)$$

$$H_1^{(F)} = H_3^{(F)} = \sum_{m=0}^{\infty} \left(\frac{(2m+1)^2}{4g} \right) \tag{15.64}$$

using the $c = 1$ Virasoro characters (12.57) and the coupling g depends on the model parameters as given in (12.46). The surface exponents can be read off

$$x_{\sigma,s} = \frac{1}{4g} \quad , \quad x_{\varepsilon,s} = 2 \tag{15.65}$$

and thus $\eta_{\parallel} = 1/2g$.

This result contains quite a few interesting special cases. These can be selected by specifying g or alternatively, via (12.46), the four-spin coupling ε. For $\varepsilon = 0$, the model decouples into two Ising models. This corresponds to $g = 1/2$ and we recover $x_{\sigma,s} = 1/2$. The four-state Potts model corresponds to $g = 1/4$ while the Kosterlitz–Thouless transition has $g = 1$. We thus have

$$\eta_{\parallel} = \begin{cases} 2 & \text{four-states Potts model} \\ 1/2 & \text{XY model} \end{cases} . \tag{15.66}$$

Finally, we mention that for peculiar values of g the spectrum of the model can be written in terms of irreducible representations of a larger algebra. We only mention here the case of superconformal invariance, which is realized for $g = 1/6, 2/3, 3/2, 6$. For example, for $g = 2/3$, one has [36]

$$H_{0,+}^{(F)} = [0] \ , \ H_{2,+}^{(F)} = [1] \ , \ H_{0,-}^{(F)} = H_{2,-}^{(F)} = [3/2] \ , \ H_1^{(F)} = H_3^{(F)} = [3/8]^R, \tag{15.67}$$

where the $[\Delta]$ are superconformal characters which decompose in terms of $c = 1$ Virasoro characters as follows

$$[0] = \{0\} + \sum_{m=0}^{\infty} (6m^2), \ [1] = \{1\} + \sum_{m=1}^{\infty} (6m^2), \tag{15.68}$$

$$[3/2] = \sum_{m=0}^{\infty} \left(\frac{3}{2}(2m+1)^2 \right), \ [3/8]^R = \sum_{m=0}^{\infty} \left(\frac{3}{8}(2m+1)^2 \right),$$

where the notation of (12.56) was used.

15.4.7 XXZ Quantum Chain

At last, we discuss the XXZ quantum chain. For the bulk, we had already seen that the unitary minimal conformal spectra could be obtained by a suitable projection. We now give the analogue for free, fixed and mixed boundary conditions [24, 280]. We repeat the quantum Hamiltonian

$$H = -\frac{\gamma}{2\pi \sin \gamma} \sum_{n=1}^{N-1} [\sigma^x(n)\sigma^x(n+1) + \sigma^y(n)\sigma^y(n+1)$$
$$- \cos \gamma \sigma^z(n)\sigma^z(n+1) + v\sigma^z(1) + v'\sigma^z(N)], \qquad (15.69)$$

where v, v' are the surface external fields and $v = v' = 0$ corresponds to free boundary conditions. These are conveniently written [24, 280]

$$v = i \sin \gamma \coth \alpha \ , \quad v' = i \sin \gamma \coth \alpha', \qquad (15.70)$$

where α and α' are complex parameters. The spectrum is real for the following choices

$$\alpha = \frac{i\pi}{2}(a - \psi) \ , \ \alpha' = -\frac{i\pi}{2}(a + \psi)$$
$$\alpha = \frac{\pi}{2}(b - i\psi) \ , \ \alpha' = -\frac{\pi}{2}(b + i\psi), \qquad (15.71)$$

where a, b, ψ are real. Free boundary conditions are recovered for $a = b = 0$, $\psi = 1$.

Denote by $E_{Q,i}(N)$ the ith level in the charge sector Q, where the U(1) charge operator is given by (12.73), obtained for N sites. Let $E_0^{(F)}(N)$ be the ground state energy for the Hamiltonian with free boundary conditions. Consider the scaled gaps

$$\mathcal{E}_{Q,i} := \frac{N}{\pi} \left(E_{Q,i}(N) - E_0^{(F)}(N) \right) \qquad (15.72)$$

and define the partition function as the $N \to \infty$ limit of the quantities [25, 297]

$$\mathcal{Z}_Q(q) := \sum_i q^{\mathcal{E}_{Q,i}(N)} = P(q)q^{(Q+\phi)^2/(4g)} \qquad (15.73)$$

for a, b finite and $\phi = 2g(1 - \psi)$, while $P(q)$ was defined in (10.31). For free boundary conditions, we have again the surface exponent $x_{\sigma,s} = 1/(4g)$. We also note that for a fixed value of Q, the operator content is given by a single irreducible representation of a so-called **shifted U(1) Kac–Moody algebra** [37], where the shift ϕ can be written in terms of the model parameters. We shall discuss in the next section shifted Kac–Moody algebras in connection with defect lines.

We now give the projectionchain to the surface operator content of the unitary minimal models. The procedure is quite analogous to the one of Chap. 12. Take $E_{0,0}(N)$ as the ground state energy and define [280]

$$\mathcal{F}_{R,i}(N) := \frac{N}{\pi}\left(E_{R,i}(N) - E_{0,0}(N)\right). \tag{15.74}$$

As before, one gets the partition functions in the $N \to \infty$ limit

$$\mathcal{Y}_R(q) := \sum_i q^{\mathcal{F}_{R,i}(N)} = \sum_{k=-\infty}^{\infty} q^{[(nk+R+\phi)^2-\phi^2]/(4g)} P(q), \tag{15.75}$$

where $R = 0, 1, \ldots, n-1$ for N even and $2R = 1, 3, \ldots, 2n-1$ for N odd. Choose

$$g = \frac{n^2}{4m(m+1)} \ , \quad \phi = \frac{n}{2m(m+1)} \tag{15.76}$$

and the partition functions $\mathcal{Y}_R$ can now be expressed in terms of the unitary Virasoro characters $\chi_{r,s}(q)$. This defines the desired projection rules.

As we did for the bulk system in Chap. 12, we now give the projection rules for the p-state Potts model and the tricritical p-states Potts model. For the p-state Potts model (which had been called the 2_R system before), one has (see (12.81)) [280]

$$\chi_{1,2u+1} = \mathcal{Y}_{-u} - \mathcal{Y}_{u+1} \quad \text{free, fixed}$$
$$\chi_{1,2u} = \mathcal{Y}_{-u+1/2} - \mathcal{Y}_{u+1/2} \quad \text{mixed.} \tag{15.77}$$

For the tricritical p-state Potts model, one has (see (12.81)) [280]

$$\chi_{2u+1,1} = \mathcal{Y}_u - \mathcal{Y}_{-u-1} \quad \text{free, fixed}$$
$$\chi_{2u,1} = \mathcal{Y}_{u-1/2} - \mathcal{Y}_{-u-1/2} \quad \text{mixed.} \tag{15.78}$$

These results reproduce all the explicit examples given above. It is now a straightforward matter to read off the surface critical exponents of any desired unitary minimal model. Having found this, the conformal techniques outlined in the context of the Ising model can be used to calculate explicitly the multipoint correlation functions.

15.4.8 Percolation

The exponent $\eta_{\parallel}$ has been conjectured by generalizing the existing results for $q = 2, 3, 4$ to the q-states Potts model with q arbitrary [133]. In the $q \to 1$ limit, the conformal weight $\Delta = \Delta_{3,1}$ can be recognized in the minimal model $\mathcal{M}_{4,6}$ and that leads to the conformal invariance prediction $\eta_{\parallel} = 2/3$. This has been checked by several numerical techniques [622, 428, 526].

First, one may study the pair connectivity which for percolation plays the role of the order parameter two-point function [622, 428]. Percolation clusters were generated through the Leath algorithm. Considering the connectivity between two points on the surface of the half-plane and relating it to the the order parameter two-point function, $\eta_{\parallel} = 0.66(1)$ was found [622]. In addition, the correct transformation of the pair connectivity under

2D conformal transformations was checked by measuring the effective scaling exponents if the boundary consists of straight line segments with corners (see (15.154)), as well as the semi-infinite strip of finite width. All results are fully consistent with the conformal predictions [622]. An extensive verification of this was presented in [428].

Second, transfer matrices on both the square and triangular lattices in the strip geometry with free boundary conditions were employed to measure $\eta_\parallel$ (and also the bulk exponent η) from the standard amplitude-exponent relations (3.49,15.26). This is illustrated in Table 15.3, where estimates (the data of [526] are truncated to six figures) are shown as a function of the width of the strip, together with the $N \to \infty$ limit as estimated from the BST algorithm. Already, the estimates from Table 15.3 are in good agreement with the expected values $\eta = 5/24$ and $\eta_\parallel = 2/3$ and universality of the exponents is explicitly reconfirmed. A further careful study [526] of the corrections to scaling gives the precise determination $\eta_\parallel = 0.6664(8)$ and also gives at the ordinary transition the leading irrelevant RG eigenvalue as $y_s = -1.001(3)$ and $-1.004(3)$ for the square and and triangular lattices, respectively and in agreement with the conformal prediction $y_s = -1$. We are not aware of a systematic numerical study of the surface operator content from the higher excitations of the transfer matrix.

Table 15.3. Exponents η and $\eta_\parallel$ for critical percolation on the square and triangular lattices, obtained from the transfer matrix on strips of width N for periodic and free boundary conditions, respectively (after [526])

N	square η	$\eta_\parallel$	triangular η	$\eta_\parallel$
2	0.216346	0.432695	0.211477	0.467384
3	0.213060	0.484810	0.211193	0.513593
4	0.212558	0.517493	0.210355	0.542353
5	0.211467	0.540029	0.209741	0.561983
6	0.210737	0.556558	0.209349	0.576243
7	0.210223	0.569219	0.209096	0.587075
8	0.209856	0.579240	0.208925	0.595584
9	0.209587	0.587373	0.208805	0.602447
10	0.209383	0.594110	0.208717	0.608102
∞	0.20835(2)	0.6664(4)	0.20833(2)	0.665(1)

In the presence of boundaries, other quantities than the operator content may be studied. Consider two disjoint segments S_1 and S_2 on the smooth boundary of a simply connected compact region $\mathcal{G}$. Let $\mathcal{Z}_{ab}$ be the partition function of the q-state Potts model with all spins on S_1 fixed in the state a and all spins on S_2 fixed in the state b. Then the **crossing probability** between S_1 and S_2 is [146, 428, 14]

$$\pi(S_1, S_2) := \lim_{q \to 1} (\mathcal{Z}_{aa} - \mathcal{Z}_{ab}) = 1 - \lim_{q \to 1} \mathcal{Z}_{ab}. \qquad (15.79)$$

Following [147], the interior of $\mathcal{G}$ may be mapped conformally to the upper half plane, so that the boundary is mapped onto the real axis. Let $\phi_{(a|f)}(r)$ denote the **boundary operator**[10] which switches a boundary state fixed at a to free boundary conditions at the point r (and similarly $\phi_{(f|a)}(r)$). It is easy to see that [146]

$$\mathcal{Z}_{ab} = \mathcal{Z}_f \langle \phi_{(f|a)}(r_1)\phi_{(a|f)}(r_2)\phi_{(f|b)}(r_3)\phi_{(b|f)}(r_4)\rangle , \qquad (15.80)$$

where for the boundary sites $r_1 < r_2 < r_3 < r_4$ may be assumed and $\mathcal{Z}_f$ is the partition function for free boundary conditions. For percolation $\mathcal{Z}_f = 1$ [146]. To calculate this, the operators $\phi_{(a|b)}$ are related to primary operators in a non-unitary minimal model. In the continuum limit, the self-duality of the critical q-states Potts model exchanges free (f) and fixed (a) boundary conditions. An insertion of the Potts spin operator on the boundary point r is mapped onto an insertion of the operator $\phi_{(a|b)}(r)$ with $b \neq a$ and it follows that the correlators of $\phi_{(a|b)}$ are simply related to the boundary spin operator for free boundary conditions, which in turn might be expressed in terms of the primary operator $\phi_{3,1}$ of the minimal model $\mathcal{M}_{4,6}$ (Chap. 12). To identify $\phi_{(f|a)}$, consider the insertion $\phi_{(a|f)}(r)\phi_{(f|b)}(r')$ for $|r - r'| \to 0$. This is given through the operator product expansion

$$\phi_{(a|f)} \cdot \phi_{(f|b)} = \delta_{a,b}\mathbf{1} + \phi_{(a|b)} + \dots . \qquad (15.81)$$

From the general form of the OPE for minimal models (Chap. 7), it follows that $\phi_{(a|f)}$ is related to the primary operator $\phi_{2,1}$ of the minimal model $\mathcal{M}_{4,6}$, with $\Delta_{2,1} = 0$. In other words, $\phi_{(a|f)}$ is a null operator at level two, which implies that $\pi(S_1, S_2)$ can be found from the solutions of a second order differential equation (see Chap. 4). If $r = [(r_4-r_3)(r_2-r_1)]/[(r_3-r_1)(r_4-r_2)]$ and $\pi(S_1, S_2) = g(r)$, this equation is [146]

$$r(1 - r)g''(r) + \frac{2}{3}(1 - 2r)g'(r) = 0 \qquad (15.82)$$

Two boundary conditions must be satisfied. For $r \to 0$, one expects that $\mathcal{Z}_{ab} \to \mathcal{Z}_{aa} = 1$ and for $r \to 1$, from the operator product expansion the solution should scale as $(1 - r)^{1/3}$. Finally [146]

$$\pi(S_1, S_2) = g(r) = \frac{3\Gamma\left(\frac{2}{3}\right)}{\Gamma\left(\frac{1}{3}\right)^2} \, _2F_1\left(\frac{1}{3}, \frac{2}{3}, \frac{4}{3}; r\right) \qquad (15.83)$$

This should still be transformed back to the interior of $\mathcal{G}$. But, since $\Delta_{\phi_{(a|f)}} = \Delta_{2,1} = 0$, the result (15.83) is truly invariant. Extensive numerical tests have fully confirmed it [428, 14].

---

[10] Boundary operators $\phi_i^{(B)}$ are defined through the OPE $\phi(z) \cdot \phi(\bar{z}) = \sum_i (z - \bar{z})^{\Delta_i - 2\Delta} \phi_i^{(B)}((z + \bar{z})/2)$ and, although restricted to the boundary, belong to the same operator product algebra as the bulk operators [145].

One may also ask how many of these **incipient spanning clusters** exist. Consider the rectangular region $\mathcal{G} = [0, kL] \times [0, L]$. Let $P(n, k, L)$ denote the probability that the strip is traversed in the direction of length kL by exactly n independent clusters. To find the asymptotic form, divide $\mathcal{G}$ into two rectangles of size $[0, kL] \times [0, L/2]$ and one expects that $P(n, k, L) \sim P(n/2, 2k, L/2)^2$ [153]. Together with the expected exponential dependence of kL for L fixed, this leads to $\ln P(n, k, L) \sim -\alpha n^2 k$, where $\alpha > 0$ is a constant. This argument can be justified rigorously through bounds on $P(n, k, L)$ [15]. Also, we point out that this contradicts the older expectation that only one such cluster should exist, see [578] for a review. P can be calculated from the partition function of the Potts model with fixed boundary conditions at both end of the strip, $\langle a | e^{-kLH(L)} | a \rangle = \sum_R e^{-\pi x_R k} \sum_{\mathcal{N}} \langle a | \mathcal{N} \rangle \langle \mathcal{N} | a \rangle e^{-\pi \mathcal{N} k}$ in a somewhat abreviated notation, where the first sum is over the primary states R and the second over the secondary states in each conformal tower. Identifying the primaries R_n with the primary operators $\phi_{1,2n+1}$, of scaling dimension $x_{R_n} = x_{1,2n+1} = n(2n - 1)/3$ for $q = 1$, one finds that, for k large and for any n [153]

$$\lim_{L \to \infty} \ln P(n, k, L) \sim -\frac{2\pi}{3} n(n-1)k \qquad (15.84)$$

(a similar result holds for an open-ended cylinder). Monte Carlo simulations for $P(n, k, L)$ with $k = 1, 2, 3$ [566] appear to be consistent with this, although for these k are not large.

15.4.9 Polymers

Surface critical phenomena in polymer systems provide a very rich field of study, with many applications of conformal invariance. We shall not discuss it here at all, but merely refer to a nice review [180] and two recent books [217, 599] for further information. The exact calculation of surface exponents at the Θ'-point and the long debate about this question (already briefly mentioned in Chap. 12) is reviewed in [579]. The extraordinary/normal transition has recently been studied in [60], where it was shown that, in the $O(n)$ model with $n < 1$, the surface critical behaviour with enhanced surface couplings differs from that occuring when fixed boundary conditions are imposed, unlike the case of $n \geq 1$ in higher dimensions.

15.5 Profiles

Since translation invariance is broken by a free surface, expectation values $\langle \phi(v) \rangle$ of some scaling operator may display a profile depending on v. For a single free surface, the form of the profile follows already from the projective Ward identities (15.15) and, as first shown by Fisher and de Gennes [229]

$$\langle \phi(v) \rangle = \phi_0 \, v^{-x}, \tag{15.85}$$

where x_ϕ is the *bulk* scaling dimension of ϕ and ϕ_0 is some constant which depends on the boundary conditions. This result, valid in the upper half-plane, may be conformally transformed into finite geometries [115]. For example, the transformation (15.25) maps the upper half-plane onto the infinitely long strip of finite width L. For a primary operator, one has

$$\langle \phi(v) \rangle_{aa} = \phi_{0,a} \left[\left(\frac{L}{\pi} \right) \sin \left(\frac{\pi v}{L} \right) \right]^{-x_\phi}, \tag{15.86}$$

where the *same* boundary condition is imposed on both surfaces.

It is also possible, to predict the profiles between surfaces with *different* boundary conditions [119]. Consider a semi-infinite system defined on the upper half-plane and such that on the real axis, the system is fixed in state a for $x > \zeta$ and fixed in state b for $x < \zeta$. Following [119], the derivation of the Ward identities for a primary operator ϕ proceeds almost as for uniform boundary conditions, with the important difference that now

$$\langle T(z) \rangle_\zeta = \frac{t_{ab}}{(z - \zeta)^2}, \tag{15.87}$$

where t_{ab} is the scaling dimension of the boundary operator which brings about the change of the boundary conditions from a to b. It is also related to the universal Casimir amplitude A_{ab} in the free energy density $f = f_{\text{bulk}} + L^{-1}(f_{\text{surf}}^{(a)} + f_{\text{surf}}^{(b)}) + L^{-2} A_{ab} + \ldots$ through [6, 93]

$$t_{ab} = \frac{1}{\pi} A_{ab} + \frac{c}{24}. \tag{15.88}$$

The conformal Ward identity becomes

$$\langle T(z) \phi_1(z_1, z_1') \cdots \rangle_\zeta = \left\{ \langle T(z) \rangle_\zeta + \frac{1}{z - \zeta} \frac{\partial}{\partial \zeta} \right.$$
$$\left. + \sum_i \left[\frac{\Delta_i}{(z - z_i)^2} + \frac{1}{z - z_i} \frac{\partial}{\partial z_i} + \frac{\bar{\Delta}_i}{(z - \bar{z}_i)^2} + \frac{1}{z - \bar{z}_i} \frac{\partial}{\partial z_i'} \right] \right\} \langle \phi_1(z_1, z_1') \cdots \rangle_\zeta. \tag{15.89}$$

From this, expressions for the operators $\mathcal{L}_k$ may be derived, as shown in Chap. 4. Using translational invariance, one may always arrange for $\zeta = 0$. For a scalar scaling operator of level 2, it follows that [119]

$$\langle \phi(z, \bar{z}) \rangle_{ab} = \left(\frac{1}{2i}(z - \bar{z}) \right)^{-2\Delta} \mathcal{F}_{ab}(X) \;, \quad X = \frac{z + \bar{z}}{2\sqrt{z\bar{z}}}, \tag{15.90}$$

where $\mathcal{F}_{ab}(X)$ satisfies the equation

$$\left[(1 - X^2) \frac{d^2}{dX^2} - \frac{5 - 8\Delta}{3} X \frac{d}{dX} + \frac{8(1 + 2\Delta)}{3} t_{ab} \right] \mathcal{F}_{ab}(X) = 0 \tag{15.91}$$

which has the solution, in terms of hypergeometric functions $_2F_1$

$$\mathcal{F}_{ab}(X) = a_1 \, _2F_1\left(a_+, a_-; \frac{1}{2}; X^2\right) + a_2 X \, _2F_1\left(a_+ + \frac{1}{2}, a_- + \frac{1}{2}; \frac{3}{2}; X^2\right)$$

$$a_\pm = \frac{1}{6}\left[1 - 4\Delta \pm \left[(1 - 4\Delta)^2 + 24t_{ab}(1 + 2\Delta)\right]^{1/2}\right], \tag{15.92}$$

where the constants $a_{1,2}$ are fixed by the boundary conditions. Through the transformation (15.25), one has on the strip

$$\langle \phi(v) \rangle_{ab} = \left(\frac{L}{\pi} \sin \frac{\pi v}{L}\right)^{-2\Delta} \mathcal{F}_{ab}\left(\cos \frac{\pi v}{L}\right) \tag{15.93}$$

which gives the desired profile for a level 2 scalar scaling operator.

In the Ising model, both the order parameter σ and the energy density ε have a null vector at level 2 of the Kac formula. The required values of t_{ab} may be read from the explicit operator content given previously. Therefore, up to a constant amplitude [119]

$$\langle \sigma(v) \rangle_{+-} \sim \left[\frac{L}{\pi} \sin \frac{\pi v}{L}\right]^{-1/8} \cos \frac{\pi v}{L} \, , \quad \langle \sigma(v) \rangle_{+f} \sim \left[\frac{L}{\pi} \sin \frac{\pi v}{L}\right]^{-1/8} \left[\cos \frac{\pi v}{2L}\right]^{1/2},$$

$$\langle \varepsilon(v) \rangle_{+-} \sim \left[\frac{L}{\pi} \sin \frac{\pi v}{L}\right]^{-1} \left[1 - 4\sin^2 \frac{\pi v}{L}\right] \, , \quad \langle \varepsilon(v) \rangle_{+f} \sim \left[\frac{L}{\pi} \sin \frac{\pi v}{L}\right]^{-1} \cos \frac{\pi v}{L}. \tag{15.94}$$

Analogous results may be derived, for example, for the p-states Potts model [119, 121, 122].

Let us return to the ordinary transition. From scaling arguments, it is easy to show that close to the surface, the local order parameter scales as $\langle \sigma(v) \rangle \sim v^{x_{\sigma,s} - x_\sigma}$ [91, 195]. For free boundary conditions, however, one might expect from (15.85,15.86) that the local order parameter vanishes. But this is not so and is seen as follows [596]. The order parameter is not obtained from a conventional average (which in fact vanishes on any finite lattice), but from a 'non-diagonal' matrix element of the form $\langle 0|\sigma(v)|\sigma \rangle$, which has the correct scaling behaviour and may be taken as a 'finite-lattice' estimate of the order parameter, see Chap. 9. The transformation of these matrix elements has not yet been taken into account and will be done now.

Consider the (connected) two-point function (15.16), as determined from covariance under the projective transformations. Using polar coordinates $r = (\rho, \theta)$,

$$G(\mathbf{r}_1, \mathbf{r}_2) \simeq (\rho_1 \rho_2)^{-x} \left(\frac{\rho_1}{\rho_2}\right)^{x_s - x} (\sin \theta_1 \sin \theta_2)^{x_s - x} \tag{15.95}$$

in the limit $\rho_1 \gg \rho_2$ and where x is the bulk and x_s the surface scaling dimension of the scalar operator ϕ. In complex coordinates, we change notation slightly with respect to previous sections and write $z = U + iV$. If ϕ is also primary, transformation onto the strip through $w = (L/\pi)\ln(U + iV) = u + iv$ gives

$$G(u_1 - u_2; v_1, v_2) \simeq \left(\frac{\pi}{L}\right)^{2x} \exp\left[-\frac{\pi x_s}{L}(u_1 - u_2)\right] \left[\sin\frac{\pi v_1}{L} \sin\frac{\pi v_2}{L}\right]^{x_s - x}.$$
(15.96)

On the other hand, this may also be written through the eigenstates of the quantum Hamiltonian H

$$G(u_1 - u_2; v_1, v_2) = \sum_{n>0} \langle 0|\phi(v_1)|n\rangle\langle n|\phi(v_2)|0\rangle \exp\left[-(E_n - E_0)(u_1 - u_2)\right]$$

$$\simeq \langle 0|\phi(v_1)|\phi\rangle\langle\phi|\phi(v_2)|0\rangle \exp\left[-(E_\phi - E_0)(u_1 - u_2)\right]$$
(15.97)

since in the limit $u_1 \gg u_2$, the dominant term is given by the lowest state with a non-vanishing matrix element. Comparison of (15.96,15.97) yields the *non-diagonal profile* (in particular, but not only, for the order parameter) on the strip [596]

$$\langle\phi(v)\rangle = \langle 0|\phi(v)|\phi\rangle \sim \left(\frac{L}{\pi}\right)^{-x} \left(\sin\frac{\pi v}{L}\right)^{x_s - x}$$
(15.98)

with symmetric boundary conditions on the strip, in terms of the bulk and surface scaling dimensions $x = x_\phi$ and $x_s = x_{\phi,s}$, respectively. At a fixed distance $\ell \ll L$ from the surface, $\langle\phi(\ell)\rangle \sim L^{-x_s}$, while deep in the bulk $v = \mathcal{O}(L/2)$, it scales as L^{-x}. Furthermore, one recovers close to the surface the correct scaling with v. In the half-plane limit $L \to \infty$, the amplitude of the off-diagonal profile vanishes. A non-vanishing profile on the semi-infinite system always gives the diagonal profile (15.86) on the strip.

For the Ising quantum chain, (15.98) is reproduced for symmetric free or fixed boundary conditions for the non-diagonal energy density profile $\langle\varepsilon(v)\rangle$ [596]. Non-symmetric non-diagonal profiles may be considered as well. For the Ising model one finds [596]

$$\langle\sigma(v)\rangle_{+-} \sim \left[\frac{L}{\pi}\right]^{-\frac{1}{8}} \left[\sin\frac{\pi v}{L}\right]^{15/8},$$

$$\langle\sigma(v)\rangle_{+f} \sim \left[\frac{L}{\pi}\right]^{-\frac{1}{8}} \left[\sin\frac{\pi v}{L}\right]^{7/8} \left[\cos\frac{\pi v}{2L}\right]^{1/2} \tan\frac{\pi v}{2L},$$
(15.99)

$$\langle\varepsilon(v)\rangle_{+-} \sim \left[\frac{L}{\pi}\right]^{-1} \sin\frac{\pi v}{L} \cos\frac{\pi v}{L}, \quad \langle\varepsilon(v)\rangle_{+f} \sim \left[\frac{L}{\pi}\right]^{-1} \sin\frac{\pi v}{L} \cos\frac{\pi v}{L}.$$

These predictions have been numerically confirmed to great precision for the order parameter and energy density profiles for the $p = 2, 3, 4$-states Potts models [154, 355], see Fig. 15.6a for the Ising quantum chain with free boundary conditions.

Remarkably, the conformal prediction (15.98) appears to remain valid [355] for the **random** quantum Ising chain

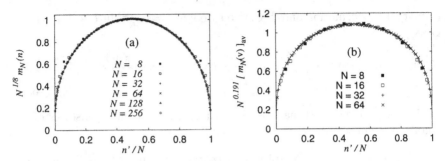

Fig. 15.6. Order parameter profile for free boundary conditions for the Ising quantum chain with N sites. **(a)** Profile for the pure quantum chain, where the full curve represents the function $1.01 \sin(\pi x)^{3/8}$ and $n' = n - 1/2$. **(b)** Profile for the random Ising chain (15.100) for a binary distribution with $\lambda = 2$ and the full curve represents $1.09 \sin(\pi x)^{0.5-0.191}$ [355, 535]

$$H = -\sum_n \left[J_n \sigma_n^x \sigma_{n+1}^x + t \sigma_n^z \right] \tag{15.100}$$

with the probability distribution for the couplings J

$$P(J) = \frac{1}{2} \left(\delta(J - \lambda) + \delta(J - 1/\lambda) \right) \tag{15.101}$$

although the randomness should break conformal invariance. At the critical point $t_c = 1$, it is known that $x_\sigma = (3 - \sqrt{5})/4 \simeq 0.191$ [233] and $x_{\sigma,s} = 1/2$ [459]. The order parameter profile is given by $m_N(n) = \langle \sigma(n) \rangle = \langle \sigma | \sigma_n^x | 0 \rangle$, where $|0, \sigma\rangle$ are two lowest eigenstates of H. As shown in Fig. 15.6b, the numerical results collapse nicely into a scaling form which is very well described by (15.98). The same observation holds true for other boundary conditions and for the energy density profile $\langle \varepsilon(n) \rangle = \langle \varepsilon | \sigma_n^z | 0 \rangle$ [355, 357].

15.6 Defect Lines

Having treated the case of a free surface in some detail, we can now ask what might happen if two critical statistical systems are weakly coupled together on their free surfaces. The line where the two systems are joined is referred to as a **defect line**. The **defect strength** κ measures the coupling between the two systems. For $\kappa = 0$, we are back to a free surface while for $\kappa = 1$, the defect line has disappeared completely. We can define local critical exponents close to the defect line in analogy to the surface exponents. We now ask for the values of these **defect line exponents**.

Here, we want to consider the effect of a single defect line. Since the defect strength couples the two systems via the interactions between neighboring spins, we obtain for the renormalization group eigenvalue y_κ, if the infinitely long defect line is treated as a perturbation of a two-dimensional system [114]

$$y_\kappa = 1 - x_\varepsilon, \tag{15.102}$$

where x_ε is the scaling dimension of the energy density. Consequently, if $x_\varepsilon > 1$, the defect perturbation is irrelevant and even close to the defect line, we shall find the critical exponents of the unperturbed system. Conversely, if $x_\varepsilon < 1$, the defect perturbation will be relevant and close to the defect line the system will be described by a new fixed point and exponents.

Let us consider the role of a defect line for a bulk system as well as for a free surface. For a free surface, we have just seen above that always $x_{\varepsilon,s} = 2$. It follows that a defect line is irrelevant close to a free surface. On the other hand, the bulk exponent $x_{\varepsilon,b} < 1$ for most of the systems considered in earlier chapters. For these, the defect line is relevant. In these cases, we expect the local exponents close to the defect line to be the *same* as the ones close to a free surface [114]. A notable exception to this is the 2D Ising model, where $x_{\varepsilon,b} = 1$. In that case, the defect perturbation is marginal and the defect exponents are expected to be functions of the defect strength κ. In fact, the defect line exponent $\beta(\kappa)$ of the local order parameter was calculated exactly for the 2D Ising model with one infinite defect line [50, 460] and found to depend continuously on κ.

Turban [591] suggested that conformal invariance might be useful to study this problem. Consider the Ising model with a single infinitely long defect line in the two-dimensional plane. By the familiar logarithmic transformation $w = (N/2\pi) \ln z$, this system can be mapped into the strip geometry, where the system has now *two* defect lines, which are $N/2$ sites apart from each other. The calculation is parallel to the one done for free boundary conditions before, if the scaling operators still transform covariantly even in the presence of the defect lines. Provided this is true, one finds again the amplitude-exponent relation between the correlation length measured parallel to the defect line and the defect exponents [591]

$$\xi_i^{-1} = L^{-1} 2\pi x_i(\kappa). \tag{15.103}$$

This relation can (and has been) explicitly checked by comparing the exactly known critical exponents as calculated in the plane [381] with the results following [286, 305, 361] from the conformal relation (15.103). Here, we shall prove the conformal invariance of the 2D Ising model with a defect line in another way [306], which has the advantage that the calculations to be done are shorter. Furthermore, this approach sheds further light on the algebraic aspects of conformal invariance.

We consider the Ising quantum chain with a single defect line, described by the quantum Hamiltonian

$$H = -\frac{1}{2} \left[\sum_{n=1}^{N} t\sigma^z(n) + \sum_{n=1}^{N-1} \sigma^x(n)\sigma^x(n+1) \right] - \frac{\kappa}{2} \sigma^x(N)\sigma^x(1). \tag{15.104}$$

In the infinite plane, this corresponds to a system with a semi-infinite defect line. The system interpolates between periodic boundary conditions (for $\kappa = 1$) and antiperiodic boundary conditions (for $\kappa = -1$) and we expect to recover in these two cases the known operator content (10.46). The off-critical behaviour of this system has received considerable attention as well and quite a complex structure for the correlation length was found [124, 651, 49, 125, 183].

The diagonalization of H [501, 2, 309] is essentially the same as for the free boundary case treated above. The matrix to be diagonalized is

$$
\widehat{M} = \begin{pmatrix}
t^2 + 1 & -t & & & & & & \mathcal{Q}\kappa t \\
-t & t^2 + 1 & -t & & & & & \\
& -t & \ddots & \ddots & & & & \\
& & \ddots & \ddots & \ddots & & & \\
& & & & \ddots & \ddots & \ddots & \\
& & & & & \ddots & t^2 + 1 & -t \\
\mathcal{Q}\kappa t & & & & & & -t & t^2 + \kappa^2
\end{pmatrix}, \qquad (15.105)
$$

where $\mathcal{Q} = 1 - 2Q$. Repeating the ansatz (15.35), we reobtain the lattice dispersion relation (15.36) and the following condition for the discretization of the lattice momenta k

$$
2\mathcal{Q}\kappa t \sin k - \kappa^2 t \sin((N-1)k) + (\kappa^2 - 1) \sin(Nk) + t \sin((N+1)k) = 0 \quad (15.106)
$$

To discuss the critical point behaviour, we put $t = 1$. Since we are merely interested in the large N limit, we make the ansatz $k = \alpha/N$ and find, to leading order in $1/N$

$$
\cos \alpha = -\frac{2\kappa Q}{1 + \kappa^2}. \qquad (15.107)
$$

Introducing this into the dispersion relation (15.36), we obtain the critical point Hamiltonian [306, 309], where we have subtracted the non-universal bulk and surface terms, of order $\mathcal{O}(N)$ and $\mathcal{O}(1)$, respectively

$$
H = \frac{2\pi}{N} \sum_{r=0}^{\infty} \left[\left(r + \frac{1}{2} - \mathcal{D}(Q, \kappa) \right) n_r^{(-)} + \left(r + \frac{1}{2} + \mathcal{D}(Q, \kappa) \right) n_r^{(+)} \right] - \frac{\pi c_{\text{eff}}}{6N},
$$
$$
(15.108)
$$

where $n_r^{(\pm)}$ are fermionic number operators and

$$
\mathcal{D}(0, \kappa) = D(\kappa) \ , \quad \mathcal{D}(1, \kappa) = \frac{1}{2} - D(\kappa) \qquad (15.109)
$$

with

$$
D(\kappa) = \left| \frac{1}{\pi} \arctan \frac{1}{\kappa} - \frac{1}{4} \right|. \qquad (15.110)
$$

Some care is necessary to select the correct branch of the inverse tangent, for example $\arctan(-1/\kappa) = \pi - \arctan(1/\kappa)$. Finally, we have [306]

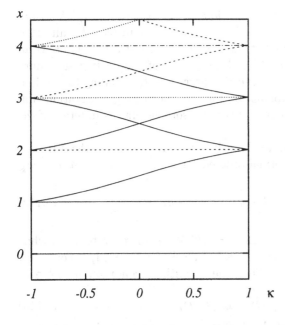

Fig. 15.7. Scaled spectrum $x(\kappa)$ of the $Q = 0$ sector of the Ising quantum Hamiltonian with a single defect line as a function of the defect strength κ. The degeneracies are as follows: 1 (full lines), 2 (dashed lines), 3 (dotted lines) and 5 (dash-dotted lines) [169]

$$c_{\text{eff}} = \frac{1}{2} - 6D(\kappa)^2. \tag{15.111}$$

In Figs 15.7 and 15.8 [306] we display the defect exponents as obtained from the Hamiltonian spectrum using the conformal invariance result (15.103). While for $\kappa = 1$ and $\kappa = -1$, respectively, the known operator

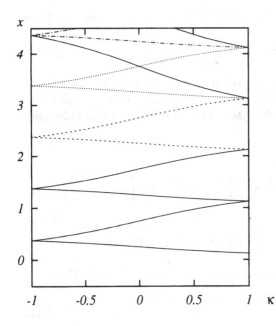

Fig. 15.8. Scaled spectrum $x(\kappa)$ of the $Q = 1$ sector of the Ising quantum Hamiltonian with a single defect line. The degeneracies are as for $Q = 0$ [169]

content as derived from the representations of the pair of $c = 1/2$ Virasoro algebras is recovered, these levels split into distinct multiplets for all other values of κ. Nevertheless, the spectrum is seen to show still a structure compatible with conformal towers. The degeneracies of the excited secondary states are those of a system with central charge $c \geq 1$, including the excited states of the vacuum. The explanation for these surprising observations follows from the explicit construction of the Virasoro generators in terms of the fermionic oscillators.

In analogy to what we have done in Chap. 10, we write the Virasoro generators in terms of the normal-ordered fermionic oscillators $a_{-k} = a_k^+$ [306]

$$L_n(\mathcal{D}) = \frac{1}{2} \sum_{k \in \mathbb{Z}+1/2} (k - \mathcal{D}) : a_{n-k+\mathcal{D}} a_{k-\mathcal{D}} : . \tag{15.112}$$

In the limit $\mathcal{D} \to +0$, we recover the $c = 1/2$ Virasoro generators L_n, while for $\mathcal{D} \to -0$, we get back to the $\bar{L}_n$, and similarly for $\mathcal{D} \to \pm 1/2$. Form now on, we always take $\mathcal{D} \neq 0, 1/2$. Then the hermiticity condition is no longer valid but is replaced by

$$L_n^+(\mathcal{D}) = L_{-n}(-\mathcal{D}). \tag{15.113}$$

The commutation relations of the $L_n(\mathcal{D})$ can now be worked out following the lines of Chap. 10, with the result [306]

$$[L_n(\mathcal{D}), L_m(\mathcal{D})] = \frac{n-m}{2} L_{n+m}(\mathcal{D}) - \frac{1}{24}(n^3 - n)\delta_{n+m,0}$$
$$+ \frac{n}{4}\left(\mathcal{D}^2 - \frac{1}{4}\right)\delta_{n+m,0}. \tag{15.114}$$

To recast this into the conventional form of the Virasoro algebra, we define

$$\tilde{L}_n(\mathcal{D}) := 2L_n(\mathcal{D}) + \frac{1}{2}\left(\mathcal{D}^2 - \frac{1}{4}\right)\delta_{n,0} \tag{15.115}$$

which satisfy a Virasoro algebra with a *negative* central charge $c = -2$, thereby illustrating the general theorem from Chap. 4 that for negative central charges hermiticity is not possible [246]. Equation (15.114) is supplemented by

$$[L_n(\mathcal{D}), L_m(-\mathcal{D})] =$$
$$\frac{n-m}{2} L_n(-\mathcal{D}) - \frac{n^2}{2} W_{n+m}(-\mathcal{D}) + \frac{1}{4}\left(\frac{n^3}{3} + 2n^2\mathcal{D} + n\left(\mathcal{D}^2 - \frac{1}{12}\right)\right)\delta_{n+m,0}$$
$$\tag{15.116}$$

where we have introduced a further set of generators $W_n(\mathcal{D})$

$$W_n(\mathcal{D}) = \frac{1}{2} \sum_{k \in \mathbb{Z}+1/2} : a_{n-k+\mathcal{D}} a_{p-\mathcal{D}} : -\frac{\mathcal{D}}{2}\delta_{n,0} = -W_n(-\mathcal{D})$$
$$W_n^+(\mathcal{D}) = W_{-n}(\mathcal{D}) \tag{15.117}$$

and, using the same techniques as before, we can verify that the algebra closes

$$[W_n(\mathcal{D}), W_m(\mathcal{D})] = \frac{n}{4}\delta_{n+m,0}$$

$$[L_n(\pm\mathcal{D}), W_m(\mathcal{D})] = -\frac{m}{2}W_{n+m}(\mathcal{D}) \pm \frac{n^2}{2}\delta_{n+m,0}. \tag{15.118}$$

We thus see that the presence of the defect lines does change the structure of the conformal algebra. We do no longer have a simple commuting pair of hermitian generators and have furthermore picked up the additional generators W_n, which form a U(1) Kac–Moody algebra (7.10) and (7.14) among themselves. However, our quantum Hamiltonian is hermitian. If the model is conformal invariant, we must be capable to express H in terms of the generators of some Virasoro algebra. To do so, we define new hermitian generators [306]

$$K_n(\mathcal{D}) = L_n(\mathcal{D}) + L_n(-\mathcal{D}) + \frac{1}{2}\mathcal{D}^2\delta_{n,0}. \tag{15.119}$$

Comparing with the spectrum of H (15.108), we can relate it to $K_0(\mathcal{D})$

$$H = \frac{2\pi}{N}\left(K_0(\mathcal{D}) - \frac{1}{24}\right) \tag{15.120}$$

and the hermitian conformal algebra becomes finally

$$[K_n(\mathcal{D}), K_m(\mathcal{D})] = (n-m)K_{n+m}(\mathcal{D}) + \frac{1}{12}\left(n^3-n\right)\delta_{n+m,0}$$

$$[K_n(\mathcal{D}), W_m(\mathcal{D})] = -mW_{n+m}(\mathcal{D})$$

$$[W_n(\mathcal{D}), W_m(\mathcal{D})] = \frac{n}{4}\delta_{n+m,0} \tag{15.121}$$

and we read off the central charge $c = 1$. This also explains the degeneracies observed in Figs 15.7 and 15.8. The infinitely many primary operators of the Virasoro algebra generated by the K_n are labeled by the eigenvalues of the generator W_0. The presence of the Kac–Moody generators leads to non-trivial relations between the scaling dimensions of the order and the disorder operators [305], the first examples being observed in [109]. We can now conclude that the 2D Ising model with a half-infinite defect line and its corresponding quantum chain are conformal invariant at the critical point. As previously seen for free boundary conditions, it follows fron the construction (15.119) that the conformal transformations permitted include those which are real analytic. This justifies the previous proposal [591] to use the conformal mapping from the two-dimensional plane onto the strip geometry and the operator content can be obtained from the Hamiltonian spectrum using (15.103) in the by now familiar way. We also see in Figs 15.7 and 15.8 that for $\kappa = 0$, corresponding to free boundary conditions, several of the U(1) primary fields recombine into conformal towers of the $c = 1/2$ Virasoro algebra. This can be understood from the relation [306]

$$K_n(1/4) = \frac{1}{2}L_{2n}(0) + \frac{1}{32}\delta_{n,0} \tag{15.122}$$

since the $L_n(0)$ satisfy a $c = 1/2$ Virasoro algebra. This relation can be used to define branching rules when going from a $c = 1$ to a $c = 1/2$ Virasoro algebra, see [38].

Before writing down the operator content, we present an alternative derivation [39] of the conformal algebra (15.121), which is close in spirit to the Coulomb gas approach of Chap. 7. At the same time, this clarifies the origin of the shift terms introduced in the definition of the K_n and the W_n. We go back to $\kappa = 1$ for a moment. As we have seen in Chap. 10, the Ising quantum Hamiltonian can be constructed in terms of fermionic oscillators. On the other hand, we have seen in Chap. 7 that the energy–momentum tensor of any minimal model could be obtained as a bilinear expression in terms of a conserved U(1) Kac–Moody current, see (7.12). The corresponding U(1) Kac–Moody algebra is

$$
\begin{aligned}
{[L_n, L_m]} &= (n - m)L_{n+m} + \frac{1}{12}\left(n^3 - n\right)\delta_{n+m,0} \\
{[L_n, J_m]} &= -mJ_{n+m} \\
{[J_n, J_m]} &= \frac{n}{4}\delta_{n+m,0},
\end{aligned}
\tag{15.123}
$$

where J_n are the components of the current J. We remark that the L_n as written here are the *sum* of the two Virasoro algebras usually encountered before, which also accounts for $c = 1$ in the Ising model context. The components L_n of the energy–momentum tensor are, as a consequence of (7.12), bilinear in the J_n and read in normal-ordered form

$$L_n = \frac{1}{2}\sum_{r\in\mathbb{Z}} : J_{n-r}J_r : \tag{15.124}$$

and the J_n have for the Ising model the fermionic representation [39]

$$J_n = \mathrm{i}\sum_{k\in\mathbb{Z}+1/2} : a_{n-k}\bar{a}_k :, \tag{15.125}$$

where $a_k^+ = a_{-k}$, $\bar{a}_k^+ = \bar{a}_{-k}$, $\{a_k, a_\ell\} = \{\bar{a}_k, \bar{a}_\ell\} = \delta_{k+\ell,0}$ and $\{a_k, \bar{a}_\ell\} = 0$. We can interpret the a_k to correspond to the excitations with k small while the $\bar{a}_k$ correspond to those with k of order N, see Chap. 10. At first sight, it might seem that the L_n should be of forth order in the fermionic operators. These terms cancel out, however, as a consequence of the identity [39]

$$\sum_{m'}\sum_{k,\ell} : a_{m-m'-k}a_{m'-\ell}a_k a_\ell : = 0 \tag{15.126}$$

and the remaining terms are

$$L_n = \frac{1}{2} \sum_{k \in \mathbb{Z}+1/2} k : a_{n-k}a_k + \bar{a}_{n-k}\bar{a}_k : . \tag{15.127}$$

Now, the commutation relations of the L_n and J_n are left invariant under the automorphism [39]

$$J_n \rightarrow W_n(\mathcal{D}) := J_n + \mathcal{D}\,\delta_{n,0}$$
$$L_n \rightarrow K_n(\mathcal{D}) := L_n + \mathcal{D}J_n + \frac{1}{2}\mathcal{D}^2\delta_{n,0} \tag{15.128}$$

which introduces the *shifted U(1) Kac-Moody algebra* [37]. The operator content is then given as follows. Starting from periodic boundary conditions in the spin variables, it follows from the Jordan-Wigner transformation that $\mathcal{D}(Q,\kappa)$ does depend on the charge sector Q. On the other hand, $\mathcal{D}$ only depends on the combination $Q + \tilde{Q}$, where $\tilde{Q}$ was used in the Chaps. 9 and 10 to specify the toroidal boundary conditions. The operator content for the Ising quantum chain with a single defect line then is [308, 39]

$$H^{(\kappa)}_{Q+\tilde{Q}} = \sum_{m=-\infty}^{\infty} \left(\frac{1}{2}\left(m + \mathcal{D}(Q + \tilde{Q}, \kappa)\right)^2\right)_{\mathrm{KM}} \tag{15.129}$$

in terms of characters of the U(1) Kac-Moody algebra (15.121). These are of the form

$$\chi_{\Delta,\mathcal{D}}(r,q) = \mathrm{tr}\left(q^{K_0(\mathcal{D})}r^{W_0(\mathcal{D})}\right) = P(q)q^{\mathcal{D}^2/2}r^{\mathcal{D}} \tag{15.130}$$

in contradistinction to the Virasoro characters for $c = 1$, which are given by (12.57). It follows that $\Delta = \mathcal{D}^2/2$. One has the following decomposition of the U(1) Kac-Moody characters (evaluated for $r = 1$) [37]

$$\left(\frac{1}{4}m^2\right)_{\mathrm{KM}} = \sum_{k=0}^{\infty} \left(\frac{1}{4}(m + 2k)^2\right)_{\mathrm{Vir}} \tag{15.131}$$

in terms of Virasoro characters (Vir). Carrying out the sum, the stated form (15.130) is obtained. This illustrates how several primary operators of the Virasoro algebra are combined into a single primary operator of the larger Kac-Moody algebra.

We close the discussion with a brief comment on the maximal spectrum generating algebra. First, the algebra can be enlarged by adding the fermionic operators a_k and $\bar{a}_k$ as generators. This then yields a supersymmetric extension of the U(1) Kac Moody algebra, whose structure is quite similar to the $N = 2$ superconformal algebra. If $\kappa = 1$, the quantum Hamiltonian provides an oscillator representation $\langle 0 \rangle^{1/2}$ of $\widehat{U(1)}$ with the decomposition

$$\langle 0 \rangle^{1/2} = \sum_{m=-\infty}^{\infty} \left(\frac{1}{2}m^2\right)_{\mathrm{KM}} , \quad \kappa = 1 \tag{15.132}$$

in terms of the U(1) Kac Moody characters. Extended symmetry algebras of this type have been studied in detail in the literature. For arbitrary κ, one has a shifted representation [39]

$$H^{(\kappa)} = \langle \mathcal{D} \rangle^{1/2} = \sum_{m=-\infty}^{\infty} \left(\frac{1}{2}(m + \mathcal{D})^2 \right)_{\mathrm{KM}} \qquad (15.133)$$

giving the complete spectrum in a *single* irreducible representation.

In a similar way, the problem of n defect lines in the (1+1)D Ising model can be studied and solved. This was done for line defects [309], extended regions of changed couplings with [77, 560] or without [77, 334, 650] an interface introduced. The isotropic 2D Ising model can be treated similarly [2]. Defects with twists have been studied [281]. One might investigate the influence of defects on surface critical phenomena [75]. Random boundary fields are treated in [144]. This is nicely reviewed in [351], where results from the mean-field approximation, scaling and conformal invariance are compared. For further illustration, consider an Ising quantum chain consisting of two coupled subsystems with spin $S = 1/2$ and $S = 1$, with Hamiltonian [391]

$$H = -\frac{1}{2\zeta} \left[\sum_{n=1}^{N-1} \sigma_n^z \sigma_{n+1}^z + \sum_{n=N+1}^{2N-1} \kappa S_n^z S_{n+1}^z \right.$$
$$\left. + t \sum_{n=1}^{N} \left(\sigma_n^x + S_{n+N}^x \right) + \gamma \left(\sigma_N^z S_{N+1}^z + \sigma_1^z S_{2N}^z \right) \right], \qquad (15.134)$$

where $\sigma^{x,z}$ are the spin 1/2 Pauli matrices and $S^{x,z}$ are spin 1 matrices. To understand the phase diagram, note that for each subsystem alone (that is, for $\gamma = 0$ and $N \to \infty$), there is a critical point at $t = t_{c,S}$ with (see Chap. 10)

$$t_{c,1/2} = 1 \ , \quad t_{c,1}/\kappa = 1.32587(1). \qquad (15.135)$$

For $\kappa > \kappa_c \simeq 0.754222$, the $S = 1$ subsystem orders first at $t = t_{c,1}$ where the system undergoes an ordinary transition, because the critical degrees of freedom are subjected to effective free boundary conditions since the $S = 1/2$ is still disordered. The $S = 1/2$ subsystem becomes critical at $t = t_{c,1/2}$ through an extraordinary (normal) transition, since the spins in the $S = 1$ are already frozen. For $\kappa < \kappa_c$, the roles of the subsystems are interchanged. Thus the meeting point $t = 1, \kappa = \kappa_c$ of these two transition lines corresponds to a *special* transition [391], see also Fig. 15.2. For both the ordinary and the extraordinary transitions, the operator content and the magnetization profiles have been checked to be in complete agreement with the predictions made earlier. Here, we concentrate on the special transition, where the parameter γ is marginal. This allows a non-trivial check of the continuum description (15.108) for a defect line perturbation, even though the finite-lattice degrees of freedom cannot be mapped one-to-one onto it. In addition, this example serves to illustrate the precision which can be achieved from purely numerical

Table 15.4. Conformal spectrum of the exponents $x_i(\gamma)$ for the $S = 1$–$S = 1/2$ Ising quantum chain (15.134) at the special transition point $t = 1, \kappa = 0.754222$. The values of $\mathcal{D}(0,1)$ used in comparing with the free fermion Hamiltonian (15.108) and the conformal normalization ζ are also given [391]

i	$\gamma = 0.5$ numerical	expected	$\gamma = 0.754222$ numerical	expected
1	0.231(1)	0.20	0.1436(6)	0.144(2)
2	0.971(3)	0.91	0.999(1)	1
3	1.034(5)	1	1.10(3)	1.072(4)
4	1.095(3)	1.20	1.12(2)	1.144(2)
5	1.72(3)	1.70	1.94(1)	1.94(1)
6	1.92(3)	1.91	1.995(5)	2
7	1.99(2)	2	2.03(3)	2.06(1)
8	-	2.20	2.11	2.072(4)
9	-	2.30	2.145(8)	2.144(2)
$\mathcal{D}(0)$	~ 0.15		0.030(5)	
$\mathcal{D}(1)$	~ 0.36		0.464(2)	
$\pi\zeta$	1.945(3)		1.8731(8)	

i	$\gamma = 0.877111$ numerical	expected	$\gamma = 1$ numerical	expected
1	0.1103(7)	0.1103(5)	0.0841(5)	0.084(5)
2	1.000(1)	1	1.00(2)	1
3	1.113(6)	1.1103(5)	1.11(2)	1.084(5)
4	1.168(5)	1.168(2)	1.232(5)	1.232(5)
5	1.92(1)	1.92(1)	1.9(1)	1.78(4)
6	2.00(2)	2	2.0(1)	2
7	2.11(1)	2.08(1)	2.1(1)	2.084(5)
8	-	2.1103(5)	2.18(3)	2.22(4)
9	-	2.168(2)	-	2.232(5)
$\mathcal{D}(0)$	0.040(5)		0.11(3)	
$\mathcal{D}(1)$	0.529(1)		0.574(4)	
$\pi\zeta$	1.8745(5)		1.88(2)	

data in testing conformal theories with a complicated operator content. This in displayed in Table 15.4, where the scaling dimensions $x_i(\gamma)$ as obtained extrapolating data up to $N = 7$ from the lowest few gaps of H are shown. The correct normalization $\zeta = 0.5964(2)$ [391] was determined from the condition $x_{\varepsilon,s} = 1$ which is necessary for the presence of a marginal operator coupled to γ. From (15.108), the two lowest values of the x_i can be expressed in terms of $\mathcal{D}(0)$ and $\mathcal{D}(1)$. Having found these, the predictions for the other gaps listed as 'expected' can be found. It can be seen that the 'numerical' estimates for the higher gaps agree within a few per cent with the conformal invariance prediction. In particular, we observe the beginning of several conformal towers, since with x_i also $x_i + 1$ and sometimes also $x_i + 2$ are found in the spectrum. While this is in agreement with the expected universality, some differences with the defects of purely spin $S = 1/2$ models show up. For example, in the purely $S = 1/2$ double defect problem, one has simply $\mathcal{D}(0) = 0$ [77, 305, 334].

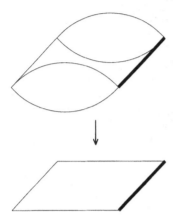

Fig. 15.9. Folding of an Ising model on a cylinder onto a $c = 1$ theory on a strip. The defect line is indicated by a thick line (after [487])

This problem has also been elegantly solved from the point of view of continuum field theory [487], which allows to understand the special cases studied in a systematic way. It turns out that in the context of boundary conformal field theory [145], one must fold [623] the systems at the defect line, as sketched in Fig. 15.9. Then one obtains a model with two Ising variables at each point, which are only coupled at the boundary. This is a special case of the Ashkin–Teller model, with an underlying $c = 1$ orbifold theory of a free boson, see Chap. 12. From the exact partition function, the boundary states can be identified in terms of the scaling operators. Since the boundary correlation functions can be determined from the boundary states and the operator product expansion [145] and the conformal blocks of the Ashkin–Teller model are known [646], explicit expressions for the spin-spin correlators in the presence of a defect line have been obtained [487]. Assuming that the defect is placed on the real axis, one has to determine, as for free boundary conditions, a scaling function which depends only on the cross-ratio $V = |(z_1 - z_2)/(z_1 - \bar{z}_2)|^2$. Define $U(V)$ through the relation $V = [\vartheta_2(U)/\vartheta_3(U)]^4$, where $\vartheta_{2,3}$ are elliptic theta functions, e.g. [1]. Writing $z = u + iv$, the spin-spin two-point function reads [487]

$$\langle \sigma(z_1)\sigma(z_2) \rangle = \left(\frac{1}{v_1 v_2 V}\right)^{1/8} \frac{1}{\vartheta_3(U(V))} \begin{cases} \vartheta_3(e^{2i\varphi_0}, \sqrt{U(V)}) \; ; \text{ same side} \\ \vartheta_2(e^{2i\varphi_0}, \sqrt{U(V)}) \; ; \text{ different sides} \end{cases}$$

$$(15.136)$$

depending on whether the two points z_1, z_2 are on the same side or on different sides of the defect line (before folding). Here $\varphi_0 = \pi/4 + \alpha/2$ is related to the momentum shift (15.107) in the free fermions due to the defect (in the $Q = 1$ sector). It is an easy exercise to check that the special cases of periodic and free boundary conditions are correctly reproduced [487]. Multipoint correlators, mostly of the local energy density, have also been studied in detail, see [409, 410, 51, 3, 120, 123]. For an S-matrix approach following the lines of Chap. 14, see [183].

Defect lines were also studied numerically in the three-states Potts [556] model and the Yang–Lee singularity [263]. The results indicate that the defect exponents are equal to the free surface exponents in agreement with the renormalization group predictions [114] discussed above. In higher dimensions, defects were studied as well in the $O(n)$ model either in the $n \to \infty$ limit [114] or via ε-expansion techniques [98]. Defect lines have also been studied in the context of conformal perturbation theory, by folding the system onto a boundary field theory of twice the central charge [431].

A related problem is posed through the **Hilhorst–van Leeuwen model** [331]. In the context of quantum chains, this may be specified in a simple way through the Hamiltonian

$$H = -t \sum_{n=1}^{N-1} \sigma^z(n) - \sum_{n=1}^{N-2} J_n \sigma^x(n) \sigma^x(n+1), \qquad (15.137)$$

where the coupling J_n is space-dependent

$$J_n = 1 - an^{-y} \qquad (15.138)$$

and a and y are constants. As reviewed in [351], it is known that the scaling behaviour at the critical point $t_c = 1$ depends on the value of y. For $y > 1$, one recovers the ordinary transition of the Ising quantum chain with a free surface. For $y < 1$, the surface orders and one has a first-order transition. Finally, for $y = 1$, the perturbation is marginal and the exponents depend continuously on the amplitude a [331, 92, 118, 350, 52, 387]. Obviously, the presence of the space-dependent coupling J_n should break conformal invariance. Remarkably, it is found that the amplitude-exponent relation (3.47) for the critical correlation length, which was derived from the assumption of conformal invariance, does remain valid. The same observation also holds when the couplings decrease towards both sides according to $J_n \sim n^{-1}$ [77, 118, 349]. This also holds true for the perturbation of the critical Ising chain with the aperiodic Fredholm sequence [387, 388]. Indeed, by considering the effect of *any* marginal perturbation of the form

$$H = H_c + g \int d\mathbf{r}\, \rho(\mathbf{r}) \varepsilon(\mathbf{r}), \qquad (15.139)$$

where $\rho(\mathbf{r})$ describes the spatial extension of the perturbation, it was shown that the amplitude-exponent relation (3.47) remains valid to first order in g [595]. One might wonder whether the amplitude-exponent relation (3.47) could have a broader range of validity than suggested through its conformal invariance derivation.

15.6.1 Aperiodically Modulated Systems

Going further the way from regular to random systems [458, 587, 621, 356], one possibility is to introduce a sequence of systems with n defect lines for a lattice of N sites and let n grow faster than N. The resulting systems, however, are no longer conformally invariant [240].

A more careful way may be to consider spin systems defined on quasiperiodic or aperiodic sequences. These may for example be generated through binary substitution sequences $A \to S\{A\}$, $B \to S\{B\}$ [525]. For example, the rules $A \to AB$ and $B \to A$, together with the starting sequence $S_1 = A$, give the celebrated **Fibonacci sequence** $ABAABABA\ldots$. Consider the quantum Hamiltonian

$$H = -\sum_{n=1}^{N} [t_n \sigma^z(n) + \sigma^x(n)\sigma^x(n+1)] \tag{15.140}$$

and let $t_n = t$ if the site n corresponds to an A and let $t_n = \kappa t$ if the site n corresponds to a B. The corresponding 2D spin model is describes an aperiodic modulation along one of the axis of the lattice while the orthogonal direction is homogeneous. The energy gaps of H are, because of the Hamiltonian limit involved, inversely proportional to the correlation length of the boundary correlators near to the alternating lines and the exponents extracted from them are surface exponents. Another observable of interest is the surface magnetization m_1 given by (Peschel [500])

$$m_1 = \left(1 + \sum_{j=1}^{\infty} \prod_{n=1}^{j} t_n^2\right)^{-1/2}. \tag{15.141}$$

For the homogeneous Ising chain, $t_n = t$ and one has directly $m_1 = (1-t^2)^{1/2}$, reconfirming $\beta_1 = 1/2$.

Considering the aperiodic modulation of the coupling t_n as a perturbation of the homogeneous system, a simple criterion formulated by Luck [442] allows to assess the relevance of that perturbation. First, one constructs the **substitution matrix**

$$\hat{M} = \begin{pmatrix} n_A^{S\{A\}} & n_A^{S\{B\}} \\ n_B^{S\{A\}} & n_B^{S\{B\}} \end{pmatrix}, \tag{15.142}$$

where $n_A^{S\{B\}}$ indicates how often the symbol A occurs in $S\{B\}$. Second, one finds the two largest eigenvalues $\Omega_1 \geq |\Omega_2|$ of $\hat{M}$. After k substitutions, the length of the sequence is $N \sim \Omega_1^k$. The **wandering exponent** is $\omega_S := \ln|\Omega_2|/\ln\Omega_1$. The aperiodicity introduces a perturbation around the mean coupling $\bar{t}$, which in turn is in the mean, after k substitutions $\overline{\delta t} = N^{-1}\sum_{n=1}^{N}(t_n - \bar{t}) \sim N^{-1}(\kappa - 1)|\Omega_2|^k \sim (\kappa - 1)N^{\omega_S - 1}$. Under a scaling transformation $r \to r/b$, the amplitude $g = \kappa - 1$ of the perturbation changes as $g \to b^{\Phi/\nu} g$, where the **cross-over exponent** [442]

$$\Phi = 1 + \nu(\omega_S - 1) \tag{15.143}$$

and ν is the correlation length exponent of the homogeneous system. The **Luck criterion** [442] is the analogue of the Harris criterion [299] for random systems.

Thus, for $\Phi < 0$, the perturbation is *irrelevant*. That is the case for the Fibonacci sequence in the 2D Ising model [441, 203, 72, 319] and for some other aperiodic sequences [157, 204, 437]. Although the long-range properties of the system are the same as for periodic ones and in particular conformal invariance holds, the excitation spectra of the quantum Hamiltonians display multifractal [292, 337, 273] features. For $\Phi > 0$, the perturbation is *relevant* and the nature of the critical point is distinct from the periodic case [352, 356]. An interesting situation occurs for the *marginal* case $\Phi = 0$, which for the Ising model is for example realized for the **period-doubling sequence** $S\{A\} = AB$ and $S\{B\} = AA$. At the critical point $t_c = \kappa^{2/3}$, the energy gaps of H scale as [78]

$$E_i - E_0 \sim N^{-\theta} , \quad \theta = \frac{\kappa^{1/3} + \kappa^{-1/3}}{\ln 2}. \tag{15.144}$$

Such a correlation length scaling is not compatible with conformal invariance but is reminiscent of strongly anisotropic scaling where in different spatial directions there are different correlation lengths $\xi_{\parallel,\perp} \sim (t - t_c)^{-\nu_{\parallel,\perp}}$ and an *anisotropy exponent* $\theta = \nu_{\parallel}/\nu_{\perp}$, which depends continuously on the modulation. Such systems will be discussed further in Chap. 16.

The physics of these systems is very rich and has been actively investigated, e.g. [40, 80, 81, 162, 329, 353, 354, 389, 505, 568]. For a review and bibliography, see [282].

15.6.2 Persistent Currents in Small Rings

We finish the discussion of defect effects by describing how some of the techniques presented can be applied to the problem of persistent currents in small rings of a mesoscopic scale penetrated by a magnetic flux Φ, first studied in the classic papers [420, 111]. We refer to the review [654] and a nice book by Imry [358] for a detailed introduction, the physical background and an extensive bibliography.

Here, we shall limit ourselves to a simple toy model which describes non-interacting (spin-less) electrons moving around a small, thin ring penetrated by an Aharonov-Bohm flux Φ, see Fig. 15.10. Experimentally, persistent currents have been observed in rings of diameters in the μm range, for example for Au [E19] or GaAS-AlGaAs [E56]. The flux $\Phi = \oint A dx = AL$, where A is the vector potential and L is the length of the ring. Assume that the ring

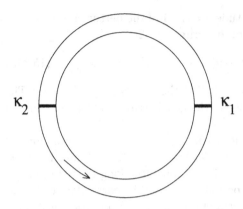

Fig. 15.10. Aharonov-Bohm geometry. The places of the defects and their couplings $\kappa_{1,2}$ are indicated. The magnetic field points out of the plane

is discretized into N cells of size a. We use a simple **tight-binding model** which describes the hopping of electrons from one cell to the next

$$H = -t_{\text{hop}} \sum_{n=1}^{N} \left[\exp\left(2\pi i \frac{\Phi}{\Phi_0} \frac{a}{N} \right) \psi^\dagger(n+1)\psi(n) + \text{h.c.} \right], \qquad (15.145)$$

where t_{hop} is the hopping rate and $\Phi_0 = hc/e$ is the flux quantum. We shall limit ourselves to the case of **half-filling**, when exactly $M = N/2$ particles are present and assume that N is even. The flux Φ can be related to the phase $\phi = 2\pi\Phi/\Phi_0$ of the wave function through the gauge transformation

$$\psi(n) \to \exp\left(-i\phi\frac{a}{L}n \right) \psi(n) \qquad (15.146)$$

leading to the conventional tight-binding Hamiltonian

$$H = -t_{\text{hop}} \sum_{n=1}^{N} \left[\psi^\dagger(n+1)\psi(n) + \psi^\dagger(n)\psi(n+1) \right] \qquad (15.147)$$

with the boundary condition $\psi(N+1) = e^{-i\phi}\psi(1)$. This is nothing but the classical Byers–Yang–Bloch result that the system is periodic in the flux Φ with period Φ_0, see [654, 358] and references therein. In particular, the persistent current is obtained from the ground state energy $E_0(\Phi)$ of H as

$$j = -\frac{\partial E_0}{\partial \Phi}. \qquad (15.148)$$

It is well known that for a perfect ring, $j = (t_{\text{hop}}/\pi)(\phi/N)$ inside the period interval $-\pi < \phi \leq \pi$ and continued periodically outside it. Note the scaling of j with the ring length $L = Na$ [654, 358] so that the current vanishes in the bulk limit.

We now consider the situation [274] where in an otherwise perfect ring there is a single link where the hopping rate is $t_{\text{hop}}\rho$, rather than t_{hop}, see Fig. 15.10. By rotational symmetry, we thus have a defect boundary condition

$\psi(N+1) = \kappa\psi(1)$ with a *complex* defect strength $\kappa = \rho e^{-i\phi}$. The free fermion Hamiltonian (15.147) can be diagonalized through the same techniques as before, leading to $H = \sum_k \Lambda_k(\eta_k^+ \eta_k - 1/2)$, where $\Lambda_k = -t_{\text{hop}} \cos k$ and the values of k are determined by solving

$$\rho^2 \sin(k(N-1)) + 2\rho \cos\phi \sin k - \sin(k(N+1)) = 0. \tag{15.149}$$

A cooperative phenomenon as the development of a persistent current should be determined by the lowest-energy modes. The physical ground state is obtained by filling up the lowest $M = N/2$ one-fermion states. In the vicinity of the Fermi energy, one has now a *linear* dispersion relation and the exact Hamiltonian may be replaced by an effective low-energy Hamiltonian H_{eff} of the form (15.108), with *two* sets of low-lying excitations and where the one-fermion energy shift $\mathcal{D} = \mathcal{D}(\rho, \phi)$. If $E_{0,\text{eff}} = -\frac{1}{2} \sum_k \Lambda_{k,\text{eff}}$ is the ground state energy of H_{eff}, the physical ground state energy is $E_0 = 2E_{0,\text{eff}}$ at half-filling.

Now, $E_{0,\text{eff}}$ is easily related to the shift $\mathcal{D}$ using a conformal invariance argument [327]. We have seen above for the Ising model that the spectrum generating algebra is spanned by the generators $K_n(\mathcal{D})$ of (15.119) and the effective Hamiltonian H_{eff} is related to $K_0(\mathcal{D})$ through (15.120). Since the bulk and surface terms not written explicitly in (15.108) do not contain the phase ϕ, the only change in the ground state energy which can give rise to a persistent current is $\delta E_0 = 2\delta E_{0,\text{eff}} = (2\pi/N)t_{\text{hop}}\mathcal{D}^2$.

To calculate $\mathcal{D}$ to leading order in $1/N$, we let $k = \alpha/N$ in (15.149), which gives $\cos\alpha = (-1)^M 2\rho/(1 + \rho^2) \cos\phi$. From (15.108), we also have $k \simeq (2\pi/N)(\ell + 1/2 \pm \mathcal{D})$, with $\ell = 0, 1, 2, \ldots$. Collecting everything, we find from (15.148) the current $j = -\partial(\delta E_0)/\partial\phi$ as a function of the phase ϕ. Explicitly [274, 327],

$$j(\phi) = t_{\text{hop}}\frac{u \arccos(u \cos\phi) \sin\phi}{\pi\sqrt{1 - u^2 \cos^2\phi}}\frac{1}{N} \quad , \quad u = \frac{2\rho}{1 + \rho^2} \tag{15.150}$$

and we find again the scaling $j \sim N^{-1}$. Note that this expression is invariant under the transformation $\rho \to 1/\rho$ which reflects that the defect, even for $\rho > 1$ disturbs the coherence of the many-particle system. As it stands, (15.150) is correct for M odd. For M even, merely replace $\phi \to \phi - \pi$. The size-independent current amplitude $j(\phi)N$ is shown in Fig. 15.11a for an entire period of the phase ϕ and for several values of the defect coupling ρ. While the periodicity with ϕ is kept unchanged with respect to the perfect case, the amplitude decreases rapidly with increasing values of $|\rho - 1|$.

For two defects which are placed opposite to each other (Fig. 15.10), we find [327]

$$j_\pm(\phi) = t_{\text{hop}}\frac{A \arccos(A \cos\phi \pm B) \sin\phi}{\pi\sqrt{1 - (A\cos\phi \pm B)^2}}\frac{1}{N}$$

$$A = \frac{4\rho_1\rho_2}{(1 + \rho_1^2)(1 + \rho_2^2)} \quad , \quad B = \frac{(1 - \rho_1^2)(1 - \rho_2^2)}{(1 + \rho_1^2)(1 + \rho_2^2)}, \tag{15.151}$$

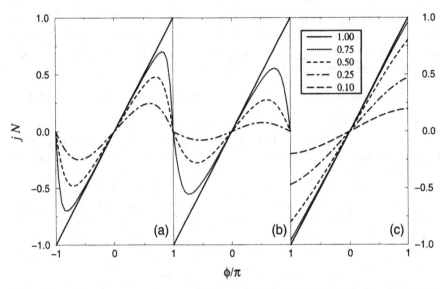

Fig. 15.11. Persistent current amplitude $j(\phi)N$ in the tight-binding model for **(a)** a single defect with defect strength $\kappa = \rho e^{-i\phi}$ for $M = N/2$ odd. For two opposite defects with strengths $\kappa_1 = \rho e^{-i\phi}$ and $\kappa_2 = \rho$ the persistent current is shown for **(b)** $N/2$ even and **(c)** $N/2$ odd. The curves correspond to different values of ρ as indicated and $t_{\mathrm{hop}} = 1$

where $+$ corresponds to $N/2$ even, $-$ corresponds to $N/2$ odd and $\rho_{1,2}$ are the couplings at the defects. For $\rho_1 = 1$ or $\rho_2 = 1$, one simply recovers (15.150). In Fig. 15.11bc, we show the current amplitude for the number of particles $M = N/2$ even and odd, respectively and for $\rho_1 = \rho_2 = \rho$. In contrast to the single-defect case, there is a clear **parity effect**. For an *even* number of particles, the amplitude is broadly similar to the single-defect case (up to a renormalization of ρ). For an *odd* number of particles, the flux-dependence of the current is almost *linear* for a wide range of values of ρ and the presence of the defects merely changes the prefactor. This is reminiscent of what happens in hopping models with interactions and boundary conditions $\sigma^{\pm}(N+1) = e^{\pm i\phi}\sigma^{\pm}(1)$. For example

$$H = -\frac{t_{\mathrm{hop}}}{4} \sum_{n=1}^{N} [\sigma^x(n)\sigma^x(n+1) + \sigma^y(n)\sigma^y(n+1) - \cos\gamma\, \sigma^z(n)\sigma^z(n+1)].$$
(15.152)

The ground state energy $E_0(\phi)$ can be found from the Bethe ansatz [22, 414] and

$$j(\phi) = t_{\mathrm{hop}} \frac{\sin\gamma}{4\gamma(1 - \gamma/\pi)} \frac{\phi}{N}$$
(15.153)

at half-filling and for $-\pi < \phi \le \pi$.

Exercises and Further Examples

1. Why does one obtain the ordinary surface critical exponents from the spectra of critical quantum Hamiltonians with free boundary conditions?

2. Consider $G = G(u_1 - u_2; v_1, v_2) = \langle \phi_1(u_1, v_1)\phi_2(u_2, v_2)\rangle$. Show that for quasi-primary operators ϕ_i and in the presence of a free surface, the scaling dimensions x_1 and x_2 which enter into the two-point function G remain independent of each other. Assuming (is this always justified ?) that the limit $v_1 \to 0$ exists, show directly that formally

$$G(u; 0, v) = G_0\, v^{x_1 - x_2} \left(u^2 + v^2\right)^{-x_1}.$$

3. Consider a corner resulting from the meeting of two straight surfaces at an angle φ. From the scaling of correlation functions near the corner, **corner exponents** $x_i^{(e)}$ may be defined [130], which in general will depend on φ. Map this problem onto the upper half-plane through the conformal transformation $w = z^{\varphi/\pi}$ to show that [133]

$$x_i^{(e)} = \frac{\pi}{\varphi}\, x_{i,s}, \qquad (15.154)$$

where $x_{i,s}$ is the surface exponent of the operator ϕ_i.[11] See [390] and Table 15.5 for a numerical verification for the Ising and Potts models, [622, 428] for percolation (with very extensive tables) and [309] for corner exponents in the presence of defect lines.

Table 15.5. Energy density corner exponent $x_\varepsilon^{(e)}$ in 2D models (after [390])

φ [degrees]	Ising	Potts-3	expected
180	2.001(1)	2.01(3)	2
90	3.999(1)	4.00(4)	4
45	7.993(7)	8.0(1)	8
26.56	13.52(2)	13.5(1)	13.5516
18.43	19.6(4)	19.7(4)	19.5281

4. Consider the two-point function of surface spins near a rectangular corner. Use the transformation $z = w^2$ to relate it to the surface two-point function. Compare your result with the exact corner two-point function $\langle \sigma(w_1)\sigma(w_2)\rangle \sim \sqrt{v_1 v_2}/(v_2^2 - v_1^2)$, with $w = u + iv$, obtained for the 2D Ising model [5].

---

[11] Since this relation fails for strongly embedded lattice animals, these cannot be conformally invariant [466]. For the Hilhorst–van Leeuwen model, supersymmetry methods allow to find the entire spectrum for surface and corner exponents, confirming (15.154) [78].

5. Consider the profile $\langle \Phi(v) \rangle$ of a quasi-primary operator with scaling dimension x. Use the conformal discrete-lattice realization from Chap. 12 to show that [326]

$$\langle \Phi(v) \rangle = \phi_0 \, a^{-x} \, \frac{\Gamma\left(\frac{v}{a} + \frac{1}{2} - \frac{x}{2}\right)}{\Gamma\left(\frac{v}{a} + \frac{1}{2} + \frac{x}{2}\right)}.$$

6. For the extraordinary transition, the magnetization has a non-vanishing value at the boundary. For the Ising model, use a duality argument to prove that $H^{(F)} = H^{(++)} + H^{(+-)}$. Generalize this to other spin systems. Show that the extraordinary surface exponent $\eta_\parallel^{\mathrm{ex}}$ can be found from the operator content of the quantum Hamiltonian H when the boundary spins are both fixed in the same orientation, e.g. $H^{(++)}$ or $H^{(00)}$. Derive the values of $\eta_\parallel^{\mathrm{ex}}$ from the operator contents given in the text.

7. Find the order-parameter two-point function for the tricritical Ising model near a free surface (see [250] for the extension to all multicritical Ising models).

8. Prove (15.141), starting from the Hamiltonian (15.140) [500] and show that the critical point of the chain (15.140) is given by

$$\lim_{L \to \infty} \prod_{k=1}^{L} (t_{k,c})^{1/L} = 1.$$

9. Consider a system with free boundaries in the form of a parabola $v^2 = 2pu + p^2$ in the complex plane $w = u + iv$. This may be related to the half-plane via the conformal map $z = i \cosh(\pi \sqrt{w/(2p)}\,)$. Analyse the two-point function and discuss the possible types of critical behaviour [502]. Quantum chains in this setting are obtained from the corner transfer matrix [G19] and leads in this case to a Hamiltonian of the form

$$H = -\sum_{n=1}^{N-1} \sqrt{n} \, \sigma^z(n) - \sum_{n=0}^{N-1} \sqrt{n+1} \, \sigma^x(n)\sigma^x(n+1).$$

Find the one-fermion energies Λ_k and use this to determine the magnetization at the tip of system $m_{\mathrm{tip}} = \prod_k \tanh(\Lambda_k/2)$ [594, 502, 179] (this is a long exercise).

10. Use (6.38) to derive the surface behaviour of $\langle \sigma\epsilon \rangle$ in the Ising model.

16. Strongly Anisotropic Scaling

The techniques of conformal invariance presented in this book are restricted to systems which are, at least in the critical regime, isotropic and static. It is of interest, however, to investigate to what extend ideas and techniques developed for conformally invariant systems can be carried over to this more general situation. At present, this field is still in its infancy and there remains plenty of scope for further investigations.

16.1 Dynamical Scaling

Consider the following scaling form of a two-point function

$$G\left(br, b^{\theta} t\right) = b^{-2x} G\left(r, t\right),\qquad(16.1)$$

where r and t denote "spatial" and "temporal" coordinates, x is the scaling dimension and θ is the **anisotropy exponent**. We are back to isotropic scaling for $\theta = 1$, but if $\theta \neq 1$, the system is called **strongly anisotropic**. For systems characterized by an isotropic fixed point, any anisotropy which might be present in a lattice formulation can be eliminated by a convenient rescaling along the axes which is independent of r and t. For strongly anisotropic systems, that is no longer possible, rather the anisotropy even increases with r, t growing, which makes any of the lattice anisotropies (including the Hamiltonian limit) considered in previous chapters look rather weak. A system which satisfies (16.1) with $\theta \neq 1$ is called a **strongly anisotropic critical system**.

Systems where some two-point function satisfies a scaling law of the form (16.1) are quite common. The *first class* of examples involves systems at thermodynamic equilibrium but with extra next-nearest neighbour interactions which compete with the usual next-nearest neighbour interactions. Consider for example the **ANNNI model**, involving classical Ising spins σ_i on a d-dimensional hypercubic lattice and the Hamiltonian

$$\mathcal{H} = -J \sum_{\langle i,j \rangle} \sigma_i \sigma_j + \kappa J \sum_{i,\|}' \sigma_i \sigma_{i'},\qquad(16.2)$$

where $J > 0$ and the second sum only runs along the next-nearest neighbours along a particular fixed axis. For a review, see [561]. For $\kappa > 0$ but small, the extra term merely represents an irrelevant perturbation of the Ising model and the system undergoes the usual ferromagnetic ordering at low temperatures. For larger values of κ, the zero-temperature ground state takes along the $\parallel$ axis the form $\uparrow\uparrow\downarrow\downarrow \cdots$. For $T > 0$, there will be (for $d > 2.5$) a meeting point of an ordered ferromagnetic phase, a paramagnetic phase and an incommensurable phase (characterized by a modulation of the order parameter). This point is called a **Lifshitz point** [341]. Identifying the "time" direction with the direction along the axis $\parallel$ and the "space" directions with the directions perpendicular to it, the *correlation functions* of the ANNNI model *precisely at* the Lifshitz point satisfy the anisotropic scaling law (16.1). In fact, for any $O(n)$ model a generalization analogous to the ANNNI model may be defined, which we shall call **ANNNO(n) model**. It is known that at the Lifshitz point, the anisotropy exponent $\theta = 1/2$ for *all* values of n [341]. We shall examine later the spherical model version $(n \to \infty)$ of this, the so-called **ANNNS model**. Anisotropic scaling has also been found in the chiral Potts model (Chap. 14).

The *second class* of systems with strongly anisotropic scaling involves the true time-dependent behaviour of spin systems. For example, one may study the time-dependence of correlators at an equilibrium critical point leading to **critical dynamics** [291] or else investigate **phase ordering kinetics**, the (out of equilibrium) time-dependent growth of domains after quenching an initially disordered system to a state in the ordered phase [90, 448]. Quantities of interest include the correlation functions $C(\boldsymbol{r}, \boldsymbol{r}'; t, t') = \langle \sigma_{\boldsymbol{r}}(t) \sigma_{\boldsymbol{r}'}(t') \rangle$ or the **response functions** $G(\boldsymbol{r}, \boldsymbol{r}'; t, , t')$, which satisfy

$$\langle \sigma_{\boldsymbol{r}}(t) \rangle = \int \mathrm{d}^d \boldsymbol{r}' \, \mathrm{d}t \, G(\boldsymbol{r}, \boldsymbol{r}'; t, t') h(\boldsymbol{r}', t') \tag{16.3}$$

and measure the response of the system to a weak time-dependent perturbation $h(\boldsymbol{r}, t)$.[1] For long times, correlators and response functions display anisotropic scaling,[2] where the anisotropy exponent $\theta = z$ is referred to as **dynamical exponent**. As reviewed in e.g. [291, 287], the value of z depends, among other things, on the conservation laws respected by the dynamics. For example, the dynamical exponents of a binary liquid-vapour transition and of a binary alloy are different, although the static critical behaviour of both is described by the Ising universality class. For a continuum field-theory description (Martin-Siggia-Rose theory [449, 64]), one may start from a phenomenological time-dependent Ginzburg–Landau equation of motion,

---

[1] Causality implies that $G = 0$ for $t < t'$. When linear-response theory is applicable, correlators can be related to response functions via the fluctuation-dissipation theorem, viz. $C(t - t') \sim \int_{-\infty}^{t'} \mathrm{d}t'' \, G(t - t'')$. In the same context, the Fourier transform $\tilde{G}(\boldsymbol{r}, \omega)$ is analytic in the frequency ω in the upper half plane, see [417, G6]. The distinction between C and G is unnecessary for the conformally invariant systems discussed so far.

[2] See [E38, E57, E80] for nice experimental demonstrations of scaling in phase ordering kinetics.

supplemented with a noise term. Introducing a **response field** $\tilde{\phi}$, conjugate to the continuum field ϕ which describes the order parameter scaling operator, the averaging over the noise can be explicitly carried out and one retains an effective field theory formulated in terms of ϕ and $\tilde{\phi}$. While correlators C are obtained as usual from averages $\langle \phi\phi \rangle$, response functions may be found in this context from $G(\boldsymbol{r}, \boldsymbol{r}'; t, t') = \langle \phi(\boldsymbol{r}, t)\tilde{\phi}(\boldsymbol{r}', t') \rangle$. Scaling dimensions and critical exponents may then be extracted through standard techniques, for example ε-expansion. Similar techniques have also been developed for domain growth problems and kinetic roughening, with scale invariance emerging far from equilibrium, see [104, 105, 419, 462] for reviews. The most thoroughly studied theory of these is given by the Kadar-Parisi-Zhang (KPZ) equation [386].[34]

Anisotropic scaling may also occur in purely geometrical problems, the simplest of which is probably **directed percolation**, see [399] and references therein. This is a percolation process on a lattice with a preferred direction $\parallel$, which takes the role of "time". By definition, a step of the directed percolation process must not proceed against the orientation of the preferred direction. As for ordinary percolation, there is a second-order phase transition between a percolating phase with an infinite cluster and a non-percolating phase. The main distinction[5] between ordinary and directed percolation is a non-trivial anisotropy exponent $\theta \neq 1$ in the latter. Contact with field-theory techniques has been achieved through a celebrated mapping of bond directed percolation onto Reggeon field theory [129]. It has been recognized that generically the continuous transitions occurring in non-equilibrium systems with a single absorbing state are in the same universality class as directed percolation [278, 369, 399].

Equation (16.1) can be cast into the form

$$G(\boldsymbol{r}, t) = t^{-2x/\theta} \, \mathcal{G}(u) \tag{16.4}$$

defining the **dynamical scaling function** $\mathcal{G}(u)$ where $u = |\boldsymbol{r}|^\theta/t$ is the scaling variable. Global scale invariance leaves $\mathcal{G}$ undetermined. It is tempting to *try to find the function $\mathcal{G}(u)$ and the anisotropy exponent θ by generalising (16.1) to local, space-time dependent rescaling factors $b = b(\boldsymbol{r}, t)$.* As for conformal invariance, the $b(\boldsymbol{r}, t)$ should satisfy some extra requirements and it is part of the problem to discover these for anisotropic scaling, if they exist at all. It is clear that from the variety of examples with anisotropic scaling

---

[3] The KPZ equation is 'Galilei'-invariant and in (1+1)D, one has $z = 3/2$.

[4] There is also a conformal KPZ equation which relates scaling dimensions of spin models coupled to 2D quantum gravity to standard universality classes [407].

[5] More subtle distinctions involve the fractal dimensions of critical percolation clusters [301, 321] and the breaking of hyperscaling of the 'surface tension' in the percolating phase [311].

that one should expect that fixing the scaling function $\mathcal{G}(u)$ will require more information than merely the value of the exponent x.[6]

The first attempt of this kind was made by Cardy [136], who considered the 2D conformal transformation $r \to b(r)r$ and $t \to b(r)^\theta t$ for the response function in critical dynamics. The response function was mapped onto the strip and found there through van Hove theory (where $\theta = 2$). For a non-conserved order parameter, this lead to $\mathcal{G}(u) \sim e^{-u}$, where a metric factor was absorbed into u and θ remained a free parameter. Later, it was realized that this scaling form is not reproduced at the (higher order) Lifshitz points of the ANNNS model [241] and for directed percolation [74, 279] (where $\theta \neq 2$). Rather, the leading behaviour for u large was either algebraic in u or of the form $\mathcal{G}(u) \sim \exp(-u^{1/(\theta-1)})$.

16.2 Schrödinger Invariance

We now try to construct more systematically groups of space-time transformations compatible with anisotropic scaling. We begin with the case $\theta = 2$. For notational simplicity, we merely consider the case $d = 1$ of one spatial dimension, but the generalisation to arbitrary d is immediate. The following set of transformations defines the **Schrödinger group** [474, 290]

$$r \to r' = \frac{r + vt + a}{\gamma t + \delta} \quad , \quad t \to t' = \frac{\alpha t + \beta}{\gamma t + \delta} \quad ; \quad \alpha\delta - \beta\gamma = 1, \qquad (16.5)$$

where $\alpha, \beta, \gamma, \delta, v, a$ are parameters. This is the maximal group [474] which transforms solutions of the free Schrödinger equation

$$\left(i\frac{\partial}{\partial t} + \frac{1}{2m}\frac{\partial^2}{\partial r^2} \right) \psi = 0 \qquad (16.6)$$

into other solutions of (16.6), viz. $(r, t) \mapsto g(r, t)$, $\psi \to T_g\psi$

$$(T_g\psi)(r, t) = f_g\left(g^{-1}(r, t)\right) \psi\left(g^{-1}(r, t)\right) \qquad (16.7)$$

and the companion function f_g is known explicitly. For diffusive processes, write $m = i/(2D)$, where D is the diffusion constant. The Schrödinger group can be obtained as the non-relativistic limit of the conformal group [54] and it is known that non-relativistic free field theory is Schrödinger invariant [290].

The set $\{X_{-1}, X_0, X_1, Y_{-1/2}, Y_{1/2}, M_0\}$ of generators spans the Lie algebra of (16.5). They read [320][7]

---

[6] While scaling functions can certainly be found in a model context, see e.g. [172, 238, 438, 506] and the examples discussed below, here the question is how much can be said in a model-independent way.

[7] For $d > 1$, there are several generators $Y_m^{(a)}$, $a = 1, \ldots, d$, straightforwardly written down and rotation generators as well.

$$X_n = -t^{n+1}\partial_t - \frac{n+1}{2}t^n r \partial_r - \frac{n(n+1)}{4}\mathcal{M}t^{n-1}r^2$$

$$Y_m = -t^{m+1/2}\partial_r - \left(m + \frac{1}{2}\right)\mathcal{M}t^{m-1/2}r$$

$$M_n = -t^n \mathcal{M}, \tag{16.8}$$

where the terms $\sim \mathcal{M}$ in X_n and Y_m come from the companion function and $n(m)$ is an integer (half-integer). The commutation relations are

$$[X_n, X_m] = (n-m)X_{n+m}$$

$$[X_n, Y_m] = \left(\frac{n}{2} - m\right)Y_{n+m}$$

$$[X_n, M_m] = -m\, M_{n+m}$$

$$[Y_n, Y_m] = (n-m)M_{n+m}$$

$$[Y_n, M_m] = [M_n, M_m] = 0 \tag{16.9}$$

which is the analogue of the loop algebra (2.19) found in Chap. 2.[8] While M_0 is sometimes referred to as central charge of the finite-dimensional subalgebra (16.9), it is clear from the infinite algebra that it is just a component of a U(1) current. From the Jacobi identities it follows that in (16.9) only the commutator $[X_n, X_m]$ admits a central extension, which must be of the familiar Virasoro form [322].

We are interested in two-point functions of quasi-primary operators $\Phi(r,t)$. Under the Schrödinger group, these should transform as $\delta_X \Phi = \epsilon[X_n, \Phi]$, $\delta_Y \Phi = \epsilon[Y_m, \Phi]$, where

$$[X_n, \Phi(r,t)] = \left(t^{n+1}\partial_t + \frac{n+1}{2}t^n r\partial_r\right.$$

$$\left. + \frac{n(n+1)}{4}\mathcal{M}t^{n-1}r^2 + (n+1)\frac{x}{2}t^n\right)\Phi(r,t)$$

$$[Y_m, \Phi(r,t)] = \left(t^{m+1/2}\partial_r + \left(m + \frac{1}{2}\right)\mathcal{M}t^{m-1/2}r\right)\Phi(r,t), \tag{16.10}$$

where $n(m)$ is integer (half-integer) and x is the scaling dimension of $\Phi(r,t)$. A further remark concerning the "masses" $\mathcal{M}$ is necessary. It is well known for non-relativistic, Galilei-invariant field theories that a 'massive' operator Φ must be *complex* [435].[9] Therefore, the operator Φ is distinguished from the operator Φ^* by the sign of $\mathcal{M}$. Galilei-invariance further implies that the total mass of a set of operators inserted into a correlator must vanish [435]. This distinction between the operators Φ and Φ^*, originally established for

---

[8] The infinite algebra (16.9) is the full dynamic symmetry algebra of the 1D Burgers equation [364].

[9] These are non-relativistic masses *not* related to $T - T_c$, as it would be the case for relativistic masses arising for perturbed conformal theories.

true masses m, must carry over to the diffusion equations studied here.[10] Following the lines of Chap. 2, the two- and three-point functions of quasi-primary operators can be found. The result is [322] (with $t, u, v > 0$)

$$\langle \Phi_a(r,t)\Phi_b^*(0,0) \rangle = \delta_{x_a,x_b} \delta_{\mathcal{M}_a,\mathcal{M}_b} \mathcal{G}_0 \, t^{-x_a} \exp\left(-\frac{\mathcal{M}_a}{2}\frac{r^2}{t}\right) \qquad (16.11)$$

$$\langle \Phi_a(r,u)\Phi_b(s,v)\Phi_c^*(0,0) \rangle = \delta_{\mathcal{M}_a+\mathcal{M}_b,\mathcal{M}_c} \exp\left[-\frac{\mathcal{M}_a}{2}\frac{r^2}{u} - \frac{\mathcal{M}_b}{2}\frac{s^2}{v}\right]$$

$$\times u^{-(x_a+x_c-x_b)/2} v^{-(x_b+x_c-x_a)/2} (u-v)^{-(x_a+x_b-x_c)/2} \, \Psi\left(\frac{(u-v)uv}{(rv-su)^2}\right),$$
$$(16.12)$$

where $\Psi = \Psi_{ab,c}$ is an arbitrary function and $\mathcal{G}_0$ a normalization constant. This result is quite similar to the corresponding result (2.36,2.38) of conformal invariance, but we stress the important role played by the (non-universal) constants $\mathcal{M}_{a,b} \geq 0$. The two-point function can also be found for the *semi-infinite geometry*, where quasi-primary fields transform covariantly under the generators $X_{\pm 1,0}$ only, which keep the surface $r = 0$ invariant. One has [322] (with $t > 0$)

$$\langle \Phi_a(r_a,t)\Phi_b^*(r_b,0) \rangle = \delta_{x_a,x_b} t^{-x_a} \chi(r_a r_b/t) \exp\left(-\frac{\mathcal{M}_a}{2}\frac{r_a^2}{t} - \frac{\mathcal{M}_b}{2}\frac{r_b^2}{t}\right),$$
$$(16.13)$$

where $\chi = \chi_{a,b}$ is an arbitrary function. Since the surface breaks Galilei invariance, there is no constraint on the masses $\mathcal{M}_{a,b}$. Finally, the two-point function for a *non-steady state* is found from the covariance under the generators $X_{0,1}, Y_{\pm 1/2}$ which keep the initial time surface $t = 0$ invariant [322] (with $t_a > t_b$)

$$\langle \Phi_a(r_a,t_a)\Phi_b^*(r_b,t_b) \rangle = \delta_{\mathcal{M}_a,\mathcal{M}_b} \mathcal{G}_0 \left(\frac{t_a}{t_b}\right)^{(x_b-x_a)/2}$$

$$\times (t_a - t_b)^{-(x_a+x_b)/2} \exp\left(-\frac{\mathcal{M}_a}{2}\frac{(r_a-r_b)^2}{t_a-t_b}\right),$$
$$(16.14)$$

where $\mathcal{G}_0$ is a normalization constant. Because there is here no time-translation invariance, there is no constraint on the exponents $x_{a,b}$.

In the same way as in Chap. 2, one may derive relations between the components of the energy–momentum tensor T_{ij} (we use $i = 0$ for the time coordinate) from the projective Ward identities. From scale invariance, it follows that[11] $2T_{00} + T_{11} = 0$ and from Galilei invariance $T_{01} = 0$. Thus [290]

---

[10] In the context of critical dynamics formulated as a continuum field theory, Φ^* is naturally interpreted as the response field associated with the physical field Φ.

[11] Here written for $d = 1$, with an obvious generalization for any d. Anomalies may modify this relation (here for Euclidean theories) in the case of interacting non-relativistic theories [86].

$$
\left.\begin{array}{c}
\text{translation invariance (in space and time)} \\
\text{rotation invariance (for } d \geq 2) \\
\text{anisotropic scale invariance} \\
\text{Galilei invariance} \\
\text{short-ranged interactions}
\end{array}\right\} \Longrightarrow \text{Schrödinger invariance}
$$

$$(16.15)$$

quite analogous to the corresponding statement for conformal invariance. We emphasize that Galilei invariance has to be required and cannot be derived from the other transformations. Certainly, the specific form of the n-point functions does depend on the way Galilei invariance is realized, in analogy with the conformal invariance examples discussed in Chap. 12.

The above results for two- and three-point functions have been reproduced in a variety of models as reviewed in [322, 323]. Here, we quote merely two examples. The first one is the 1D **kinetic Ising model** with Glauber dynamics which is defined as follows. Consider the time-dependent probability distribution $P(s_1, \ldots, s_N; t)$ of the spin configuration $\{s\} = (s_1, \ldots, s_N)$. It is convenient to describe the evolution of $P(\{s\}; t)$ in terms of a **master equation** describing the rates by which a spin configuration $\{s\}$ changes into another configuration $\{s'\}$

$$
\partial_t P(\{s\}; t) = \sum_{\{s'\} \neq \{s\}} [w(\{s'\} \to \{s\}) P(\{s'\}; t) - w(\{s\} \to \{s'\}) P(\{s\}; t)],
$$

$$(16.16)$$

where $w(\{s\} \to \{s'\})$ are the phenomenologically given transition rates between the two spin configurations $\{s\}$ and $\{s'\}$.[12] **Glauber dynamics** [269] only allows a single spin flip per time with the rate[13]

$$
w_i(s_i) = \frac{\alpha}{2} \left(1 - \frac{1}{2} \tanh\left(\frac{2J}{T}\right) s_i (s_{i-1} + s_{i+1}) \right), \tag{16.17}
$$

where α is the transition rate. This automatically satisfies the two essential conditions to be required, namely (i) probability conservation

$$
\sum_{\{s\}} P(\{s\}; t) = 1 \tag{16.18}
$$

at all times, and (ii) the **detailed balance condition**

$$
\frac{w(\{s'\} \to \{s\})}{w(\{s\} \to \{s'\})} = \exp\left[-\frac{1}{T} (\mathcal{H}[\{s\}] - \mathcal{H}[\{s'\}]) \right] \tag{16.19}
$$

which makes the equilibrium distribution $P_{\text{eq}} = e^{-\mathcal{H}/T}/\mathcal{Z}$ a stationary solution of (16.16), where $\mathcal{H}[\{s\}] = -J \sum_i s_i s_{i+1}$ is the energy. Time-dependent averages are obtained from

---

[12] Deriving the master equation from quantum mechanics is far from trivial [417].
[13] An experimental example is given by 1D Fe stripes on a Cu(111) surface [E81].

$$\langle X \rangle(t) = \sum_{\{s\}} X(\{s\}) \, P(\{s\}; t). \qquad (16.20)$$

The model so defined is completely integrable [269]. The critical point is at zero temperature. From the exact solution, it is easily verified that, where $r = i - j$, $t = t_1 - t_2$

$$C(i - j, t_1 - t_2) = \langle s_i(t_1) s_j(t_2) \rangle_T = (2\pi\alpha t)^{-1/2} \exp\left(-\frac{1}{2\alpha} \frac{r^2}{t}\right) \qquad (16.21)$$

at thermal equilibrium, in the scaling limit $r \to \infty$, $t \to \infty$ with $u = r^2/t$ fixed and where finally the $T \to 0$ limit was also taken [320, 322]. The scaling form (16.11) is clearly recovered and the remaining parameters x and $\mathcal{M}$ can be identified.

As a second example, we consider the *response functions* obtained in the late and intermediate stages of the time evolution of a spin system quenched from an initially uncorrelated state to a temperature $T \leq T_c$ and then allowed to relax. For the spherical model with a non-conserved order parameter, the dynamical exponent $z = 2$ [370, 472]. The existing exact results in d dimensions thus allow to test (16.14). Calculating the two-time response function which measures the response of the order parameter (Φ_a) at time t to a thermal fluctuation (Φ_b^*) at time $t' \leq t$, complete consistency is found if [322]

$$x_a = \frac{d}{4} \quad , \quad x_b = \frac{3d}{4} \quad ; \text{ if } T = 0 \; [472] \qquad (16.22)$$

$$x_a = \frac{3d}{4} - 1 \quad , \quad x_b = 1 + \frac{d}{4} \quad ; \text{ if } T = T_c \; [370] \qquad (16.23)$$

and we see that the scaling dimensions only depend on T and are in general different from each other, $x_a \neq x_b$.[14] This is related to the existence of universal time-dependent scaling at *intermediate* times first described in [370]. Finally, if we have that $x_a = x_b$, we may take the limit $t_b \to 0$ in (16.14) and find (with $t > 0$)

$$\langle \Phi_a(r_a, t) \Phi_b^*(r_b, 0) \rangle = \delta_{\mathcal{M}_a, \mathcal{M}_b} \, \mathcal{G}_0 \, t^{-x_a} \exp\left(-\frac{\mathcal{M}_a}{2} \frac{(r_a - r_b)^2}{t}\right) \quad ; \; x_a = x_b. \qquad (16.24)$$

This form should be obeyed by single-time response function measuring the response of the order parameter at time t to a fluctuation of itself at time zero. The explicit results in the spherical model at $T = 0$ are consistent with this and do reproduce $x_a = d/4$ [472], as expected from (16.22).

Schrödinger invariance in the ANNNS model will be discussed at the end of the next section.

---

[14] Thermal fluctuations are expected to be irrelevant for $T < T_c$. Correlations in the initial states only affect the values of $x_{a,b}$ [472].

16.3 Towards Local Scale Invariance for General θ

Schrödinger invariance has provided us with a useful case study of a dynamical symmetry compatible with scaling, yet different from conformal invariance. We shall now describe a recent attempt to construct a set of local, space-time-dependent scale transformations for more general values of the anisotropy exponent [325]. This will be done through specifying a set of conditions on the these transformations, which are formulated such that one remains as closely as possible to the known cases of conformal and Schrödinger invariance. We do so for sheer conservatism. Let us require that

1. Möbius transformations occur in both conformal and Schrödinger transformations, see (2.22,16.5). We want to keep them for the time coordinate.
2. The dilatation generator is $X_0 = -t\partial_t - \frac{1}{\theta}r\partial_r$.
3. Spatial translation invariance is satisfied.
4. 'Mass' terms built in analogy with the Schrödinger generators must be present.
5. When applied to two-point functions, through the projective Ward identities we will obtain a set of differential equations. We require that there will be at most a finite number of conditions. This means that the infinitesimal generators *when applied to the states* should provide a realization of a finite-dimensional Lie algebra.

From the first condition, the time-changing generators X_n, $n = 0, \pm 1$, must have the form $X_n = -t^{n+1}\partial_t + \ldots$, where the dots stand for the other terms we still have to construct. The first term alone fixes the commutation relation $[X_n, X_m] = (n - m)X_{n+m}$. In the sequel, *any extra term in X_n must be built such that these commutation relations are kept.*[15] Now, from the explicit form of X_0, it is easy to show that

$$X_n = -t^{n+1}\frac{\partial}{\partial t} - \frac{n+1}{\theta}t^n r\frac{\partial}{\partial r} + \ldots, \qquad (16.25)$$

where the dots now stand for any mass term still to be constructed. Let us write

$$\theta = \frac{2}{N}. \qquad (16.26)$$

We now digress to discuss spatial translation invariance, generated by ∂_r. Letting

$$Y_m = -t^{N/2+m}\frac{\partial}{\partial r} + \ldots \qquad (16.27)$$

with $m = -N/2 + k$ and $k = 0, 1, 2, \ldots$, we easily find

$$[X_n, X_m] = (n - m)X_{n+m} \ , \ \ [X_n, Y_m] = \left(N\frac{n}{2} - m\right)Y_{n+m} \qquad (16.28)$$

---

[15] If the reader should have a better idea, he is urged to write it up and let me know.

and in particular

$$[X_1, Y_{-N/2+k}] = (N - k) Y_{-N/2+k+1} \tag{16.29}$$

which shows that the repeated action of X_1 on $Y_{-N/2} = -\partial_r$ is creating a potentially infinite set of generators. The only way to truncate this to a finite number of generators is through the requirement that $N = 1, 2, \ldots$ is a positive integer. In other words, only if

$$\theta = \frac{2}{N} = 2, 1, \frac{2}{3}, \frac{1}{2}, \frac{2}{5}, \ldots \tag{16.30}$$

then is it possible to retain a *finite* number of the Y_m generators. For all other values of θ, the set of the Y_m is infinite. At this exploratory stage, it is not really clear what the meaning of this large set of conditions on the two-point functions might be and we shall not follow this possibility any further here.

Since the 'conformal' structure of these transformations always involve time, this might suggest that the temporal degrees of freedom are enough to render the system critical. Results to be obtained for the two-point function should be valid independently of whether the spatial degrees of freedom are critical by themselves or not. Of course, one might formally interchange the roles of space and time and find then the sequence $\theta = 1/2, 1, 3/2, 2, \ldots$ of anisotropy exponents. In that case, however, the spatial degrees of freedom must be critical by themselves, as is for example realized for critical dynamics. In (1+1)D however, it should be possible to permute space and time quite freely and there should be two infinite sequences of values of θ where the local scale invariance, under construction, might apply.

To complete the construction, the mass terms must be determined. Their form determines to a large extent to what type of physical system the final results can be applied. For the determination of the scaling function $\mathcal{G}(u)$ of the two-point function, the 'Galilei'-subalgebra spanned by $\{X_{-1,0}, Y_{-N/2,-N/2+1}\}$ is enough. Its Casimir operator, when formally rewritten in energy–momentum space, provides information about those lattice models which might possibly realize that symmetry. Since a full derivation takes too much space, we shall here simply quote the result [325] and discuss some of its consequences. The generators $X_{-1,0}, Y_{-N/2}$ do not contain any mass term. All information required is contained in the form of the generators

$$X_1 = -t^2 \partial_t - Ntr\partial_r - \alpha r^2 \partial_t^{N-1} \ , \quad Y_{-N/2+1} = -t\partial_r - \frac{2\alpha}{N} r \partial_t^{N-1}, \tag{16.31}$$

where the dimensionful constant α parametrizes the mass term. Obviously, commuting the Y's with each other will generate an infinity of extra generators when $\alpha \neq 0$ and unless $N = 1, 2$, there does not appear a simple way to close the algebra. However, all what we really want is to be able to find the two-point functions $G = \langle \Phi_a \Phi_b \rangle$. If we can arrange that for any such two-point function the action of any commutator $[Y_m, Y_{m'}]$ vanishes, only a finite number of conditions will be imposed on G. It is easy to see that

$$\alpha_a = (-1)^N \alpha_b \qquad (16.32)$$

does meet this requirement. On the other hand, because of the commutator (16.28), any further action with the generators X_n cannot invalidate this result. Thus, although there is no finite-dimensional Lie algebra in an operator sense, the calculation of the two-point functions can formally proceed by just considering the projective Ward identities obtained from the action of the X_n, Y_m, provided the condition (16.32) is met. Then the effective commutator algebra acting on the states $|\Phi_a \Phi_b\rangle$ is given by (16.28).

In consequence, if N is even, $\alpha_a = \alpha_b$ for any pair of operators. An extreme way of satisfying this is by introducing an universal mass constant α, which is the same for all operators (and $\Phi = \Phi^*$). For N odd, one has for each scaling operator Φ, characterized by the pair of numbers (x, α), a conjugate scaling operator Φ^*, characterized by the pair $(x, -\alpha)$. This general structure is reproduced in the two known special cases $N = 1, 2$. For $N = 1$, we recover exactly the Lie algebra generators (16.8) of Schrödinger invariance and we identify $\alpha_i = \mathcal{M}_i / 2$, the masses of the scaling operators. For $N = 2$, we get back the conformal algebra in the form presented in Table 15.1, with $X_n = l_n + \bar{l}_n$ and $Y_n = i(l_n - \bar{l}_n)$. This becomes obvious when introducing the light-cone coordinates $z, \bar{z} = t \pm \sqrt{\alpha}\, r$, where $\alpha = -1/c^2$ and c can be interpreted as the speed of light, usually set to unity.

We now determine the two-point function ($r = r_a - r_b, t = t_a - t_b > 0$)

$$G(r_a - r_b; t_a - t_b) = \langle \Phi_a(r_a, t_a)\Phi_b^*(r_b, t_b)\rangle = \delta_{x_a, x_b}\delta_{\alpha_a, \alpha_b} r^{-2x_a} \Omega\left(\frac{t}{r^{2/N}}\right),$$
$$(16.33)$$

where translation and dilatation invariance are already used. Local scale invariance yields the single extra condition

$$\left(\alpha_1 \frac{d^{N-1}}{dv^{N-1}} - v^2 \frac{d}{dv} - \zeta v\right)\Omega(v) = 0, \qquad (16.34)$$

where

$$\zeta := \frac{N}{2}(x_a + x_b) \qquad (16.35)$$

characterizes the scaling operator under consideration. This must be supplemented by some boundary conditions. If we let either $t = 0$ or $r = 0$, we have from (16.1) that G should decay with a power law. That leads to the requirements

$$\Omega(0) = \text{cste.} , \quad \Omega(v) \sim v^{-\zeta} \text{ for } v \to \infty. \qquad (16.36)$$

Since the cases $N = 1, 2$ are already known, we concentrate on $N \geq 3$. Then the general solution of (16.34) is[16]

---

[16] We leave it as an exercise to check that for $N = 2$, the conformal invariance result is recovered.

$$\Omega(v) = \sum_{p=0}^{N-2} b_p v^p \, {}_2F_{N-1}\left(\frac{\zeta+p}{N}, 1; 1 + \frac{p}{N}, 1 + \frac{p-1}{N}, \ldots, \frac{p+2}{N}; \frac{v^N}{N^{N-2}\alpha_1}\right),$$

$$(16.37)$$

where $_2F_{N-1}$ is a generalized hypergeometric function and the b_p are free parameters. In order to check the boundary conditions, the asymptotic behaviour of $\Omega(v)$ for $v \to \infty$ is required. The asymptotics of the generalized hypergeometric functions had been discussed in detail by Wright [624]. It can be shown that each of the terms occurring in (16.37) consists of two terms. The first term is an infinite series which grows exponentially for v large and the second term shows algebraic behaviour for v large [624]. At first sight, the presence of infinitely many exponentially growing terms might exclude the compatibility of the solution (16.37) with the boundary condition (16.36), but very remarkably, it turns out that a *single* condition on the b_p, namely

$$\sum_{p=0}^{N-2} b_p \frac{\Gamma(p+1)}{\Gamma\left(\frac{p+1}{N}\right)\Gamma\left(\frac{p+\zeta}{N}\right)} \left(\frac{\alpha_1}{N^2}\right)^{p/N} = 0$$

$$(16.38)$$

is enough to cancel the *entire* exponential series.[17] The final result for the scaling form of the two-point function, eliminating b_{N-2}, is $\Omega(v) = \sum_{p=0}^{N-3} b_p \Omega_p(v)$. We simply give here the asymptotics [325]

$$\Omega_p(v) \simeq \begin{cases} v^p & ; \ v \to 0 \\ \Omega_\infty v^{-\zeta} & ; \ v \to \infty \end{cases}$$

$$(16.39)$$

which is found to be in agreement with the physically requested boundary condition! We illustrate this remarkable and unexpected cancellation in Fig. 16.1 for $N = 4$, where the functions $\Psi_p(v) = v^\zeta \Omega_p(v)$ are shown, which for $v \to \infty$ should tend towards a constant Ω_∞, which finally is

$$\Omega_\infty = -\left(\frac{\alpha_1}{N^2}\right)^{(\zeta+p)/N} \frac{\Gamma(\frac{1-\zeta}{N})}{\Gamma(1-\zeta)} \frac{\Gamma(p+1)}{\Gamma(\frac{p+1}{N})}$$

$$(16.40)$$

$$\times \frac{\pi \sin\left(\frac{\pi}{N}(p+2)\right)}{\Gamma(\frac{p+\zeta}{N}) \sin\left(\frac{\pi}{N}(p+\zeta)\right) \sin\left(\frac{\pi}{N}(\zeta-2)\right)}.$$

Thus, the solution to our problem is given by (16.37,16.38). After fixing an arbitrary normalization constant, $N-3$ of the parameters b_p are still arbitrary and there is at present no known criterion to fix their values in a model-independent way. Finally, by considering the Casimir operator of the 'Galilei'-subalgebra, one may show that the underlying dispersion relation is of the form $E^N = k^2$. That indicates that the class of models to which (16.37,16.38) apply are those of the first class described at the beginning of this chapter. In

---

[17] For $N = 3$, this is a known relation between Kummer and Tricomi functions [1].

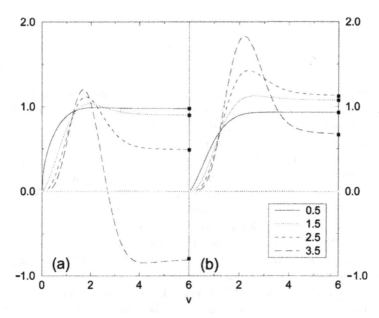

Fig. 16.1. Scaling functions $\Psi_p(v) = v^\zeta \Omega_p(v)$ for $N = 4$ and (a) $p = 0$ and (b) $p = 1$. The different curves correspond to the indicated values of ζ and the 'mass' $\alpha_1 = 1$. The full squares indicate the values for $v \to \infty$ according to (16.40)

particular, since for *Lifshitz points* the anisotropy exponent is always $\theta = 1/2$ [341], these are ideal candidates on which to try the predictions obtained.

In the *ANNNS model*, the spin-spin correlation function *at* the Lifshitz point is known exactly in d dimensions [241]. From a straightforward calculation we have

$$C(r_\|, r_\perp) = \langle s_{r_\|, r_\perp} s_{0,0} \rangle$$

$$= \frac{T_c}{2(2\pi)^d J} \int_{-\pi}^{\pi} \cdots \int_{-\pi}^{\pi} dk_1 \ldots dk_d \frac{\cos(\boldsymbol{k} \cdot \boldsymbol{r})}{d + \kappa - \sum_{j=1}^{d} \cos k_j - \kappa \cos(2k_1)}$$

$$= C_0 r_\perp^{-d-d_*} \Psi \left(\frac{d - d_*}{2}, \sqrt{\frac{1}{32c_2}} \frac{r_\|^2}{r_\perp} \right), \tag{16.41}$$

where $\boldsymbol{r} = (r_\|, r_\perp)$ and

$$\Psi(a, x) := \sum_{k=0}^{\infty} \frac{(-1)^k}{k!} \frac{\Gamma(k/2 + a)}{\Gamma(k/2 + 3/4)} x^k \tag{16.42}$$

while $d_* = 5/2$ is the lower critical dimension and C_0, c_2 are known non-universal constants. On the other hand, since $\theta = 1/2$ at a Lifshitz point, this corresponds to local scaling with $N = 4$. Then, the two-point function is

$$G(r, t) = r^{-\zeta/2} \Omega(v) = r^{-\zeta/2} \left(b_0 \Omega_0(v) + b_1 \Omega_1(v) \right). \tag{16.43}$$

Comparison with the $\Omega_p(v)$ calculated above shows that for $N = 4$, $v = t/\sqrt{r}$ and

$$\Omega_0(v) = \frac{\Gamma(3/4)}{\Gamma(\zeta/4)} \, \Psi\left(\frac{\zeta}{4}, \sqrt{\frac{1}{4\alpha_1}} \, v^2\right) \quad ; \quad N = 4. \qquad (16.44)$$

Thus, with the correspondence $t \leftrightarrow r_{\parallel}$, $r \leftrightarrow r_{\perp}$, $\zeta = 2(d - d_*)$ and $\alpha_1 = 8c_2$, the scaling function for the spin-spin correlation function in the ANNNS model precisely at the Lifshitz point is exactly reproduced by the local scaling form (16.37,16.38) for the parameter value $b_1 = 0$.

It is tempting to speculate that precisely at the Lifshitz point in any ANNNO(n) model with $d > 5/2$ the correlators might be of the form (16.43). At the very least, these models all obey anisotropic scaling with $N = 4$ [341]. The question can probably be decided through direct numerical simulation, e.g. of the 3D ANNNI model. Further examples with an anisotropy exponent falling into the list (16.30) include the superintegrable chiral Potts model (Chap. 14), the asymmetric six-vertex model (with $\theta = 1/2$) [19, 398] and a certain non-hermitian quantum chain (with $\theta \simeq 2/3$) [603].

Also, comparing with what has been achieved in the conformal case $\theta = 1$, it is obvious how much more remains to be done before local anisotropic scaling may eventually be turned into a predictive tool. Finally, we point out that the entire discussion in this chapter was centered on systems with dispersion relations of the form $E^N = k^2$ and thus does not tell anything yet about stochastic differential equations, with dispersion relations of the form $E = k^{2/N}$, with $N \neq 1$, as they may arise in critical dynamics, the KPZ equation or directed percolation. The analysis of the generators of local scale invariance in that case is an open problem under investigation.

To close the discussion, let us reconsider the scaling function (16.42). Obviously, $\Psi(3/4, x) = e^{-x}$, reminiscent of the scaling function $\mathcal{G}(u)$ for Schrödinger-invariant systems. In fact, the ANNNS model is easily modified to allow for an additional test of Schrödinger invariance. First, next-nearest neighbour interactions are introduced in all but a single direction, which will play the role of "time", while the other direction play the role of "space". Then the condition $a = 3/4$ corresponds to $d = 6$ dimensions. Second, the model must be extended to complex spins in order to be able to implement Galilei-invariance. Then the Hamiltonian of the modified complex ANNNS' model is

$$\mathcal{H} = -\sum_{i,j} \frac{1}{2} J_{i,j} \left(s_i^* s_j + s_i s_j^*\right) + \lambda \sum_i |s_i|^2, \qquad (16.45)$$

where the $J_{i,j}$ describe the nearest- and next-nearest neighbour interactions and the s_i are complex numbers, subject to the (mean) spherical constraint

$$\sum_i \langle |s_i|^2 \rangle = 2\mathcal{N}, \qquad (16.46)$$

where $\mathcal{N}$ is the total number of sites, which determines λ. It is easy to show that the free energy $F = 2F_{\mathrm{SM}}$, where F_{SM} is the free energy of the ordinary

ANNNS model and consequently the thermodynamics is unchanged [322]. Lattice realizations of various scaling operators are now readily constructed, for example the order parameter at site i is simply $\sigma(i) := s_i$. For the energy density, however, there are two distinct operators, namely $\varepsilon(i) := s_i s_{i'}$ and $\eta(i) := s_i^* s_{i'}$, where i' stands for a nearest neighbor of the site i. While the scaling dimensions $x_\varepsilon = x_\eta = 3$, ε is a massive scaling operator while η is massless.[18] It has been explicitly checked that the two- and three-point functions of these and some other, similarly built, operators are in full agreement with the predictions (16.11,16.12) of Schrödinger invariance [322].

16.4 Some Remarks on Reaction–Diffusion Processes

For numerical studies, it is sometimes useful to relate non-equilibrium systems to the quantum chains which were encountered in the earlier chapters. As an example, consider the 1D kinetic Ising model with Glauber dynamics (16.17). The probability $P(s_1, \ldots, s_N; t)$ of a certain configuration is taken to be an element of the state vector $|\mathcal{P}(t)\rangle$, of 2^N elements. The master equation takes the form $\partial_t |\mathcal{P}\rangle = -H|\mathcal{P}\rangle$, where the elements of the stochastic **quantum Hamiltonian** H are given through the transition rates

$$\langle s'|H|s\rangle = -w(\{s\} \to \{s'\}) \text{ if } \{s\} \neq \{s'\}$$
$$\langle s|H|s\rangle = \sum_{\{s'\} \neq \{s\}} w(\{s\} \to \{s'\}). \tag{16.47}$$

For the kinetic Ising model with Glauber dynamics, H is nothing else than the Ising model quantum chain (10.1) with the model parameters [226, 567]

$$t = \tanh(2J/T) \ , \ \eta^{-1} = \cosh(2J/T) \tag{16.48}$$

(in this context, t is a transverse field not to be confused with time). In other words, as T is varied, the model moves on the **disorder line** $t^2 + \eta^2 = 1$. This is a general property of stochastic systems studied through the master equation, e.g. [266, 545]. From the spectrum of H Glauber's solution [269] can be completely recovered [226]. In general, the time-dependent 1D Ising model with a more general dynamics than (16.17) corresponds to the XYZ chain in a magnetic field [372].

As a second example, we briefly mention **reaction–diffusion processes**. In a simple way, these may be defined by considering particles moving on a lattice. Diffusion is modeled by allowing the particles to hop to a nearby empty lattice site. Two colliding particles (or two particles on nearby sites) may undergo a reaction, for example of the simple types $A+A \to A$, $A+A \to \emptyset$ or $A + B \to \emptyset$, which means that the reaction products disappear from the

---

[18] If real spins were considered, the physical energy density would be a combination of ε and η.

system (or become inert). For these simple cases, the mean particle density decreases monotonously towards zero. For late times, one typically has a power law

$$\rho_A(t) \sim t^{-\alpha} \; ; \; t \to \infty. \tag{16.49}$$

Here, we consider only situations involving a single species A of particles undergoing two-body interactions $A + A \to \emptyset, A$. In low dimensions, strong fluctuation effects reduce the value of α from unity, as obtained from standard kinetic (mean field) equations to $\alpha = 1/2$ for 1D lattices. That is known from numerous exact solutions or simulations (see [523] for a collection of reviews), but we merely list in Table 16.1 some experimental results which clearly show that α is close to $1/2$ and far from the mean-field value of unity.[19] In these experiments, the particles are excitons moving along polymer chains whose only role is a carrier for the excitons. Schematically, the reactions of the excitons can be read from Table 16.1. We observe that the value of α is universal in the sense that it does not depend on the branching ratio of the two reactions $A + A \to \emptyset$ and $A + A \to A$. In [E48] it is also shown that α is independent of the initial density of particles.[20] It transpires that apparently also for non-equilibrium systems there is a sort of 'critical' behaviour governed by power laws which enjoys universality and can thus be found from rather simple models.

Table 16.1. Exciton kinetics on polymer chains and the exponent α in (16.49).

substance	α	source	reaction(s)
$C_{10}H_8$	0.52 - 0.59	[E67]	$A + A \to \emptyset, A$
P1VN/PMMA film	0.47(3)	[E47]	$A + A \to \emptyset, A$
TMMC	~ 0.48	[E48]	$A + A \to A$

In the 1D examples at hand, the associated quantum Hamiltonian $H = H_0 + H_1$, where H_0 turns out to be the XXZ Heisenberg quantum chain, describing essentially the diffusion process, whereas the terms in H_1 describe the chemical reactions. For the cases considered here H_1 has no influence the spectrum of the total Hamiltonian H, viz. spec (H) = spec (H_0) [27]. The spectrum of H is thus the one of a diffusive process, independently of the precise form the reactions. From this the observed universality can be derived. Of course, this statement can be made more precise and general, but that is a very long and fascinating story by itself.

---

[19] For multispecies reactions of the type $A + B \to \emptyset$, more exponents describing the movement of reaction fronts can be defined. For 1D systems, experiments demonstrate again the importance of fluctuation effects for the values of these exponents [E46].

[20] In addition, the results of [E48] imply that the scaling amplitude occuring in (16.49) is independent of the initial density.

Anhang/Annexe

Diplomarbeiten und Dissertationen werden an deutschsprachigen Universitäten weiterhin in Deutsch geschrieben. Da die Fachliteratur heute fast ausschließlich auf Englisch erscheint, ist die Übertragung spezifischer Terminologie ins Deutsche manchmal mühsam. Dieses kurze Verzeichnis möchte dabei helfen.

Les DEAs et les thèses universitaires s'écrivent toujours en français. Comme la littérature spécialisée apparaît aujourd'hui presque exclusivement en anglais, la traduction en français des termes techniques est parfois difficile. Cette liste brève pourra aider.

We compile a short list of specialised terms, translated into German and French.

Deutsch	English	Français
Primäroperator, m	primary operator	opérateur primaire, m
Sekundäroperator, m	secondary operator	opérateur secondaire, m
Operatorinhalt, m	operator content	contenu opératoriel, m
relevant	relevant	pertinent
irrelevant	irrelevant	non pertinent
Eigenwert, m	eigenvalue	valeur propre, f
Eigenvektor, m	eigenvector	vecteur propre, m
Entartung, f	degeneracy	dégénérescence, f
Energielücke, f	energy gap	lacune d'énergie, f
Renormierungsgruppe, f	renormalization group	groupe de renormalisation, m
Skalenverhalten, n	scaling (behaviour)	comportement d'échelle, m
Hyperskalengesetz, n	hyperscaling (law)	loi de hyperéchelle, f
Gitterskalenverhalten, n	finite-size scaling	comportement d'échelle de taille finie, m
Gitterskalenregion, f	finite-size scaling region	région du comportement d'échelle sur réseau, f
Gitterkorrektur, f	finite-size correction	correction de taille finie, f
Gitterpunkt, -platz, m	site	site, m
Abrundungstemperatur, f	rounding temperature	température d'arondissement, f
Überkeuzen, n	cross–over	changement de comportment, m
Kreuzexponent, m	crossing exponent	exposant de changement de comportement, m
Verschiebungsexponent, m	shift exponent	exposant de déplacement, m
Wanderexponent, m	wandering exponent	exposant de divagation, m
Volumenexponent, m	bulk exponent	exposant de volume, m
Oberflächenexponent, m	surface exponent	exposant de surface, m

Eckenexponent, m	corner exponent	exposant de coin, m
Eckentransfermatrix, f	corner transfer matrix	matrice (de transfert) de coin, f
Eckensingularität, f	edge singularity	singularité de seuil, f
Schleife, f	loop	boucle, f
Zopf, m	braid	tresse, f
Abbildung, f	mapping	application, f
Menge, f	set	ensemble, m
Teilmenge, f	subset	sous-ensemble, m
selbstdual	self-dual	auto-dual
Dualität, f	duality	dualité, f
Tröpfchenbild, n	droplet picture	image de gouttes, m
Auswahlregel, f	selection rule	règle de sélection, f
Zustandssumme, f	partition function	fonction de partition, f
Hamiltonoperator, m	Hamiltonian	l'hamiltonien, m
Transfermatrix, f	transfer matrix	matrice de transfert, f
Transversalfeld, n	transverse field	champ transverse, m
Isingmodell, n	Ising model	modèle d'Ising, m
Uhrmodell, n	clock model	modèle d'horloge, m
Starkbindungsmodell, n	tight-binding model	modèle de liaisons fortes, m
Punktperkolation, f	site percolation	percolation de sites, f
Kantenperkolation, f	bond percolation	percolation de liens, f
Spannungstensor, m	stress-energy tensor	tenseur d'énergie-contrainte, m
Energieimpulstensor, m	energy–momentum tensor	tenseur impulsion-énergie, m
selbstkonsistent	self-consistent	auto-cohérent
Betheansatz, m	Bethe ansatz	ansatz de Bethe, m
Abschneidemethode, f	truncation method	méthode de troncation, f
Restfunktion, f	remnant function	fonction du reste, f
Anfangsbedingung, f	initial condition	condition initiale, f
Randbedingung, f	boundary condition	condition de bord, f
Dauerstrom, m	persistent current	courant permanent, m
Magnet, m	magnet	aimant, m
Magnetisierung, f	magnetization	aimantation, f
Saitentheorie, f	string theory	théorie de cordes, f
Urknall, m	big bang	'big bang'
Irrfahrt, f; Zufallsweg, m	random walk	marche aléatoire, f
Haufen, m	cluster	amas, m
verzweigtes Polymer, n	lattice animal	animal sur réseau, m
selbstvermeidend	self-avoiding	auto-évitant
selbstmittelnd	self-averaging	auto-moyennant
Zufallszahl, f	random number	nombre aléatoire, f
Rechner, m	computer	ordinateur, m
Datei, f	file	fichier, m
Dateienverzeichnis, n	directory	répertoire, m
Internetz, n	internet	réseau inter, m
Elektropost, f	e-mail	courrier électronique, m (courriel)
Register, n	index	index, m
Anhang, m	appendix	annexe, m
aufzählen	to list	énumérer
(nach-, über-) prüfen	to test	vérifier, contrôler
verschieben	to shift	déplacer, glisser

List of Tables

1.1   Possible melting transitions of adatoms   20
1.2   Coxeter number and exponents of the simply-laced Lie algebras   30
1.3   Curie temperatures and susceptibility exponents for Gd/W(110)   31
1.4   Critical point and exponent β for ultrathin Fe or Ni layers   32
1.5   Experimentally measured critical exponents of 2D systems, I . .   34
1.6   Experimentally measured critical exponents of 2D systems, II . .   35
1.7   Experimentally measured critical exponents of 2D systems, III .   35
1.8   Analogies between quantum theory and statistical mechanics . .   39
1.9   Correspondence of physical quantities of statistical
      and quantum systems   39

3.1   Bulk scaling and finite-size scaling close to T_c with $\tilde{z} = 0$ 68
3.2   Experimental results for the shift exponent   80
3.3   Effective shift exponent λ_{eff} and comparison with experiment . .   81

4.1   Kac table for the Ising model.   95
4.2   Kac table for the A_5 RSOS model.   96

7.1   Numerical check of OPE coefficients in the Ising
      and Potts models   140

8.1   Conformal primary operators and the LGW fields ϕ^n   155

9.1   Symmetries of a $\mathbb{Z}_p$ symmetric quantum Hamiltonian   160
9.2   Convergence of the Lanczos algorithm.   168
9.3   Extrapolation with the VBS algorithm   176
9.4   Extrapolation with the BST algorithm   176
9.5   Infinite-system DMRG algorithm   178
9.6   Finite-system DMRG algorithm   179
9.7   Convergence of the DMRG algorithm   181

10.1   Calculation of the character $d(0, I)$ of $\phi_{1,1}$ for $m = 3$ 190
10.2   Virasoro character of the primaries for $m = 3$ 191
10.3   Critical points of the spin-1 quantum Ising chain   199
10.4   Estimates for the central charge of the quantum Ising chain . . .   200
10.5   Scaled energies for the spin-1 Ising chain and $Q = 0$ 200
10.6   Scaled energies for the spin-1 Ising chain and $Q = 1$ 201

11.1   Critical partition functions of the unitary ADE series   208
11.2   Critical partition function of the $\mathbb{Z}_2$ symmetric A series
       with antiperiodic boundary conditions   210
11.3   Physical branches and central charges of the dilute ADE models   217

12.1   Virasoro character of the primaries for $m = 5$ 220
12.2   Virasoro character of the primaries for $m = 4$ 222
12.3   Discrete series of some extended conformal algebras   226
12.4   Location of the Yang–Lee singularity
       and conformal normalization   227
12.5   Virasoro character of the primaries for $p = 2, p' = 5$ 228
12.6   Sector equivalence of the XXZ chain
       with the Ashkin–Teller chain   231
12.7   Symmetries of the Ashkin–Teller model
       and boundary conditions   232
12.8   Operator content of the Ashkin–Teller model   235
12.9   Scalar primaries of the Ashkin–Teller quantum chain   235
12.10 Exponent η for the (1+1)D XY model   238
12.11 Some scaling dimensions in 2D percolation   246
12.12 Virasoro characters of the primaries for $p = 4, p' = 6$ 246
12.13 Finite-lattice estimates for the critical parameters of a SAW . . .   249
12.14 Critical point t_c and amplitude ratio Ξ
       in the (2+1)D Ising model   259
12.15 Critical point K_c and amplitude ratio Ξ in the 3D O(n) model .   259

13.1   Finite-size correction coefficients
       for the three-states Potts model   268
13.2   Quasiprimary states up to level four
       for the Yang–Lee singularity   277
13.3   Normalized scalar states for the truncated space   277

14.1   Counting criterion for the Ising model perturbed with $\phi_{1,2}$. . . .   288
14.2   Central charge and scaling dimensions for the diluted A_3 model.   303
14.3   Mass ratios for the diluted A_3 model.   303
14.4   Universal amplitude ratios for the q-states Potts model   305
14.5   Mass ratios for the Ashkin–Teller model.   314
14.6   Numerical values for μ_{abc} and ρ_{abc} in the 2D Ising model   317
14.7   Mass shifts δG_i for the Ising model in a magnetic field.   319

15.1   Real projective conformal generators   327
15.2   Bulk and surface free energies of the p-states Potts chain   340
15.3   Exponents η and $\eta_{\parallel}$ for critical percolation   344
15.4   Conformal spectrum of the $S = 1$–$S = 1/2$ Ising quantum chain   359
15.5   Energy density corner exponent in models   367

16.1   Exciton kinetics on polymer chains   384

List of Figures

1.1   Phase diagram of an Ising ferromagnet.   2
1.2   Magnetization as a function of B, for several values of T   3
1.3   Droplet picture for the Ising model   4
1.4   Antiferromagnetic and metamagnetic ground states   15
1.5   Phase diagram of the tricritical Ising model   16
1.6   Duality in the Potts model   18
1.7   Dynkin diagrams of the simply-laced Lie algebras.   29
1.8   Classical path joining two space-time points   38
1.9   Hypercubic lattice with "time" and "space" directions   38

2.1   Conformal and non-conformal transformations   45
2.2   Bounded coordinate transformation   53
2.3   Contours defining the Virasoro generators   58
2.4   Commutation contours for the Virasoro algebra   59

3.1   Finite-size effects illustrated for the specific heat   65
3.2   Finite-size scaling of the critical point   66
3.3   Solving the phenomenological renormalization condition
      to find T^*   73
3.4   Logarithmic transformation onto the strip   74
3.5   Finite-size scaling in Ni films   79

4.1   Vanishing curves for level 2   91
4.2   Vanishing curves for the first four levels   93
4.3   Sequence of Verma submodules.   98

5.1   Deformation of an oriented contour   102
5.2   Crossing symmetry condition for the four-point functions   109
5.3   Decomposition of a n-point function   109
5.4   Monodromy transformations.   110

8.1   Effective ϕ^4 potential   147
8.2   Relationship between systems in the Ising universality class . . .   148
8.3   Transformation of contours under $w = \ln z$ 150

9.1   Finite-size estimates of x_ε in the (1+1)D Ising model   171

9.2 Finite-size estimates for K_c in the 3D spherical model 172
9.3 Two-point function of the four-states Potts model 182

10.1 Phase diagram of the quantum Ising chain 188
10.2 Critical spectrum for $H_0^{(0)}$ in the 2D Ising model 194
10.3 Critical spectrum for $H_1^{(0)} = H_0^{(1)}$ in the 2D Ising model 195
10.4 Critical spectrum for $H_1^{(1)}$ in the 2D Ising model 195

11.1 Modular transformations of the torus 206
11.2 The manifold S^1 and the orbifold $S^1/\mathbb{Z}_2$ 214
11.3 Known conformal modular invariant systems with $c = 1$. 215

12.1 Phase diagram of the Ashkin–Teller quantum chain 233
12.2 Watermelon topology. 248
12.3 Some examples of 2D lattice animals. 251
12.4 Phase diagram of the infinite lattice animal 253

13.1 Finite-size scaling functions for the quantum Ising chain. 275

14.1 Mapping the s-plane to the θ-plane. 290
14.2 Factorizable scattering 291
14.3 Anomalous threshold 293
14.4 Mass ratios for the Ising model 299
14.5 Interacting oriented self-avoiding walk 308
14.6 Manhattan lattice 310
14.7 Mass ratios for the tricritical Ising model 312
14.8 Masses for the Ashkin–Teller model 315
14.9 $\mathbb{Z}_2$ breaking perturbations in multicritical Ising
       and Ashkin–Teller models. 315
14.10 Mass shift in the (1+1)D Ising model with a magnetic field. . . . 318

15.1 Schematic local order parameter profiles. 322
15.2 Schematic surface phase diagram 323
15.3 Semi-infinite geometry 327
15.4 Contours for the half-plane 329
15.5 Surfaces (10) and (11) for square lattice 336
15.6 Order parameter profile for the Ising quantum chain 350
15.7 Spectrum of the Ising model with a defect line, $Q = 0$ 353
15.8 Spectrum of the Ising model with a defect line, $Q = 1$ 353
15.9 Folding of an Ising model onto a $c = 1$ theory 360
15.10 Aharanov-Bohm geometry 364
15.11 Persistent current amplitude in the tight-binding model 366

16.1 Scaling functions for an anisotropy exponent $\theta = 1/2$ 381

References

The references are arranged as follows. First we give a list of general introductions to the fields of critical phenomena and conformal invariance. The works cited are roughly ordered in increasing technical complexity. This is followed by a short list of reprint volumes. The original papers used in the main text then follow in alphabetical order, separated into experimental and theoretical papers.

Many of these papers are also available in electronic form, for example in the archives **cond-mat, hep-lat, hep-th**.

General Overviews and Introductory Texts

[G1] M.E. Fisher in F.J.W. Hahne (Ed), *Critical Phenomena*, Springer Lecture Notes in Physics, Vol. 186 (Heidelberg 1983), p. 1

[G2] J.M. Yeomans, *Statistical Mechanics of Phase Transitions*, Oxford University Press (Oxford 1992)

[G3] N. Boccara, *Symétries brisées: théorie des transitions avec paramètre d'ordre*, Herrman (Paris 1976)

[G4] S.K. Ma, *Modern Theory of Critical Phenomena*, Benjamin (Reading 1976)

[G5] G.A. Baker Jr., *Quantitative Theory of Critical Phenomena*, Academic Press (New York 1990)

[G6] J.L. Cardy, *Scaling and Renormalization in Statistical Physics*, Cambridge University Press (Cambridge 1996)

[G7] C. Domb and M.S. Green (Eds.), *Phase Transitions and Critical Phenomena*, Vols. 1-6, Academic Press (New York 1972-1976)

[G8] C. Domb and J. Lebowitz (Eds.), *Phase Transitions and Critical Phenomena*, Vols. 7-17, Academic Press (New York 1983-1996)

[G9] D.J. Amit, *Field Theory, the Renormalization Group and Critical Phenomena*, 2nd ed., World Scientific (Singapore 1984)

[G10] V. Privman (Ed.), *Finite Size Scaling and Numerical Simulation of Statistical Systems*, World Scientific (Singapore 1990)

[G11] J.L. Cardy in [G8], Vol. 11, p. 55

[G12] C. Itzykson and J.M. Drouffe, *Statistical Field Theory*, Vols. 1,2, Cambridge University Press (Cambridge 1988)

[G13] E. Brézin and J. Zinn-Justin (Eds), *Fields, Strings and Critical Phenomena*, Les Houches XLIX, North Holland (Amsterdam 1990)

[G14] J.L. Cardy in [G13], p. 169

[G15] P. Ginsparg in [G13], p. 1

[G16] I. Affleck in [G13], p. 563

[G17] M. Lüscher in [G13], p. 451

[G18] P. di Francesco, P. Mathieu and D. Sénéchal, *Conformal Field Theory*, Springer Verlag (Heidelberg 1997)

[G19] R.J. Baxter, *Exactly Solved Models in Statistical Mechanics*, Academic Press (London 1982)

[G20] J. Zinn-Justin, *Quantum Field Theory and Critical Phenomena*, Clarendon Press (Oxford 1989)

[G21] A.M. Polyakov, *Gauge Fields and Strings*, Harwood (New York 1987)

[G22] B.M. McCoy and T.T. Wu, *The Two-dimensional Ising Model*, Harvard University Press (Cambridge, Mass. (USA) 1973)

[G23] V.E. Korepin, N.M. Bogoliubov and A.G. Izergin, *Quantum Inverse Scattering Method and Correlation Functions*, Cambridge University Press (Cambridge 1993)

[G24] J. Fröhlich, G. 't Hooft, A. Jaffe, G. Mack, P.K. Mitter and R. Stora (Eds), *New Symmetry Principles in Quantum Field Theory*, NATO-ASI, Vol. B295, Plenum Press (New York 1992)

[G25] C. Gómez, M. Ruiz-Altaba and G. Sierra, *Quantum Groups in Two-dimensional Physics*, Cambridge University Press (Cambridge 1996)

[G26] J.A. de Azcárraga and J.M. Izquiero, *Lie groups, Lie algebras, cohomology and some applications in physics*, Cambridge University Press (Cambridge 1995)

[G27] M. Schottenloher, *Eine mathematische Einführung in die konforme Feldtheorie*, Lecture Notes Hamburg University (1994)

Reprint Volumes

[R1] J.L. Cardy (Ed), *Finite-size Scaling*, North Holland (Amsterdam 1988)

[R2] C. Itzykson, H. Saleur and J.-B. Zuber (Eds), *Conformal Invariance and Applications to Statistical Mechanics*, World Scientific (Singapore 1988)

[R3] P. Bouwknegt and K. Schoutens (Eds), *W-Symmetry*, World Scientific (Singapore 1995)

Experimental Work

[E1] D.L. Abernathy, S. Song, K.I. Blum, R.J. Birgenau and S.G.J. Mochrie, Physica **B221**, 126 (1996)

[E2] T. Ambrose and C.L. Chien, Phys. Rev. Lett. **76**, 1743 (1996)

[E3] T. Ambrose and C.L. Chien, J. Appl. Phys. **79**, 5920 (1996)

[E4] I. Affleck, in K. Dietz and V. Rittenberg (Eds) *Infinite-dimensional Lie Algebras and Conformal Invariance in Condensed Matter and Particle Physics*, World Scientific (Singapore 1987), p. 1

[E5] S.F. Alvarado, M. Campagna and H. Hopster, Phys. Rev. Lett. **48**, 51 (1982)

[E6] D. Arndt, St. Faßbender, M. Enderle and K. Knorr, Phys. Rev. Lett. **80**, 1686 (1998)

[E7] A.S. Arrott and B. Heinrich, in M. Hong, S. Wolf and D.C. Gruber (Eds) *Metallic Multilayers and Epitaxy*, The Metallurgical Society (1988), p. 188

[E8] C.H. Back, Ch. Würsch, D. Kerkmann and D. Pescia, Z. Phys. **B96**, 1 (1994)

[E9] C.H. Back, Ch. Würsch, A. Vaterlaus, U. Ramsperger, U. Maier and D. Pescia, Nature **378**, 597 (1995)

[E10] C.A. Ballentine, R.L. Fink, J. Araya-Pochet and J.L. Erskine, Phys. Rev. **B41**, 2631 (1990)

[E11] A. Baraldi, V.R. Dhanak, G. Comelli, K.C. Prince and R. Rosei, Phys. Rev. **B53**, 4073 (1996); **B56**, 10511 (1997)

[E12] D. Bitko, T.F. Rosenbaum and G. Aeppli, Phys. Rev. Lett. **77**, 940 (1996)

[E13] S.T. Bramwell and P.C.W. Holdsworth, J. Appl. Phys. **73**, 6096 (1993)

[E14] K. Brennan, C. Hohenemser and M. Eibschütz, J. Appl. Phys. **73**, 5500 (1993)

[E15] M. Bretz, Phys. Rev. Lett. **38**, 501 (1977)

[E16] K. Budde, L. Schwenger, C. Voges and H. Pfnür, Phys. Rev. **B52**, 9275 (1995)

[E17] W. Cai, M. Zhang, S. Arjas and H.F. Helbig, J. Phys. Chem. Solids **57**, 175 (1996)

[E18] J.C. Campuzano, M.S. Foster, G. Jennings, R.F. Willis and W.N. Unertl, Phys. Rev. Lett. **54**, 2684 (1985)

[E19] V. Chadrasekhar, R.A. Webb, M.J. Brady, M.B. Ketchen, W.J. Gallagher and A. Kleinsasser, Phys. Rev. Lett. **67**, 3578 (1991)

[E20] P. Chieux and M.J. Sienko, J. Chem. Phys. **53**, 566 (1970)

[E21] M.F. Collins, *Magnetic Critical Scattering*, Oxford University Press (Oxford 1989)

[E22] R.A. Cowley, M. Hagen and D.P. Belanger, J. Phys. **C17**, 3763 (1984)

[E23] W. Dürr, M. Taborelli, O. Paul, T. Germar, W. Gudat, D. Pescia and M. Landolt, Phys. Rev. Lett. **62**, 206 (1989)

[E24] W. Dürr, D. Kerkmann and D. Pescia, Int. J. Mod. Phys. **B4**, 401 (1990)

[E25] R.A. Dunlap, N.M. Fujiki, P. Hargraves and D.J.W. Geldart, J. Appl. Phys. **76**, 6338 (1994)

[E26] C.A. Eckert, B.L. Knutson and P.G. Debenedetti, Nature **383**, 313 (1996)

[E27] T.L. Einstein, in K. Dietz and V. Rittenberg (Eds) *Infinite-dimensional Lie Algebras and Conformal Invariance in Condensed Matter and Particle Physics*, World Scientific (Singapore 1987), p. 17

[E28] H.J. Elmers, J. Hauschild, H. Höche, U. Gradmann, H. Bethge, D. Heuer and U. Köhler, Phys. Rev. Lett. **73**, 898 (1994)

[E29] H.J. Elmers, J. Hauschild, G.H. Liu and U. Gradmann, J. Appl. Phys. **79**, 4984 (1996)

[E30] M. Farle and K. Baberschke, Phys. Rev. Lett. **58**, 511 (1987)

[E31] M. Farle, A. Berghaus and K. Baberschke, Phys. Rev. **B39**, 4838 (1989)

[E32] M. Farle, K. Baberschke, U. Stetter, A. Aspelmeier and F. Gerhardter, Phys. Rev. **B47**, 11571 (1993)

[E33] Y.P. Feng and M.H.W. Chan, Phys. Rev. Lett. **71**, 3822 (1993)

[E34] H. Freimuth and H. Wiechert, Surf. Sci. **162**, 432 (1985)

[E35] H. Freimuth and H. Wiechert, Surf. Sci. **178**, 716 (1986)

[E36] E.E. Fullerton, S. Adenwalla, G.P. Felcher, K.T. Riggs, C.H. Sowers, S.D. Bader and J.L. Robertson, Physica **B221**, 370 (1996)

[E37] G. Garreau, M. Farle, E. Beaurepaire and K. Baberschke, Phys. Rev. **B55**, 330 (1997)

[E38] B.D. Gaulin, S. Spooner and Y. Morii, Phys. Rev. Lett. **59**, 668 (1987)

[E39] M. Hagen and D. McK. Paul, J. Phys. **C17**, 5605 (1984)

[E40] M. Henkel, S. Andrieu, Ph. Bauer and M. Piecuch, Phys. Rev. Lett. **80**, 4783 (1998)

[E41] F. Huang, G.J. Mankey, M.T. Kief and R.F. Willis, J. Appl. Phys. **73**, 6760 (1993); Phys. Rev. **B49**, 3962 (1994)

[E42] H. Ikeda and K. Hirakawa, Solid State Comm. **14**, 529 (1974)

[E43] H. Ikeda, M. Suzuki and M.T. Hutchings, J. Phys. Soc. Japan **46**, 1153 (1979)

[E44] J.S. Jiang and C.L. Chien, J. Appl. Phys. **79**, 5615 (1996)

[E45] H.K. Kim and M.H.W. Chan, Phys. Rev. Lett. **53**, 170 (1984)

[E46] Y-E. L. Koo and R. Kopelman, J. Stat. Phys. **65**, 893 (1991)

[E47] R. Kopelman, C.S. Li and Z.-Y.Shi, J. Luminescence **45**, 40 (1990)

[E48] R. Kroon, H.Fleurent and R. Sprik, Phys. Rev. **E47**, 2462 (1993)

[E49] D. Lederman, C.A. Ramos, V. Jaccarino and J.L. Cardy, Phys. Rev. **B48**, 8365 (1993)

[E50] Y. Li, K. Baberschke and M. Farle, J. Appl. Phys. **69**, 4992 (1991)
[E51] Y. Li and K. Baberschke, Phys. Rev. Lett. **68**, 1208 (1992)
[E52] J.A. Lipa, D.R. Swanson, J.A. Nissen, T.C.P. Churi and U.E. Israelsson, Phys. Rev. Lett. **76**, 944 (1996)
[E53] C. Liu and S.D. Bader, J. Appl. Phys. **73**, 5758 (1993)
[E54] M.D. Lumsden, B.D. Gaulin, H. Dabkowska and M.L. Plumer, Phys. Rev. Lett. **76**, 4919 (1996)
[E55] X. Mailänder, H. Dosch, J. Peisl and R.L. Johnson, Phys. Rev. Lett. **64**, 2527 (1990)
[E56] D. Mailly, C. Chapelier and A. Benoit, Phys. Rev. Lett. **70**, 2020 (1993)
[E57] N. Mason, A.N. Pargellis and B. Yurke, Phys. Rev. Lett. **70**, 190 (1993)
[E58] S. Mehta and F.M. Gasparini, Phys. Rev. Lett. **78**, 2596 (1997)
[E59] K.D. Miner, M.H.W. Chan and A.D. Migone, Phys. Rev. Lett. **51**, 1465 (1983)
[E60] Ch.V. Mohan, H. Kronmüller and M. Kelsch, Phys. Rev. **B57**, 2701 (1998)
[E61] F.C. Motteler and J.G. Dash, Phys. Rev. **B31**, 346 (1985)
[E62] G.A. Mulhollan, Mod. Phys. Lett. **B7**, 655 (1993)
[E63] B.M. Ocko, J.X. Wang and T. Wandlowski, Phys. Rev. Lett. **79**, 1511 (1997)
[E64] V. Pasler, P. Schweiss, C. Meingast, B. Obst, H. Wühl, A.I. Rykov and S. Tajima, Phys. Rev. Lett. **81**, 1094 (1998)
[E65] P. Piercy and H. Pfnür, Phys. Rev. Lett. **59**, 1124 (1987)
[E66] C.D. Potter, M. Swiatek, S.D. Bader, D.N. Argyriou, J.F. Mitchell, D.J. Miller, D.G. Hinks and J.D. Jorgensen, Phys. Rev. **B57**, 72 (1998)
[E67] J. Prasad and R. Kopelman, Chem. Phys. Lett. **157**, 535 (1989)
[E68] G.A. Prinz and J.J. Krebs, Appl. Phys. Lett. **39**, 397 (1981)
[E69] G.A. Prinz, in R.F.C. Farrow, B. Dieney, M. Donath, A. Fert and B.D. Hermsmeier (Eds) *Magnetism and Structure in Systems of Reduced Dimensions*, NATO-ASI, Vol. B309, Plenum (New York 1993), p. 1
[E70] Z.Q. Qiu, J. Pearson and S.D. Bader, Phys. Rev. Lett. **67**, 1646 (1991)
[E71] Z.Q. Qiu, J. Pearson and S.D. Bader, Phys. Rev. **B49**, 8797 (1994)
[E72] C. Rau and C. Jin, J. Physique, Colloque C8, **49**, C8-1627 (1988)
[E73] C. Rau, G. Xing, C. Liu and M. Robert, Phys. Lett. **135A**, 227 (1989)
[E74] C. Rau, P. Mahavadi and M. Lu, J. Appl. Phys. **73**, 6757 (1993)
[E75] L.D. Roelofs, A.R. Kortan, T.L. Einstein and R.L. Park, Phys. Rev. Lett. **46**, 1465 (1981)
[E76] J. Rossat-Mignod, P. Burlet, L.P. Regnault and C. Vettier, J. Mag. Mag. Mat. **90& 91**, 5 (1990)
[E77] E.J. Samuelsen, Phys. Rev. Lett. **31**, 936 (1973)
[E78] F.O. Schumann, M.E. Buckley and J.A.C. Bland, Phys. Rev. **B50**, 16424 (1994)
[E79] F.O. Schumann and J.A.C. Bland, J. Appl. Phys. **73**, 5945 (1993)
[E80] R.F. Shannon, S.E. Nagler and C.R. Harkless, Phys. Rev. **B46**, 40 (1992)
[E81] J. Shen, R. Skomski, M. Klaua, S.S. Manoharan and J. Kirschner, Phys. Rev. **B56**, 2340 (1997)
[E82] M. Sokolowski and H. Pfnür, Phys. Rev. **B49**, 7716 (1994)
[E83] E.D. Specht, M. Sutton, R.J. Birgenau, D.E. Moncton and P.M. Horn, Phys. Rev. **B30**, 1589 (1984)
[E84] M. Tejwani, O. Ferreira and O.E. Vilches, Phys. Rev. Lett. **44**, 152 (1980)
[E85] M. Thommes and G.H. Findenegg, Langmuir **10**, 4270 (1994)
[E86] A. Thomy and X. Duval, J. Chimie Physique **67**, 1101 (1970)
[E87] A. Thomy, X. Duval and J. Regnier, Surf. Sci. Rep. **1**, 1 (1981)
[E88] A. Thomy and X. Duval, Surf. Sci. **299/300**, 415 (1994)
[E89] R.M. Tromp, W. Theis and N.C. Bartelt, Phys. Rev. Lett. **77**, 2522 (1996)
[E90] W.N. Unertl, Comments Cond. Matt. Phys. **12**, 289 (1986)

[E91] R. Vilanove and F. Rondelez, Phys. Rev. Lett. **45**, 1502 (1980)
[E92] C. Voges and H. Pfnür, Phys. Rev. **B57**, 3345 (1998)
[E93] G.C. Wang and T.M. Lu, Phys. Rev. **B31**, 5918 (1985)
[E94] H. Wiechert, Physica **169B**, 144 (1991)
[E95] H. Wiechert and S.-A. Arlt, Phys. Rev. Lett. **71**, 2090 (1993)
[E96] Ch. Binek, preprint `cond-mat/9811157`

Research Papers

1. M. Abramowitz and I.A. Stegun (Eds), *Handbook of Mathematical Functions*, Dover (New York 1965)
2. D.B. Abraham, L.F. Ko and N.M. Švrakić, Phys. Rev. Lett. **61**, 2393 (1988); J. Stat. Phys. **56**, 563 (1989)
3. D.B. Abraham, N.M. Švrakić and P. Upton, Phys. Rev. Lett. **68**, 423 (1992)
4. D.B. Abraham, D. O'Connor, A.O. Parry and P.J. Upton, Phys. Rev. Lett. **73**, 1742 (1994)
5. D.B. Abraham and F.T. Latrémolière, Phys. Rev. Lett. **76**, 4813 (1996)
6. I. Affleck, Phys. Rev. Lett. **56**, 746 (1984)
7. I. Affleck, Phys. Rev. Lett. **56**, 2763 (1986)
8. I. Affleck and A.W.W. Ludwig, Phys. Rev. Lett. **67**, 191, 3160 (1991); Nucl. Phys. **B360**, 641 (1991)
9. I. Affleck, Acta Phys. Pol. **B26**, 1869 (1995)
10. I. Affleck, M. Oshikawa and H. Saleur, preprint `condmat/9804117`
11. A. Aharony, Phys. Rev. **B8**, 3349 and 3363 (1973)
12. A. Aharony, in F.J.W. Hahne (Ed) *Critical Phenomena*, Springer Lecture Notes in Physics, Vol. 186 (Heidelberg 1983), p. 210
13. L.V. Ahlfors, *Complex Analysis*, 3rd edition, McGraw-Hill (London 1979)
14. M. Aizenman, in H. Bailin (Ed) *Statphys 19*, World Scientific (Singapore 1996), p. 104
15. M. Aizenman, Nucl. Phys. **B485**, 551 (1997)
16. G. Albertini, B.M. McCoy and J.H.H. Perk, Phys. Lett. **135A**, 159 (1989); **139A**, 204 (1989); Adv. Stud. Pure Math. **19**, 1 (1989)
17. G. Albertini, B.M. McCoy, J.H.H. Perk and S. Tang, Nucl. Phys. **B314**, 741 (1989)
18. G. Albertini and B.M. McCoy, Nucl Phys. **B350**, 745 (1991)
19. G. Albertini, S.R. Dahmen and B. Wehefritz, J. Phys. **A29**, L369 (1996); Nucl. Phys **B493**, 541 (1997)
20. F.C. Alcaraz, J.R. Drugowich de Felicio, R. Köberle and J.F. Stilck, Phys. Rev. **B32**, 7469 (1985)
21. F.C. Alcaraz and H.J. Herrmann, J. Phys. **A20**, 5735 (1987)
22. F.C. Alcaraz, M.N. Barber and M.T. Batchelor, Ann. of Phys. **182**, 280 (1988)
23. F.C. Alcaraz, M. Baake, U. Grimm and V. Rittenberg, J. Phys. **A21**, L117 (1988)
24. F.C. Alcaraz, U. Grimm and V. Rittenberg, Nucl. Phys. **B316**, 735 (1989)
25. F.C. Alcaraz, M. Baake, U. Grimm and V. Rittenberg, J. Phys. **A22**, L5 (1989)
26. F.C. Alcaraz, J. Phys. **A23**, L1105 (1990)
27. F.C. Alcaraz, M. Droz, M. Henkel and V. Rittenberg, Ann. of Phys. **230**, 250 (1994)
28. F.C. Alcaraz and A.L. Malvezzi, J. Phys. **A28**, 1521 (1995); **A29**, 2283 (1996)
29. F.C. Alcaraz and J.C. Xavier, J. Phys. **A29**, 3329 (1996)
30. C.R. Allton and C.J. Hamer, J. Phys. **A21**, 2417 (1988)

31. L. Alvarez-Gaumé and P. Zaugg, Ann. of Phys. **215**, 171 (1992)
32. G.E. Andrews, R.J. Baxter and P.J. Forrester, J. Stat. Phys. **35**, 193 (1984)
33. W.E. Arnoldi, Quart. Appl. Math. **9**, 17 (1951)
34. N.W. Ashcroft and N.D. Mermin, *Solid State Physics*, Saunders College Publishing (New York 1976)
35. H. Au-Yang and J.H.H. Perk, J. Stat. Phys. **78**, 17 (1995)
36. M. Baake, G.v. Gehlen and V. Rittenberg, J. Phys. **A20**, L479, L487 (1987)
37. M. Baake, P. Christe and V. Rittenberg, Nucl. Phys. **B300**[FS22], 637 (1988)
38. M. Baake, J. Math. Phys. **29**, 1753 (1989)
39. M. Baake, P. Chaselon and M. Schlottmann, Nucl. Phys. **B314**, 625 (1989)
40. M. Baake, U. Grimm and C. Pisani, J. Stat. Phys. **78**, 285 (1995)
41. R. Badke, P. Reinicke and V. Rittenberg, J. Phys. **A18**, 653 (1985)
42. R. Badke, V. Rittenberg and H. Ruegg, J. Phys. **A18**, L867 (1985)
43. D.B. Balbão and J.R. Drugowich de Felicio, J. Phys. **A20**, L207 (1987)
44. M. Bander and D.L. Mills, Phys. Rev. **B38**, 12015 (1988)
45. Yu. A. Bashilov and S.V. Pokrovsky, Comm. Math. Phys. **113**, 115 (1987)
46. M.N. Barber and M.E. Fisher, Ann. of Phys. **77**, 1 (1973)
47. M.N. Barber and C.J. Hamer, J. Austr. Math. Soc. **B23**, 229 (1982)
48. M.N. Barber in [G8], Vol. 8, p. 145
49. M.N. Barber and M.E. Cates, Phys. Rev. **B36**, 2024 (1987)
50. R.Z. Bariev, Sov. Phys. JETP **50**, 613 (1979)
51. R.Z. Bariev, J. Phys. **A22**, L397 (1989)
52. R.Z. Bariev and I. Peschel, Phys. Lett. **153A**, 166 (1991); J. Phys. **A24**, L87 (1991)
53. E. Barouch and B.M. McCoy, Phys. Rev. **A3**, 786 (1971)
54. A.O. Barut, Helv. Phys. Acta **46**, 496 (1973)
55. G.T. Barkema and S. Flesia, J. Stat. Phys. **85**, 363 (1996)
56. G.T. Barkema and J. McCabe, J. Stat. Phys. **84**, 1067 (1996)
57. G.T. Barkema, U. Bastolla and P. Grassberger, preprint `cond-mat/9707312`
58. G.K. Batchelor, Phys. Fluid Suppl. II, **12**, 233 (1969)
59. M.T. Batchelor and C.J. Hamer, J. Phys. **A23**, 761 (1990)
60. M.T. Batchelor and J.L. Cardy, Nucl. Phys. **B506**, 553 (1997)
61. M.T. Batchelor and K.A. Seaton, preprint `hep-th/9712121`
62. M.T. Batchelor and K.A. Seaton, preprint `cond-mat/9803206`
63. M. Bauer, P. diFrancesco, C. Itzykson and J.-B. Zuber, Nucl. Phys. **B362**, 515 (1991)
64. R. Bausch, H.K. Janssen and H. Wagner, Z. Phys. **B24**, 113 (1976)
65. R.J. Baxter, Phys. Lett. **133A**, 185 (1988); J. Stat. Phys. **57**, 1 (1989)
66. V.V. Bazhanov, B. Nienhuis, S.O. Warnaar, Phys. Lett. **322B**, 198 (1994)
67. P.D. Beale, J. Phys. **A17**, L335 (1984)
68. A.A. Belavin, A.M. Polyakov and A.B. Zamolodchikov, Nucl. Phys. **B241**, 333 (1984); J. Stat. Phys. **34**, 763 (1984)
69. D. Bennett-Wood, J.L. Cardy, S. Flesia, A.J. Guttmann and A.L. Owczarek, J. Phys. **A28**, 5143 (1995)
70. D. Bennett-Wood, J.L. Cardy, I.G. Enting, A.J. Guttmann and A.L. Owczarek, Oxford preprint (1997)
71. L. Benoit and Y. Saint-Aubin, Phys. Lett. **215B**, 517 (1988)
72. V.G. Benza, Europhys. Lett. **8**, 321 (1989)
73. R. Benzi, B. Legras, G. Parisi and R. Scardovelli, Europhys. Lett. **29**, 203 (1995)
74. J. Benzoni, J. Phys. **A17**, 2651 (1984)
75. B. Berche and L. Turban, J. Stat. Phys. **56**, 589 (1989)
76. B. Berche and L. Turban, J. Phys. **A23**, 3029 (1990)
77. B. Berche and L. Turban, J. Phys. **A24**, 245 (1991)

78. B. Berche, P.-E. Berche, M. Henkel, F. Iglói, P. Lajkó, S. Morgan and L. Turban, J. Phys. **A28**, L165 (1995)
79. B. Berche and F Iglói, J. Phys. **A28**, 3579 (1995)
80. P.-E. Berche, B. Berche and L. Turban, J. Physique **I6**, 621 (1996)
81. P.-E. Berche and B. Berche, Phys. Rev. **B56**, 5276 (1997)
82. P.-E. Berche, C. Chatelain and B. Berche, Phys. Rev. Lett. **80**, 297 (1998)
83. V.L. Berezinskii, Sov. Phys. JETP **32**, 493 (1971)
84. B. Berg, M. Karowski and P. Weisz, Phys. Rev. **D19**, 2477 (1979)
85. W.A. Berger, H.G. Müller, K.G. Kreuzer and R.M. Dreizler, J. Phys. **A10**, 1089 (1977)
86. O. Bergman, Phys. Rev. **D46**, 5474 (1992)
87. E. Berkcan, Nucl. Phys. **B215**[FS7], 68 (1983)
88. T.H. Berlin and M. Kac, Phys. Rev. **86**, 821 (1952)
89. M.A. Bershadsky, V.G. Knizhnik and M.G. Teitelman, Phys. Lett. **151B**, 31 (1985)
90. K. Binder and D. Stauffer, Phys. Rev. Lett. **33**, 1006 (1974)
91. K. Binder, in [G8], Vol. 8, p. 1
92. H.W. Blöte and H.J. Hilhorst, Phys. Rev. Lett. **51**, 2015 (1983)
93. H.W. Blöte, J.L. Cardy and M.P. Nightingale, Phys. Rev. Lett. **56**, 742 (1984)
94. H.W. Blöte and M.P.M. den Nijs, Phys. Rev. **B37**, 1766 (1988)
95. H.W.J. Blöte, to be published
96. N. Boccara and G. Sarma, J. Physique Lett. **35**, L95 (1974)
97. P. Bouwknegt, Jim McCarthy and K. Pilch, Prog. Theor. Phys. Suppl. **102**, 67 (1990)
98. D. Boyanovsky and J.L. Cardy, Phys. Rev. **B26**, 154 (1982)
99. H.W. Braden, E. Corrigan, P.E. Dorey, and R. Sasaki, Phys. Lett. **227B**, 441 (1989); Nucl. Phys. **B338**, 689 (1990); **B356**, 469 (1991)
100. S.T. Bramwell and P.C.W. Holdsworth, J. Phys. Condens. Matt. **5**, L53 (1993)
101. S.T. Bramwell, P.C.W. Holdsworth and J. Rothman, Mod. Phys. Lett. **B11**, 139 (1997)
102. J.G. Brankov and V.B. Priezzhev, J. Phys. **A25**, 4297 (1992)
103. P. Brax, J. Physique **I7**, 759 (1997)
104. A.J. Bray, Physica **A194**, 41 (1993)
105. A.J. Bray, Adv. Phys. **43**, 357 (1994)
106. E. Brézin, J. Physique **43**, 15 (1982); Ann. New York Acad. Sciences **410**, 339 (1983)
107. E. Brézin and J. Zinn-Justin, Nucl. Phys. **B257**[FS17], 867 (1985)
108. R.C. Brower, M.A. Furman and K. Subbarao, Phys. Rev. **D15**, 1756 (1977)
109. A.C. Brown, Phys. Rev. **B25**, 331 (1982)
110. S.G. Brush, *Statistical Physics and the Atomic Theory of Matter*, Princeton University Press (Princeton 1983)
111. M. Büttiker, Y. Imry and R. Landauer, Phys. Lett. **96A**, 365 (1983)
112. A.I. Bugrij and V.N. Shadura, Phys. Lett. **150A**, 171 (1990)
113. R. Bulirsch and J. Stoer, Num. Math. **6**, 413 (1964)
114. T.W. Burkhardt and E. Eisenriegler, Phys. Rev. **B24**, 1236 (1981); **B25**, 3283 (1982)
115. T.W. Burkhardt and I. Guim, J. Phys. **A18**, L33 (1985)
116. T.W. Burkhardt and E. Eisenriegler, J. Phys. **A18**, L83 (1985)
117. T.W. Burkhardt and I. Guim, Phys. Rev. **B35**, 1799 (1987)
118. T.W. Burkhardt and F. Iglói, J. Phys. **A23**, L633 (1990)
119. T.W. Burkhardt and T. Xue, Phys. Rev. Lett. **66**, 895 (1991); Nucl. Phys. **B354**, 653 (1991)
120. T.W. Burkhardt and J.-Y. Choi, Nucl. Phys. **B376**, 447 (1992)
121. T.W. Burkhardt and E. Eisenriegler, Nucl. Phys. **B424**, 487 (1994)

122. T.W. Burkhardt and H.W. Diehl, Phys. Rev. **B50**, 3894 (1994)
123. D. Cabra and C. Naón, Mod. Phys. Lett. **A9**, 2107 (1994)
124. G.G. Cabrera and R. Jullien, Phys. Rev. Lett. **57**, 393 (1986); Phys. Rev. **B35**, 7062 (1987)
125. G.G. Cabrera, Int. J. Mod. Phys. **B4**, 1671 (1990)
126. C.G. Callan, S. Coleman and R. Jackiw, Ann. of Phys. **59**, 42 (1970)
127. A. Cappelli, C. Itzykson and J.B. Zuber, Nucl. Phys. **B280**, 445 (1987); Comm. Math. Phys. **113**, 1 (1987)
128. A. Cappelli, Phys. Lett. **185B**, 82 (1987)
129. J.L. Cardy and R.L. Sugar, J. Phys. **A13**, L423 (1980)
130. J.L. Cardy, J. Phys. **A16**, 3617 (1983)
131. J.L. Cardy, J. Phys. **A17**, L385 (1984)
132. J.L. Cardy, J. Phys. **A17**, L961 (1984)
133. J.L. Cardy, Nucl. Phys. **B240**[FS12], 514 (1984)
134. J.L. Cardy, Phys. Rev. Lett. **54**, 1354 (1985)
135. J.L. Cardy, J. Phys. **A18**, L757 (1985)
136. J.L. Cardy, J. Phys. **A18**, 2771 (1985)
137. J.L. Cardy, Nucl. Phys. **B270**[FS16], 186 (1986)
138. J.L. Cardy, Nucl. Phys. **B275**[FS17], 200 (1986)
139. J.L. Cardy, J. Phys. **A20**, L891 (1987)
140. J.L. Cardy and I. Peschel, Nucl. Phys. **B300**[FS22], 377 (1988)
141. J.L. Cardy, Phys. Rev. Lett. **60**, 2709 (1988)
142. J.L. Cardy and G. Mussardo, Phys. Lett. **225B**, 275 (1989)
143. J.L. Cardy, Nucl. Phys. **B366**, 403 (1991)
144. J.L. Cardy, J. Phys. **A24**, L1315 (1991)
145. J.L. Cardy and D.C. Lewellen, Phys. Lett. **259B**, 274 (1991)
146. J.L. Cardy, J. Phys. **A25**, L201 (1992)
147. J.L. Cardy, Nucl. Phys. **B389**, 577 (1993)
148. J.L. Cardy and A.J. Guttmann, J. Phys. **A26**, 2485 (1993)
149. J.L. Cardy, in *Fluctuating Geometries in Statistical Mechanics and Field Theory*, Les Houches Summer School 2 Aug - 9 Sep 1994, preprint `cond-mat/9409094`
150. J.L. Cardy, Phys. Rev. Lett. **72**, 1580 (1994)
151. J.L. Cardy, Nucl. Phys. **B419**, 411 (1994)
152. J.L. Cardy, J. Phys. **A26**, 1897 (1996)
153. J.L. Cardy, preprint `cond-mat/9705137`
154. E. Carlon and F. Iglói, Phys. Rev. **B57**, 7877 (1998)
155. M. Caselle and M. Hasenbusch, J. Phys. **A30**, 4963 (1997)
156. J.-S. Caux, I.I. Kogan and A.M. Tsvelik, Nucl. Phys. **B466**, 444 (1996)
157. H.A. Cecatto, Z. Phys. **B75**, 253 (1989)
158. B.K. Chakrabarti, A. Dutta and P. Sen, *Quantum Ising Phases and Transitions in Transverse Ising Models*, Lecture Notes in Physics, vol. **m41**, Springer (Heidelberg 1996)
159. I. Chang and H. Meirovitch, Phys. Rev. Lett. **69**, 2232 (1992)
160. C. Chatelain, private communication
161. S. Chaudhuri and J.A. Schwartz, Phys. Lett. **219B**, 291 (1989)
162. J.C. Chaves and I.I. Satija, Phys. Rev. **B55**, 14076 (1997)
163. L. Chim and A.B. Zamolodchikov, Int. J. Mod. Phys. **A7**, 5317 (1992)
164. P. Christe and R. Flume, Phys. Lett. **188B**, 219 (1987)
165. P. Christe and F. Ravanini, Phys. Lett. **217B**, 252 (1989)
166. P. Christe and G. Mussardo, Nucl. Phys. **B330** 465 (1990)
167. P. Christe and G. Mussardo, Int. J. Mod. Phys. **A5** 4581 (1990)
168. P. Christe and M.J. Martins, Mod. Phys. Lett. **A5** 2189 (1990)

169. P. Christe and M. Henkel, *Introduction to Conformal Invariance and its Applications to Critical Phenomena*, Springer Lecture Notes in Physics m16, Springer (Heidelberg 1993)
170. A. Coniglio, N. Jan, I. Maijd and H.E. Stanley, Phys. Rev. **B35**, 3617 (1987)
171. F. Constantinescu and R. Flume, Phys. Lett. **326B**, 101 (1994)
172. S. Cornell, B. Chopard, and M. Droz, Phys. Rev. **A44**, 4826 (1991); Physica **A188**, 322 (1992)
173. J.F. Cornwell, *Group Theory in Physics*, Vol. 3, Academic Press (New York 1989)
174. M. de Crombrugghe and V. Rittenberg, Ann. of Phys. **151**, 99 (1983)
175. M.S.L. du Croo de Jongh and J.M.J. van Leeuwen, preprint cond-mat/9709103
176. E. Dagotto, Rev. Mod. Phys. **66**, 763 (1994)
177. C. Dasgupta and B.I. Halperin, Phys. Rev. Lett. **47**, 1556 (1981)
178. E.R. Davidson, J. Comp. Phys. **17**, 87 (1975)
179. B. Davies and I. Peschel, Ann. der Physik **2**, 79 (1992)
180. K. De'Bell and T. Lookman, Rev. Mod. Phys. **65**, 87 (1993)
181. J.M. Debierre and L. Turban, J. Phys. **A20**, 1819 (1987)
182. P.G. de Gennes, *Scaling Concepts in Polymer Physics*, Cornell University Press (Ithaca 1979)
183. G. Delfino, G. Mussardo and P. Simonetti, Nucl. Phys. **B432**, 518 (1994)
184. G. Delfino, P. Simonetti and J.L. Cardy, Phys. Lett. **387B**, 327 (1996)
185. G. Delfino, G. Mussardo and P. Simonetti, Nucl. Phys. **B473**, 469 (1996)
186. G. Delfino, Phys. Lett. **B**, preprint hep-th/9710019
187. G. Delfino and J.L. Cardy, Nucl. Phys. **B519**, 551 (1998)
188. G.W. Delius, M.T. Grisaru, and D. Zanon, Phys. Lett. **277B**, 414 (1992); Nucl. Phys. **B382**, 365 (1992)
189. J.W. Demmel, M.T. Heath and H. A. van der Vorst, *Parallel Numerical Linear Algebra*, in Acta Numerica (Cambridge University Press 1993), p. 111
190. B. Derrida and L. de Seze, J. Physique **43**, 475 (1982)
191. B. Derrida and H.J. Herrmann, J. Physique **44**, 1365 (1983)
192. B. Derrida and D. Stauffer, J. Physique **46**, 1623 (1985)
193. G. Deutscher, R. Zallen and J. Adler (Eds), *Percolation Structures and Processes*, Adam Hilger (Bristol 1983)
194. H.W. Diehl and S. Dietrich, Phys. Lett. **80A**, 408 (1980)
195. H.W. Diehl, in [G8], Vol. 10, p. 75
196. H.W. Diehl and A. Nüsser, Phys. Rev. Lett. **56**, 2834 (1986)
197. H.W. Diehl, Phys. Rev. **B49**, 2846 (1994)
198. H.W. Diehl, in *Renormalization group - 96*, World Scientific (Singapore 1997), preprint cond-mat/9610143
199. S. Dietrich and H.W. Diehl, Z. Phys. **B43**, 315 (1981)
200. S. Dietrich, in [G8], Vol. 12, p. 1
201. M. Doi, *Introduction to Polymer Physics*, Oxford University Press (Oxford 1996)
202. E. Domany, M. Schick, J.S. Walker and R.B. Griffiths, Phys. Rev. **B18**, 2209 (1978)
203. M.M. Doria and I. Satija, Phys. Rev. Lett. **60**, 444 (1988)
204. M.M. Doria, F. Nori and I. Satija, Phys. Rev. **B39**, 6802 (1988)
205. Vl. S. Dotsenko and V.A. Fateev, Nucl. Phys. **B240**, 312 (1984)
206. Vl. S. Dotsenko and V.A. Fateev, Nucl. Phys. **B251**, 691 (1985)
207. Vl. S. Dotsenko and V.A. Fateev, Phys. Lett. **154B**, 291 (1985)
208. Vl. S. Dotsenko, Int. J. Mod. Phys. **B4**, 1039 (1990)
209. A. Drewitz, R. Leidl, T.W. Burkhardt and H.W. Diehl, Phys. Rev. Lett. **78**, 1090 (1997)
210. M. Droz, A. Malaspinas and A.L. Stella, J. Phys. **C18**, L245 (1985)

211. B. Duplantier and H. Saleur, Phys. Rev. Lett. **59**, 539 (1987)
212. B. Duplantier and H. Saleur, Phys. Rev. Lett. **61**, 1521 (1988); **62**, 2641 (1989)
213. B. Duplantier and F. David, J. Stat. Phys. **51**, 327 (1988)
214. B. Duplantier, in [G13], p. 393
215. T. Eguchi and S.-K. Yang, Phys. Lett. **235B**, 282 (1990)
216. H. Eichenherr, Phys. Lett. **151B**, 26 (1985)
217. E. Eisenriegler, *Polymers near surfaces*, World Scientific (Singapore 1993)
218. e.g. S. Elitzur, R.B. Pearson and J. Shigemitsu, Phys. Rev. **D19**, 3698 (1979)
219. J. Erdmenger and H. Osborn, Nucl. Phys. **B483**, 431 (1997)
220. G. Falkovich and A. Hanany, Phys. Rev. Lett. **71**, 3454 (1993)
221. V.A. Fateev and A.B. Zamolodchikov, Int. J. Mod. Phys. **A5**, 1025 (1990)
222. V.A. Fateev, Phys. Lett. **B324**, 45 (1994)
223. B.L. Feigin and D.B. Fuks, Funct. Anal. and Appl. **16**, 114 (1982); **17**, 241 (1983)
224. G. Felder, Nucl. Phys. **B317**, 215 (1989), err. **B324**, 548 (1989)
225. G. Felder, J. Fröhlich, and G. Keller, Comm. Math. Phys. **124**, 417 (1989); **130**, 1 (1990)
226. B.U. Felderhof, Rep. on Math. Phys. **1**, 215 (1971)
227. M.E. Fisher and M.N. Barber, Phys. Rev. Lett. **28**, 1516 (1972)
228. M.E. Fisher and M.N. Barber, Arch. Rat. Mech. Anal. **47**, 205 (1972)
229. M.E. Fisher, Phys. Rev. Lett. **40**, 1610 (1978)
230. M.E. Fisher and P.G. de Gennes, C.R. Acad. Sci. Paris **B287**, 207 (1978)
231. see M.E. Fisher in [G1], Appendix D
232. M.E. Fisher in *Collective Properties of Physical Systems*, Proc. 24th Nobel Symposium, Academic Press (New York 1973), p. 16
233. D.S. Fisher, Phys. Rev. Lett. **69**, 534 (1992); Phys. Rev. **B51**, 6411 (1995)
234. S. Flesia, D.S. Gaunt, C.E. Soteros and S.G. Whittington, J. Phys. **A25**, L1169 (1992)
235. S. Flesia, Europhys. Lett. **32**, 149 (1995)
236. M.A.I. Flohr, Int. J. Mod. Phys. **A11**, 4147, (1996)
237. M.A.I. Flohr, Nucl. Phys. **B482**, 567 (1996)
238. G. Foltin, K.Oerding, Z. Rácz, R.L. Workman and R.K.P. Zia, Phys. Rev. **E50**, R639 (1994)
239. D.P. Foster, E. Orlandini and M.C. Tesi, J. Phys. **A25**, 1211 (1992)
240. L. Frachebourg and M. Henkel, J. Phys. **A24**, 5121 (1991)
241. L. Frachebourg and M. Henkel, Physica **A195**, 577 (1993)
242. E. Fradkin and L. Susskind, Phys. Rev. **D17**, 2637 (1978)
243. P. di Francesco, H. Saleur and J.B. Zuber, J. Stat. Phys. **49**, 57 (1987)
244. P. di Francesco, H. Saleur and J.B. Zuber, Nucl. Phys. **B290**, 527 (1987)
245. P. di Francesco, Int. J. Mod. Phys. **A7**, 407 (1992)
246. D. Friedan, Z. Qiu and S. Shenker, Phys. Rev. Lett. **52**, 1575 (1984); Comm. Math. Phys. **107**, 535 (1986)
247. D. Friedan, Z. Qiu and S. Shenker, Phys. Lett. **151B**, 37 (1985)
248. U. Frisch, *Turbulence*, Cambridge University Press (Cambridge 1995)
249. J. Fröhlich and F. Gabbiani, Rev. Math. Phys. **2**, 251 (1990)
250. P. Fröjdh, J. Phys. **A27**, 41 (1994)
251. J. Fuchs and C. Schweigert, preprint `hep-th/9806121`
252. K. Gadwedzki, *Turbulence under a magnifying glass*, Cargèse Summer Institute 1996, preprint `chao-dyn/9610003`
253. D.S. Gaunt and C.J. Domb, J. Phys. **C3**, 1442 (1970)
254. Y. Gefen, Y. Imry and D. Mukamel, Phys. Rev. **B23**, 6099 (1981)
255. G.v. Gehlen, C. Hoeger and V. Rittenberg, J. Phys. **A17**, L469 (1984); err. **A18**, 187 (1985)

256. G.v. Gehlen, V. Rittenberg and H. Ruegg, J. Phys. **A18**, 107 (1985)
257. G.v. Gehlen and V. Rittenberg, Nucl. Phys. **B257**[FS14], 351 (1985)
258. G.v. Gehlen and V. Rittenberg, J. Phys. **A19**, L625 (1986)
259. G.v. Gehlen and V. Rittenberg, J. Phys. **A19**, L631 (1986)
260. G.v. Gehlen and V. Rittenberg, J. Phys. **A20**, 227 (1987)
261. G.v. Gehlen, V. Rittenberg and T. Vescan, J. Phys. **A20**, 2577 (1987)
262. G.v. Gehlen, Nucl. Phys. **B330**, 741 (1990)
263. G.v. Gehlen, J. Phys. **A24**, 5371 (1991)
264. G.v. Gehlen, Int. J. Mod. Phys. **B8**, 3507 (1994)
265. G.A. Gehring, R.J. Bursill and T. Xiang, Acta Phys. Pol. **A91**, 105 (1997)
266. A. Georges and P. Le Doussal, J. Stat. Phys. **54**, 1011 (1989)
267. B. Germain-Bonne, Rev. Fran. Autom. Inform. Rech. Operat. **7**(R-1), 84 (1973)
268. P. Ginsparg, Nucl. Phys. **B295**, 153 (1988)
269. R. Glauber, J. Math. Phys. **4**, 294 (1963)
270. P. Goddard and D. Olive, Int. J. Mod. Phys. **A1**, 303 (1983)
271. P. Goddard, A. Kent and D. Olive, Comm. Math. Phys. **103**, 105 (1986)
272. P. Goddard, *Conformal symmetry and its extensions*, IXth IAMP Congress, Swansea (1988)
273. C. Godrèche and J.M. Luck, J. Phys. **A23**, 3769 (1990)
274. A.O. Gogolin, Phys. Rev. Lett. **71**, 2995 (1993)
275. S.G. Gorishny, S.A. Larin and F.V. Tkachov, Phys. Lett. **101A**, 120 (1984)
276. I.S. Gradshteyn and I.M. Ryzhik, *Table of Integrals, Series, and Products*, 4th edition, Academic Press (New-York 1965)
277. Y. Grandati, Ann. de Physique **17**, 159 (1992)
278. P. Grassberger, Z. Phys. **B47**, 365 (1982)
279. P. Grassberger, private communication of improvement of results in [74] by an order of magnitude.
280. U. Grimm and V. Rittenberg, Int. J. Mod. Phys. **B4**, 969 (1990)
281. U. Grimm, Nucl. Phys. **B340**, 633 (1990)
282. U. Grimm and M. Baake, in R.V. Moody (Ed) *The Mathematics of Long-Range Aperiodic Order*, Kluwer (Amsterdam 1997), p. 199
283. U. Grimm and B. Nienhuis, Phys. Rev. **E55**, 5011 (1997)
284. R. Guida and N. Magnoli, Int. J. Mod. Phys. **A13**, 1145 (1998)
285. R. Guida and J. Zinn-Justin, preprint **cond-mat/9803240**
286. L.G. Guimarães and J.R. Drugowich de Felicio, J. Phys. **A19**, L341 (1986)
287. J.D. Gunton and M. Droz, *Introduction to the Theory of Metastable and Unstable States*, Springer Lecture Notes in Physics, Vol. 183, Springer (Heidelberg 1983)
288. V. Gurarie, Nucl. Phys. **B410**, 535 (1993)
289. A.J. Guttmann in [G8], Vol. 13, p. 1
290. C.R. Hagen, Phys. Rev. **D5**, 377 (1972)
291. B.I. Halperin and P.C. Hohenberg, Rev. Mod. Phys. **49**, 435 (1977)
292. T.C. Halsey, M.H. Jensen, L.P. Kadanoff, I.Procaccia and B.I. Shraiman, Phys. Rev. **A33**, 1141 (1986)
293. M. Hamermesh, *Group Theory and its Applications to Physical Problems*, Academic Press (New York 1962)
294. C.J. Hamer and M.N. Barber, J. Phys. **A14**, 2009 (1981)
295. C.J. Hamer, J. Phys. **A15**, L675 (1982)
296. C.J. Hamer, J. Phys. **A16**, 3085 (1983)
297. C.J. Hamer and M.T. Batchelor, J. Phys. **A21**, L173 (1988)
298. E.R. Hansen, *A Table of Series and Products*, Prentice Hall (Englewood Cliffs, 1975)

299. A.B. Harris, J. Phys. **C7**, 1671 (1974)
300. A.B. Harris and T.C. Lubensky, Phys. Rev. **B24**, 2656 (1981)
301. B. Hede, J. Kertész and T. Vicsek, J. Stat. Phys. **64**, 829 (1991)
302. M. Henkel, J. Phys. **A20**, 995 (1987)
303. M. Henkel, J. Phys. **A20**, 3569 (1987)
304. M. Henkel, J. Phys. **A20**, L769 (1987)
305. M. Henkel and A. Patkós, J. Phys. **A20**, 2199 (1987)
306. M. Henkel and A. Patkós, Nucl. Phys. **B285**[FS19], 29 (1987)
307. M. Henkel and G. Schütz, J. Phys. **A21**, 2617 (1988)
308. M. Henkel and H. Saleur, J. Phys. **A22**, L513 (1989)
309. M. Henkel, A. Patkós and M. Schlottmann, Nucl. Phys. **B314**, 609 (1989)
310. M. Henkel and J. Lacki, Phys. Lett. **138A**, 105 (1989)
311. M. Henkel and V. Privman, Phys. Rev. Lett. **65**, 1777 (1990)
312. M. Henkel in [G10], ch. VIII, p. 353
313. M. Henkel and H.J. Herrmann, J. Phys. **A23**, 3719 (1990)
314. M. Henkel, J. Phys. **A23**, 4369 (1990); Phys. Lett. **247B**, 567 (1990)
315. M. Henkel and A.W.W. Ludwig, Phys. Lett. **249B**, 463 (1990)
316. M. Henkel and H. Saleur, J. Phys. **A23**, 791 (1990)
317. M. Henkel, J. Phys. **A24**, L133 (1991)
318. M. Henkel and R.A. Weston, J. Phys. **A25**, L207 (1992)
319. M. Henkel and A. Patkós, J. Phys. **A25**, 5223 (1992)
320. M. Henkel, Int. J. Mod. Phys. **C3**, 1011 (1992)
321. M. Henkel and R. Peschanski, Nucl. Phys. **B390**, 637 (1993)
322. M. Henkel, J. Stat. Phys. **75**, 1023 (1994)
323. M. Henkel and G.M. Schütz, Int. J. Mod. Phys. **B8**, 3487 (1994)
324. M. Henkel and F. Seno, Phys. Rev. **E53**, 3662 (1996)
325. M. Henkel, Phys. Rev. Lett. **78**, 1940 (1997)
326. M. Henkel and D. Karevski, J. Phys. **A31**, 2503 (1998)
327. M. Henkel and D. Karevski, Eur. Phys. J. **B5**, 787 (1998)
328. H.J. Herrmann, Phys. Lett. **100A**, 256 (1984)
329. J. Hermisson, U. Grimm and M. Baake, J. Phys. **A30**, 7315 (1997)
330. Y. Hieda, preprint `cond-mat/9711072`
331. H.J. Hilhorst and J.M.J. van Leeuwen, Phys. Rev. Lett. **47**, 1188 (1981)
332. E. Hille, *Lectures on Ordinary Differential Equations*, Addison-Wesley (New-York 1969)
333. E. Hille, *Ordinary Differential Equations in the Complex Domain*, Wiley (New-York 1976)
334. H. Hinrichsen, Nucl. Phys. **B336**, 377 (1990)
335. H. Hinrichsen and V. Rittenberg, Phys. Lett. **275B**, 350 (1992)
336. H. Hinrichsen, J. Phys. **A27**, 1121 (1994)
337. H. Hiramoto and M. Kohmoto, Int. J. Mod. Phys. **B6**, 281 (1992)
338. W. Hofstetter and M. Henkel, J.Phys. **A29**, 1359 (1996)
339. T.J. Hollowood and P. Mansfield, Phys. Lett. **226B**, 73 (1989)
340. A. Honecker, J. Stat. Phys. **82**, 687 (1996)
341. R.M. Hornreich, M. Luban and S. Strikman, Phys. Rev. Lett. **35**, 1678 (1975); Phys. Lett. **55A**, 269 (1975)
342. J.-P. Hovi and A. Aharony, Phys. Rev. Lett. **76**, 3874 (1996); **72**, 1941 (1994)
343. S. Howes, L.P. Kadanoff and M. den Nijs, Nucl. Phys. **B215**[FS7], 169 (1983)
344. C.-K. Hu, C.-Y. Lin and J.-A. Chen, Phys. Rev. Lett. **75**, 193 (1995); err. 2786 (1995); **77**, 8 (1996)
345. M.-C. Huang and H.-P. Hsu, J. Phys. **A31**, (1998), to be published
346. D.A. Huse, Phys. Rev. **B24**, 5180 (1981)
347. D.A. Huse, Phys. Rev. **B30**, 3908 (1984)
348. D. Iagolnitzer, Phys. Rev. **D18**, 1275 (1978)

349. F. Iglói, Phys. Rev. Lett. **64**, 3035 (1990)
350. F. Iglói, B. Berche and L. Turban, Phys. Rev. Lett. **65**, 1773 (1990)
351. F. Iglói, I. Peschel and L. Turban, Adv. Phys. **42**, 683 (1993)
352. F. Iglói and L. Turban, Europhys. Lett. **27**, 91 (1994)
353. F. Iglói and P. Lajkó, J. Phys. **A29**, 4803 (1996)
354. F. Iglói, L. Turban, D. Karevski and F. Szalma, Phys. Rev. **B56**, 11031 (1997)
355. F. Iglói and H. Rieger, Phys. Rev. Lett. **78**, 2473 (1997)
356. F. Iglói, D. Karevski and H. Rieger, Eur. Phys. J. **B1**, 513 (1998)
357. F. Iglói and H. Rieger, preprint `cond-mat/9709260`
358. Y. Imry, *Introduction to Mesoscopic Physics*, Oxford University Press (Oxford 1997)
359. J.O. Indekeu, M.P. Nightingale and W.V. Wang, Phys. Rev. **B34**, 330 (1986)
360. E. Ising, Z. Phys. **31**, 253 (1925)
361. A.C. Irving, G. Odor and A. Patkós, J. Phys. **A22**, 4665 (1989)
362. C. Itzykson, H. Saleur and J.-B. Zuber, Europhys. Lett. **2**, 91 (1986)
363. V.I. Ivanov and M.K. Trubetskov, *Handbook of conformal mapping*, CRC Press (Boca Raton 1995)
364. E.V. Ivashkevich, J. Phys. **A30**, L525 (1997)
365. E. Jäger and R. Perthel, *Magnetische Eigenschaften von Festkörpern*, 2nd edition, Akademie Verlag (Berlin 1996)
366. W. Janke and K. Nather, Phys. Rev. **B48**, 7419 (1993); Phys. Lett. **157A**, 11 (1991)
367. W. Janke, Phys. Rev. **B55**, 3580 (1997)
368. W. Janke, private communication
369. H.K. Janssen, Z. Phys. **B42**, 151 (1981)
370. H.K. Janssen, B. Schaub and B. Schmittmann, Z. Phys. **B73**, 539 (1989)
371. A. Jennings, *Matrix Computation for Engineers and Scientists*, Wiley (New York 1977)
372. J.D. Johnson and B.M. McCoy, Phys. Rev. **A6**, 1613 (1973)
373. D. Johnston, preprint `cond-mat/9807091`
374. V.R. Jones, Int. J. Mod. Phys. **B4**, 701 (1990)
375. J.V. José, L.P. Kadanoff, S. Kirkpatrick and D.R. Nelson, Phys. Rev. **B16**, 1217 (1977)
376. V.G. Kac, Springer Lecture Notes in Physics, Vol. 94, Springer (Berlin 1979), p. 441
377. V.G. Kac *Infinite Dimensional Lie Algebras*, 2nd ed. Cambridge University Press (Cambridge 1985)
378. V.G. Kac and A.K. Raina, *Heighest Weight Representations of Infinite-Dimensional Lie Algebras*, World Scientific (Singapore 1987)
379. L.P. Kadanoff in [G7], Vol. 5A, p. 1
380. L.P. Kadanoff and A.C. Brown, Ann. of Phys. **121**, 318 (1979)
381. L.P. Kadanoff, Phys. Rev. **B24**, 5382 (1981)
382. G. Kamieniarz, P. Kozlowski and R. Dekeyser, preprint `cond-mat/9803277`
383. E. Kamke, *Differentialgleichungen: Lösungsmethoden und Lösungen*, Vol. 2 (Partielle Differentialgleichungen), 6th edition, Teubner (Leipzig 1979)
384. M. Karowski and P. Weisz, Nucl. Phys. **B139**, 445 (1978)
385. M. Karowski, Phys. Rep. **49C**, 229 (1979)
386. M. Kardar, G. Parisi and Y.C. Zhang, Phys. Rev. Lett. **56**, 889 (1986)
387. D. Karevski, G. Palágyi and L. Turban, J. Phys. **A28**, 45 (1995)
388. D. Karevski, L. Turban and F. Iglói, J. Phys. **A28**, 3925 (1995)
389. D. Karevski and L. Turban, J. Phys. **A29**, 3461 (1996)
390. D. Karevski, P. Lajkó and L. Turban, J. Stat. Phys. **86**, 1153 (1996)
391. D. Karevski and M. Henkel, Phys. Rev. **B55**, 6429 (1997)

392. D. Kastor, Nucl. Phys. **B280**[FS18], 304 (1987)
393. S. Katsura, Phys. Rev. **127**, 1508 (1962)
394. A. Kato, Mod. Phys. Lett. **A2**, 585 (1987)
395. M. Kaulke and I. Peschel, preprint `cond-mat/9802175`
396. A. Kent, Phys. Lett. **273B**, 56 (1991)
397. W. Kerner, J. Comp. Phys. **85**, 1 (1989)
398. D. Kim, J. Phys. **A30**, 3817 (1997)
399. W. Kinzel, in G. Deutscher, R. Zallen and J. Adler (Eds) *Percolation Structures and Processes*, Adam Hilger (Bristol 1983), p. 425
400. Y. Kitazawa, N. Ishibashi, A. Kato, K. Kobayashi, Y. Matsuo and S. Odake, Nucl. Phys. **B306**, 425 (1988)
401. T. Klassen and E. Melzer, Nucl. Phys. **B338**, 485 (1990); **B350**, 635 (1991)
402. T. Klassen and E. Melzer, Nucl. Phys. **B362**, 329 (1991)
403. T. Klassen and E. Melzer, Nucl. Phys. **B370**, 511 (1992)
404. P. Kleban and I. Vassileva, J. Phys. **A25**, 5779 (1992); **A24**, 3407 (1991)
405. P. Kleban and I. Peschel, Z. Phys. **B101**, 447 (1996)
406. H. Kleinert and V. Schulte-Frohlinde, Phys. Lett. **342B**, 284 (1995)
407. V.G. Knizhnik, A.M. Polyakov and A.B. Zamolodchikov, Mod. Phys. Lett. **A3**, 819 (1988)
408. C.M. Knobler and R.L. Scott, in [G8], Vol. 9, p. 164
409. L.-F. Ko, H. Au-Yang and J.H.H. Perk, Phys. Rev. Lett. **54**, 1091 (1985)
410. L.-F. Ko, Phys. Lett. **131A**, 285 (1988)
411. J.B. Kogut, Rev. Mod. Phys. **51**, 659 (1979)
412. M. Kohmoto, M. den Nijs and L.P. Kadanoff, Phys. Rev. **B24**, 5229 (1981)
413. W.M. Koo, J. Stat. Phys. **81**, 561 (1995)
414. V.E. Korepin and A.C.T. Wu, Int. J. Mod. Phys. **B5**, 497 (1991)
415. J.M. Kosterlitz and D.J. Thouless, J. Phys. **C6**, 118 (1973)
416. R.H. Kraichnan, Phys. of Fluids **10**, 1417 (1967); J. of Fluid Mech. **47**, 525 (1971); J. of Fluid Mech. **67**, 155 (1975)
417. H.J. Kreuzer, *Nonequilibrium Thermodynamics and its Statistical Foundations*, Clarendon Press (Oxford 1981)
418. S. Krivonos and K. Thielemans, Class. Quant. Grav. **13**, 2899 (1996)
419. J. Krug, Adv. Phys. **46**, 141 (1997)
420. I.O. Kulik, Sov. Phys. JETP Lett. **11**, 275 (1970)
421. J. Kurmann, H. Thomas and G. Müller, Physica **A112**, 235 (1982)
422. M. Lässig, G. Mussardo and J.L. Cardy, Nucl. Phys. **B348**, 591 (1991)
423. M. Lässig and G. Mussardo, Comp. Phys. Comm. **66**, 71 (1991)
424. M. Lässig, Phys. Lett. **278B**, 439 (1992)
425. D.P. Landau, Phys. Rev. Lett. **28**, 449 (1972)
426. D.P. Landau and R.H. Swendsen, Phys. Rev. Lett. **46**, 1437 (1981)
427. K. Lang and W. Rühl, Nucl. Phys. **B377**, 371 (1992); Z. Phys. **C50**, 285 (1991)
428. R. Langlands, P. Pouliot and Y. Saint-Aubin, Bull. Am. Math. Soc. **30**, 1 (1994)
429. P.G.M. Lauwers and V. Rittenberg, Phys. Lett. **233B**, 197 (1989)
430. I.D. Lawrie and S. Sarbach in [G8], Vol. 9, p. 1
431. A. LeClair and A.W.W. Ludwig, preprint `hep-th/9708135`
432. K.-C. Lee, Phys. Rev. Lett. **69**, 9 (1992)
433. Ö. Legeza and G. Fáth, Phys. Rev. **B53**, 14349 (1996); preprint `cond-mat/9809035`
434. R. Leidl, A. Drewitz and H.W. Diehl, preprint `cond-mat/9704215`; R. Leidl and H.W. Diehl, preprint `cond-mat/9707345`
435. J.-M. Levy-Leblond, Comm. Math. Phys. **4**, 157 (1967); **6**, 286 (1968)

436. E. Lieb, T. Schultz and D. Mattis, Ann. of Phys. **16**, 407 (1961)
437. Z. Lin and R. Tao, Phys. Lett. **150A**, 11 (1990)
438. K. Lindenberg, W.-S. Sheu and R. Kopelman, J. Stat. Phys. **65**, 1269 (1991)
439. J.M. Luck, J. Phys. **A15**, L169 (1982)
440. J.M. Luck, Phys. Rev. **B31**, 3069 (1985)
441. J.M. Luck and T.M. Nieuwenhuizen, Europhys. Lett. **2**, 257 (1986)
442. J.M. Luck, J. Stat. Phys. **72**, 417 (1993); Europhys. Lett. **24**, 359 (1993)
443. A.W.W. Ludwig and J.L. Cardy, Nucl. Phys. **B285**[FS19], 687 (1987)
444. M. Lüscher and U. Wolff, Nucl. Phys. **B339**, 222 (1991)
445. A.A. Lushnikov, Sov. Phys. JETP **64**, 811 (1986); Phys. Lett. **120A**, 135 (1987)
446. V. Lvov and I. Procaccia, Physica Scripta **T67**, 131 (1996)
447. D. Maes and C. Vanderzande, Phys. Rev. **A41**, 1808 (1990)
448. J. Marro, J. Lebowitz and M.H. Kalos, Phys. Rev. Lett. **43**, 282 (1979)
449. P.C. Martin, E.D. Siggia and H.H. Rose, Phys. Rev. **A8**, 423 (1973)
450. P.P. Martin, *Potts Model and Related Problems in Statistical Mechanics*, World Scientific (Singapore 1990)
451. P.P. Martin and V. Rittenberg, Int. J. Mod. Phys. **A7**, Suppl. 1B, 707 (1992)
452. M.J. Martins, Phys. Rev. Lett. **65**, 2091 (1990); Phys. Lett. **257B**, 317 (1991)
453. M.J. Martins, Santa Barbara preprint UCSBTH-91-45
454. Y. Matsuo and S. Yahikozawa, Phys. Lett. **178B**, 211 (1986)
455. N.E. Mavromatos and R.J. Szabo, preprint `hep-th/9803092`
456. D.M. McAvity and H. Osborn, Nucl. Phys. **B455**, 522 (1995)
457. B.M. McCoy and T.T. Wu, Phys. Rev. **162**, 436 (1967)
458. B.M. McCoy and T.T. Wu, Phys. Rev. **176**, 631 (1968)
459. B.M. McCoy, Phys. Rev. **188**, 1014 (1969)
460. B.M. McCoy and J.H.H. Perk, Phys. Rev. Lett. **44**, 840 (1980)
461. B.M. McCoy and W.P. Orrick, Phys. Lett. **230A**, 24 (1997)
462. A. McKane, M. Droz, J. Vannimenus and D. Wolf (Eds), *Scale Invariance, Interfaces and Non-Equilibrium Dynamics*, NATO-ASI, vol. B344, Plenum (New York 1995)
463. H. Meirovitch and H.A. Lim, Phys. Rev. Lett. **62**, 2640 (1989)
464. N.D. Mermin and H. Wagner, Phys. Rev. Lett. **17**, 1133 (1966)
465. A. Meurman and A. Rocha-Caridi, Comm. Math. Phys. **107**, 263 (1986)
466. J.D. Miller and K. De'Bell, J. Physique **I3**, 1717 (1993)
467. H.K. Moffat, in G. Comte-Bellot, J. Mathieu (Eds) *Advances in Turbulence*, Springer (Heidelberg 1986), p. 284
468. A.I. Mudrov and K.B. Varnashev, Phys. Rev. **B57**, 3562 (1998)
469. D.B. Murray and B.G. Nickel, unpublished. (quoted in [285])
470. G. Mussardo, Phys. Rep. **218C**, 215 (1992)
471. J.P. Nadal, B. Derrida and J. Vannimenus, J. Physique **43**, 1561 (1982)
472. T.J. Newman and A.J. Bray, J. Phys. **A23**, 4491 (1990)
473. Th. Niemeijer, Physica **36**, 377 (1967)
474. U. Niederer, Helv. Phys. Acta **45**, 802 (1972); **46**, 191 (1973); **47**, 119, 167 (1974); **51**, 220 (1978)
475. B. Nienhuis, E.K. Riedel and M. Schick, J. Phys. **A13**, L18 (1980)
476. B. Nienhuis, Phys. Rev. Lett. **49**, 1062 (1982)
477. B. Nienhuis, J. Stat. Phys. **34**, 731 (1984); in [G8], Vol. 11, p. 1
478. M.P. Nightingale, Physica **83A**, 561 (1976)
479. M.P. Nightingale and H.W. Blöte, J. Phys. **A16**, L657 (1983)
480. M.P. Nightingale in [G10], ch. VII, p. 287
481. D.L. O'Brien and P.A. Pearce, J. Phys. **A28**, 4891 (1995)
482. G. Odor, Int. J. Mod. Phys. **C3**, 1195 (1992)
483. Y. Okabe and M. Kikuchi, Int. J. Mod. Phys. **C7**, 287 (1996)

484. K. Okamoto and K. Nomura, J. Phys. **A29**, 2279 (1996)
485. P. Olsson and S. Teitel, Phys. Rev. Lett. **80**, 1964 (1998)
486. H. Osborn and A. Petkou, Ann. of Phys. **231**, 311 (1994)
487. M. Oshikawa and I. Affleck, Phys. Rev. Lett. **77**, 2604 (1996); Nucl. Phys. **B495**, 533 (1997)
488. S. Östlund, Phys. Rev. **B24**, 398 (1981)
489. S. Östlund and S. Rommer, Phys. Rev. Lett. **75**, 3537 (1995)
490. G. Parisi and N. Sourlas, Phys. Rev. Lett. **46**, 871 (1981)
491. A.O. Parry, J. Phys. **A25**, L1015 (1992)
492. V. Pasquier, Nucl. Phys. **B285**, 162 (1987)
493. V. Pasquier, J. Phys. **A20**, 5707 (1987); Nucl. Phys. **B295**[FS21], 491 (1988)
494. W. Pauli, Helv. Phys. Acta, **13**, 204 (1940)
495. P.A. Pearce and K.A. Seaton, J. Phys. **A23**, 1191 (1990)
496. P.A. Pearce, Physica **A205**, 15 (1994)
497. P.J. Peard and D.S. Gaunt, J. Phys. **A28**, 6109 (1995)
498. K.A. Penson, R. Jullien and P. Pfeuty, Phys. Rev. **B26**, 6334 (1982)
499. K.A. Penson and M. Kolb, Phys. Rev. **B29**, 2854 (1983)
500. I. Peschel, Phys. Rev. **B30**, 6783 (1984)
501. I. Peschel and K.D. Schotte, Z. Phys. **B54**, 305 (1984)
502. I. Peschel, L. Turban and F. Iglói, J. Phys. **A24**, L1229 (1991)
503. P. Pfeuty, Ann. of Phys. **57**, 79 (1970)
504. J.L. Pichard and G. Sarma, J. Phys. **C14**, L617 (1981)
505. F. Piechon, M. Benakli and A. Jagannathan, Phys. Rev. Lett. **74**, 5248 (1995)
506. M. Plischke, Z. Rácz and R.K.P. Zia, Phys. Rev. **E50**, 3589 (1994)
507. A.M. Polyakov, Sov. Phys. JETP Lett. **12**, 381 (1970)
508. A.M. Polyakov, Nucl. Phys. **B396**, 367 (1993)
509. A.M. Polyakov, Phys. Rev. **E52**, 6183 (1995)
510. P.H. Poole, A. Coniglio, N. Jan and H.E. Stanley, Phys. Rev. Lett. **60**, 1203 (1988); Phys. Rev. **B39**, 495 (1989)
511. R.B. Potts, Proc. Cam. Phil. Soc. **48**, 106 (1952)
512. T. Prellberg and B. Drossel, preprint `cond-mat/9704100`
513. P.F. Price, C.J. Hamer and D. O'Shaughnessy, J. Phys. **A26**, 2855 (1993)
514. V. Privman and M.E. Fisher, J. Stat. Phys. **33**, 385 (1983)
515. V. Privman and M.E. Fisher, J. Phys. **A16**, L295 (1983)
516. V. Privman and M.E. Fisher, Phys. Rev. **B30**, 322 (1984)
517. V. Privman, Phys. Rev. **B32**, 6089 (1985)
518. V. Privman and S. Redner, J. Phys. **A18**, L781 (1985)
519. V. Privman, Phys. Rev. **B38**, 9261 (1988)
520. V. Privman in [G10], ch. I, p. 1
521. V. Privman, P.C. Hohenberg and A. Aharony in [G8], Vol. 14, p. 1
522. V. Privman, Int. J. Mod. Phys. **C3**, 857 (1992)
523. V. Privman (Ed), *Nonequilibrium Statistical Mechanics in One Dimension*, Cambridge University Press (Cambridge 1996)
524. P. Pujol, Europhys. Lett. **35**, 283 (1996)
525. M. Quéffelec, in A. Dold and B. Eckmann (Eds) *Substitution Dynamical Systems - Spectral Analysis*, Lecture Notes in Math., vol. 1294, Springer (Heidelberg 1987), p. 97
526. S.L.A. de Queiroz, J. Phys. **A28**, L363 (1995)
527. M.R. Rahimi Tabar and S. Rouhani, Ann. of Phys. **246**, 446 (1996)
528. M.R. Rahimi Tabar, A. Aghamohammadi and M. Khorrami, Nucl. Phys. **B497**, 555 (1997)
529. M.R. Rahimi Tabar and S. Rouhani, Phys. Lett. **224A**, 331 (1997)
530. F. Ravanini, Phys. Lett. **274B**, 345 (1992)

531. F. Ravanini, P. Tateo and A. Valleriani, Phys. Lett. **293B**, 361 (1992)
532. J.S. Reeve and A.J. Guttmann, Phys. Rev. Lett. **45**, 1581 (1980)
533. P. Reinicke, J. Phys. **A20**, 4501 and 5325 (1987)
534. P. Reinicke and T. Vescan, J. Phys. **A20**, L653 (1987)
535. H. Rieger, private communication
536. P.A. Rikvold, W. Kinzel, J.D. Gunton and K. Kaski, Phys. Rev. **B28**, 2686 (1983)
537. U. Ritschel and P. Czerner, Phys. Rev. Lett. **77**, 3645 (1996)
538. V. Rittenberg in P. Dita and V. Georgescu (Eds) *Conformal Invariance and String Theory*, Academic Press (San Diego 1989), p. 37
539. A. Rocha-Caridi in J. Lepowski, S. Mandelstam and I.M. Singer (Eds) *Vertex Operators in Mathematics and Physics*, MSRI Publications #3, Springer (New York 1985), p. 451
540. P. Roche, Phys. Lett. **285B**, 49 (1992)
541. H.H. Roomany and H.W. Wyld, Phys. Rev. **D21**, 3341 (1980)
542. J.S. Rowlinson, in *J.D. van der Waals: On the continuity of the gaseous and liquid states*, Studies in Statistical Mechanics (edited by J.L. Lebowitz), Vol. XIV, North Holland (Amsterdam 1988)
543. P. Ruelle and O. Verhoeven, preprint `hep-th/9803129`
544. A. Ruhe in V.A. Barker (Ed) *Sparse Matrix Techniques*, Springer Lecture Notes in Mathematics, Vol. 572, Springer (Heidelberg 1977), p. 130
545. P. Rujàn, J. Stat. Phys. **49**, 139 (1987)
546. P.G. Saffman, Stud. Appl. Math. **50**, 277 (1971)
547. H. Saleur, J. Phys. **A19**, L807 (1986)
548. H. Saleur and B. Duplantier, Phys. Rev. Lett. **57**, 3179 (1986)
549. H. Saleur, J. Phys. **A20**, L1127 (1987)
550. H. Saleur and C. Itzykson, J. Stat. Phys. **48**, 449 (1987)
551. H. Saleur, J. Phys. **A20**, 455 (1987)
552. H. Saleur, Nucl. Phys. **B382**, 486 (1992)
553. J. Sasaki and F. Matsubara, J. Phys. Soc. Japan **66**, 2138 (1997)
554. M. Scheunert, *The Theory of Lie Superalgebras*, Springer Lecture Notes in Mathematics, Vol. 716, Springer (Heidelberg 1979)
555. M. Schick, Prog. Surf. Sci. **11**, 245 (1981)
556. M. Schlottmann, Diplomarbeit Bonn-IR-88-41
557. K. Schoutens, Phys. Rep. **223C**, 183 (1993)
558. K. Schoutens, Phys. Rev. Lett. **79**, 2608 (1997)
559. H.J. Schulz and T.A.L. Ziman, Europhys. Lett. **18**, 355 (1992)
560. W. Selke, N.M. Švrakić and P.J. Upton, Z. Phys. **B89**, 231 (1992)
561. W. Selke, in [G8], Vol. 15, p. 1
562. F. Seno and A.L. Stella, J. Physique **49**, 739 (1988); Europhys. Lett. **7**, 605 (1988)
563. F. Seno, A.L. Stella and C. Vanderzande, Phys. Rev. Lett. **61**, 1520 (1988)
564. F. Seno and C. Vanderzande, J. Phys. **A27**, 5813 (1994); err. **A27**, 7937 (1994)
565. F. Seno, private communication
566. L.N. Shchur and S.S. Kosyakov, Int. J. Mod. Phys. **C8**, 473 (1997)
567. E. Siggia, Phys. Rev. **B16**, 2319 (1977)
568. H. Simon and M. Baake, J. Phys. **A30**, 5319 (1997)
569. P. Simon, Europhys. Lett. **41**, 605 (1998); Nucl. Phys. **B515**, 624 (1998)
570. S. Singh and R.K. Pathria, Phys. Rev. **B31**, 4483 (1983)
571. F.A. Smirnov, *Form Factors in Completely Integrable Models of Quantum Field Theory*, World Scientific (Singapore 1992)
572. J. Sólyom and P. Pfeuty, Phys. Rev. **B24**, 218 (1981)
573. C. Sommerfield, Phys. Rev. **176**, 2019 (1968)

574. B. Spain and M.G. Smith, *Functions of Mathematical Physics*, Van Nostrand Reinhold (New-York 1970)
575. H.E. Stanley, Phys. Rev. **176**, 718 (1968)
576. H.E. Stanley, Nature **378**, 554 (1995)
577. D. Stauffer and A. Aharony, *Introduction to Percolation Theory*, 2nd edition, Taylor and Francis (London 1991)
578. D. Stauffer, Physica **A242**, 1 (1997)
579. A. Stella, F. Seno and C. Vanderzande, J. Stat. Phys. **73**, 21 (1993)
580. L.M. Stratychuk and C.E. Soteros, J. Phys. **A29**, 7067 (1996)
581. H. Sugawara, Phys. Rev. **170**, 1659 (1968)
582. M. Suzuki, Prog. Theor. Phys. **46**, 1337 (1971)
583. J. Suzuki, T. Nagao and M. Wadati, Int. J. Mod. Phys. **B6**, 1119 (1992)
584. H.N.V. Temperley and E.H. Lieb, Proc. Roy. Soc. (London) **A322**, 251 (1971)
585. K. Thielemans, Int. J. Mod. Phys. **C2**, 787 (1991)
586. K. Thielemans, *An algorithmic approach to operator product expansions, W-algebras and W-strings*, PhD thesis Leuven (1994), preprint `hep-th/9506159`
587. C.A. Tracy, J. Phys. **A21**, L603 (1988)
588. A. Trovato, tesi di laurea, Universitá di Padova (1996)
589. A. Trovato and F. Seno, Phys. Rev. **E56**, 131 (1997)
590. L. Turban, J. Phys. **C15**, L65 (1982)
591. L. Turban, J. Phys. **A18**, L325 (1985)
592. L. Turban and J.-M. Debierre, J. Phys. **A19**, 1033 (1986)
593. L. Turban, Phys. Rev. **B44**, 7051 (1991)
594. L. Turban, J. Phys. **A25**, L127 (1992)
595. L. Turban and B. Berche, J. Phys. **A26**, 3131 (1993)
596. L. Turban and F. Iglói, J. Phys. **A30**, L105 (1997)
597. K. Uzelac and R. Jullien, J. Phys. **A14**, L151 (1981)
598. C. Vanderzande, A.L. Stella and F. Seno, Phys. Rev. Lett. **67**, 2757 (1991)
599. C. Vanderzande, *Lattice models of polymers*, Cambridge University Press (Cambridge 1998)
600. A.R. Veal, J.M. Yeomans and G. Jug, J. Phys. **A24**, 827 (1991)
601. J.M. van den Broeck and L.W. Schwartz, SIAM J. Math. Anal. **10**, 658 (1979)
602. E. Verlinde, Nucl. Phys. **B300**, 360 (1988)
603. T. Vescan, G.v. Gehlen and V. Rittenberg, J. Phys. **19**, 1957 (1986)
604. M.A. Virasoro, Phys. Rev. **D1**, 2933 (1970)
605. S.O. Warnaar, B. Nienhuis and K.A. Seaton, Phys. Rev. Lett. **69**, 710 (1992)
606. S.O. Warnaar, M.T. Batchelor and B. Nienhuis, J. Phys. **A25**, 3077 (1992)
607. S.O. Warnaar and B. Nienhuis, J. Phys. **A26**, 2301 (1993)
608. S.O. Warnaar, P.A. Pearce, K.A. Seaton and B. Nienhuis, J. Stat. Phys. **74**, 469 (1994)
609. G.M.T. Watts, preprint `hep-th/9708167`
610. F. Wegner in [G7], Vol. 6, p. 8
611. M. Weigel, Diplomarbeit, Mainz (1998); M. Weigel and W. Janke, preprint `cond-mat/9809253`
612. E.J. Weniger, Comp. Phys. Rep. **10**, 189 (1989)
613. J. Wess and J. Bagger, *Supersymmetry and Supergravity*, Princeton University Press (Princeton 1983)
614. R.A. Weston, Phys. Lett. **248B**, 340 (1990)
615. S.R. White, Phys. Rev. Lett. **69**, 2863 (1992)
616. S.R. White, Phys. Rev. **B48**, 10345 (1993)
617. S.R. White, Phys. Rev. Lett. **77**, 3633 (1996)
618. H. Whitney, Ann. Math. **33**, 688 (1932)
619. K.G. Wilson, Phys. Rev. **179**, 1499 (1969)

620. T. Wittlich, J. Phys. **A23**, 3825 (1990)
621. W.F. Wolff, P. Hoever and J. Zittartz, Z. Phys. **B42**, 259 (1981)
622. T. Wolf, R. Blender and W. Dietrich, J. Phys. **A23**, L153 (1990)
623. E. Wong and I. Affleck, Nucl. Phys. **B417**, 403 (1994)
624. E.M. Wright, Proc. London Math. Soc. **46**, 389 (1940); err. J. London Math. Soc. **27**, 256 (1952)
625. F.Y. Wu, Rev. Mod. Phys. **54**, 235 (1982)
626. F.Y. Wu, Rev. Mod. Phys. **64**, 1099 (1992)
627. T.T. Wu, B.M. McCoy, C.A. Tracy and E. Barouch, Phys. Rev. **B13**, 316 (1976)
628. C.N. Yang and T.D. Lee, Phys. Rev. **87**, 404, 410 (1952)
629. C.N. Yang, Phys. Rev. **85**, 808 (1952)
630. C.N. Yang and C.P. Yang, J. Math. Phys. **10**, 1115 (1969)
631. S.K. Yang, Nucl. Phys. **B285**[FS19], 183 (1987)
632. S.K. Yang and H.B. Zheng, Nucl. Phys. **B285**[FS19], 410 (1987)
633. K. Yildrim, Diplomarbeit Bonn-IB-95-02
634. V.P. Yurov and Al.B. Zamolodchikov, Int. J. Mod. Phys. **A5**, 3221 (1990)
635. V.P. Yurov and Al.B. Zamolodchikov, Int. J. Mod. Phys. **A6**, 4557 (1991)
636. M.A. Yurishchev, Phys. Rev. **B50**, 13533 (1994); **E55**, 3915 (1997)
637. A.B. Zamolodchikov and Al.B. Zamolodchikov, Ann. of Phys. **120**, 253 (1979)
638. A.B. Zamolodchikov and V.A. Fateev, Sov. Phys. JETP, **62**, 215 (1985)
639. A.B. Zamolodchikov, Sov. J. Nucl. Phys. **44**, 529 (1986)
640. A.B. Zamolodchikov, Sov. Phys. JETP Lett. **43**, 730 (1986)
641. A.B. Zamolodchikov, Int. J. Mod. Phys. **A3**, 743 (1988)
642. A.B. Zamolodchikov, Int. J. Mod. Phys. **A4**, 4235 (1989)
643. A.B. Zamolodchikov, Adv. Stud. Pure Math. **19**, 641 (1989)
644. A.B. Zamolodchikov, Int. J. Mod. Phys. **A4**, 2371 (1989)
645. A.B. Zamolodchikov, Nucl. Phys. **B358**, 497, 524 (1991)
646. Al.B. Zamolodchikov, Sov. Phys. JETP **63**, 1061 (1986)
647. Al.B. Zamolodchikov, Nucl. Phys. **B342**, 695 (1990)
648. Al.B. Zamolodchikov, Nucl.Phys. **B348**, 619 (1991)
649. Al.B. Zamolodchikov, Phys. Lett. **253B**, 391 (1991)
650. D-G. Zhang, B.-Z. Li and M.-G. Zhao, Phys. Rev. **B53**, 8161 (1996)
651. J. Zinn-Justin, Phys. Rev. Lett. **57**, 3296 (1986)
652. J.-B. Zuber, Phys. Lett. **176B**, 127 (1986)
653. J.-B. Zuber, Acta Phys. Pol. **B26**, 1785 (1995)
654. A.A. Zvyagin and I.V. Krive, Low Temp. Phys. **21**, 533 (1995)
655. R.E. Behrend, P.A. Pearce, V.B. Petkova and J.-B. Zuber, preprint **hep-th/9809097**
656. V. Rittenberg, private communication
657. S. Caracciolo, M.S. Causo, P. Grassberger and A. Pelissetto, preprint **cond-mat/9812267**

Index

ADE classification 207, 223
ADE model 216, 300
- dilute 217, 301
-- free energy for A_3 model 301
-- order parameter 301
- from Dynkin diagram 29
- order parameter 300
- related models 29
adjacency matrix 28, 216
affine extension *see*
  • Kac–Moody algebra
affine Lie algebra *see*
  • Kac–Moody algebra
affine Toda theory 296
amplitude *see*
  • critical amplitude
amplitude-exponent relation 75, 331,
  351, 361
- in 3D 258, 259
- non-linear 283
anisotropic local scaling 377
anisotropy exponent 20, 42, 369
ANNNI model 369
ANNNO(n) model 370
ANNNS model 370, 381, 382
anomalous thresholds 292
anomaly 57
antiferromagnet 336
aperiodic modulation 362
Arnoldi algorithm 170
Ashkin–Teller model 230, 313, 340
- operator content 234
- quantum chain phase diagram 233
- relation to defect line 360
- surface operator content 340
associativity of the OPE 108
average value 5, 375

beta function 236
Bethe ansatz equations 302
block spin 11

Blume–Capel model 16, 221
BMKD algorithm 169
Boltzmann constant 4
Boltzmann weight 28
bootstrap equation 292
bound states 292
boundary operator 345
BST algorithm 174
bulk scaling amplitudes 279

Callan–Symanzik beta function 236,
  281
Cappelli–Itzykson–Zuber–Kato
  theorem 207
Casimir effect 61, 77
central charge 56, 77, 78, 331, 354
- c, from specific heat 77
- k, from susceptibility 241
- effective 228
- free boson 78, 130
- Ising model 193, 335
- of Coulomb gas 133
central extension *see*
  • central charge
character 87, 88, 97–99, 189
charge conjugation operator 158, 230
charge operator 158
charge sector 158
charge U(1) 131, 248, 308
chemisorption 31
circle models 212
classification of modular invariant
  partition functions
- minimal models with $c < 1$ 207
- models with $c = 1$ 215
clock model 19
column geometry 259
compactification
- circle 212
- orbifold 213, 230
compactification radius 212

conformal algebra
- generators   47, 62, 242, 327
conformal anomaly   *see*
   • central charge
conformal block   108   *see also*
   • conformal tower
conformal group   45
- analytic functions   46
conformal invariance   4, 49, 54, 326, 351
conformal normalization   163, 188, 199, 262, 332, 339, 359
conformal perturbation theory   263
conformal tower   85, 86, 163, 194, 263, 266, 354, 359
conformal transformations   45
- angle-preserving   45
- projective   48, 51, 244, 326
conformal turbulence   254
- minimal model   257
conformal weights   48, 213
conservation law   287, 288
- absence for ϕ_{22} perturbation   313
- energy–momentum tensor   55
convergence
- of DMRG algorithm   181
- of finite-size amplitudes   171
- of Lanczos algorithm   168
convergence of finite-size sequences
- linear   173
- logarithmic   173
corner exponent   367
correction exponent ω 72, 81, 156, 264
correlation function   *see*
   • two-, three-, four-point function
correlation functions
- covariance   7, 44, 49
- differential equation   103
correlation length   2, 5, 41, 68, 222, 279
Coulomb gas   127, 248, 309
- and minimal models   134
- correlators   134
- screened   132, 134
Coxeter exponents   29, 209, 296
Coxeter number   29, 209, 216, 296
C-parity   159
critical amplitude   67, 304
- universal ratio   304
critical dynamics   370
critical exponents   6, 21, 220, 324
- bulk   6
- definition   6
- finite-size estimates   162, 331
- non-equilibrium   384

- surface   324
- tricritical   15
critical point   3
cross-over exponent   363
crossing condition   107, 108, 120, 292
c-theorem   280, 281

dangerous irrelevant scaling field   68
defect line   350
defect line exponents   350
defect strength   350
Δ-theorem   283
density matrix renormalization group   177
detailed balance   375
dihedral group   158
dihedral group $\mathbb{D}_4$ 158, 230, 264
- and breaking of rotation invariance on square lattice   278
- irreducible representations   232
dilatation   11, 45, 46
dipolar interactions   13, 32
directed percolation   176, 371
discretized energy and momentum   148, 163, 191
disorder line   187, 199, 259, 383
dispersion relation   77, 163, 188, 196, 244, 380, 382
DMRG algorithm   177
- finite system   179
- infinite system   178
droplet picture   3
dual bond   18
dual coupling   18
dual lattice   18
dual site   18
duality   17, 20, 67, 151, 153, 159, 182, 188
dynamical exponent   370   *see also*
   • anisotropy exponent
dynamical scaling function   371
dynamical symmetry   83, 148, 225, 331
Dynkin diagram   28

eigenvalues   167
eigenvectors   168
elastic scattering   289
energy density
- on the lattice   242
- surface exponent   332
energy spectrum
- turbulence   255
energy–momentum tensor   54, 86, 129, 133, 280, 356, 374

enstrophy   255–257
ϵ-expansion   22, 24, 326
Euler pentagonal theorem   209
Euler relation   18
experiment   2, 3, 13, 14, 21
– bulk exponents in 2D   30
– central charge   241
– classical 2D and quantum (1+1)D
  systems   14
– finite-size scaling in 2D   81
– finite-size scaling in 3D   78
– reaction–diffusion in 1D   375, 384
– surface exponents in 3D   326
– Yang–Lee zeroes   23
exponent   see
  • critical exponents
exponent inequalities   10
extended conformal algebras   226
extraordinary transition   323, 333, 335,
  368
extrapolation algorithms   see
  • BST-, VBS-algorithm
extrapolation length   322
extreme anisotropic limit   see
  • Hamiltonian limit

fermion number operator   184
Fibonacci sequence   362
finite group
– cyclic $\mathbb{Z}_p$ 19
– dihedral $\mathbb{D}_p$ 158
– permutation $\mathbb{S}_p$ 17
finite sizes
– relation to inverse temperature   40
finite-lattice extrapolation   173
finite-size corrections   71, 197, 263, 268
– from conformal tower of the identity
  266
finite-size scaling   63–68, 75, 77, 161,
  162, 274, 330
finite-size scaling amplitude   70, 75,
  191, 259, 331, 334, 337
finite-size scaling functions   70, 197,
  198, 270–272, 274, 298, 312, 316
finite-size scaling limit   67, 264
finite-size scaling region   65, 67
finite-size scaling variables   70, 264
first intersection   94
fixed point   11
fluctuation-dissipation theorem   7,
  370
form factor   304
four-point function   52, 103, 107, 117,
  120, 122, 124, 134, 136, 137

free boson   78, 127, 209   see also
  • spherical model
free boundary conditions   64, 78, 162,
  333
free energy   5, 40, 63, 68, 76, 191, 279,
  301, 322, 325
– surface   64, 322
free energy density   7
free fermion   82   see also
  • Ising model
free Parisi–Sourlas field theory   254
Friedan–Qiu–Shenker theorem   94
Frobénius series   116
Fuchs differential operator   114
Fuchs theorem   116

Galilei invariance   373
generating function
– of lattice animals   251
– of polymers   27, 250
– of the partitions of the integers   189
geometric phase transition   24
Glauber dynamics   375, 383
global symmetry   17, 19, 21, 158, 160,
  184, 221, 229, 230, 232, 338
Goddard–Kent–Olive theorem   97

half-filling   364
half-partition function   240
Hamiltonian limit   141, 143, 147, 157,
  230
Heisenberg model   22, 80, 241, 260
highest weight
– of Kac–Moody algebra   131
– of superconformal algebra   225
– of Virasoro algebra representations
  85
Hilhorst–van Leeuwen model   361, 367
history   13, 31
Hubbard–Stratonovich transformation
  144
hypergeometric function   115, 116,
  118, 126, 135, 140, 269
– asymptotics   380
hyperscaling   10, 20, 69, 70, 325
hyperuniversality   70

ideal Bose gas   78
ideal Fermi gas   82
integrability   288
IRF model   28, 216
irrelevant   see
  • scaling field
irrelevant scaling field   71, 263

Ising lattice gas   13
Ising metamagnet   16, 221
Ising model   1, 13, 34, 35, 95, 117, 139, 141, 171, 183, 210, 240, 242, 258, 264, 267, 271, 272, 288, 294, 297, 298, 305, 317, 332, 336, 351
- (1+1)D   see
   • Ising quantum chain
- (2+1)D   161, 258
- antiferromagnetic   336
- in magnetic field at T_c 288, 296, 298, 301
- operator content   196
- operator content with a defect line   357
- random   349
- relation with diluted A_3 model   301
- relation with RSOS model   30
- surface operator content   335
Ising quantum chain   14, 143, 147, 183, 245, 298, 351
- aperiodic   362
- diagonalization   184, 333, 352
- Glauber dynamics in 1D   383
- multispin interactions   181
- phase diagram   187
- spin $S = 1$ 198
Ising uniaxial magnet   13

Jordan-Wigner transformation   184

Kac formula   93, 96, 97, 155, 156, 189, 227
- from Coulomb gas   133
- non-unitary   227
- superconformal   225
Kac table   see
   • Kac formula
Kac–Moody algebra   28, 129, 216, 241, 355, 356
- shifted   342, 357
kinetic Ising model   375, 383
kinetic spherical model   376

Lanczos algorithm   163–169
- convergence   167
- Fortran code   164
- re-orthogonalization   168
Landau–Ginzburg–Wilson   146, 154
latent heat   68, 161
lattice animal   251, 367
lattice momentum   149, 162
level
- of Kac–Moody algebra   129

- of Virasoro algebra representation   85, 87, 189
Lie superalgebra   224
Lieb, Schultz, Mattis technique   185, 333, 352
Lifshitz point   370, 381
local scale invariance   371, 377
logarithmic operators   243
logarithmic transformation   74, 149, 331, 351
loop algebra   47
Luck criterion   362, 363

magnetic field   1, 298, 311
- complex   23
- staggered   14
magnetization   see
   • order parameter
Manhattan lattice   310
marginal   see
   • scaling field
marginal scaling field   286, 308, 351
mass ratios
- from factorizable S-matrix   294–297
- numerical   300, 302, 313, 316
mass term   373, 378
master equation   375
metamagnet   16
metric factor   68, 198, 331
minimal S-matrix   293, 317
minimal models   90, 93, 96, 134
- non-unitary   227, 245, 257
- unitary   94, 207, 240
Möbius transformation   48
modular group   205
modular invariance   206, 258
modular transformation   209
monodromy   110, 112, 124, 135
monodromy matrix   110, 111, 119, 126
monodromy transformation   119, 126
multicritical Ising models   154, 313

Navier–Stokes equation   254
neutrality condition
- Coulomb gas   131
non-linear scaling field   71
normal ordering   129, 201, 354
normal transition   335   see also
   • extraordinary transition
N-shift technique   174
null operator   87, 92, 97, 101, 287
- differential equation for level 2   103
- differential equation for level 4   116
- explicit for $\phi_{1,2}$, $\phi_{1,3}$, $\phi_{2,2}$ 286

– explicit for $\phi_{1,s}$ 287
null vector
– null operator   92

$O(n)$ model   22, 240, 259, 260, 326, 332
– $n \to 0$ limit and polymers   26, 284
one-fermion energies   186
OPE   see
    • operator product expansion
operator product algebra   270 see also
    • operator product expansion
operator product algebra hypothesis
    105
operator product expansion   84, 101,
    118, 125, 135, 156, 211, 256, 262, 274,
    283
– coefficients   52, 120, 124, 125, 137,
    140, 272
orbifold   213, 360
orbifold models   212, 215, 230 see also
    • Ashkin–Teller model
order parameter   5, 68, 160, 349
– staggered   15
ordinary transition   322, 332
orthosymplectic Lie superalgebra   225

parastatistics   82
parity operator   158
parity sector   158
partition function   5, 206, 208, 209,
    214, 215, 239
– modular invariant   207
Pauli matrices   143
percolation   24, 34, 81, 245, 250, 273,
    344
– bond   25
– critical exponents   26
– crossing probability   344
– directed   176, 371
– incipient spanning cluster   346
– operator content   247
– relation to Potts model   25
– site   25
– surface operator content   343
period-doubling sequence   363
periodic boundary conditions   64, 74,
    162
persistence length   222
persistent current   363, 364
– defect   364
– parity effect   366
phase ordering kinetics   370
phenomenological renormalization   72,
    222

physisorption   31
Poisson resummation formula   209
polymer   26, 35, 247, 250, 284, 346
– branched   see
    • lattice animal
– relation to $O(n)$ spin model   26, 284
Potts model   17, 19, 35, 95, 96, 139,
    151, 153, 157, 214, 219, 229, 240, 264,
    268, 273, 304, 337–339, 341, 343
– $q \to 1$ limit   25
– chiral   19, 20, 35, 306
– operator content   219
– relation to lattice animals   252
– relation with RSOS model   30
– surface operator content   337
– two-point function   182
– vector   19
primary   50, 72, 84, 131
profile   346
– non-diagonal   349
projective invariant   52, 328

quantum chain   14, 143, 147
– at finite temperature   77, 82, 278
– central charge   77, 241
quantum Hamiltonian   14, 39, 141,
    143, 383
quasi-primary   49, 86, 88, 244, 288, 373
quasiperiodic modulation   362

radial quantization   89
radius of gyration   27, 284
random Ising model   349
rapidity   290
reaction–diffusion process   383
reduced magnetic field   4
reduced surface magnetic field   324
reduced temperature   4
relevant   see
    • scaling field
relevant scaling field   264
remnant function   198, 272
renormalization group   11, 52
renormalization group eigenvalue   7
renormalization group trajectory   52,
    280, 314
response field   371
response function   370, 376
Reynolds number   254
Rocha–Caridi formula   99, 189, 246
Roomany–Wyld approximant   237
rotation   45, 46
rounding exponent   65
rounding temperature   65

RSOS model   27, 28, 95, 96, 156, 300

σ-Γ algebra   159
SAW   *see*
     • self-avoiding walk
scalar ϕ^4 theory   145, 146   *see also*
     • Ising model
scale invariance   4, 6, 328, 375, 377
scaling amplitude ratios   304
scaling dimension   7, 12, 48
scaling field   7
– dangerously irrelevant   10
– irrelevant   9
– marginal   9
– relevant   9
scaling function   9
scaling operator   7, 12
– covariance   43, 49
scaling relations   10
– anisotropic   20, 42
– surface   326
– tricritical   42
Schrödinger group   372
Schrödinger invariance   373–376, 382
Schwarzian derivative   60
screening charges   133
screening current
– Coulomb gas   133
secondary   50, 71
self-avoiding walk   26, 248, 249
– oriented   307
self-duality   18, 67, 153, 159, 181, 188,
     305
sequence extrapolation   173
shift exponent   65, 79, 80, 82
– effective   81
singular exponent (ODE)   112
singular regular point (ODE)   114
slab geometry   321
S-matrix   289, 316
special conformal transformation   45,
     46, 327
special transition   322, 358
specific heat   5, 68, 78, 160
– amplitudes and self-duality   67, 305
spectral parameter   216
spherical model   22, 82, 171, 245, 260
spin   48
stream function   255
stress-energy tensor   *see*
     • energy–momentum tensor
strip geometry   61, 74, 76, 82, 157, 331
strong embedding   251, 252
strongly anisotropic system   369

substitution matrix   362
substitution sequence   362
Sugawara form   129
sum rule
– central charge   282
– conformal weights   114, 283
summary of steps of conformal
     invariance applied to a given model
     203
superconformal algebra   225
superconformal characters   225
superconformal invariance   224, 341
– and percolation   246
supersymmetry   224, 253
surface exponent   324
– energy density   332
surface magnetic field   322
surface magnetization   362
surface thermodynamics
– excess quantity   324
– local quantity   324
– mixed quantity   324
surface transition   323
susceptibility   5, 68, 160, 237, 241, 283

Temperley–Lieb algebra   150, 153, 338
theta point   27, 35, 250–252
three-dimensional systems   258
three-point function   52, 106, 228, 243,
     262, 265, 374
tight-binding model   364
toroidal boundary conditions   162,
     183, 232
torus   205
transfer matrix   38, 142, 218, 245, 249,
     344
translation   45, 46, 181
translation invariance   162
transverse field   14, 143
tricritical exponents   42
tricritical Ising model   14–16, 221, 264,
     273, 311
– operator content   223
– surface operator content   340
tricritical point   15
– non-unitary   229
tricritical Potts model   19, 35
truncation method   275, 318
two-point function   5, 51, 74, 82, 90,
     127, 128, 131, 228, 242, 243, 245, 261,
     265, 325, 326, 328, 333, 360, 367, 370,
     374, 379, 381
– for quantum chains at finite
     temperature   278

– scaling   12, 327, 369

unitarity   90, 94, 97, 225
– scattering   291
unitary minimal series   94
universality   4, 33, 37, 68, 70, 148, 192, 198, 258, 272, 274, 384
universality classes   4, 20 *see also*
    • ADE classification
upper critical dimension   10

vacancy operator   270
vacuum   88
vanishing curves   91, 94
VBS algorithm   174
Verlinde formula   211
Verma module   85, 86, 97
– degenerate   87, 92
vertex operator   131, 211
Virasoro algebra   60, 83, 133
Virasoro character   97–99, 189, 228
– for $c = 1$ 234
– generic   87
– modular transformation   207
– numerical computation   190
Virasoro generators   57, 201, 204, 354, 356
– hermiticity   89
vorticity   254

wandering exponent   362
Ward identity   44, 51, 54, 55, 131, 329, 347
– projective   51, 243, 244, 327, 373, 379
watermelon topology   247
Wright formula   380
W symmetry   221

XXZ quantum chain   238, 338, 342
– operator content   239
– relation to nonequilibrium systems   384
– relation to other spin systems   231, 239, 339, 342
– surface operator content   342
XY model   21, 22, 34, 212, 213, 236, 260, 341
XY quantum spin model   183, 188, 191, 198, 258, 267, 272, 383

Yang–Baxter equation   291
Yang–Lee singularity   23, 24, 96, 227, 229, 253, 256, 276
– operator content   229
– surface operator content   340

Zamolodchikov counting criterion   287, 288